Springer Tracts in Advanced Robotics

Volume 47

Editors: Bruno Siciliano · Oussama Khatib · Frans Groen

Srinivas Akella, Nancy M. Amato,
Wesley H. Huang, Bud Mishra (Eds.)

Algorithmic Foundation of Robotics VII

Selected Contributions of the Seventh International Workshop on the Algorithmic Foundations of Robotics

Editors

Srinivas Akella
Department of Computer Science
Rensselaer Polytechnic Institute
110 Eighth Street
Troy, New York 12180
USA
E-Mail: sakella@cs.rpi.edu

Nancy M. Amato
Department of Computer Science
Texas A&M University
College Station, Texas 77843
USA
E-Mail: amato@cs.tamu.edu

Wesley H. Huang
Applied Perception, Inc.
220 Executive Drive, Suite 400
Cranberry Township, PA 16066
E-Mail: wes@appliedperception.com

Bud Mishra
Department of Computer Science
New York University
Courant Inst, 251 Mercer St
New York, NY 10012
USA
E-Mail: mishra@nyu.edu

ISBN 978-3-642-08798-1 e-ISBN 978-3-540-68405-3

DOI 10.1007/978-3-540-68405-3

Springer Tracts in Advanced Robotics ISSN 1610-7438

Printed in acid-free paper

5 4 3 2 1 0

springer.com

STAR (Springer Tracts in Advanced Robotics) has been promoted under the auspices of EURON (European Robotics Research Network)

Foreword

By the dawn of the new millennium, robotics has undergone a major transformation in scope and dimensions. This expansion has been brought about by the maturity of the field and the advances in its related technologies. From a largely dominant industrial focus, robotics has been rapidly expanding into the challenges of the human world. The new generation of robots is expected to safely and dependably co-habitat with humans in homes, workplaces, and communities, providing support in services, entertainment, education, healthcare, manufacturing, and assistance.

Beyond its impact on physical robots, the body of knowledge robotics has produced is revealing a much wider range of applications reaching across diverse research areas and scientific disciplines, such as: biomechanics, haptics, neurosciences, virtual prototyping, animation, surgery, and sensor networks among others. In return, the challenges of the new emerging areas are proving an abundant source of stimulation and insights for the field of robotics. It is indeed at the intersection of disciplines that the most striking advances happen.

The goal of the series of Springer Tracts in Advanced Robotics (STAR) is to bring, in a timely fashion, the latest advances and developments in robotics on the basis of their significance and quality. It is our hope that the wider dissemination of research developments will stimulate more exchanges and collaborations among the research community and contribute to further advancement of this rapidly growing field.

This volume is the outcome of the seventh edition of the biennial Workshop Algorithmic Foundations of Robotics (WAFR). Edited by S. Akella, N.M. Amato, W.H. Huang, and B. Mishra, the book offers a collection of a broad range of topics in advanced robotics. The contents of these contributions represent a cross-section of the current state of research from one particular aspect: algorithms, and how they reflect on the theoretical basis of subsequent developments. Validation of algorithms, design concepts, or techniques is the common thread running through this focused collection.

Rich by topics and authoritative contributors, WAFR culminates with this unique reference on the current developments and new directions in the field of algorithmic foundations. A fine addition to the series!

Naples, Italy
April 2008

Bruno Siciliano
STAR Editor

Preface

Algorithms are a fundamental component of robotic systems: they control or reason about motion and perception in the physical world. They receive input from noisy sensors, consider geometric and physical constraints, and operate on the world through imprecise actuators. The design and analysis of robot algorithms therefore raises a unique combination of questions in control theory, computational and differential geometry, and computer science.

The Workshop on the Algorithmic Foundations of Robotics (WAFR) is a multi-disciplinary single-track workshop with submitted papers and invited talks on advances on algorithmic problems in robotics. It has been held every other year since 1994 and has an established reputation as one of the most (if not the most) important venues for presenting algorithmic work related to robotics.

As you will see, the topics of interest in WAFR are very broad since the focus is on algorithm development and analysis rather than on specific problems or applications. Increasingly, robotics algorithms are finding use in areas far beyond the traditional scope of robots. One of the most important aspects of WAFR is its informal atmosphere which allows a frank exchange of new, previously unpublished ideas. In particular, WAFR has been an occasion for graduate students to meet and interact with more senior researchers who many times are not accessible to students at the larger robotics conferences.

The seventh WAFR was held July 16–18, 2006, in New York City at the Tribeca Grand Hotel in lower Manhattan.WAFR 2006 had a record number of submissions and a record attendance with 106 registrants, just over 50 students. In addition to the 32 contributed papers contained in this volume, the workshop featured six invited speakers, including both researchers who defined the field and who are today defining the frontiers of the field – in several cases the same people: James Gimzewski (UCLA), Jessica K. Hodgins (CMU), Jean-Claude Latombe (Stanford), Tomás Lozano-Pérez (MIT), Jacob Schwartz (NYU), and Sebastian Thrun (Stanford).

WAFR 2006 had a very strong program of 32 contributed technical papers. These papers were selected from 62 submissions by a rigorous evaluation process, with each submission being reviewed by at least 3 members of the program committee. The authors of selected papers were invited to submit expanded versions of their WAFR 2006 papers to a special issue of the International Journal of Robotics Research.

We are extremely grateful to the program committee for their careful and insightful reviews. The program committee members were: O. Burchan Bayazit, Antonio Bicchi, Greg Chirikjian, Mike Erdmann, Dan Halperin, Hirohisa Hirukawa, Seth Hutchinson, Lydia Kavraki, James Kuffner, Vijay Kumar, Jean-Paul Laumond, Steve LaValle, Ming Lin, Yoshi Nakamura, Dinesh Pai, Elon Rimon, Jack Snoeyink, Dezhen Song, Frank van der Stappen, and Gaurav Sukhatme.

This meeting would not have been possible without the dedicated work and assistance of many individuals and organizations. We have many thanks to give: to Kay Jones from Texas A&M for overall support and logistics; to the student volunteers from Rensselaer and from Texas A&M for their diligent work; to our institutions (NYU, Rensselaer, and Texas A&M) for their support; to the National Science Foundation for the student travel grants that provided support to 38 students; to Microsoft for sponsoring the banquet cruise; and, of course, to the WAFR steering committee for their advice and suggestions.

Thank you all for making WAFR 2006 a WAFR to remember!

Srinivas Akella
Nancy M. Amato
Wesley H. Huang
Bud Mishra

Contents

Part I

Probabilistic Roadmap Methods (PRMs)

Quantitative Analysis of Nearest-Neighbors Search in High-Dimensional Sampling-Based Motion Planning

Erion Plaku and Lydia E. Kavraki

Department of Computer Science, Rice University
{plakue,kavraki}@cs.rice.edu

Abstract. We quantitatively analyze the performance of exact and approximate nearest-neighbors algorithms on increasingly high-dimensional problems in the context of sampling-based motion planning. We study the impact of the dimension, number of samples, distance metrics, and sampling schemes on the efficiency and accuracy of nearest-neighbors algorithms. Efficiency measures computation time and accuracy indicates similarity between exact and approximate nearest neighbors.

Our analysis indicates that after a critical dimension, which varies between 15 and 30, exact nearest-neighbors algorithms examine almost all the samples. As a result, exact nearest-neighbors algorithms become impractical for sampling-based motion planners when a considerably large number of samples needs to be generated. The impracticality of exact nearest-neighbors algorithms motivates the use of approximate algorithms, which trade off accuracy for efficiency. We propose a simple algorithm, termed Distance-based Projection onto Euclidean Space (DPES), which computes approximate nearest neighbors by using a distance-based projection of high-dimensional metric spaces onto low-dimensional Euclidean spaces. Our results indicate DPES achieves high efficiency and only a negligible loss in accuracy.

1 Introduction

Research in motion planning has in recent years focused on sampling-based algorithms [1,5,9,13,14,17,20,22] for solving problems involving multiple and highly complex robots. Such algorithms rely on an efficient sampling of the configuration space and compute nearest neighbors for the sampled points. In general, the k nearest neighbors of a point in a data set are defined as the k closest points in the data set according to a distance metric.

As research in motion planning progressively addresses problems of unprecedented complexity, nearest-neighbors computations based on arbitrary distance metrics and large high-dimensional data sets become increasingly challenging. Researchers have developed many nearest-neighbors algorithms, such as the kd-tree, R-tree, X-tree, M-tree, VP-tree, Gnat, iDistance, surveyed in [7, 10, 11], and others [3]. Analysis has shown that for certain distance metrics and data distributions, the computational efficiency of such algorithms decreases as the dimension increases [4,6,12,18,23]. As summarized in [15], although the nearest

S. Akella et al. (Eds.): Algorithmic Foundation of Robotics VII, STAR 47, pp. 3–18, 2008.
springerlink.com

neighbor of a point according to L_2 from an n-point d-dimensional data set can be computed in $O(d^{O(1)} \log n)$ time, the associated $n^{O(d)}$ space requirement is impractical. Reducing the space requirement to what is practical, i.e., $O(dn)$, increases the query time to $\min(2^{O(d)}, dn)$. In fact, after a critical dimension, the brute-force linear method, which examines the entire data set, is computationally faster than other exact nearest-neighbors algorithms. The work in [23] shows 610 as the theoretical bound on the critical dimension for L_2 and uniformly distributed points in $[0, 1]^d$. The experiments in [23] however indicate 10 as the critical dimension, a much lower estimate than the theoretical bound. In [4, 12], the critical dimension is estimated between 10–15 for L_p and several synthetic and image data sets.

Another viable approach for nearest-neighbors computations is to use approximate algorithms which trade off accuracy for efficiency [2,19,21,10], where accuracy indicates similarity between exact and approximate nearest neighbors. As summarized in [15], in the case of L_2, approximate nearest neighbors can be computed probabilistically in $dn^{1/1+\epsilon}$ time and $O(dn)$ space or deterministically in $(d \log n/\epsilon)^{O(1)}$ time and $n^{1/\epsilon^{O(1)}}$ space. Such algorithms gain efficiency by projecting the data set onto low-dimensional spaces and achieve high accuracy when the projection results in low distortion of distances. The computational advantages of approximate nearest-neighbors algorithms are more evident when the dimension d of the data set is high. The problem however remains challenging for general metrics. As summarized in [16], any n-point metric space can be projected onto $\mathbb{R}^{O(\log^2 n)}$ with only $O(\log n)$ distortion. However, solving high-dimensional motion planning problems requires generating millions of samples, which makes $O(\log^2 n)$ impractical. By increasing the distortion to $O(n^{2/d} \log^{3/2} n)$, and thus reducing the accuracy, any n-point d-dimensional metric space could be projected onto $\mathbb{R}^d$. Therefore, the efficiency or accuracy of approximate nearest-neighbors algorithms is typically reduced when general metrics are used instead of L_2.

The analysis of nearest-neighbors algorithms generally assume a uniform distribution of points and the use of L_2. In motion planning, the distribution is impacted by the sampling scheme. Samples satisfy certain criteria, such as representing collision-free configurations, and, consequently, the distribution is usually non-uniform. Furthermore, distances between configurations are not necessarily defined by L_2, but instead attempt to capture the success of the local planner. Motion planners therefore exhibit a degree of flexibility which can be exploited to compute approximate instead of exact nearest neighbors. Research in [22] shows that in certain high-dimensional problems using random neighbors actually improves the performance of motion planners. Therefore, an understanding of the impact of these factors on the efficiency and accuracy of nearest-neighbors algorithms employed by motion planners could provide valuable insight in addressing high-dimensional motion planning problems.

In this work, we quantitatively analyze exact and approximate nearest-neighbors algorithms in the context of high-dimensional sampling-based motion planning. We focus on roadmap-based algorithms, such as the Probabilistic

RoadMap (PRM) method with uniform [17], bridge [13], Gaussian [5], and obstacle [1] sampling, and tree-based algorithms, such as the Rapidly-exploring Random Tree (RRT) [20] and the Expansive-Spaces Tree (EST) [14].

We address the following questions: (i) under what conditions, if any, should motion planners use the brute-force linear method instead of other exact nearest-neighbors algorithms? (ii) do approximate nearest-neighbors algorithms compute more efficiently nearest neighbors that are similar to exact nearest neighbors on high-dimensional motion planning problems? We study the impact of the dimension, number of samples, distance metrics, and sampling schemes on the efficiency and accuracy of nearest-neighbors algorithms.

Our analysis indicates that after a critical dimension the brute-force linear method is computationally more efficient than other exact nearest-neighbors algorithms. The critical dimension however depends on the number of samples, distance metric, and sampling scheme. We present results that quantify these dependencies.

Motivated by the impracticality of exact nearest-neighbors algorithms, we propose the use of approximate algorithms for the computation of neighbors in high-dimensional motion planning problems. In this work, we develop a simple algorithm, termed Distance-based Projection onto Euclidean Space (DPES), which computes approximate nearest neighbors by projecting high-dimensional metric spaces onto low-dimensional Euclidean spaces. The projection is based on distances between a set of selected points and points in the data set. Our experiments indicate DPES achieves high computational efficiency and only a negligible loss in accuracy.

The rest of the paper is organized as follows. In Sect. 2 we describe the methodology we use in our analysis and the DPES algorithm. In Sect. 3 we describe the experimental setup. In Sect. 4 we present the results of our analysis of nearest-neighbors algorithms. We conclude in Sect. 5 with a discussion.

2 Methods

In this section, we describe the nearest-neighbors algorithms we use in this paper including DPES. We also outline the motion planners and distance metrics we use in this study. We denote the data set, number of nearest neighbors, and distance metric by S, k, and $\rho : S \times S \rightarrow \mathbb{R}^{\geq 0}$, respectively.

2.1 Exact k Nearest-Neighbors Algorithms

We define the k nearest neighbors (kNN) of a point $s_i \in S$, denoted by $\mathrm{NN}_S(s_i, k)$, as the k closest points to s_i from $S - \{s_i\}$ according to ρ.

Linear. This is a brute-force approach which resolves $\mathrm{NN}_S(s_i, k)$ by computing the distance from s_i to each point in S. The Linear method provides the basis for the comparison with other more sophisticated kNN algorithms.

Gnat. Gnat [7] constructs a tree recursively by partitioning S into smaller sets and associating each set with a branch in the tree. Gnat then uses the triangle inequality to prune certain branches of the tree in order to compute kNN more efficiently. We choose Gnat in our analysis, since the results in [7] and our experiments in the context of motion planning with several kNN algorithms, such as kd-tree, M-tree, VP-tree, [3], etc., indicate Gnat to be more efficient especially on large data sets and metric spaces.

2.2 Approximate k Nearest-Neighbors Algorithms

We define approximate k nearest neighbors (kANN) of a point $s_i \in S$, denoted by $\text{ANN}_S(s_i, k)$, as a subset of $S - \{s_i\}$ of cardinality k that according to certain measures is similar to $\text{NN}_S(s_i, k)$.

Random. This method selects $S' \subset S$ uniformly at random, $|S'| \gg k$, and computes $\text{ANN}_S(s_i, k)$ as $\text{NN}_{S'}(s_i, k)$. Random provides a basis for evaluating the quality of other kANN algorithms.

Distance-based Projection onto Euclidean Space (DPES). Our kANN algorithm is based on projecting each point $s_i \in S$ to a point $v(s_i) \in \mathbb{R}^m$, for some fixed $m > 0$. We then use L_2 to define distances between projected points and compute $\text{ANN}_S(s_i, k)$ as

$$\text{ANN}_S(s_i, k) = \{s' : v(s') \in \text{NN}_{V(S)}(v(s_i), k)\}, \quad V(S) = \{v(s_i) : s_i \in S\}.$$

We thus compute $\text{ANN}_S(s_i, k)$ according to the distance metric ρ by computing $\text{NN}_{V(S)}(v(s_i), k)$ according to L_2. Any kNN data structure $\mathcal{A}$ can be used to compute $\text{NN}_{V(S)}(v(s_i), k)$. DPES supports dynamic addition and removal of points. When a point s is added to or removed from S, the corresponding projection $v(s)$ is added to or removed from $\mathcal{A}$, respectively.

We obtain the projection by selecting m pivots $\{p_1, p_2, \cdots, p_m\} \subset S$ and setting each $v(s_i) \in \mathbb{R}^m$ to $v(s_i)[j] = \rho(s_i, p_j)$, $1 \leq j \leq m$. We select p_1 uniformly at random in S and each p_j, $2 \leq j \leq m$, as the point in S that maximizes $\min_{i=1}^{j-1} \rho(p_i, p_j)$. The objective is to select pivots that preserve relative distances between points in S when projected onto $\mathbb{R}^m$, e.g., projections of points in S that are close according to ρ should be close according to L_2.

DPES has certain computational advantages. The projection of S onto $\mathbb{R}^m$ greatly improves the efficiency, since, as shown in Sect. 4.2, typically fewer distance evaluations are necessary for computing nearest neighbors. In addition, evaluating L_2 is more efficient than evaluating distance metrics commonly used in motion planning, such as those in Sect. 2.3.

Quality Evaluation. We determine the quality of $\text{ANN}_S(s_i, k)$ by using two common measures based on distances between points in $\text{ANN}_S(s_i, k)$ and $\text{NN}_S(s_i, k)$. Similar to [10], we use the ratio of false dismissals:

$$\text{rfd}_\epsilon = \frac{1}{k} \sum_{s \in \text{ANN}_S(s_i, k)} \begin{cases} 1, & \rho(s, s_i) > (1 + \epsilon) \max_{s' \in \text{NN}_S(s_i, k)} \rho(s_i, s'), \\ 0, & \text{otherwise.} \end{cases}$$

The rfd_ϵ error, $\epsilon \geq 0$, indicates the fraction of points in $\text{ANN}_S(s_i, k)$ that are $(1+\epsilon)$-times farther away from the k-th nearest neighbor of s_i. Note however that some $s', s'' \in \text{ANN}_S(s_i, k)$ could contribute the same value to rfd_ϵ even when $\rho(s', s_i) \gg \rho(s'', s_i)$. Thus, two different sets could have the same rfd_ϵ value even when points in one set are farther away from s_i than points in the other set. Therefore, as in [10], we also use the ratio of distance errors:

$$\text{rde} = 1 - \sum_{s \in \text{NN}_S(s_i,k)} \rho(s, s_i) / \sum_{s \in \text{ANN}_S(s_i,k)} \rho(s, s_i).$$

The range of rfd_ϵ and rde is $[0, 1]$ and smaller values indicate high quality.

2.3 Sampling-Based Motion Planning and Distance Metrics

In this study, we use roadmap-based algorithms, such as `PRM` with uniform (`PRMu`) [17], bridge (`PRMb`) [13], Gaussian (`PRMg`) [5], and obstacle (`PRMo`) [1] sampling, and tree-based algorithms, such as bi-directional `RRT` [20] and `EST` [14]. We follow standard implementations as in [22, 9]. We consider problems with multiple robots moving freely in 2D or 3D workspaces with static obstacles. We gradually increase the number of robots until we reach the critical dimension. We create data sets using configurations of the roadmap in `PRM` and the initial tree in `RRT` and `EST`.

In 2D workspaces, we use $\rho_{\text{SE}(2)}$, the geodesic distance in SE(2) [8], as the distance between any two single robot configurations a and b, i.e., length of shortest path in SE(2) from a to b. We also use $\rho_{w\text{SE}(2)}$, which weighs, as discussed below, the geodesic distances in $\mathbb{R}^2$ and SO(2). Similarly, in 3D workspaces, we use $\rho_{\text{SE}(3)}$ [8], the geodesic distance in SE(3), and $\rho_{w\text{SE}(3)}$, which weighs the geodesic distances in $\mathbb{R}^3$ and SO(3). We also use ρ_{L_2}, which approximates the volume of the workspace region swept by the robot [9] . We experimented with several weighting schemes for $\rho_{w\text{SE}(2)}$ and $\rho_{w\text{SE}(3)}$, but found little variation in the results of nearest-neighbors algorithms. Therefore in this study we set the weights to one. In the case of multiple robots, we sum up $\rho_{\text{SE}(2)}$, $\rho_{w\text{SE}(2)}$, $\rho_{\text{SE}(3)}$, $\rho_{w\text{SE}(3)}$, and ρ_{L_2} distances between configurations for each robot to obtain $\rho^*_{\text{SE}(2)}$, $\rho^*_{w\text{SE}(2)}$, $\rho^*_{\text{SE}(3)}$, $\rho^*_{w\text{SE}(3)}$, and $\rho^*_{\text{L}_2}$, respectively.

3 Experimental Setup

Data Sets. We use 2D and 3D workspaces, shown in Fig. 1, that provide a representative benchmark for motion planners. The "maze2d" workspace is a 2D maze, as in Fig. 1(a). Robots must move from one of the borders of the maze to the opposite border. The "narrow2d" workspace has several narrow passages, as in Fig. 1(b). Robots must move from the left side to the right side of the box. In the 3D workspaces, robots are objects in the shape of letters, as in Fig. 1(c). The "maze3d" workspace is a 3D maze, as in Fig. 1(d). Robots must move from one corner of the maze to the other and always remain inside the maze.

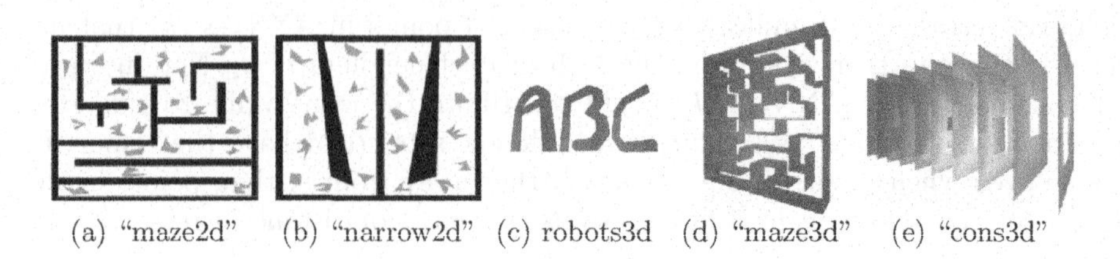

(a) "maze2d" (b) "narrow2d" (c) robots3d (d) "maze3d" (e) "cons3d"

Fig. 1. Workspaces. (a), (b) The black and gray polygons indicate obstacles and robots, respectively. (c),(d),(e) In the 3D workspaces, robots consist of 3D renderings of English letters and the i-th robot corresponds to the i-th letter.

Table 1. Summary of data sets

motion planner	PRM[uniform, bridge, Gaussian, obstacle], RRT, EST	
number of points (n)	10000, 50000, 100000	
workspace	maze2d, narrow2d	maze3d, cons3d
distance	$\rho^*_{\mathrm{SE}(2)}$, $\rho^*_{w\mathrm{SE}(2)}$	$\rho^*_{\mathrm{SE}(3)}$, $\rho^*_{w\mathrm{SE}(3)}$, $\rho^*_{\mathrm{L}_2}$
dimension (d)	$3, 6, 9, \ldots, 60$	$6, 12, 18, \ldots, 60$

The "cons3d" workspace has ten consecutive walls each with a small hole, as in Fig. 1(e). Robots must move through all the ten holes.

We created many data sets as summarized in Table 1. We use 1, 2, ..., 20 and 1, 2, ..., 10 robots in each 2D and 3D workspace to obtain configurations with 3, 6, ..., 60 and 6, 12, ..., 60 dimensions, respectively. As an example, a 60-dimensional "maze3d" problem is obtained by placing 10 robots, consisting of 3D renderings of letters A through I as in Fig. 1(c), in the "maze3d" workspace. We note that the choice of letters for the robots does not affect the results of our experiments. For each dimension, we generate data sets with 10000, 50000, and 100000 points. For each dimension and number of points, we use each motion planner to generate data sets using $\rho^*_{\mathrm{SE}(2)}$ and $\rho^*_{w\mathrm{SE}(2)}$ in each 2D workspace and $\rho^*_{\mathrm{SE}(3)}$, $\rho^*_{w\mathrm{SE}(3)}$, and $\rho^*_{\mathrm{L}_2}$ in each 3D workspace. During data generation, each motion planner uses Linear for the kNN computations.

Experiments. For each data set, we use Gnat and Linear to compute kNN queries and DPES and Random to compute kANN queries for various values of $k \in \{15, 45, 150\}$. We report only results obtained for $k = 45$, since the results for the other values of k are similar. In each case, we measure the time and distance evaluations required to compute nearest neighbors of a point $s \in S$ selected uniformly at random. In addition, for kANN algorithms, we measure the rfd_ϵ, $\epsilon \in \{0.00, 0.05, 0.10\}$, and rde errors. We obtain averages of these quantities by repeating the above step 100 times. We choose $|S'|$ such that the running time of Random is the same as that of DPES.

Platform. We utilized three high-performance computing clusters, Rice Terascale Cluster, PBC Cluster, and Rice Cray XD1 Cluster ADA.

4 Results

We compare the computational efficiency of Gnat and DPES relative to Linear for kNN and kANN computations, respectively. We also focus on the accuracy of DPES and Random. The use of Linear and Random provides a normalization of the results obtained for data sets generated using various motion planners, distance metrics, number of points, dimensions, and workspaces. We present results for "maze2d" and "maze3d" workspaces, since the correlation with results for "narrow2d" and "cons3d" workspaces is above 90%.

4.1 Exact k Nearest-Neighbors Algorithms

We compare Gnat to Linear for various distance metrics.

Using $\rho^*_{\mathbf{SE(2)}}$. We present the results in Fig. 2. These results are indicative of other distances and illustrate general trends observed in kNN algorithms. We indicate the workspace, motion planner, distance metric, and number of points at the top and legend of each figure. In Fig. 2(a), we compare the computation time of Gnat relative to Linear on data sets generated using `PRM` with uniform sampling. We observe that Gnat is more efficient than Linear on low-dimensional data sets. The efficiency of Gnat increases even more when the number of points increases. However, as the dimension increases, the efficiency of Gnat deteriorates rapidly. In fact, after a certain dimension, $d > 18$, Gnat is even less efficient than Linear.

In Fig. 2(b), we focus on the number of distance evaluations. We observe trends similar to Fig. 2(a). We note that Gnat evaluates far fewer distances than Linear on low-dimensional data sets, especially on large low-dimensional data sets. However, as in Fig. 2(a), the number of distance evaluations by Gnat relative to Linear increases rapidly with the dimension and even approaches 1.0 when $d > 18$. Since Gnat has more computational overhead than Linear, we observed in Fig. 2(a), that for $d > 18$, Gnat is less efficient than Linear.

In Fig. 2(c), we compare the computation time of Gnat relative to Linear on data sets with $n = 100000$ points generated using `PRM` with different sampling schemes. We observe that Gnat is unable to take advantage of the different distributions that result from changing the sampling in `PRM`. There is almost no variation in the efficiency of Gnat when the sampling in `PRM` is changed from uniform to bridge, Gaussian, or obstacle. Similar observations also hold for the smaller data sets.

In Fig. 2(d) and (e), we focus on `RRT` and `EST`, respectively. As in Fig. 2(a), the efficiency of Gnat relative to Linear is significantly better on low-dimensional data sets, but quickly deteriorates as the dimension increases, and becomes worse after the critical dimension. However, the critical dimension is higher for `RRT` and `EST` than for `PRM` due to the local nature of `RRT` and `EST` which create samples that are more distinctly clustered and, consequently, can be used by Gnat to eliminate certain distance computations.

Using $\rho^*_{w\mathbf{SE(2)}}$, $\rho^*_{\mathbf{SE(3)}}$, $\rho^*_{w\mathbf{SE(3)}}$, and $\rho^*_{\mathbf{L_2}}$. We present the results in Fig. 3. We compare the computation time of Gnat relative to Linear. The filled region

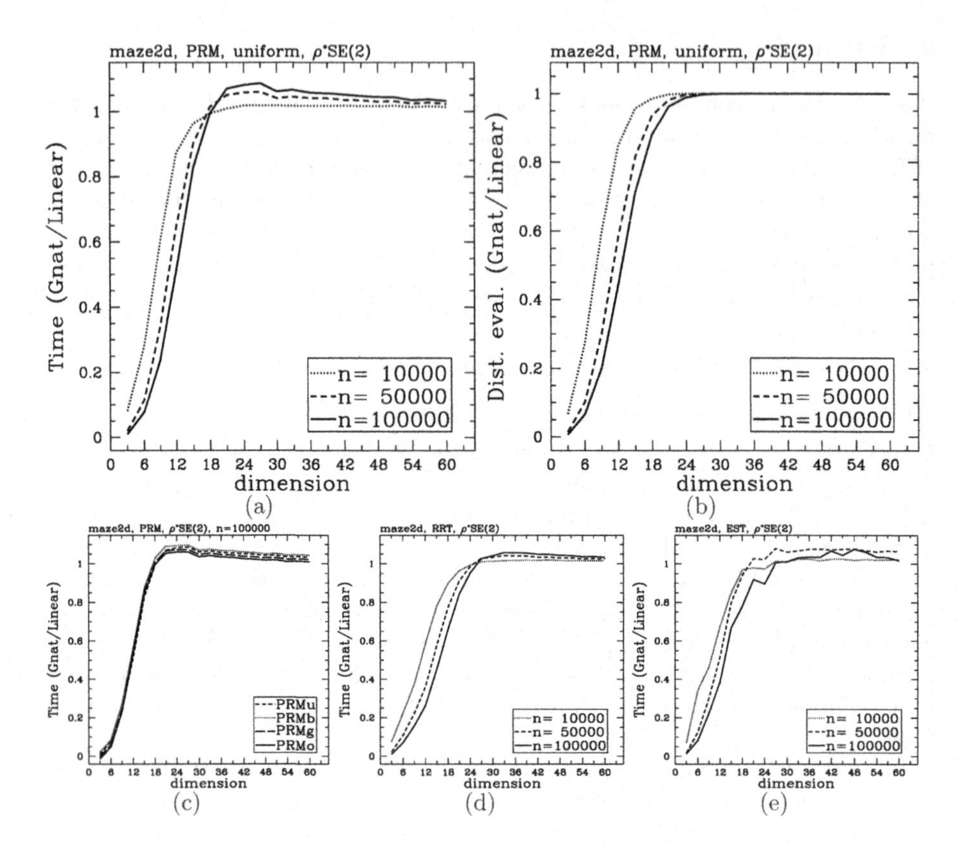

Fig. 2. Comparing Gnat to Linear when using $\rho^*_{\mathrm{SE}(2)}$

indicates the variation in the efficiency of Gnat for the different PRM versions, while the dashed and dotted lines indicate the results obtained for EST and RRT, respectively. We show results only for data sets with $n = 100000$ points, since we obtain similar results for the smaller data sets.

In Fig. 3(a), we focus on $\rho^*_{w\mathrm{SE}(2)}$. As in the case of $\rho^*_{\mathrm{SE}(2)}$, the efficiency of Gnat relative to Linear is at least one order of magnitude better on low-dimensional data sets, but rapidly decreases with the dimension. We note there is almost no variation in the efficiency of Gnat when the sampling in PRM is changed from uniform to bridge, Gaussian, or obstacle.

Similarly, in Fig. 3(b) and (c), we observe that when using $\rho^*_{\mathrm{SE}(3)}$ or $\rho^*_{w\mathrm{SE}(3)}$, the efficiency of Gnat remains the same for PRM variants, but improves for RRT and EST due to the local sampling.

In Fig. 3(d), we focus on $\rho^*_{\mathrm{L}_2}$. The efficiency of Gnat remains the same when PRM uses uniform, bridge, or obstacle sampling, as indicated by the small area of the shaded region, but decreases when Gaussian sampling is used. As before, due to the local sampling Gnat is more efficient when RRT and EST are used.

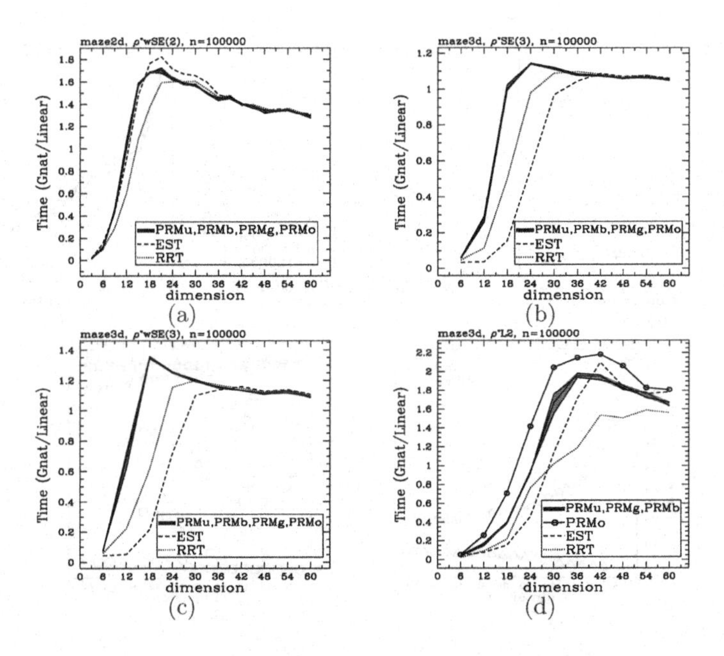

Fig. 3. Comparing Gnat to Linear when using $\rho^*_{w\mathrm{SE}(2)}$, $\rho^*_{\mathrm{SE}(3)}$, $\rho^*_{w\mathrm{SE}(3)}$, and $\rho^*_{\mathrm{L}_2}$

A comparison between Fig. 2(c, d, e) and Fig. 3(a) and between Fig. 3(b) and Fig. 3(c) indicates that in general the efficiency of Gnat is better when geodesic distances are used instead of weighted distances. Our intuition is that this is due to the decoupling of translational and rotational components which reduces the number of distinct clusters in the data set. As mentioned in Sect. 2.3, we obtained similar results for several different weighting schemes.

4.2 Approximate k Nearest-Neighbors Algorithms

In this section, we analyze the efficiency and accuracy of DPES using Linear and Random as the basis of comparison, respectively. We present results for various distance metrics. In the experiments presented in this section, DPES uses $m = 15$ pivots for the projection. These results are indicative of the behavior of DPES. In the next section, we present results where we vary m.

Using $\rho^*_{\mathrm{SE}(2)}$. We focus on PRM variants in Fig. 4 and RRT and EST in Fig. 5. In Fig. 4(a), we compare the computation time of DPES relative to Linear on data sets generated using PRM with uniform sampling. We observe that for low-dimensional and small data sets, DPES is less efficient than Linear, since data sets are projected onto $\mathbb{R}^{15}$. However, as the dimension increases, the efficiency of DPES relative to Linear improves rapidly. The improvement is even greater on the larger data sets.

In Fig. 4(b), we focus on distance evaluations for the same data sets as in Fig. 4(a). Recall that DPES uses L_2 in $\mathbb{R}^{15}$, while Linear uses $\rho^*_{\mathrm{SE}(2)}$. We note

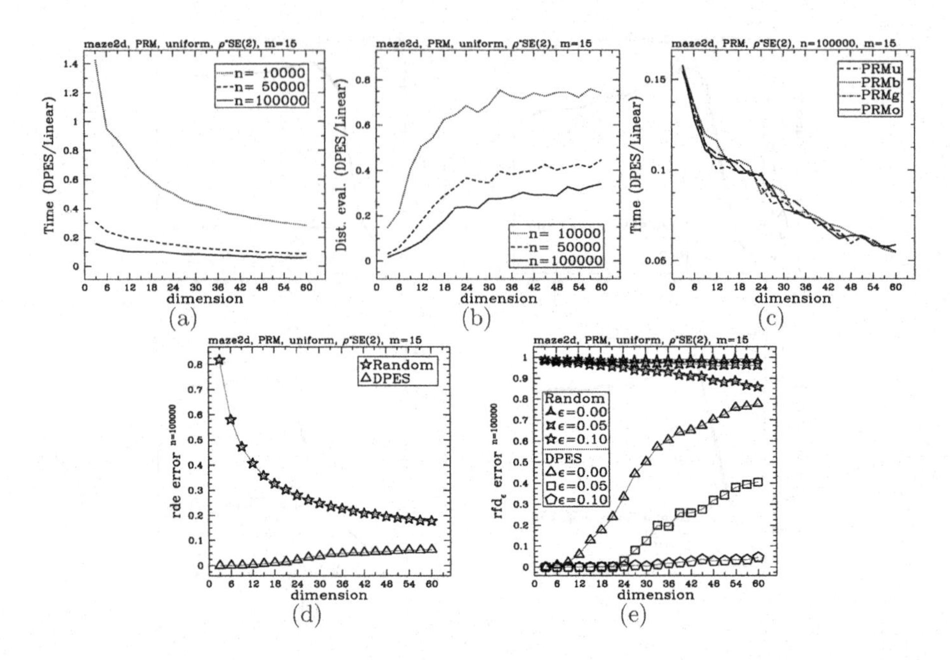

Fig. 4. kANN results for PRM and the $\rho^*_{\mathrm{SE}(2)}$ distance metric

that although the number of distance evaluations by DPES relative to Linear increases with the dimension, it decreases with the number of points. In general, DPES evaluates only a fraction of distances to the query point.

In Fig. 4(c), we compare the computation time of DPES relative to Linear on data sets with $n = 100000$ points generated using different PRM variants. We observe only small changes in the computational time of DPES when the sampling in PRM is changed from uniform to bridge, Gaussian, and obstacle. We obtain similar results for the smaller data sets as well.

In Fig. 4(d), we compare the rde error of Random and DPES on data sets with $n = 100000$ points generated using PRM with uniform sampling. The rde error of Random is high on low-dimensional data sets but decreases with the dimension. This is due to the sparsity of data on high-dimensional spaces which as shown in [4] implies that the relative distances between points decrease as the dimension increases. On the other hand, the rde error of DPES is small on low-dimensional data sets but increases, although only slightly, with the dimension. The increase in the rde error of DPES is a consequence of the projection onto a low-dimensional Euclidean space, i.e., $\mathbb{R}^{15}$. In all cases however the rde error of DPES relative to Random is at least 2.5 times smaller.

In Fig. 4(e), we focus on the rfd error using the same data sets as in Fig. 4(d). The rfd error of Random remains very high even when the dimension increases. Such high values indicate that even though relative distances between points decrease with the dimension, as seen in Fig. 4(d), there is a clear distinction between the nearest neighbors and other points in the data set. A similar

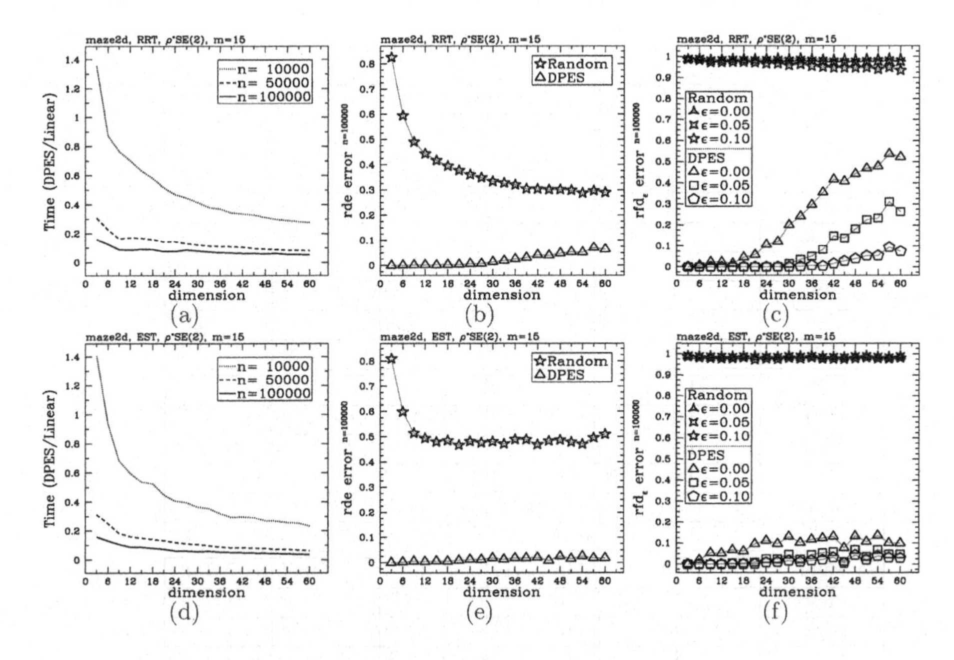

Fig. 5. kANN results for RRT and EST and the $\rho^*_{\mathrm{SE}(2)}$ distance metric

observation has also been made in [12], where it is shown that under certain conditions nearest neighbors are meaningful on high-dimensional data sets. In the case of DPES, we observe that the rfd error increases with the dimension. The increase is more rapid when $\epsilon = 0.00$. This is expected since $\mathrm{rfd}_{0.00}$ indicates how many points in $\mathrm{ANN}_S(s_i, k)$ are not in $\mathrm{NN}_S(s_i, k)$. However, as ϵ increases, the rfd error of DPES decreases significantly and for $\epsilon = 0.10$ comes close to zero even on the high-dimensional data sets.

The accuracy results in Fig. 4(d) and (e) indicate that although the approximate nearest neighbors computed by DPES are not the same as the exact nearest neighbors, the differences between them are small. In Sect. 4.3, we show how to further improve the accuracy of DPES.

In Fig. 5(a) and (d), we compare the efficiency of DPES relative to Linear on data sets generated using RRT and EST, respectively. As in the case of PRM in Fig. 4(a), the relative efficiency improves rapidly with the dimension and for $d > 12$, DPES is several times faster than Linear.

In Fig. 5(b) and (e), we compare the rde error of Random and DPES on data sets with $n = 100000$ points generated using RRT and EST, respectively. We obtain similar results on the smaller data sets. In addition to the observations made for Fig. 4(d), we note that for RRT and especially EST the rde error of Random is larger while the rde error of DPES is smaller than for PRM. This is due to the local sampling of RRT and EST, which generate data sets where relative distances between points are more distinct, especially in the case of EST which expands slower than RRT. Consequently, the likelihood that a random point is a

Table 2. Summary of kANN results for various distance metrics

$n = 100000$, $d = 60$	Efficiency ($t_{\text{DPES}}/t_{\text{Linear}}$)	Accuracy (•DPES ⋆Random) rde •	rde ⋆	rfd $\epsilon=0.00$ •	$\epsilon=0.00$ ⋆	$\epsilon=0.05$ •	$\epsilon=0.05$ ⋆	$\epsilon=0.10$ •	$\epsilon=0.10$ ⋆	
PRMu	0.40	0.07	0.17	0.80	0.99	0.46	0.95	0.04	0.84	$\rho^*_{w\text{SE}(2)}$
PRMb	0.38	0.06	0.17	0.79	0.99	0.42	0.95	0.04	0.84	
PRMg	0.42	0.07	0.17	0.80	0.99	0.45	0.95	0.05	0.84	
PRMo	0.39	0.06	0.17	0.79	0.99	0.42	0.96	0.03	0.85	
RRT	0.36	0.06	0.28	0.54	0.99	0.30	0.97	0.06	0.93	
EST	0.24	0.03	0.49	0.14	0.99	0.08	0.98	0.06	0.98	
PRMu	0.08	0.07	0.19	0.78	0.99	0.45	0.96	0.08	0.88	$\rho^*_{\text{SE}(3)}$
PRMb	0.08	0.07	0.19	0.79	0.98	0.46	0.96	0.08	0.88	
PRMg	0.08	0.07	0.19	0.78	0.99	0.46	0.96	0.09	0.89	
PRMo	0.08	0.07	0.19	0.80	0.99	0.51	0.96	0.11	0.88	
RRT	0.06	0.05	0.25	0.55	0.99	0.20	0.97	0.02	0.94	
EST	0.07	0.06	0.25	0.58	0.99	0.27	0.97	0.06	0.94	
PRMu	0.14	0.07	0.17	0.81	0.99	0.48	0.95	0.07	0.84	$\rho^*_{w\text{SE}(3)}$
PRMb	0.15	0.07	0.17	0.81	0.99	0.49	0.96	0.08	0.85	
PRMg	0.14	0.07	0.17	0.82	0.99	0.49	0.96	0.09	0.86	
PRMo	0.14	0.07	0.17	0.82	0.99	0.50	0.95	0.10	0.85	
RRT	0.11	0.05	0.25	0.56	0.99	0.22	0.97	0.03	0.95	
EST	0.11	0.05	0.25	0.57	0.99	0.25	0.97	0.04	0.94	
PRMu	0.50	0.02	0.24	0.39	0.99	0.04	0.97	0.00	0.94	$\rho^*_{\text{L}_2}$
PRMb	0.46	0.02	0.24	0.38	0.99	0.03	0.97	0.00	0.94	
PRMg	0.49	0.02	0.24	0.37	0.99	0.04	0.97	0.00	0.94	
PRMo	0.46	0.03	0.22	0.45	0.99	0.07	0.97	0.01	0.93	
RRT	0.29	0.01	0.29	0.20	0.99	0.00	0.98	0.00	0.96	
EST	0.44	0.01	0.28	0.14	0.99	0.00	0.98	0.00	0.95	

nearest neighbor decreases while the projection done by DPES better preserves the relative distances between points.

In Fig. 5(c) and (f), we focus on the rfd error using the same data sets as in Fig. 5(b) and (e). In addition to the observations made for Fig. 4(e), we note that the rfd error of Random remains high, while the rfd error of DPES decreases when RRT and especially EST are used instead of PRM. In fact, in the case of EST, even the $\text{rfd}_{0.00}$ error of DPES is less than 0.1, which indicates that DPES computes above 90% of the exact nearest neighbors.

Using $\rho^*_{w\text{SE}(2)}$, $\rho^*_{\text{SE}(3)}$, $\rho^*_{w\text{SE}(3)}$, and $\rho^*_{\text{L}_2}$. We summarize the results obtained for the other distance metrics in Table 2. We focus on data sets with $n = 100000$ points and $d = 60$ dimensions generated using the "maze2d" and "maze3d" workspaces. The results for the other data sets are similar. For each motion planner, we present the computation time of DPES relative to Linear and the rde and rfd_ϵ, $\epsilon \in \{0.00, 0.05, 0.10\}$, errors of DPES and Random.

As in the case of $\rho^*_{\mathrm{SE}(2)}$, DPES is more efficient than Linear. The improvements vary between 2–4 and 12–16 times on the high-dimensional data sets, in the worst and best cases, corresponding to $\rho^*_{\mathrm{L}_2}$ and $\rho^*_{\mathrm{SE}(3)}$, respectively. In addition, the efficiency of DPES remains almost the same when the sampling in PRM is changed from uniform to bridge, Gaussian, or obstacle. However, DPES is generally more efficient in the case of RRT and EST.

We observe in Table 2 that DPES achieves high quality especially in the case of $\rho^*_{\mathrm{L}_2}$. The small values of rde indicate that approximate nearest neighbors computed by DPES are very close to exact nearest neighbors. This is further confirmed by the small values of the rfd error of DPES for $\epsilon \geq 0.05$ when $\rho^*_{\mathrm{L}_2}$ is used and $\epsilon \geq 0.10$ when the other distance metrics are used.

4.3 Improving the Quality of *k*ANN Queries Computed by DPES

The quality of DPES can be improved by increasing the dimension of the Euclidean space onto which data sets are projected. The results in Sect. 4.2 are obtained by using $m = 15$ pivots. In Fig. 6, we present results where we vary the number of pivots $m \in \{10, 30, 50\}$. We focus on large data sets generated using $\rho^*_{\mathrm{SE}(2)}$ and PRM with uniform sampling, since we obtain similar results with the other data sets and distance metrics.

In Fig. 6(a), we compare the computation time of DPES relative to Linear. As expected, the computation time of DPES relative to Linear increases as the number of pivots increases. However, even for $m = 50$, DPES is still several

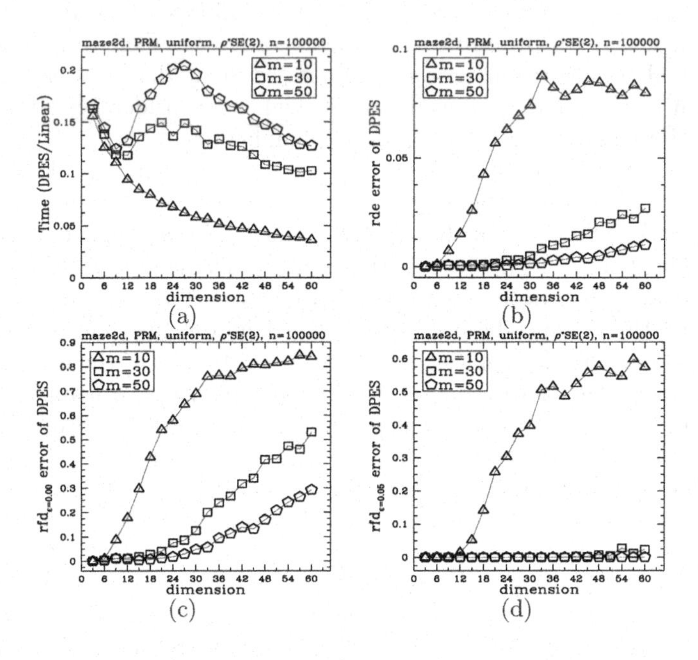

Fig. 6. Improved *k*ANN results for PRM and the $\rho^*_{\mathrm{SE}(2)}$ distance metric

times faster than Linear. As the dimension increases, the improvement in the computation time of DPES relative to Linear increases as well.

In Fig. 6(b), we focus on the rde error of DPES. As before, we observe that the rde error of DPES increases with the dimension but still remains small even when $d = 60$ and $m = 10$. Furthermore, as the number of pivots increases, the rde error of DPES quickly approaches zero.

In Fig. 6(c), we focus on the rfd error for $\epsilon = 0.00$. We note that the $\text{rfd}_{0.00}$ error of DPES increases with the dimension but decreases rapidly as the number of pivots increases. In fact, when $d = 54$ and $m = 50$, the $\text{rfd}_{0.00}$ error is less than 0.20. This indicates that DPES includes at least 80% of the exact k nearest neighbors in the computed approximate k nearest neighbors.

In Fig. 6(d), we focus on the rfd error for $\epsilon = 0.05$. We again note that the $\text{rfd}_{0.05}$ error of DPES increases with the dimension but decreases rapidly and approaches 0 as the number of pivots increases. In fact, when $m = 50$, the $\text{rfd}_{0.05}$ error of DPES is 0.00, i.e., all the approximate nearest neighbors are no more than 1.05 times farther away from the k-th nearest neighbor.

5 Discussion

In this work, we quantitatively analyzed exact and approximate nearest-neighbors algorithms for points obtained from sampling-based motion planning methods in high-dimensional problems.

Our analysis indicates that the computational efficiency of exact nearest-neighbors algorithms deteriorates rapidly as the dimension increases. After a critical dimension, which in our experiments varied between 15 and 30, exact nearest-neighbors algorithms evaluate almost as many distances as the brute-force Linear method and are thus impractical when a considerably large number of samples is necessary for solving motion planning problems.

Motivated by the impracticality of exact nearest-neighbors algorithms on high-dimensional motion planning problems, we developed a simple approximate nearest-neighbors algorithm, DPES, which achieves high computational efficiency and only a negligible loss in accuracy. The computational efficiency of DPES relative to Linear improves rapidly as the dimension increases. This is due to (i) the distance-based projection of high-dimensional data sets onto low-dimensional Euclidean spaces reduces to a certain extent the computational dependencies on the dimension (ii) the number of distance computations by DPES relative to Linear is only a small fraction; and (iii) DPES uses L_2 which is computationally more efficient than $\rho^*_{\mathrm{SE}(2)}$, $\rho^*_{\mathrm{SE}(2)}$, $\rho^*_{w\mathrm{SE}(3)}$, $\rho^*_{w\mathrm{SE}(3)}$, or $\rho^*_{\mathrm{L}_2}$. Our analysis also shows that DPES is highly accurate. In the computed queries, DPES includes many of the exact nearest neighbors and the rest are close to the exact nearest neighbors.

Since in motion planning the purpose of nearest neighbors is to provide candidates which the local planner can connect to the query point, using approximate nearest neighbors that are similar to exact nearest neighbors may indeed be sufficient. This paper shows that in high-dimensional motion planning problems nearest neighbors can be computed more efficiently by using highly accurate approximate nearest-neighbors algorithms, such as DPES, instead of exact nearest-neighbors algorithms.

Acknowledgement. Work on this paper by the authors has been supported in part by NSF 0205671, NSF 0308237, ATP 003604-0010-2003, NIH GM078988, and a Sloan Fellowship to LK. The Rice Terascale Cluster, PBC Cluster, and Rice Cray XD1 Cluster ADA used in this work are supported by EIA 0216467, CNS 0454333, CNS 0421109, AMD, Cray, Intel, and Hewlett Packard.

References

1. Amato, N.M., Bayazit, B., Dale, L., Jones, C., Vallejo, D.: OBPRM: An obstacle-based PRM for 3d workspaces. In: Agarwal, P., Kavraki, L.E., Mason, M. (eds.) Robotics: The Algorithmic Perspective, pp. 156–168. AK Peters (1998)
2. Arya, S., Mount, D.M., Nathan, S.: An optimal algorithm for approximate nearest neighbor searching in fixed dimensions. Journal of the ACM 45(6), 891–923 (1998)
3. Atramentov, A., LaValle, S.M.: Efficient nearest neighbor searching for motion planning. In: IEEE International Conference on Robotics and Automation, Washington, DC, pp. 632–637 (2002)
4. Beyer, K., Goldstein, J., Ramakrishnan, R., Shaft, U.: When is "nearest neighbor" meaningful? In: Beeri, C., Bruneman, P. (eds.) ICDT 1999. LNCS, vol. 1540, pp. 217–235. Springer, Heidelberg (1998)
5. Boor, V., Overmars, M.H., van der Stappen, A.F.: The gaussian sampling strategy for probabilistic roadmap planners. In: IEEE International Conference on Robotics and Automation, Detroit, MI, pp. 1018–1023 (1999)
6. Borodin, A., Ostrovsky, R., Rabani, Y.: Lower bounds for high dimensional nearest neighbor search and related problems. In: ACM Symposium on Theory of Computing, Atlanta, GA, pp. 312–321 (1999)
7. Brin, S.: Near neighbor search in large metric spaces. In: International Conference on Very Large Data Bases, San Francisco, California, pp. 574–584 (1995)
8. Bullo, F., Murray, R.M.: Proportional derivative (PD) control on the Euclidean group. In: European Control Conference, Rome, Italy, pp. 1091–1097 (1995)
9. Choset, H., Lynch, K.M., Hutchinson, S., Kantor, G., Burgard, W., Kavraki, L.E., Thrun, S.: Principles of Robot Motion: Theory, Algorithms, and Implementations. MIT Press, Cambridge (2005)
10. Cui, B., Shen, H.T., Shen, J., Tan, K.-L.: Exploring bit-difference for approximate knn search in high-dimensional databases. In: Australasian Database Conference, Newcastle, Australia, pp. 165–174 (2005)
11. Gaede, V., Günther, O.: Multidimensional access methods. ACM Computing Surveys 30(2), 170–231 (1998)
12. Hinneburg, A., Aggarwal, C.C., Keim, D.A.: What is the nearest neighbor in high dimensional spaces? In: International Conference on Very Large Data Bases, Cairo, Egypt, pp. 506–515 (2000)
13. Hsu, D., Jiang, T., Reif, J., Sun, Z.: The bridge test for sampling narrow passages with probabilistic roadmap planners. In: IEEE International Conference on Robotics and Automation, Taipei, Taiwan, pp. 4420–4442 (2003)

14. Hsu, D., Kindel, R., Latombe, J.-C., Rock, S.: Randomized kinodynamic motion planning with moving obstacles. International Journal of Robotics Research 21(3), 233–255 (2002)
15. Indyk, P.: Nearest neighbors in high-dimensional spaces. In: Goodman, J.E., O'Rourke, J. (eds.) Handbook of Discrete and Computational Geometry, pp. 877–892. CRC Press, Boca Raton (2004)
16. Indyk, P., Matoušek, J.: Low-distortion embeddings of finite metric spaces. In: Goodman, J.E., O'Rourke, J. (eds.) Handbook of Discrete and Computational Geometry, pp. 177–196. CRC Press, Boca Raton (2004)
17. Kavraki, L.E., Švestka, P., Latombe, J.-C., Overmars, M.H.: Probabilistic roadmaps for path planning in high-dimensional configuration spaces. IEEE Transactions on Robotics and Automation 12(4), 566–580 (1996)
18. Korn, F., Pagel, B.-U., Faloutsos, C.: On the 'dimensionality curse' and the 'self-similarity blessing'. IEEE Transactions on Knowledge and Data Engineering 13(1), 96–111 (2001)
19. Kushilevitz, E., Ostrovsky, R., Rabani, Y.: Efficient search for approximate nearest neighbor in high dimensional spaces. SIAM Journal of Computing 30(2), 457–474 (2000)
20. LaValle, S.M., Kuffner, J.J.: Randomized kinodynamic planning. International Journal of Robotics Research 20(5), 378–400 (2001)
21. Liu, T., Moore, A.W., Gray, A., Yang, K.: An investigation of practical approximate nearest neighbor algorithms. In: Saul, L.K., Weiss, Y., Bottou, L. (eds.) Advances in Neural Information Processing Systems, pp. 825–832. MIT Press, Cambridge (2005)
22. Plaku, E., Bekris, K.E., Chen, B.Y., Ladd, A.M., Kavraki, L.E.: Sampling-based roadmap of trees for parallel motion planning. IEEE Transactions on Robotics 21(4), 597–608 (2005)
23. Weber, R., Schek, H.-J., Blott, S.: A quantitative analysis and performance study for similarity-search methods in high-dimensional spaces. In: International Conference on Very Large Data Bases, pp. 194–205. New York (1998)

Path Deformation Roadmaps

Léonard Jaillet and Thierry Siméon

LAAS-CNRS, Toulouse, France
{ljaillet,nic}@laas.fr

Abstract. This paper describes a new approach to sampling-based motion planning with PRM methods. Our aim is to compute good quality roadmaps that encode the multiple connectedness of the Cspace inside small but yet representative graphs, that capture well the different varieties of free paths. The proposed approach relies on a notion of path deformability indicating whether or not a given path can be continuously deformed into another existing one. By considering a simpler form of deformation than the one allowed between homotopic paths, we propose a method that extends the Visibility-PRM technique [12] to constructing compact roadmaps that encode a richer and more suitable information than representative paths of the homotopy classes. The Path Deformation Roadmaps also contain additional useful cycles between paths in the same homotopy class that can be hardly deformed into each other. First experiments presented in the paper show that our technique enables small roadmaps to reliably and efficiently capture the multiple connectedness of the space in various problems.

1 Introduction

Robot motion planning has led to active research over the past decades [5] and sampling-based planning techniques have now emerged as a general and effective framework for solving challenging problems that remained out of reach of the previously existing complete algorithms. Today they make it possible to handle the complexity of many practical problems arising in such diverse fields as robotics, graphics animation, virtual prototyping and computational biology. In particular, the Probabilistic RoadMap planner (PRM) introduced in [4, 8] and further developed in many other works (see [2, 6] for a survey) has been conceived to solve multiple-query problems.

While most of the PRM variants focus on the fast computation of roadmaps reflecting the connectivity of the free configuration space, only few works [7, 9] address the problem of computing good quality roadmaps that encode the multiple connectedness of the space inside small graphs containing only useful cycles, ie. cycles representative of the varieties of free paths. Introducing such cycles is important for getting higher quality solutions when postprocessing queries, thus avoiding the computation of unnecessarily long paths, difficult to shorten by the smoothing techniques (e.g. [10, 13]).

S. Akella et al. (Eds.): Algorithmic Foundation of Robotics VII, STAR 47, pp. 19–34, 2008.
springerlink.com

Intuitively, the probability that a roadmap captures well the different paths varieties of $\mathcal{C}_{free}$ increases with its degree of redundancy. However, a direct approach attempting connections between all pair of nodes is far too costly and several heuristic-based connection strategies have been proposed to limit the number of redundant connections. A first way (e.g. [4]) is to limit the connection attempts of new samples to the k nearest nodes of the roadmap (or of each connected component). Another variant is to only consider nodes within a ball of radius r centered at the new sampled configuration (e.g. [1]). A more recent technique proposed in [7] only creates cycles between already connected nodes if they are k times more distant in the roadmap than in the configuration space. In all cases, the chances of capturing the different path varieties of $\mathcal{C}_{free}$ notably varies depending on the choice of the k or r parameter. Moreover it is difficult to choose with these heuristic sampling strategies the good parameter values for a given environment. This may result in a significant loss of performance regarding the roadmap construction process.

In this paper we present an alternative method to building compact roadmaps that are yet representative of the different varieties of free paths. The method only generates a limited number of useful cycles in the roadmap. Moreover it stops automatically when most of the relevant alternative paths have been found. Our approach relies on a notion of path deformability indicating whether or not a given path can be continuously deformed into another existing one. Compared with the standard notion of homotopy which is not directly suitable for our purpose because it relies on too complicated deformations (Sect. 2), we consider simpler and more easily computable deformations between paths (Sect. 3). This results in compact roadmaps capturing a richer set of paths than homotopy (Sect. 4). We describe in Section 5 a two-stage algorithm for constructing such (easy) path deformation roadmaps. The first stage uses Visibility-PRM [12] to construct a small tree covering the space and capturing its connected components as well as possible. The second stage aims at enriching the roadmap with new nodes involved in the creation of useful cycles. The key ingredient of this step is an efficient path visibility test used for the filtering of useless cycles that can be easily deformed into existing roadmap paths. Following the philosophy of Visibility-PRM, the second stage integrates a stop condition based on the difficulty of finding new useful cycles. Finally, our first experiments (Sect. 6) show that the technique enables small roadmaps to reliably capture the multiple connectedness of configuration spaces in various problems involving free flying or articulated robots.

2 Homotopy Versus Useful Roadmap Paths

First we informally discuss the relation between homotopy and the representative path varieties that it would be desirable to store in the roadmap. The capture of the homotopy classes of $\mathcal{C}_{free}$ corresponds to a stronger property than connectivity. Two paths are called homotopic (with fixed end points) if one can be "continuously deformed" into the other (see section 3.1). Homotopy defines an equivalence relation on the set of all paths of $\mathcal{C}_{free}$. A roadmap capturing the homotopy classes means that every valid path (even cyclic paths) can be continuously deformed into a path of the roadmap. PRM methods usually do not ensure this property. Only the work of Schmitzberger [9] considers the problem formally and sketches a method for encoding the set of homotopy classes inside a probabilistic roadmap. However, the approach is only applied on two-dimensional problems and its extension is limited by the difficulty of characterizing homotopic deformations in higher dimensions.

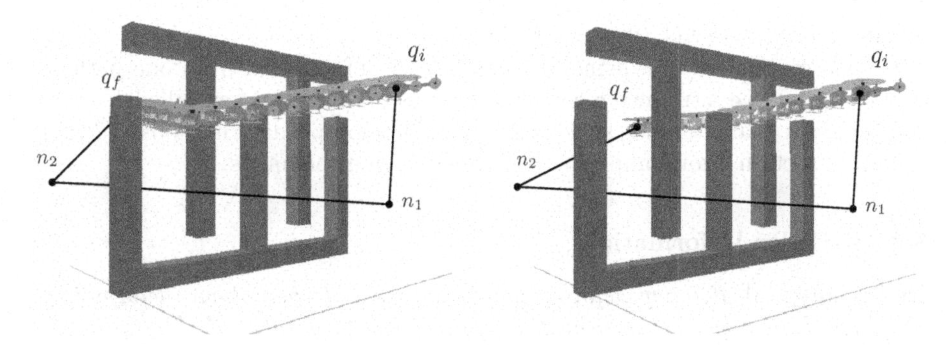

Fig. 1. Two examples of query for a 2 nodes graph (n_1-n_2). In the left picture, the solution path (q_i-n_1-n_2-q_f) extracted from the graph could be easily deformed into the displayed short path connecting query configurations (q_i, q_g) whereas a deformation in $\mathcal{C}_{free}$ would be much complex in the case of the right picture.

Moreover, as it was noted in [7] capturing the homotopy classes in higher dimensions may not be sufficient to encode the set of representative paths since homotopic paths (i.e. paths in the same homotopy class) may be too hard to deform into each other. This problem is illustrated by the example in Figure 1. Here $\mathcal{C}_{free}$ contains only one homotopy class. Therefore, an homotopy-based roadmap would have a tree structure, such as the simple 2 nodes (n_1,n_2) tree shown in the figure. While for the left query example, the solution path (q_i-n_1-n_2-q_f) found in the roadmap could be easily deformed into the displayed short path connecting query configurations (q_i, q_g), a free deformation would be much difficult to compute for the right example. Even if the topological nature of the two displayed paths is the same, their difference is such that it is preferable to store a representation of both paths in the roadmap. Generalizing this idea, we say that a roadmap is a good representation of the varieties of free paths if any path can be "easily" deformed into a path of the roadmap. This notion of simple path deformation is formalized below.

3 Complexity of a Path Deformation

In this section, after a brief reminder of the definition of a homotopic deformation, we propose a way to characterize classes of path deformations according to their complexity.

3.1 Homotopy

The homotopy between two paths is a standard notion from Topology (see [3] for a complete definition). Two paths τ and τ' in a topological space X are *homotopic* (with fixed end points) if there exists a continuous map $h : [0,1] \times [0,1] \to X$ with $h(s,0) = \tau(s)$ and $h(s,1) = \tau'(s)$ for all $s \in [0,1]$ and $h(0,t) = h(0,0)$ and $h(1,t) = h(1,0)$ for all $t \in [0,1]$.

Homotopy is a way to define any continuous deformation from one path to another. Next, we introduce a less general class of deformations, called *K-order deformations* characterizing specific subsets of homotopic deformations and that is used in section 4 for computing path deformation roadmaps.

3.2 K-Order Deformation

Definition 1. *A K-order deformation is a particular homotopic deformation such that each curve transforming a point of τ into a point of τ' is an angle line of K segments.*

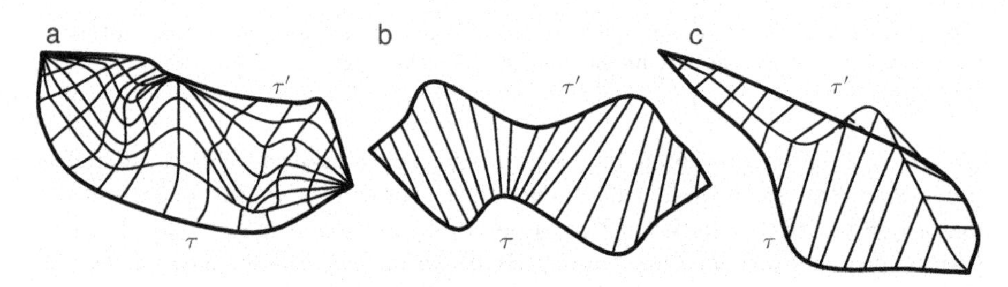

Fig. 2. (a) General homotopic deformation. (b) first order deformation: the deformation surface is a ruled surface. (c) Second order deformation: the deformation surface is obtained by concatenating two ruled surfaces.

Therefore, a first-order deformation surface describes a ruled surface and a K-order deformation is obtained by concatenation of K ruled surfaces. This is illustrated by Figure 2, which shows different types of path deformations: (a) is a general homotopic deformation whereas (b) and (c) respectively show 1st-order and a 2nd-order deformations.

Let D_i denote the set of i-order deformations. We clearly have $D^i \subset D^j$ for all $i < j$. Thus, the value K of the smallest K-order deformation existing between two paths is a good measure of the difficulty to deform one path into the other.

3.3 Visibility Diagram of Paths

It is important to note that a first-order deformation between two paths exists if and only if it is possible to simultaneously go through the two paths while maintaining a visibility constraint between the points of each path (see Figure 3). This formulation provides a computational way to test the existence of a first-order deformation, also called *visibility deformation* between two paths.

Let $\mathcal{L}_{lin}$ be the straight line segment between two configurations of $\mathcal{C}$. The parametric visibility function Vis of two paths (τ, τ') is defined as follows:

$$Vis : \begin{cases} [0,1] \times [0,1] \rightarrow \{0,1\} \\ Vis(t,t') \quad = 1 \text{ if } \mathcal{L}_{lin}(\tau(t), \tau'(t')) \in \mathcal{C}_{free} \\ Vis(t,t') \quad = 0 \text{ otherwise} \end{cases}$$

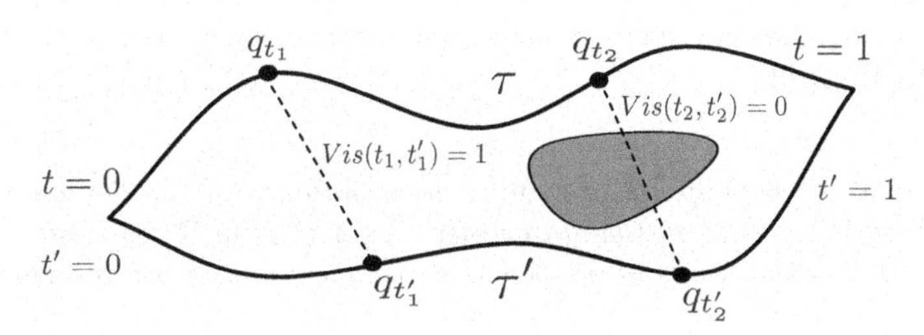

Fig. 3. The parametric visibility function of two paths evaluates the visibility between the points of each path

Then, the visibility diagram of paths (τ, τ') is defined as the two-dimensional diagram of the Vis function. It is illustrated by Figure 4 showing several examples of computed visibility diagrams with the corresponding paths.

Thanks to the visibility diagram, the visibility (i.e. first-order) deformation between two paths can now be expressed as follows: two paths $(\tau,\ \tau')$ (with the same endpoints) are *visibility deformable* one into the other if and only if there is a path in their visibility diagram linking the points of parameters $(0, 0)$ and $(1, 1)$. Therefore it is possible to test the visibility deformation between two paths by computing their visibility diagram and then searching for a path in the diagram linking the points $(0, 0)$ and $(1, 1)$. In Figure 4, such a deformation is only possible for the last example (d).

4 K-Order Deformation Roadmap

In the previous section we have defined a way to characterize the complexity for two paths to be deformed one into the other. This formalism is now used to define, for a given roadmap, its ability to capture the different varieties of free paths of the configuration space.

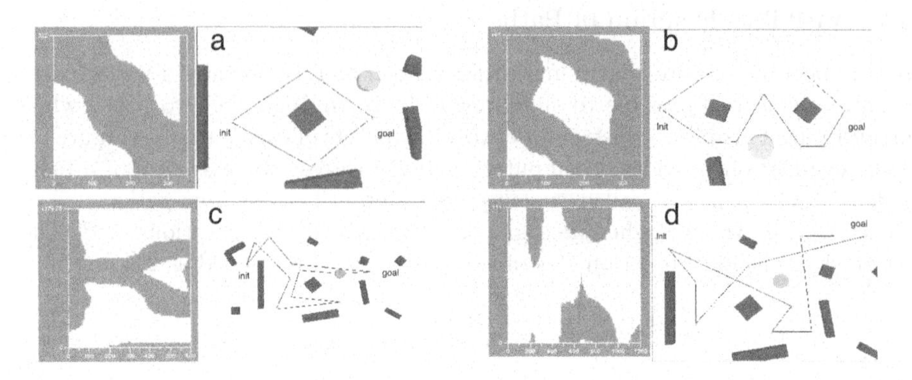

Fig. 4. Visibility diagrams for pairs of paths with the same endpoints. White areas represent regions where $Vis(t, t') = 1$. A visibility deformation is only possible in the last example (d), where a valid path linking the points (0,0) and (1,1) can be found in the visibility diagram.

Definition 2. *A roadmap R is a K-order deformation roadmap if and only if for any path τ of $\mathcal{C}_{free}$ it is possible to extract a path τ' from R (by connecting the two endpoint configurations of the paths) such that τ and τ' are K-deformable.*

This definition establishes a strong criterion specifying how the different varieties of free paths are captured inside the roadmap. One can also note that since a K-order deformation is a specific kind of homotopic deformation, any deformation roadmap captures the homotopy classes of $\mathcal{C}_{free}$. The following subsections present a computational method to construct such roadmaps.

4.1 Visibility Deformation Roadmap

We first define the notion of *Roadmap Connected from any Point of View* (called *RCPV* roadmaps) previously introduced in [9]. Then we establish that RCPV roadmaps are visibility (i.e. first-order) deformation roadmaps.

Visible Subroadmap

Let R be a roadmap with a set N of nodes and a set E of edges. If R covers $\mathcal{C}_{free}$, we can extract a set of nodes N_g (called guards) maintaining this coverage. Then, we can define for a free configuration q_v, the *Visible Subroadmap* $R_v = (N_v, E_v)$, as follows :

- N_v sublist of guards visible from q_v: $N_v = \{n \in N_g / \mathcal{L}_{lin}(q_v, n) \in \mathcal{C}_{free}\}$
- E_v, sublist of edges visible from q_v: $E_v = \{e \in E / \mathcal{L}_{lin}(q_v, e) \in \mathcal{C}_{free}\}$

Note that the notation $\mathcal{L}_{lin}(q_v, e) \in \mathcal{C}_{free}$ means that $\{\forall q \in e,\ \mathcal{L}(q_v, q) \in \mathcal{C}_{free}\}$. Examples of visible subroadmaps are presented in Figure 5.

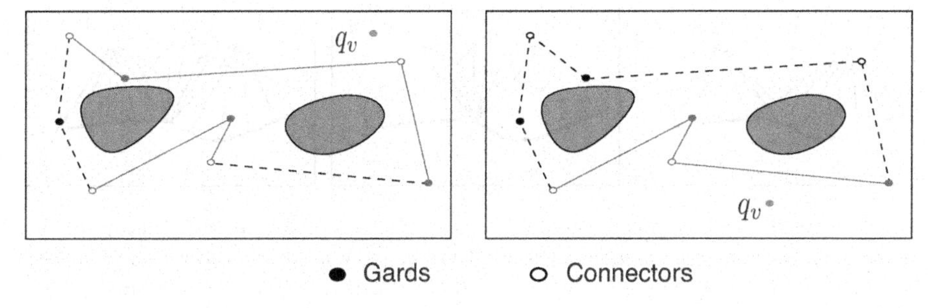

Fig. 5. Two examples of visible subroadmap from a given configuration q_v. On the left, the visible subroadmap is disconnected whereas it is connected on the right.

RCPV Roadmaps

Definition 3. *A Roadmap Connected from any Point of View (or RCPV roadmap) is such that for any configuration of $\mathcal{C}_{free}$, the visible subroadmap is connected.*

The following property establishes the link between RCPV roadmaps and visibility deformation roadmaps.

Property: A RCPV roadmap is a particular case of visibility deformation roadmap.

Sketch of proof: Let R be a RCPV roadmap and τ, a path of $\mathcal{C}_{free}$. τ can be partitioned into $2n-1$ successive paths:

$$\tau = \{\tau_{g_1} \oplus \tau_{g_1 \cap g_2} \oplus ... \oplus \tau_{g_i} \oplus \tau_{g_i \cap g_{i+1}} \oplus \tau_{g_{i+1}} \oplus ...\tau_{g_{n-1}} \oplus \tau_{g_{n-1} \cap g_n} \oplus \tau_{g_n}\}$$

with τ_{g_i} denoting the portion of path only visible from the g_i guard and $\tau_{g_i \cap g_{i+1}}$ the portion visible simultaneously from g_i and g_{i+1}. Since τ_{g_i} and g_i are by definition visible, it is possible to build a patch of ruled surface between them (Figure 6.a). Similarly, there is a patch of ruled surface between $\tau_{g_{i+1}}$ and g_{i+1}. Because R is a RCPV roadmap, any configuration $q_v \in \tau_{g_i} \cap \tau_{g_{i+1}}$ sees a path τ' connecting g_i to g_{i+1}. This property makes it possible to build a third patch of ruled surface between q_v and τ' (Figure 6.b). Finally, it is possible to fuse these three patches into a single ruled surface between $\tau_{g_i} \cap \tau_{g_{i+1}}$ and τ' (Figure 6.c). Thus, there exists a ruled surface (i.e. a visibility deformation surface) between the totality of τ and a path of the roadmap.

RCPV roadmaps are first-order deformation roadmaps. However, these roadmaps involve a high level of redundancy (see results section 6) and yet contain many useless cycles, especially in constrained situations. Therefore, to keep a compact structure we filter a part of the redundancy as explained in the following section. We will show that this filtering leads to a second-order deformation roadmap.

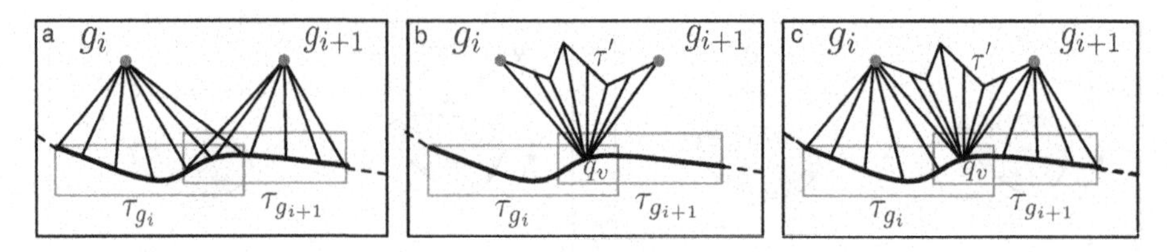

Fig. 6. A RCPV roadmap is a visibility deformation roadmap. (a) the visibility of the guards gives first patches of ruled surfaces. (b) the RCPV roadmap property guarantees the visibility of a roadmap path connecting two guards. (c) By construction, a global visibility deformation surface can be built.

4.2 Second-Order Deformation Roadmaps

Let R be a RCPV roadmap, $N_g \in R$ be a set of guard nodes ensuring the $\mathcal{C}_{free}$ coverage. Let us consider a pair of guards and τ, τ' two paths of the roadmap linking theses guards (i.e. creating a cycle) and visibility deformable one into the other. Then we have the following property:

Property: From a RCPV roadmap, the deletion of redundant paths τ' (i.e. visibility deformable into paths τ and connecting the same guards) leads to a second-order deformation roadmap.

Sketch of proof: Let us consider the partition of a free path τ, as defined in section 4.1. In that section we have shown that with a RCPV roadmap, one can extract a roadmap path τ' such that $\tau_{g_i} \cap \tau_{g_{i+1}}$ is visibility deformable into τ' (Figure 7.a). Now suppose that the redundant path τ' has been deleted as proposed above. It means that τ' was visibility deformable into another path τ'' which remains in the roadmap (Figure 7 b). Thus, by concatenation of the two ruled surfaces it is possible to build a second-order deformation surface between any path τ of $\mathcal{C}_{free}$ and a path of the roadmap (Figure 7 c).

Based on this notion of deformation roadmap, we describe below an algorithm for constructing such roadmaps.

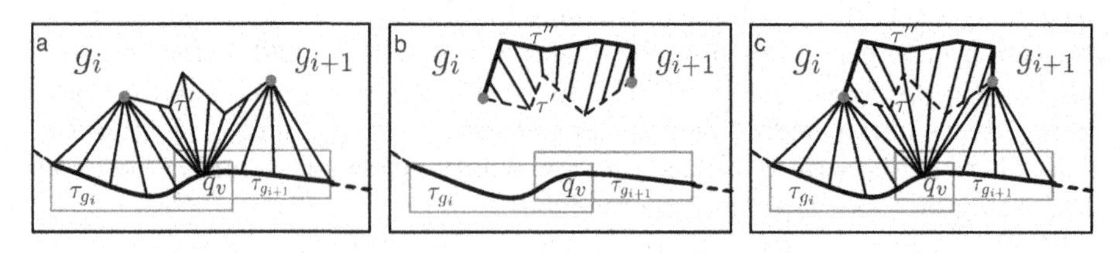

Fig. 7. Deleting redundant paths in a RCPV roadmap leads to a second-order deformation roadmap. (a) Visibility deformation for a RCPV roadmap. (b) A filtered path is itself deformable by visibility into a roadmap path. (c) By construction, there is a second-order deformation surface between a free path and a portion of roadmap.

5 Algorithm for Building Deformation Roadmaps

First, the roadmap is initialized with a tree structure computed with the Visibility-PRM method [12]. This ensures the coverage of the free space with a limited number of nodes and edges (i.e. no cycles). Then, instead of first building a RCPV roadmap and filtering in a second step the redundant cycles (as defined in section 4.2), the redundancy test is directly performed for efficiency purposes before each addition of a new cycle to the roadmap.

The pseudo-code of the algorithm used to build a second-order deformation roadmaps is shown in Figure 8. At each iteration a free configuration q_v is randomly sampled and the connectivity of the visible subroadmap is computed (*TestVisibSubRoadmap* function line 6). The evaluation of its connectivity is performed avoiding as much as possible the whole computation of the subroadmap. The redundancy test is only performed when the visible subroadmap is disconnected. For this test, we randomly choose two disconnected components of the subroadmap and pick among them the nearest guards n_1, n_2, from q_v. Then, we test whether there is a visibility deformation between the path $\tau = n_1 - q_v - n_2$ and a path of the roadmap (*TestRedundancy* function line 10). If such a visibility deformation exists, the configuration is useless with regards to the construction of a second-order deformation roadmap and is therefore rejected. The algorithm memorizes the number of successive failure since the last useful cycle inserted. This information is used to stop the iterations when the insertion of a new cycle becomes too difficult, meaning that most of the useful cycles are already captured by the roadmap.

PATH-DEFORMATION-PRM
input : the robot A, the environment B, $ntry_{max}$, $ntry\_cycl_{max}$
output : a Path Deformation Roadmap

1 $G \leftarrow$ Visibility-PRM$(A, B, ntry_{max})$
2 $ntry \leftarrow 0$
3 **While** $ntry < ntry\_cycl_{max}$
4     $q_v \leftarrow$ RandomFreeConfig(A, B)
5     $ntry \leftarrow ntry + 1$
6   **If** TestVisibSubRoadmap$(G, q_v) = Disconnected$
7         $n_1 \leftarrow$ NearestGuard$(q_v, Comp_1(G_v))$
8         $n_2 \leftarrow$ NearestGuard$(q_v, Comp_2(G_v))$
9         $\tau \leftarrow$ BuildPath(n_1, q_v, n_2)
10   **If** TestRedundancy $(\tau, n_1, n_2, G) = False$
11   CreateCyclicPath(τ, G)
12             $ntry \leftarrow 0$
13   **End If**
14   **End If**
15 **End While**

Fig. 8. General algorithm for building a Path Deformation Roadmap

We next detail the algorithms used to establish the subroadmap connectivity (*TestVisibSubRoadmap* function) and to test the visibility deformation between pairs of paths (*TestRedundancy* function).

5.1 Visible Subroadmap

The method used to check the connectivity of a visible subroadmap from a given configuration q_v (*TestVisibSubRoadmap* function in the *Path-Deformation-PRM* algorithm) corresponds to the pseudo-code of Figure 9. First, the set of nodes visible from q_v is computed by testing whether the straight line segments linking q_v to each of the roadmap nodes are free. Then, we test in two phases the connectivity of these nodes from the point of view of q_v. First, we evaluate all the roadmap edges as potentially visible. Thus, two nodes are detected as disconnected if all the paths of the roadmap connecting them pass through at least one invisible node (*VisibleConnectivity* function line 8). If this fast test is not sufficient to establish the connectivity of the visible subroadmap, we establish it by computing the visibility of the edges linking the visible nodes (*VisibleConnectivity* function line 12). We describe in the next section how the visibility of an edge from a given configuration can be tested.

```
TestVisibSubRoadmap(G, q_v)
1  N_vis ← EmptyList
2  For all node n ∈ G
3       If VisibleNode(n, q_v)
4            AddToList(n, N_vis)
5       End If
6  Endfor
7  TestEdges ← False
8  If VisibleConnectivity(q_v, N_vis, G,TestEdges) = False
9       Return Disconnected
10 End If
11 TestEdges ← True
12 If VisibleConnectivity(q_v, N_vis, G ,TestEdges) = False
13      Return Disconnected
14 End If
15 Return Connected
```

Fig. 9. Algorithm testing the visible subroadmap connectivity from a given configuration q_v

5.2 Edge Visibility

Testing the visibility of an edge from a configuration q_v is equivalent to checking the validity of triangular configuration-space facets, defined by q_v and the two edge's endpoints (c.f. Figure 10). This visibility test possibly involves one or several facets depending on the topological nature of $\mathcal{C}$:

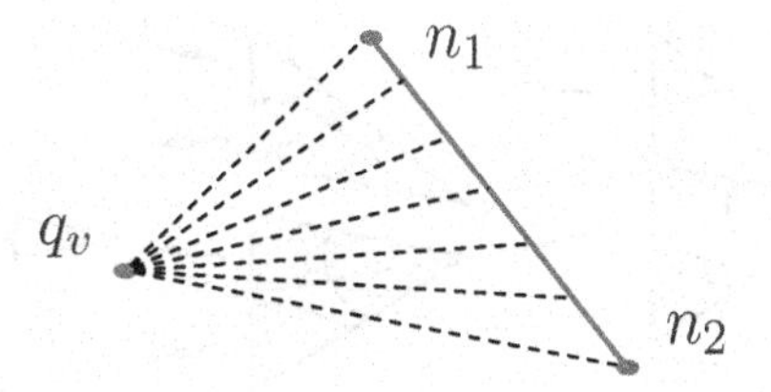

Fig. 10. Edge visibility: $n_1 - n_2$ is visible from q_v if the facet $\{q_v, n_1, n_2\}$ is valid

- If $\mathcal{C}$ is isomorphic to $[0,1]^n$ (the robot's degrees of freedom are only translations and/or bounded rotations) then the visibility test can be done by testing only a single facet in $\mathcal{C}$ (Figure 11.a).
- If $\mathcal{C}$ is isomorphic to $[0,1]^n \times SO(d)^m$ with $m > 0$ (one or more degrees of freedom are cyclic), the visibility test of an edge can lead to test several facets (Figure 11.b). In fact, a discontinuity occurs each time the distance between q_v and a configuration on the edge is equal to π according to a given degree of freedom.

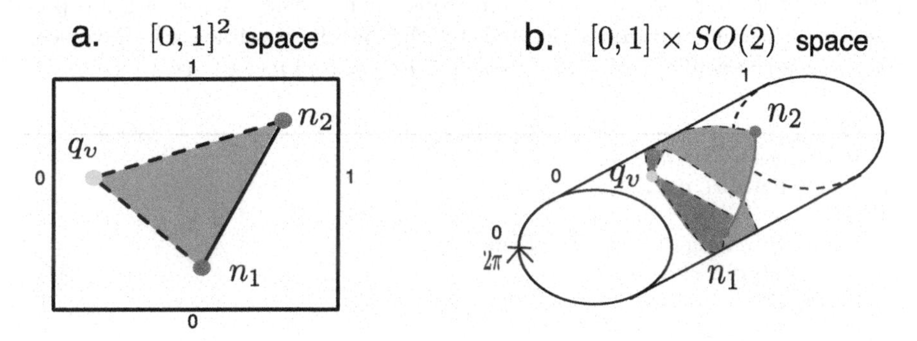

Fig. 11. Testing the visibility of an edge can lead to test one (a) or several (b) facets, depending on the topological nature of the configuration space

5.3 Elementary Facet Test

To test the validity of a facet we try to cover it entirely with free balls of $\mathcal{C}$ (Figure 12). First, the radii of the balls centered on each vertex of the facet are computed using a conservative method based on the robot kinematics and the distance of its bodies to the obstacles. If the balls are sufficient for covering the facet, then the algorithm returns that the facet is valid. Otherwise it is split into two sub-facets computed such that their common vertex is as far as possible from the regions already covered by the balls. The radius of the ball centered on this vertex is then computed. This dichotomic process is performed until the entire facet is covered or one vertex is tested as invalid.

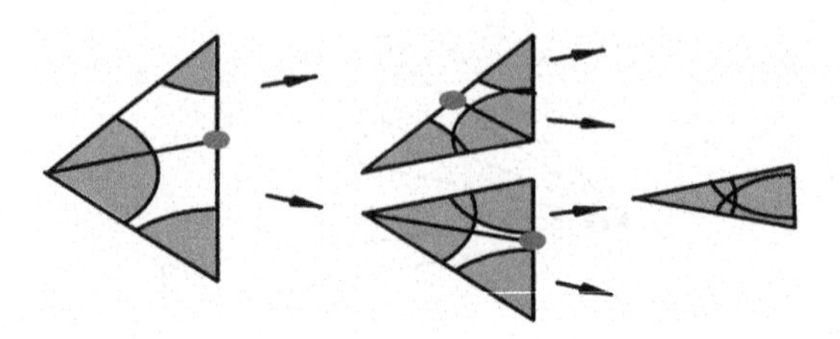

Fig. 12. Dichotomic covering of a valid facet with $\mathcal{C}_{free}$ balls

5.4 Redundancy Test

A disconnected subroadmap from the point of view of a configuration q_v can be reconnected by linking two of the subcomponents through q_v. Before performing such connection attempt, we test whether it may lead to a redundant path which could be filtered. To do so, first we build the path $\tau = n_1 - q_v - n_2$ with n_1, n_2 belonging to two distinct subcomponents. Then we test its visibility deformation into a roadmap path thanks to the *TestRedundancy* algorithm (line 10 in Figure 8). This algorithm is shown in Figure 13. Roadmap paths are iteratively extracted and tested according to their visibility deformation relatively to τ. This process starts with the shortest path found and stops when a visible deformation is possible (then the configuration is rejected) or when all the possible paths have been tested (then the configuration and the edges $n_1 - q_v$ and $n_2 - q_v$ are inserted).

TestRedundancy(τ, n_1, n_2, G)
1 $\tau' \leftarrow$ BestPath(n_1, n_2, G)
2 **While** $\tau' \neq \varnothing$
3   **If** VisibDeformation(τ, τ') $= True$
4   Return $True$
5   **End If**
6     $\tau' \leftarrow$ BestPath(n_1, n_2, G)
7 **End While**
8 Return False

Fig. 13. Visibility deformation test between a path τ and a roadmap path

The *VisibDeformation* function (line 3 of algorithm 13) tests whether two paths τ and τ' can be visibility deformed one into the other. This function is based on the grid based computation of the visibility diagram associated to the two paths. The deformation is only possible when there exists a path between the $(0, 0)$ and $(1, 1)$ points in this diagram (c.f. section 3.3). In practice, the whole diagram is not computed. The tests are limited to the grid cells visited during the A^* search of a valid path in the visibility diagram, incrementally developed during the search. This implicit search of the diagram notably limits the number of visibility tests to be performed (Figure 14) and highly speeds up the redundancy test.

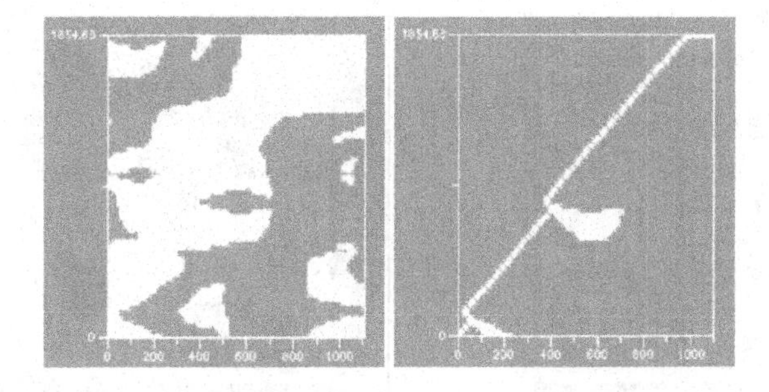

Fig. 14. Visibility diagram (left) and cells explored during the visibility deformation test (right)

6 Experimental Results

We implemented the algorithm for constructing (second-order) deformation roadmaps in the Move3D software platform [11]. The experiments reported below were performed on a 1.2GHz G4 PowerPC running on Mac OS-X. The performance results summarized in Table 2 correspond to average values computed over several runs of the algorithm.

The first experiment shown on Figure 15 compares the level of redundancy obtained in function of the algorithm used: (a), a minimum tree structure obtained with the Visibility-PRM, (b) a first-order roadmap (built without the filtering process) and (c) a second-order deformation roadmap (PDR) that captures the different varieties of paths while maintaining a compact structure.

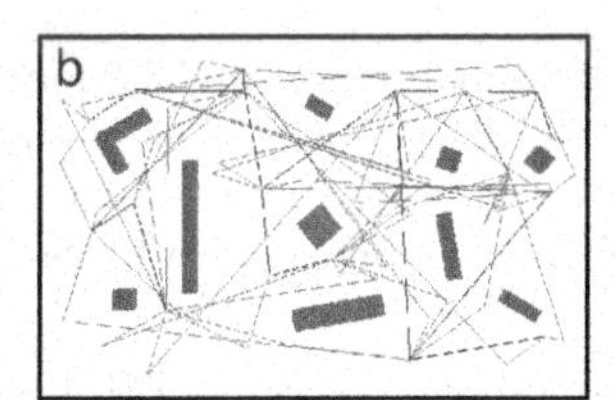

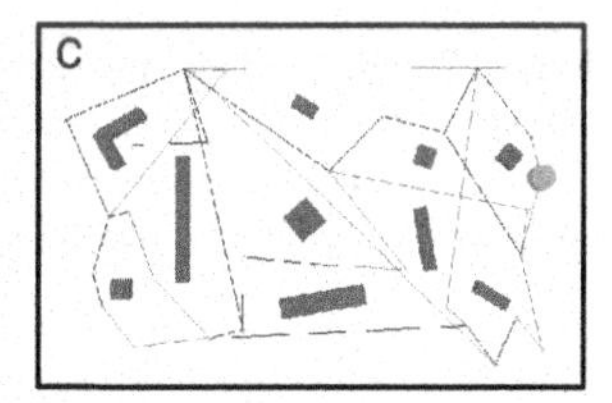

Fig. 15. Comparison between three algorithms of roadmap construction. (a) Visibility-PRM. (b), first-order and (c), second-order deformation roadmap.

The next set of experiments (Figure 16) presents the path deformation roadmaps obtained for a 2-dof robot evolving in complex environments. The first scene (a) requires 25 elementary cycles to capture the homotopy. Our method builds a roadmap capturing these cycles in only 109 seconds. The second scene (b) has a higher geometrical complexity (70 000 facets). The computing time (164 secs) reported in Table 2 shows that the algorithm can efficiently handle such geometrically complex scenes. One can also note that the resulting 2D roadmaps contain a very limited number of additional cycles compared to homotopy.

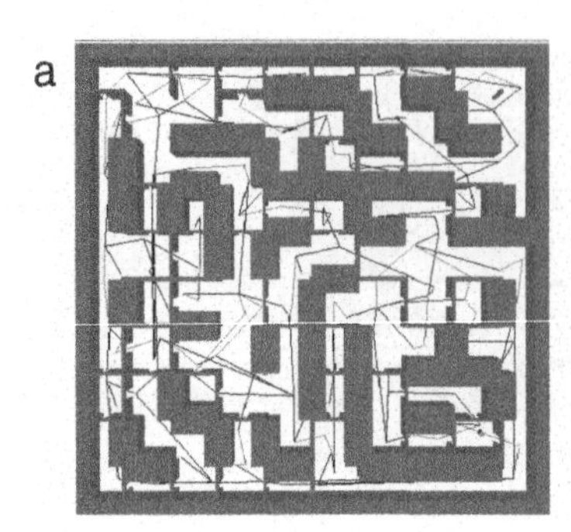

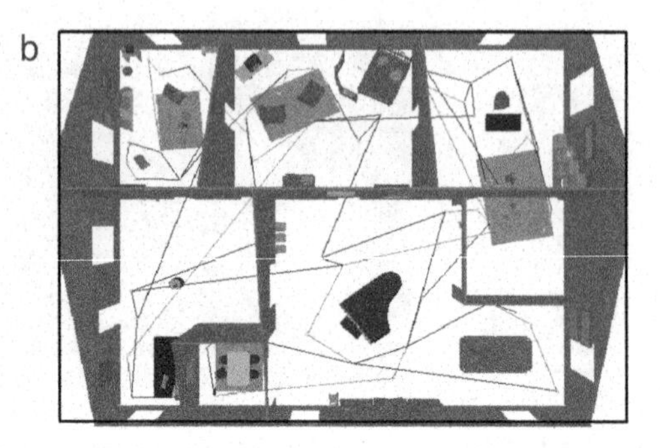

Fig. 16. Path Deformation Roadmaps for 2D environments: (a) a labyrinth with many homotopy classes. (b) an indoor environment with a complex geometry.

The third experiment (Figure 17) involves a narrow passage problem for a squared robot with 3-dof (two translations and one rotation). The robot has four ways to go through the narrow passage, depending on its orientation. Therefore the narrow passage corresponds to four homotopy classes in the configuration space.

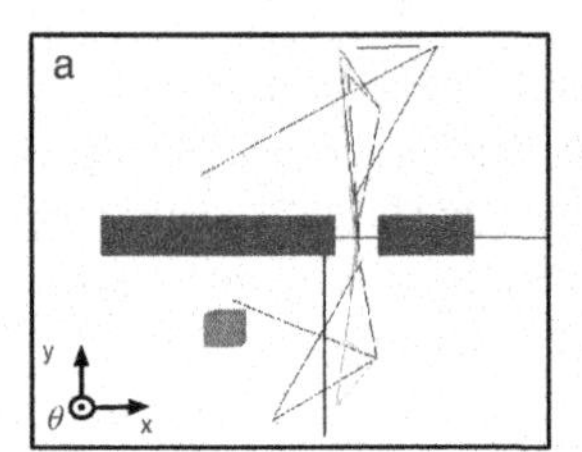

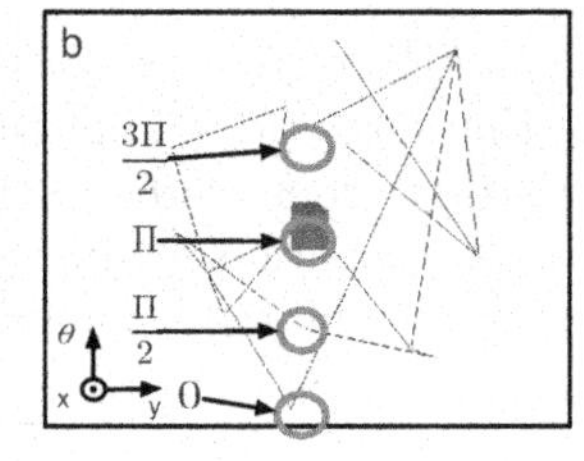

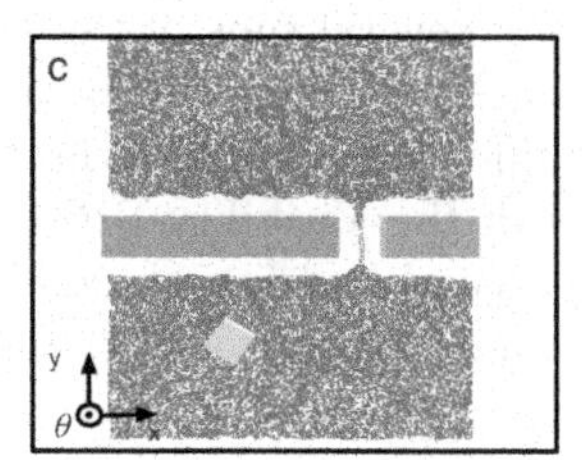

Fig. 17. Path Deformation Roadmap capturing the four homotopy classes for a rotating square and a narrow passage. (a) (x,y) view of the deformation roadmap, (b) (y,θ) view of the same roadmap showing the four kinds of passages found in $\mathcal{C}$, (c) comparison with the dense roadmap obtained with a classic k-nearest PRM.

Table 1. Homotopy classes found by a k-nearest PRM for the problem of Figure 17

		n_classes			time (s)		
	k	10	20	100	10	20	100
N	1000	0.1	0.2	1.2	6.4	9.3	33.2
	2000	0.1	0.6	1.6	33.2	43.5	110.0
	4000	0.8	1.0	2.8	246	336	455
	8000	1.4	2.4	**3.2**	2947	3295	**3819**

Table 1 presents results obtained with a traditional k-nearest PRM [4] for different couples (N, k) (with N, the number of roadmap nodes). The reported results (averaged over several runs) show that even for the densest and most redundant case $(N = 8000, k = 100)$, the homotopy is not well captured (n_classes =

3.2/4) by the k-nearest PRM. Moreover, the large size of the computed roadmap results in a significant computing time (3819 secs) due to the amount of collision tests required for adding new nodes and edges. Comparatively, our method captures the four homotopy classes in only 37 secs. The high speed-up comes from the very compact size of the path deformation roadmap (only 12 nodes) which largely compensates the additional cost of filtering the useless redundant cycles.

The last set of experiments (Figure 18) involves 6-dof robots in 3D environments. In the first case (free flying robot), the free space has only one homotopy class. Thus, a roadmap based on homotopy would have a tree structure. The results show that our method makes it possible to build a compact roadmap (in 56 secs) while capturing a richer variety of paths than the homotopy. The second scene concerns a 6-dof manipulator arm where 6 additional nodes (and 12 edges) are added to the visibility roadmap (total time of 99 secs) to represent the complexity of the space. In both cases, the number of roadmap cycles, although limited, results into shorter paths during the query phase.

Finally, the performance results are summarized in Table 2 which also provides a break-up of the total computing time showing the respective contributions of the visibility tree building and the cycle addition stages.

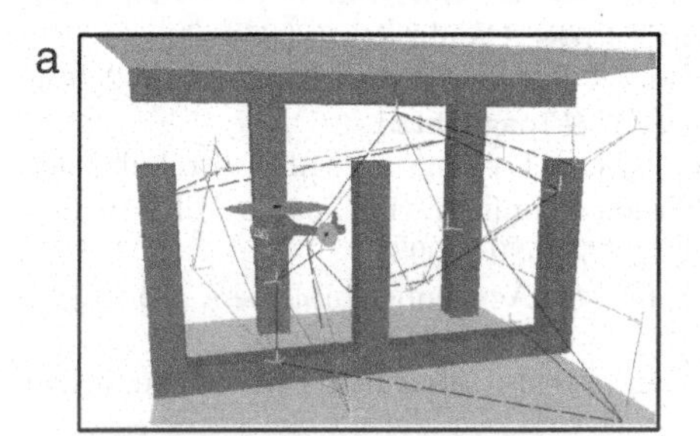

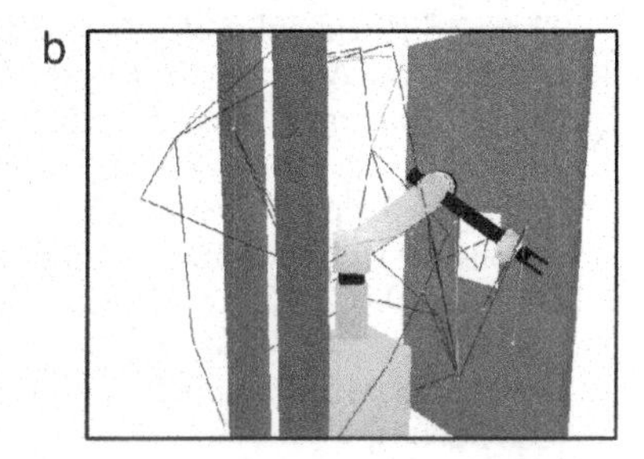

Fig. 18. Path Deformation Roadmaps for complex environments: (a) free flying robot, (b) 6-dof manipulator arm

Table 2. Computing time of the Deformation Roadmaps

						time repartition (%)			
	dof	nodes	edges	cycles	time (s)	Vis-PRM	SubRoadmap	Redundancy	Other
Laby	2	149	177	29	109	19	32	35	14
Indoor	2	66	83	18	164	25	20	49	6
Square	3	12	14	3	37	24	61	11	4
Helico	6	30	39	10	56	5	9	80	6
Arm	6	41	46	6	99	12	70	13	5

7 Conclusion

We have presented a general method to build compact PDR roadmaps with useful cycles representative of the different varieties of free paths of the configuration space. The introduction of these cycles is important for obtaining higher quality solutions when postprocessing queries inside the roadmap. Our approach is based on the notion of path deformability indicating whether or not a given path can be easily deformed into another one. Our experiments show that the method enables small roadmaps to reliably capture the multiple connectedness of possibly complex configuration spaces. Several improvements remain for future work. First, the method has so far been tested for free flying and articulated robots with up to 6 dof. We need to further evaluate its performance for higher dof articulated robots. We would also like to further investigate the link between the varieties of free paths stored in the roadmap and the smoothing method used to shorten the solution paths when postprocessing queries. Finally, another improvement concerns the extension to robots with kinematically constrained motions requiring the use of a non-linear local method.

References

1. Bohlin, R., Kavraki, L.E.: Path planning using lazy prm. In: IEEE Int. Conf. on Robotics and Automation, pp. 521–528 (2000)
2. Choset, H., Lynch, K.M., Hutchinson, S., Kantor, G., Burgard, W., Kavraki, L.E., Thrun, S.: Principles of robot motion. MIT Press, Cambridge (2005)
3. Hatcher, A.: Algebraic Topology. Cambridge University Press, Cambridge (2002), `http://www.math.cornell.edu/~hatcher/AT/ATpage.html`
4. Kavraki, L.E., Svestka, P., Latombe, J.-C., Overmars, M.H.: Probabilistic roadmaps for path planning in high-dimensional configuration spaces. IEEE Transactions on Robotics and Automation 12(4), 566–580 (1996)
5. Latombe, J.-C.: Robot Motion Planning. Kluwer Academic Publishers, Dordrecht (1991)
6. LaValle, S.M.: Planning Algorithms. Cambridge University Press, Cambridge (2004-2005), `http://msl.cs.uiuc.edu/planning/`
7. Nieuwenhuisen, D., Overmars, M.H.: Useful cycles in probabilistic roadmap graphs. In: IEEE Int. Conf. on Robotics and Automation, pp. 446–452 (2004)
8. Overmars, M.H., Svestka, P.: A probabilistic learning approach to motion planning. In: Goldberg, K., et al. (eds.) Algorithmic Foundations of Robotics (WAFR 1994), pp. 19–37. A.K. Peters (1995)
9. Schmitzberger, E., Bouchet, J.L., Dufaut, M., Didier, W., Husson, R.: Capture of homotopy classes with probabilistic road map. In: IEEE/RSJ Int. Conf. on Robots and Systems (2002)
10. Sekhavat, S., Svestka, P., Laumond, J.-P., Overmars, M.H.: Multi-level path planning for nonholonomic robots using semi-holonomic subsystems. International Journal of Robotics Research 17(8), 840–857 (1998)
11. Siméon, T., Laumond, J.-P., Lamiraux, F.: Move3d: a generic platform for path planning. In: IEEE Int. Symp. on Assembly and Task Planning (2001)
12. Siméon, T., Laumond, J.-P., Nissoux, C.: Visibility-based probabilistic roadmaps for motion planning. Advanced Robotics Journal 14(6), 477–494 (2000)
13. Sánchez, G., Latombe, J.-C.: On delaying collision checking in prm planning - application to multi-robot coordination. International Journal of Robotics Research 21(1), 5–26 (2002)

Workspace-Based Connectivity Oracle: An Adaptive Sampling Strategy for PRM Planning

Hanna Kurniawati and David Hsu

National University of Singapore, Singapore 117543, Singapore
{hannakur,dyhsu}@comp.nus.edu.sg

Summary. This paper presents *Workspace-based Connectivity Oracle* (WCO), a dynamic sampling strategy for probabilistic roadmap planning. WCO uses both domain knowledge—specifically, workspace geometry—and sampling history to construct dynamic sampling distributions. It is composed of many component samplers, each based on a geometric feature of a robot. A component sampler updates its distribution, using information from the workspace geometry and the current state of the roadmap being constructed. These component samplers are combined through the adaptive hybrid sampling approach, based on their sampling histories. In the tests on rigid and articulated robots in 2-D and 3-D workspaces, WCO showed strong performance, compared with sampling strategies that use dynamic sampling or workspace information alone.

1 Introduction

Probabilistic roadmap (PRM) planning [6] is currently the most successful approach for motion planning of robots with many degrees of freedom (DOFs). PRM planners sample a robot's configuration space $\mathcal{C}$ according to a suitable probability distribution and capture the connectivity of $\mathcal{C}$ in a graph, called a roadmap, which is an extremely simplified representation of $\mathcal{C}$.

Despite their successes, PRM planners behave poorly when $\mathcal{C}$ contains narrow passages. A narrow passage is a small region whose removal changes the connectivity of $\mathcal{C}$. The probability of sampling in narrow passages is low, because of their small volumes. In such spaces, it is difficult for PRM planners to build roadmaps with good connectivity. Although many PRM planners have been proposed (e.g., [4, 9, 12, 16, 17, 18, 21, 22]), narrow passages remain a bottleneck for PRM planning.

With few exceptions, most PRM planners use *static* sampling distributions based on *a priori* assumed geometric properties of the configuration space or the workspace. Interestingly, the first PRM planner [16], which consists of two sampling stages, uses *dynamic* sampling: the second stage exploits information gathered in the first stage to update the sampling distribution and resample $\mathcal{C}$. Recently, with the use of machine learning techniques in PRM planning [5, 13, 19], dynamic sampling has again gained popularity. Dynamic sampling incrementally infers partial knowledge of key geometric properties of $\mathcal{C}$ during the

S. Akella et al. (Eds.): Algorithmic Foundation of Robotics VII, STAR 47, pp. 35–51, 2008.
springerlink.com

roadmap construction and uses this knowledge to adapt the sampling distribution. It reveals and exploits the probabilistic foundations of PRM and is a promising way of speeding up PRM planners [11].

To infer geometric properties of $\mathcal{C}$, existing PRM planners with dynamic sampling use sampling history. This is, however, inadequate. For example, to learn the usefulness of sampling a particular region of $\mathcal{C}$, we need many samples in that region. This is difficult to achieve in narrow passages. One way of addressing this issue is to use domain knowledge such as the geometry of robots and obstacles in the workspace $\mathcal{W}$. Compared with $\mathcal{C}$, $\mathcal{W}$ has low dimensionality and an explicit geometric representation, which make it easy to find narrow passages. Narrow passages in $\mathcal{W}$ often suggest the presence and the location of narrow passages in $\mathcal{C}$. Furthermore, workspace geometry provides information complementary to that from sampling history.

In this paper, we present a new PRM planner that combines information from both workspace geometry and sampling history to construct a dynamic sampling distribution. Core to our new planner is a new sampling strategy called *Workspace-based Connectivity Oracle* (WCO). WCO is an ensemble sampler composed of many component samplers. They are all based on a key observation: a collision-free path between a start configuration s and a goal configuration g in a robot's configuration space $\mathcal{C}$ implies a collision-free path in $\mathcal{W}$ for *every* point in the robot between the corresponding start and goal positions of the point. So, if we find a collision-free path in $\mathcal{W}$ for every point in the robot and all these paths correspond to the same path γ in $\mathcal{C}$, then γ is indeed a collision-free path in $\mathcal{C}$ for the robot to move from s to g. Finding a path for every point is, of course, impractical. Nevertheless, we can use the paths of a set of geometric features in $\mathcal{W}$—points, line segments, triangles, etc.— to predict regions of $\mathcal{C}$ that are likely to be useful for connecting disconnected components of a roadmap. Each WCO component sampler is based on a single geometric feature. They are then combined, based on their sampling histories, through the adaptive hybrid sampling (AHS) approach [13], which is a restricted form of reinforcement learning.

2 Background

2.1 PRM Basics

The standard multi-query PRM approach consists of two phases, roadmap construction and roadmap query. In the roadmap construction phase, the planner samples $\mathcal{C}$ according to a suitable probability distribution and approximates the connectivity of $\mathcal{C}$ with a roadmap graph $\mathcal{R}$. The nodes of $\mathcal{R}$ represent sampled collision-free configurations, called *milestones*. An edge exists between two milestones if they can be connected with a collision-free straight-line path. In the query phase, the planner is given a start and a goal configuration. It tries to connect the two query configurations to two corresponding milestones in $\mathcal{R}$ and then searches for a path in $\mathcal{R}$ between these two milestones, using standard graph search algorithms. See [6] for details.

The performance of PRM planners depends critically on the quality of roadmaps constructed. A good roadmap has two important properties: *coverage* and *connectivity*. Denote by $\mathcal{F}_\mathcal{C}$ the collision-free subset of $\mathcal{C}$. Good coverage means that for any configuration $q \in \mathcal{F}_\mathcal{C}$, there is a collision-free straight-line path between q and a milestone in $\mathcal{R}$ with high probability. Good connectivity means that any two milestones in the same connected component of $\mathcal{F}_\mathcal{C}$ are also connected by a path in $\mathcal{R}$. Good coverage is relatively easy to achieve through uniform sampling [15]. Good connectivity is more difficult to achieve, especially when $\mathcal{F}_\mathcal{C}$ contains narrow passages.

2.2 PRM Planners That Generate Dynamic Sampling Distributions

In addition to the early PRM planner in [16], the planner in [19] also uses a two-stage sampling strategy, but employs more sophisticated techniques to generate the distribution for resampling $\mathcal{C}$ in the second stage. Instead of breaking sampling into two stages, some recent planners use on-line machine learning to update sampling distributions incrementally. In adaptive hybrid sampling [13], the sample distribution is constructed as a linearly weighted combination of component distributions. To adapt the distribution, the weights are updated after each sampling operation during roadmap construction to favor component distributions having the most promising results. In entropy-guided planning [5], an approximate model of $\mathcal{C}$ is built and used to sample $\mathcal{C}$ so that the expected value of a utility function is maximized.

2.3 PRM Planners That Use Workspace Information

Several PRM planners use workspace geometry as a guide for sampling $\mathcal{C}$. Let $\mathcal{F}_\mathcal{W}$ denote the subset of $\mathcal{W}$ that is not occupied by obstacles. Some planners bias sampling by focusing a fixed set of workspace paths, e.g., paths on the medial axis of $\mathcal{F}_\mathcal{W}$ [9, 10, 22]. Other planners bias sampling by identifying important regions in $\mathcal{W}$. For instance, the watershed algorithm focuses on small regions connecting large open regions [21]. Workspace importance sampling (WIS) focuses on regions with small local feature size [17]. These planners all use static sampling distributions.

In summary, workspace geometry has been used in earlier work to construct static sampling distributions for PRM planners. Sampling history has been used to update sampling distributions dynamically. Our new sampling strategy combines the information from both workspace geometry and sampling history to construct a dynamic sampling distribution.

3 Overview

3.1 The WCO Planner

Our planner adopts the standard multi-query PRM approach described in Section 2.1 and uses WCO for sampling $\mathcal{C}$. Since there is no confusion, we use WCO to refer to both the sampling strategy and the planner.

Before describing WCO, let us first consider the relationship between $\mathcal{C}$ and $\mathcal{W}$. For a point f in the robot, let $P_f(q)$ be the position in $\mathcal{W}$ of f when the robot is placed at configuration $q \in \mathcal{C}$. We call the mapping $P_f : \mathcal{C} \rightarrow \mathcal{W}$ a projection, as $\mathcal{C}$ has higher dimensionality than $\mathcal{W}$. Similarly, we define the lift mapping $L_f : \mathcal{W} \rightarrow 2^{\mathcal{C}}$. For any $x \in \mathcal{W}$, $L_f(x)$ is the subset of $\mathcal{C}$ such that each configuration in $L_f(x)$ places f at x. For convenience, we extend the definitions of P_f and L_f to subsets of $\mathcal{C}$ and $\mathcal{W}$, respectively, by taking set union. Using this notation, we can state the observation in Section 1 formally:

Proposition 1. *If two configurations $q, q' \in \mathcal{C}$ are connected by a path in $\mathcal{F}_{\mathcal{C}}$, then for any point f in a robot, $P_f(q)$ and $P_f(q')$, the projections of q and q' in $\mathcal{W}$, are connected by a path in $\mathcal{F}_{\mathcal{W}}$.*

During the roadmap construction, WCO maintains a partially constructed roadmap $\mathcal{R}$. Distinct connected components of $\mathcal{R}$ may in fact lie in the same connected component of $\mathcal{F}_{\mathcal{C}}$, due to inadequate sampling of certain critical regions. To sample such regions, WCO examines the workspace paths of a set of *feature points* in the robot and constructs a sampler for each feature point f. To connect two components $\mathcal{R}_1$ and $\mathcal{R}_2$ of $\mathcal{R}$, we use P_f to project the milestones of $\mathcal{R}$ into $\mathcal{W}$ and search for "channels" in $\mathcal{W}$ that connect the projected milestones of $\mathcal{R}_1$ and $\mathcal{R}_2$. These channels suggest the regions of $\mathcal{C}$ that may connect $\mathcal{R}_1$ and $\mathcal{R}_2$. So, we use L_f to lift the channels into $\mathcal{C}$ and adapt the distribution to sample more densely in the regions covered by the lifted channels. To be sensitive to the changes in $\mathcal{R}$, WCO adapts its sampling distribution incrementally whenever a new milestone is added to $\mathcal{R}$.

Although workspace-based PRM planners often consider only a single feature point, this is inadequate. By Proposition 1, a collision-free path in $\mathcal{C}$ implies a collision-free path in $\mathcal{W}$ for every point in the robot. So, we use a set of pre-selected feature points and construct an independent sampler s_i for each feature point f_i, $i = 1, 2, \ldots$. We make two simplifying assumptions. First, a finite number of feature points are sufficient to indicate the important regions of $\mathcal{C}$ for sampling. Second, we can treat the feature points independently. These two assumptions reduce the computational cost and are shown to be effective in identifying important regions of $\mathcal{C}$ (see Section 6). Despite the independence assumption, the kinematic constraints of a robot are not entirely ignored. Implicitly, WCO assigns higher sampling density to regions obeying such constraints. To provide roadmap coverage, we add a uniform sampler s_0 to the WCO samplers $s_1, s_2, \ldots$ and form the set $S = \{s_0, s_1, s_2, \ldots\}$. The component samplers in S are combined through the AHS approach to form an ensemble sampler: each component sampler has an associated weight proportional to the probability of it being used, and the weights are adjusted to reflect the success of the sampler according to the sampling history.

3.2 When Is Workspace Connectivity Information Useful?

To represent $\mathcal{F}_{\mathcal{W}}$, the collision-free subset of $\mathcal{W}$, WCO computes a decomposition $\mathcal{T}$ of $\mathcal{F}_{\mathcal{W}}$ into non-overlapping cells. It represents the connectivity of $\mathcal{F}_{\mathcal{W}}$ as an

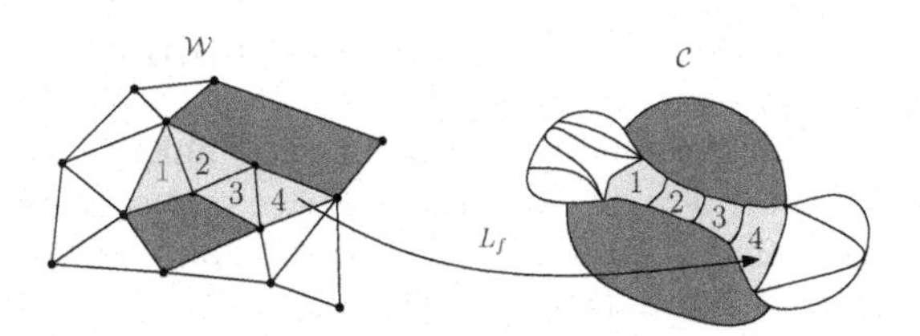

Fig. 1. A partition of $\mathcal{W}$ induces a partition of $\mathcal{C}$. Obstacles are shaded in dark color. A workspace channel and its lifted version are shaded in light color. In general, L_f is a one to many mapping. It maps a region of $\mathcal{W}$ to several regions of $\mathcal{C}$.

adjacency graph $G_\mathcal{T}$ for the cell decomposition $\mathcal{T}$. Each node of $G_\mathcal{T}$ represents a cell in $\mathcal{T}$, and two nodes are connected by an edge if the corresponding cells are adjacent. By Proposition 1, if two cells $t, t' \in \mathcal{T}$ are disconnected in $\mathcal{F}_\mathcal{W}$, then $L_f(t)$ and $L_f(t')$ are disconnected in $\mathcal{F}_\mathcal{C}$ for any feature point f. Thus, the connectivity information encoded in $G_\mathcal{T}$ can help in capturing the connectivity of $\mathcal{F}_\mathcal{C}$ during the roadmap construction.

Although we often think of $\mathcal{W}$ and $\mathcal{C}$ as two distinct spaces, they are closely related. For a fixed feature point f, $\mathcal{T}$ induces a partition of the collision-free subset of $\mathcal{C}$ into equivalent classes: $\mathcal{F}_\mathcal{C} = \bigcup_{t\in\mathcal{T}}(L_f(t) \cap \mathcal{F}_\mathcal{C})$ and for all $t, t' \in \mathcal{T}$ and $t \neq t'$, $L_f(t) \cap L_f(t') = \emptyset$ unless when t and t' share a boundary, in which case $L_f(t)$ and $L_f(t')$ share a boundary too (Fig. 1). Two configurations are in the same equivalent class if they project to the same cell in $\mathcal{T}$. WCO exploits this connection extensively to integrate the information from both $\mathcal{W}$ and $\mathcal{C}$.

To connect two milestones m and m' of a roadmap, consider their projections. Suppose that $P_f(m) \in t$ and $P_f(m') \in t'$, where $t, t' \in \mathcal{T}$. A *workspace channel* λ is the set of cells corresponding to the nodes on a path in $G_\mathcal{T}$ between t and t'. The lifted channel $L_f(\lambda)$ suggests a region of $\mathcal{C}$ for sampling in order to connect m and m'. Of course, no particular $L_f(\lambda)$ guarantees that a path between m and m' can be found within it, as the converse of Proposition 1 is not true in general. Nevertheless, a channel helps to improve sampling efficiency by narrowing down the sampling domain to a relevant subset of $\mathcal{C}$.

The usefulness of a workspace channel can be defined formally:

Definition 1. *Let m and m' be two milestones of a roadmap and λ be a workspace channel between the cells containing $P_f(m)$ and $P_f(m')$ for some feature point f. The channel λ has the (n,p)-property if n samples drawn from $L_f(\lambda)$ are sufficient to find a path in $\mathcal{F}_\mathcal{C}$ between m and m' with probability at least p, provided such a path exists.*

A channel $L_f(\lambda)$ has good (n,p)-property if n is small and p is large. It is known that $L_f(\lambda)$ has good (n,p)-property under various conditions, e.g., path clearance [14], ε-complexity [20], and expansiveness [12]. The effectiveness in finding useful workspace channels depends on the cell decomposition and the method of searching for channels. These issues are detailed in the next section.

4 Constructing a WCO Component Sampler

We now describe the construction of a component sampler of WCO for a fixed feature point, specifically, how to extract workspace connectivity (Section 4.1), how to adapt the sampling distribution (Section 4.2), and how to take a sample for rigid and articulated robots (Section 4.3).

4.1 Extracting Workspace Connectivity

WCO computes a cell decomposition $\mathcal{T}$ of $\mathcal{F}_{\mathcal{W}}$ and represents the connectivity of $\mathcal{F}_{\mathcal{W}}$ in the adjacency graph of $\mathcal{T}$. This decomposition is computed only once and used by all WCO component samplers. Many spatial decomposition methods can be used here, e.g., triangulations and quadtrees for 2-D workspaces and their counterparts for 3-D workspaces.

Building on our earlier work [17], we have chosen to sample the boundary of obstacles in $\mathcal{W}$ and construct a Delaunay triangulation [8] over the sampled points. Under reasonable assumptions, the constructed triangulation is conforming [1], meaning that every triangle in the resulting triangulation lies either entirely in $\mathcal{F}_{\mathcal{W}}$ or its complement. Although helpful, this property is not required for our purposes. Throughout the rest of the paper, triangles refer to triangles in 2-D workspaces and tetrahedra in 3-D workspaces.

4.2 Adapting the Sampling Distribution

A skeleton of a WCO component sampler is shown in Algorithm 1. Let us now look at how it represents and updates the sampling distribution based on workspace channels. During the roadmap construction, WCO maintains a partially constructed roadmap $\mathcal{R}$. To sample a new milestone, each component sampler maintains a separate sampling distribution $\pi_{\mathcal{T}}$ defined over $\mathcal{T}$. The distribution $\pi_{\mathcal{T}}$ assigns equal probabilities to all triangles of $\mathcal{T}$ inside workspace channels and zero probabilities to all other triangles.

To find workspace channels, we first project milestones of $\mathcal{R}$ to $\mathcal{W}$ (Algorithm 1, line 8). Suppose that a milestone m belongs to a roadmap component $\mathcal{R}_i$ of $\mathcal{R}$. We associate $\mathcal{R}_i$ with the triangle $t \in \mathcal{T}$ that contains $P_f(m)$. Thus, each triangle t contains a set of labels that indicates the roadmap components which t is associated with. A triangle t is a *terminal* if its label set is non-empty, meaning that t contains at least one projected milestone. See Fig. 2*a* for an example.

Next, we find channels that connect terminals with different label sets by considering the adjacency graph $G_{\mathcal{T}}$. We compute a subgraph of $G_{\mathcal{T}}$, called a *channel graph* G', that spans all the terminals and connect them together. The workspace channels are then the triangles corresponding to the vertices of G'. The intuition behind the channel graph is very much like that of a roadmap in the configuration space: it uses simple paths, in this case, the shortest paths to connect every pair of two terminals that are close to each other and have different label sets. See Fig. 2 for an example.

Algorithm 1. A WCO component sampler.

1: Given a feature point f, sample a configuration q, based on the sampling distribution defined over the decomposition $\mathcal{T}$.
2: **if** q is collision-free **then**
3: Insert q into the roadmap $\mathcal{R}$ as a new milestone m.
4: $N_m \leftarrow$ the set of at most M milestones closest to m among all existing milestones of $\mathcal{R}$ within a distance of $D_{\max}$ from m. M and $D_{\max}$ are fixed constants.
5: **for** each $m' \in N_m$ **do**
6: **if** m' and m lie in different connected components of $\mathcal{R}$ **then**
7: Check whether there is a collision-free straight-line path between m and m'. If so, insert an edge between m and m' into $\mathcal{R}$,
8: Project m to $\mathcal{W}$.
9: Update the label sets for all affected triangles in $\mathcal{T}$.
10: Let $t \in \mathcal{T}$ be the triangle that contains $P_f(m)$. Perform a breadth-first search from t and stop when reaching the first terminal t' other than t.
11: Add the path between t and t' to G' if t and t' hold different label sets and delete the path from G' if t and t' hold the same label sets.
12: Update the sampling distribution.

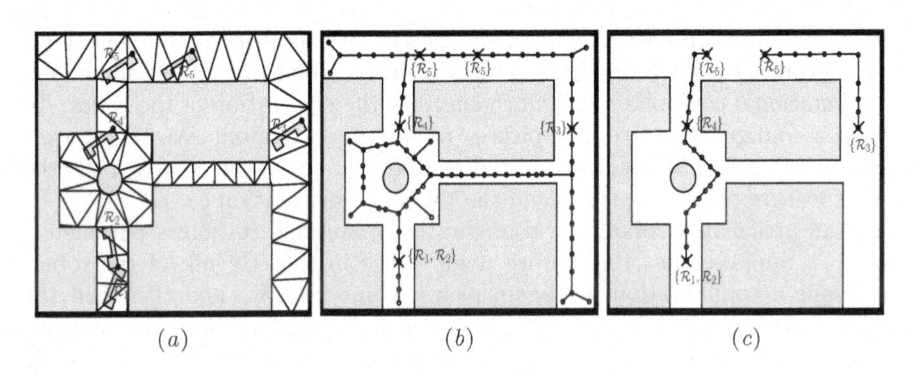

Fig. 2. (a) Milestones projected to the triangulated workspace. The labels indicate the roadmap components to which the milestones belong. The feature point of the rigid robot is marked by a black dot. (b) The adjacency graph $G_{\mathcal{T}}$. Terminals are marked by crosses. (c) The channel graph G' . Paths that connect terminals with the same label set (e.g., the path between the two terminals labelled $\{\mathcal{R}_5\}$) are not in G', as they connect terminals corresponding to milestones in the same connected components of $\mathcal{R}$ and hence unlikely to help in improving the connectivity of $\mathcal{R}$.

The channel graph is computed incrementally, as new milestones are added (Algorithm 1, lines 10–11). The incremental construction allows WCO to respond to changes in $\mathcal{R}$ and simplifies computation. To see that G' indeed "connects" all the terminals together, note that the channel graph G' clearly contains all the terminals. Furthermore, it is *weakly connected* in the sense that between every pair of terminals t and t', there is a sequence of terminals $t_i, i = 1, 2, \ldots, n$ with $t_1 = t$ and $t_n = t'$ such that every adjacent pair t_i and t_{i+1} either have exactly the same label set or have a path between them in G'. In the example shown in Fig. 2c, the two terminals $\{\mathcal{R}_3\}$ and $\{\mathcal{R}_4\}$ are weakly connected.

The incremental construction of the channel graph is quite efficient. Using the union-find data structure [7], we can project a milestone and update the label sets (Algorithm 1, lines 8–9) in $O(|\mathcal{R}|)$ worst-case time, where $|\mathcal{R}|$ is the number of different labels and is equal to the number of roadmap components that has been constructed so far. A loose upper bound for updating G' (Algorithm 1, lines 10–11) is $O(|\mathcal{T}|)$, where $|\mathcal{T}|$ is the number of triangles in $\mathcal{T}$. In practice, the upper bound is rarely reached. The entire update takes little time, compared with other parts of the planner (see Section 6.1).

4.3 Sampling a Configuration

To generate a sample from $\mathcal{C}$ (Algorithm 1, line 1), we perform two simple steps. First, we sample a point $x \in \mathcal{F}_{\mathcal{W}}$ by picking a triangle $t \in \mathcal{T}$ according to the distribution $\pi_{\mathcal{T}}$ and then picking a point $x \in t$ uniformly at random. Next, we sample a configuration from $L_f(x)$. The details depend on the specifics of the robot's kinematics and are described below separately for rigid and articulated robots.

The configuration q of a rigid robot consists of a positional component q_x, which specifies the position of the robot's reference point in the workspace, and an orientational component q_θ, which specifies the orientation of the robot. To sample a configuration, we first pick q_θ uniformly at random. We then pick a point $x \in \mathcal{F}_{\mathcal{W}}$, as described above, and compute q_x so that at $q = (q_x, q_\theta)$, the robot's feature point f lies at x and the robot has orientation q_θ.

For an articulated robot, the configuration q specifies its joints parameters $q_1, q_2, \ldots$. Suppose that the feature point f lies in the ℓth link of the robot. To sample a configuration, we again pick a point $x \in \mathcal{F}_{\mathcal{W}}$ and then find the joint parameters $q_1, q_2, \ldots, q_\ell$ that place f at x by solving the robot's inverse kinematics (IK) equations. If IK has no solution, we must pick another x. If IK has more than one solution, we pick one at random. We then sample the other joint parameters $q_{\ell+1}, q_{\ell+2}, \ldots$ uniformly at random. Various improvements can be made to speed-up this process. For instance, we may restrict the sampling domain according to the reachability of each feature point.

5 Constructing the Ensemble Sampler

WCO is an ensemble sampler composed of many component samplers, which all have different distributions due to the different feature points used. If WCO uses a single feature point, i.e., a single component sampler, then to perform well, this component sampler must generate a good distribution everywhere in $\mathcal{C}$. In general, such a sampler is difficult to construct. Using multiple feature points simplifies the task. It is sufficient for a component sampler to work well in only part of $\mathcal{C}$, provided that several component samplers can be combined effectively to generate a distribution good in entire $\mathcal{C}$.

Algorithm 2. Workspace-based Connectivity Oracle.

1: Let p_i be the probability of picking a component sampler s_i. Initialize $p_i = 1/K$ for $i = 0, 1, \ldots, K-1$.
2: **for** $t = 1, 2, \ldots$ **do**
3: Pick a component sampler s_i from $S = \{s_0, s_1, \ldots, s_{K-1}\}$ with probability p_i.
4: Sample a new configuration q using the component sampler picked.
5: **if** a new milestone m is added to the roadmap $\mathcal{R}$ **then**
6: Update the distributions for each component sampler $s_i, i = 1, \ldots, K-1$.
7: Update the probabilities $p_i, i = 0, 1, \ldots, K-1$.

5.1 Combining Samplers through AHS

Recall from Section 3.1 that WCO uses a set of component samplers, $S = \{s_0, s_1, \ldots, s_{K-1}\}$, where s_0 is a special, uniform sampler and each $s_i, i = 1, 2, \ldots, K-1$ is based on a feature point of the robot. We combine the component samplers through AHS. Each sampler s_i has an associated weight w_i, which reflects the success of s_i according to its sampling history. The sampler s_i is chosen to run with probability p_i that depends on w_i. To adapt the ensemble distribution, we adjust the weights so that the component samplers with better performance have higher weights. See Algorithm 2 for an outline of the algorithm.

In iteration t of Algorithm 2, we choose s_i with probability

$$p_i = (1-\eta)\frac{w_i(t)}{\sum_{i=0}^{K-1} w_i(t)} + \frac{\eta}{K}, \tag{1}$$

where $w_i(t)$ is the weight of s_i in iteration t and $\eta \in (0, 1]$ is a small fixed constant. We use the chosen s_i to sample a new milestone m and assign to s_i a reward r that depends on the effect of m on the roadmap $\mathcal{R}$:

- The milestone m reduces the number of connected components of $\mathcal{R}$. In this case, m merges two or more connected components and improves its connectivity. We set $r = 1$.
- The milestone m increases the number of connected components of $\mathcal{R}$. In this case, m creates a new connected component and potentially improves the coverage of $\mathcal{R}$. We also set $r = 1$.
- Otherwise, $r = 0$.

We then update the weight of s_i:

$$w_i(t+1) = w_i(t)\exp\left((r/p_i)\eta/K\right). \tag{2}$$

Note that the exponent depends on the received reward r weighted by the probability p_i of choosing s_i. If a sampler is not chosen, then its weight remains the same as before: $w_i(t+1) = w_i(t)$. More details on AHS are available in [13].

Although there are many possible schemes for updating the weights, AHS has an important advantage. It can be shown that under suitable assumptions, the ensemble sampler generated by AHS is competitive against the best component sampler [13]. More precisely, the following competitive ratio holds:

$$R_{\max} - R \leq (\mathrm{e} - 1)\eta R_{\max} + \frac{K \ln K}{\eta}, \quad (3)$$

where R is the expected total reward received by the ensemble sampler and $R_{\max}$ is the total reward received by the best component sampler if it is always chosen to run. This result can be interpreted as saying that the ensemble sampler performs almost as well as the best component sampler, without knowing in advance which component sampler is the best. With some small variations on the scheme for updating the weights, one can also show that the modified ensemble sampler is competitive against any linearly weighted combination of component samplers, an even stronger result theoretically [2]. This is one reason why we choose AHS for combining component samplers.

The ensemble sampling distribution π is a linearly weighted combination of component distributions: $\pi = \sum_{i=0}^{K-1} p_i.\pi_i$, where p_i is the probability for choosing s_i and π_i is the distribution for s_i. Each WCO component sampler maintains its own workspace channels and only samples in the lifted channels in $\mathcal{C}$. Since the component sampling distributions are combined linearly, the intersections of lifted channels from different component samplers have higher probability of being sampled. These intersections contain the configurations that simultaneously place multiple feature points in their respective workspace channels. Thus, although each component sampler operates independently, the ensemble sampler automatically takes into account a robot's kinematic constraints on the feature points.

5.2 Choosing Feature Points

We must still choose a set of feature points. By Proposition 1, a collision-free path in $\mathcal{C}$ implies a collision-free path in $\mathcal{W}$ for every point in the robot. So, to infer the configuration-space path from the workspace paths of a finite set of feature points, these workspace paths must be *representative.* Ideally, the feature set is small, because we construct a component sampler for each feature point. A large number of component samplers increase both the difficulty of identifying the good ones through AHS and the computational cost.

Since small feature sets are preferred, we choose feature points to be spaced far apart. The reason is that feature points close together generate similar sampling distributions. Below we give specific choices for rigid and articulated robots. These heuristics worked well in our experiments, but more research is needed to develop a principled method for selecting feature points.

For a rigid robot, the feature set is the union of two point sets, CH and MP. CH consists of the vertices on the convex hull of the robot. MP contains a single point in the middle of the robot, e.g., the centroid. For an articulated robot, we take CH and MP with respect to each rigid link of the robot and take their union.

6 Implementation and Experimental Results

We implemented the new planner in C++, using the Qhull [3] library for workspace triangulation. We tested the planner and compared it with other

PRM planners. For each planner, we set the required parameters by performing 10 trial runs and choosing the values that gave good results. We ran the planner 30 times independently on each test environment. Each run was terminated once the query was solved. Note that we did not insert the query configurations into the roadmap as milestones or used any information from the query configurations to bias sampling. The experiments were conducted on a PC with a 3 GHz processor and 1 GB memory. The results reported below are the averages of 30 runs.

6.1 Comparison with Other PRM Planners

Since WCO uses workspace information and generates a dynamic sampling distribution, we compared it with PRM planners of these two classes. For the former, we compared with workspace importance sampling (WIS) [17]. For the latter, we compared with the original AHS [13], whose component samplers consist of a uniform sampler, several Gaussian samplers, and several bridge tests. These two planners were chosen because they are closely related to our work, and both have shown strong performance in narrow passages sampling. We also ran a PRM planner with uniform sampling to benchmark the difficulty of test environments.

We tested our planner on Tests 1–4. In all tests, WCO uses CH $\cup$ MP as the feature set. However, since Test 4 uses a common articulated robot with a fixed base and all rotational joints, the workspace displacement of the robot's links near the base is very limited. To improve computational efficiency, we consider only the feature points in the furthest link, which contains the end-effector and the large plate.

WIS uses a single feature point. In our tests, it used MP for a rigid robot (Tests 1–3) and MP of the furthest link for an articulated robot (Test 4).

Overall, WCO performed much better than the original AHS and WIS (see Table 1). Although WCO incurs the additional costs of processing the workspace geometry and updating the sampling distribution, it uses fewer milestones and places them in strategic locations. It improves the overall performance by reducing the total number of collision checks needed for sampling new milestones and connecting milestones in the roadmap. See Fig. 4 for an illustration of the differences between WCO and the other planners.

For comparison between WCO and the original AHS, it is especially interesting to consider Test 2. The start configuration s and the goal configuration g, when projected to $\mathcal{W}$, are very close; however, to go from s to g, the robot must go out of the narrow tunnel, reorient, and then go back to the tunnel again. Regardless of which feature point f is chosen, it may potentially mislead the planner, because all short paths in $\mathcal{W}$ between $P_f(s)$ and $P_f(g)$ give little information on the correct configuration-space path that connects s and g. Nevertheless, WCO performed well here, because it combines information from both $\mathcal{W}$ and $\mathcal{C}$. It dynamically updates the workspace channels, which provide information for connecting distinct roadmap components. By doing so, as soon as $\mathcal{F}_c$

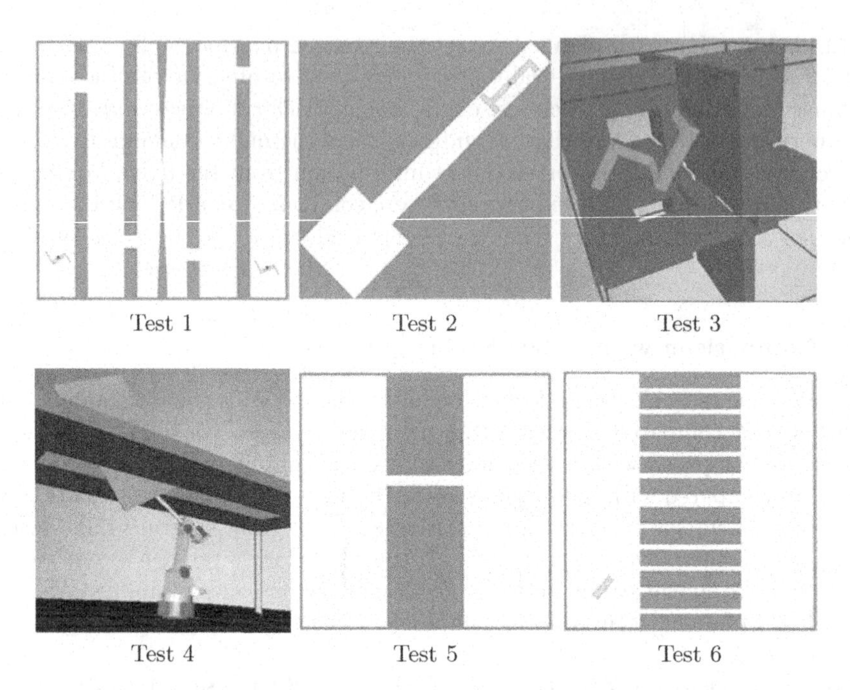

Fig. 3. Test 1: A 3-DOFs rigid-body robot moves from the lower left corner to the lower right corner by passing through five narrow openings. **Test 2**: A 3-DOFs rigid-body robot turns 180 degrees in a narrow deadend. **Test 3**: A 6-DOFs rigid-body robot must pass through 6 out of 7 narrow openings in order to answer the given query. **Test 4**: A 6-DOFs robot manipulator with its end-effector holding a large plate maneuvers through a narrow slot. **Test 5**: The robot is a planar articulated arm with a free-flying base. The dimensionality of $\mathcal{C}$ is increased by adding up to 8 links to the robot, resulting in a maximum of 10 DOFs. The robot must move through the narrow passage in the middle. **Test 6**: A 3-DOFs rigid-body robot moves from the left to the right wide-open space. It only fits through the passage in the middle. The number of false passages increases from 2 to 10.

is covered adequately by $\mathcal{R}$, WCO can potentially identify the correct regions of $\mathcal{C}$ to sample.

Compared with WIS, WCO performed significantly better except for Test 2. This is expected, because WIS uses a single feature point (MP) and a static sampling distribution, which does not respond to changes in $\mathcal{R}$ and wastes lots of effort in sampling regions of $\mathcal{C}$ already well covered. In Test 3 and 4, WIS performed as badly as uniform sampling. The reason is that in both cases, the solution path requires the robot to rotate and translate in a coordinated fashion. A single feature, used by WIS, is incapable of representing such complex motion and generates a suitable sampling distribution for solving the problem. Furthermore, WIS does not update the distribution dynamically to improve the performance. In Test 2, to solve the query, the entire $\mathcal{F}_c$ must be adequately covered, whether a static or a dynamic sampling distribution is used. WIS has

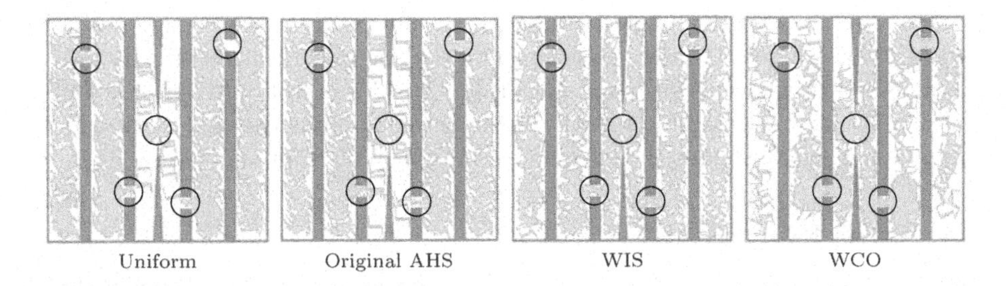

Fig. 4. 700 milestones generated by different planners. The pictures show that WCO increases the number of milestones in important regions that improve the connectivity of the roadmap, without generating too many unnecessary milestones in unimportant regions.

Table 1. Performance comparison of several PRM planners. All times are measured in seconds.

Sampler	**Test 1**					**Test 2**				
	T_{pre}	T_{upd}	T_{tol}	N_{mil}	N_{sam}	T_{pre}	T_{upd}	T_{tol}	N_{mil}	N_{sam}
Uniform			75.9	13,540	52,687			4.1	601	53,616
Original AHS			23.0	3,477	164,776			3.3	163	76,742
WIS	0.034		6.6	1,660	7,024	0.007		0.7	154	11,521
WCO	0.045	0.072	2.7	650	2,448	0.008	0.012	0.8	170	5,531
Sampler	**Test 3**					**Test 4**				
	T_{pre}	T_{upd}	T_{tol}	N_{mil}	N_{sam}	T_{pre}	T_{upd}	T_{tol}	N_{mil}	N_{sam}
Uniform			94.6	9,011	36,594			69.8	9,246	35,878
Original AHS			56.7	1,669	198,313			56.0	2,672	168,013
WIS	0.607		80.3	5,723	160,686	0.071		200.7	14,423	961,613
WCO	0.942	2.408	25.9	2,080	22,811	0.244	0.993	31.1	3,211	62,405

T_{pre}: time for triangulating $F_{\mathcal{W}}$. T_{upd}: time for updating component sampling distributions (Algorithm 1, lines 8–12). T_{tot}: total running time. N_{mil}: number of milestones required for answering the query. N_{sam}: number of configurations sampled.

Table 2. The effect of feature points on the running times of WCO. All times are measured in seconds. |CH| denotes the number of feature points in CH.

Test Env.	|CH|	MP	CH	CH ∪MP
Test 1	6	2.2	4.7	2.7
Test 2	5	1.3	0.7	0.8
Test 3	13	40.8	28.9	25.9
Test 4	8	154.3	62.0	31.1

an advantage, because it is simpler and does not incur the cost of updating the sampling distribution. Even so, the performance of WCO is comparable.

6.2 The Choice of Feature Points

Different feature points are good for sampling different regions of $\mathcal{C}$. To assess the benefits of multiple feature points, we ran WCO on Tests 1–4 with different feature sets. The experimental results show that although the performance of CH and MP varies across the test environments, the combined feature set CH $\cup$MP has consistently good performance (see Table 2). This corroborates the theoretical result that the ensemble sampler is almost as good as the best component sampler and demonstrates the effectiveness of the AHS approach.

6.3 Other Experiments

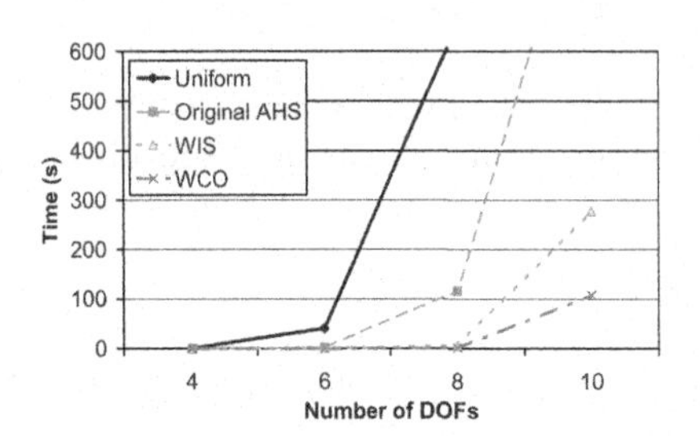

Fig. 5. The performance of WCO, as $\dim(\mathcal{C})$ increases (Test 5)

One concern of using workspace information to guide sampling in $\mathcal{C}$ is that as the dimensionality of $\mathcal{C}$ increases, workspace information becomes less useful. For this, we constructed a test environment with increasing dimensionality of $\mathcal{C}$ (Test 5). The results indicate that workspace information still has its merit (Fig. 5). The usefulness of workspace information does not directly depend on the dimensionality of $\mathcal{C}$ [11]. Instead, it depends more on whether there are workspace channels with good (n, p)-property.

One drawback of WCO is that it may find false workspace passages as channels, i.e., workspace passages that the robot can not pass through. It seems plausible that as the number of false passages grows, the performance of WCO will keep on worsening. So, we performed a test with an increasing number of false workspace passages (Test 6). The results indicate that this trend happens, but only to a limited extent (Fig. 6). The reason is that by construction, the number of channels in a channel graph G' is linear in the number of terminals in G'. Hence, after a certain limit, increasing the number of invalid workspace passages does not increase the number of channels or affect WCO's performance.

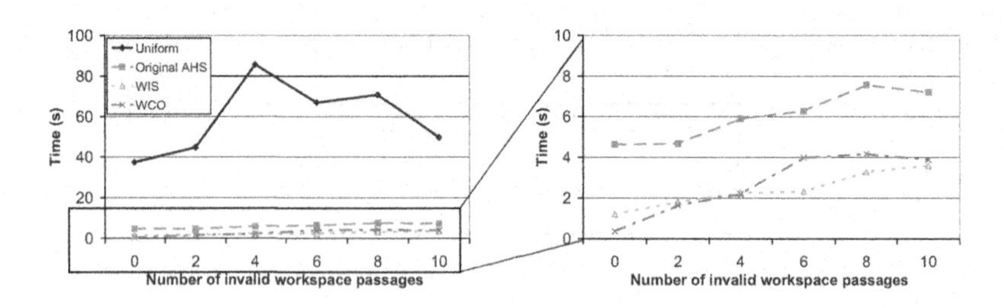

Fig. 6. The performance of WCO, as the number of false passages increases (Test 6)

7 Discussion

The WCO component samplers treat the feature points independently. Only the ensemble sampler implicitly accounts for the robot's kinematic constraints on the feature points. To further improve sampling efficiency, we may explicitly incorporate such constraints. We start with the simplest type, namely, the distance between a pair of feature points, as the distance between two feature points of a rigid robot or a rigid link of an articulated robot is fixed. One way of imposing such a constraint is to find the workspace channels for all feature points, as explained in Section 4.2. However, instead of immediately updating the sampling distribution, we check whether two channels violate the distance constraint for the corresponding feature points and ignore the channels if they do. One way to check distance constraint is by enlarging each channel according to the given distance constraint and check whether the enlarged channels intersect. There are also many other types of kinematic constraints, which can possibly be incorporated through more sophisticated geometric features, such as line segments and surface triangles.

Although the above idea is simple and intuitively should improve WCO's performance, more thoughts and experiments are needed for finding good methods to incorporate and combine the information from multiple kinematic constraints. Treating each geometric feature as a "robot" and viewing WCO as multi-robot planning give us a spectrum of options. One extreme is decoupled planning with no coordination among robots. This approach is computationally efficient, because it ignores all constraints, but is not complete. WCO is somewhat similar to this approach, though it actually handles kinematic constraints implicitly. The other extreme is centralized configuration-space planning, which accounts for all the constraints, but is slower. Between the extremes, there are decoupled planning with pairwise coordination, global coordination, and prioritized planning. Each can translate to a method for incorporating kinematic constraints. For instance, the idea in the previous paragraph is similar to pairwise coordination.

8 Conclusion

This paper presents WCO, an adaptive sampling strategy for PRM planning. WCO is composed of many component samplers, each based on a geometric feature of a robot. Using the adaptive hybrid sampling approach, it combines information from both workspace geometry and sampling history to construct a dynamic sampling distribution. In our experiments, WCO significantly outperformed two recently proposed sampling strategies, which use, respectively, workspace information and dynamic sampling alone.

For future work, it would be interesting to extend WCO by relaxing the independence assumption on the component samplers and developing good methods to incorporate a robot's kinematic constraints *explicitly*. Viewing WCO as multi-robot planning is a promising direction. Another interesting extension is to use sampling history to improve the search for workspace channels, instead of just relying on the channel graph.

References

1. Amenta, N., Choi, S., Kolluri, R.: The power crust. In: Proc. ACM Symp. on Solid Modeling, pp. 249–260 (2001)
2. Auer, P., Cesa-Bianchi, N., Freund, Y., Schapire, R.: The nonstochastic multiarmed bandit problem. SIAM J. Computing 32(1), 48–77 (2002)
3. Barber, C., Dobkin, D., Huhdanpaa, H.: The quickhull algorithm for convex hulls. ACM Trans. on Mathematical Software 22(4), 469–483 (1996)
4. Boor, V., Overmars, M., van der Stappen, A.: The gaussian sampling strategy for probabilistic roadmap planners. In: Proc. IEEE Int. Conf. on Robotics & Automation, pp. 1018–1023 (1999)
5. Burns, B., Brock, O.: Toward optimal configuration space sampling. In: Robotics: Science and Systems (2005)
6. Choset, H., Lynch, K., Hutchinson, S., Kantor, G., Burgard, W., Kavraki, L., Thrun, S.: Principles of Robot Motion: Theory, Algorithms, and Implementations, ch.7. The MIT Press, Cambridge (2005)
7. Cormen, T., Leiserson, C., Rivest, R., Stein, C.: Introduction to Algorithms, 2nd edn. The MIT Press, Cambridge (2001)
8. de Berg, M., van Kreveld, M., Overmars, M., Schwarzkopf, O.: Computational Geometry: Algorithms and Applications, 2nd edn. Springer, Heidelberg (2000)
9. Foskey, M., Garber, M., Lin, M., Manocha, D.: A Voronoi-based hybrid motion planner. In: Proc. IEEE/RSJ Int. Conf. on Intelligent Robots & Systems, pp. 55–60 (2001)
10. Holleman, C., Kavraki, L.: A framework for using the workspace medial axis in PRM planners. In: Proc. IEEE Int. Conf. on Robotics & Automation, pp. 1408–1413 (2000)
11. Hsu, D., Latombe, J.-C., Kurniawati, H.: On the probabilistic foundations of probabilistic roadmap planning. Int. J. Robotics Research 25(7), 627–643 (2006)
12. Hsu, D., Latombe, J.-C., Motwani, R.: Path planning in expansive configuration spaces. Int. J. Comp. Geometry & Applications 9(4–5), 495–512 (1999)
13. Hsu, D., Sánchez-Ante, G., Sun, Z.: Hybrid PRM sampling with a cost-sensitive adaptive strategy. In: Proc. IEEE Int. Conf. on Robotics & Automation, pp. 3885–3891 (2005)
14. Kavraki, L., Kolountzakis, M., Latombe, J.-C.: Analysis of probabilistic roadmaps for path planning. In: Proc. IEEE Int. Conf. on Robotics & Automation, pp. 3020–3025 (1996)
15. Kavraki, L., Latombe, J.-C., Motwani, R., Raghavan, P.: Randomized query processing in robot path planning. J. Comp. & Syst. Sci. 57(1), 50–60 (1998)
16. Kavraki, L., Švestka, P., Latombe, J.-C., Overmars, M.: Probabilistic roadmaps for path planning in high-dimensional configuration space. IEEE Trans. on Robotics & Automation 12(4), 566–580 (1996)
17. Kurniawati, H., Hsu, D.: Workspace importance sampling for probabilistic roadmap planning. In: Proc. IEEE/RSJ Int. Conf. on Intelligent Robots & Systems, pp. 1618–1623 (2004)
18. LaValle, S., Kuffner, J.: Randomized kinodynamic planning. Int. J. Robotics Research 20(5), 378–400 (2001)

19. Morales, M., Tapia, L., Pearce, R., Rodriguez, S., Amato, N.: A machine learning approach for feature-sensitive motion planning. In: Algorithmic Foundations of Robotics VI, pp. 361–376 (2004)
20. Švestka, P.: On probabilistic completeness and expected complexity for probabilistic path planning. Technical Report UU-CS-1996-08, Utrecht University, Dept. of Information & Computing Sciences, Utrecht, the Netherlands (1996)
21. van den Berg, J., Overmars, M.: Using workspace information as a guide to non-uniform sampling in probabilistic roadmap planners. Int. J. Robotics Research 24(12), 1055–1071 (2005)
22. Yang, Y., Brock, O.: Adapting the sampling distribution in PRM planners based on an approximated medial axis. In: Proc. IEEE Int. Conf. on Robotics & Automation, pp. 4405–4410 (2004)

Incremental Map Generation (IMG)

Dawen Xie, Marco Morales, Roger Pearce, Shawna Thomas, Jyh-Ming Lien, and Nancy M. Amato

Parasol Lab, Department of Computer Science,
Texas A&M University, College Station, TX USA
{dawenx,marcom,rap2317,sthomas,neilien,amato}@cs.tamu.edu

Abstract. Probabilistic roadmap methods (PRMs) have been highly successful in solving many high degree of freedom motion planning problems arising in diverse application domains such as traditional robotics, computer-aided design, and computational biology and chemistry. One important practical issue with PRMs is that they do not provide an automated mechanism to determine how large a roadmap is needed for a given problem. Instead, users typically determine this by trial and error and as a consequence often construct larger roadmaps than are needed. In this paper, we propose a new PRM-based framework called Incremental Map Generation (IMG) to address this problem. Our strategy is to break the map generation into several processes, each of which generates samples and connections, and to continue adding the next increment of samples and connections to the evolving roadmap until it stops improving. In particular, the process continues until a set of evaluation criteria determine that the planning strategy is no longer effective at improving the roadmap. We propose some general evaluation criteria and show how to apply them to construct different types of roadmaps, e.g., roadmaps that coarsely or more finely map the space. In addition, we show how IMG can be integrated with previously proposed adaptive strategies for selecting sampling methods. We provide results illustrating the power of IMG.

1 Introduction

Automatic motion planning has applications in many areas such as robotics, computer animation, computer-aided design (CAD), virtual prototyping, and computational biology and chemistry. Although many deterministic motion planning methods have been proposed, most are not used in practice because they are computationally infeasible except for some restricted cases, e.g., when the robot has few degrees of freedom (dof) [14]. Indeed, there is strong evidence that any complete planner (one that is guaranteed to find a solution or determine that none exists) requires time exponential in the robot's dof [21].

For this reason, attention has focused on randomized approaches that sample and connect points in the robot's configuration space (C-space). Such methods include graph-based methods such as the *probabilistic roadmap methods* (PRMs) [13] (along with their various extensions and variants [1,4,5,9,25]) and tree-based methods such as Ariadne's Clew algorithm [16], RRT [15], and Hsu's expansive

S. Akella et al. (Eds.): Algorithmic Foundation of Robotics VII, STAR 47, pp. 53–68, 2008.
springerlink.com

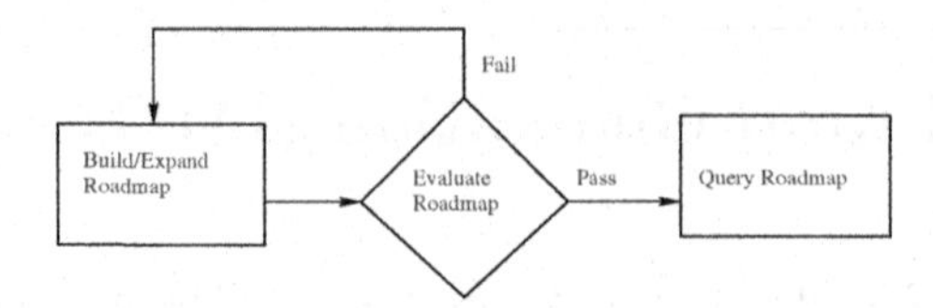

Fig. 1. Flow diagram for Incremental Map Generation (IMG)

planner [10]. These methods have been highly successful in solving challenging problems with many dof that were previously unsolvable and thus have become the method of choice for a wide range of applications.

One important practical issue not addressed by the PRM framework is that it does not provide an automated mechanism to determine when to stop. Ideally, planning should be terminated when the planner is no longer adding useful information to the roadmap. In practice, users select a roadmap size they believe is appropriate, usually by trial and error. This often results in larger maps than needed or in the construction of several maps before obtaining one that meets the user's needs. While there are a number of reasons for this disconnect between the ideal and practice, perhaps the most important has been the lack of effective techniques for measuring roadmap improvement.

In this paper, we propose a PRM-based framework called Incremental Map Generation (IMG) that addresses this issue. In particular, we advocate a strategy that measures the improvement achieved over time in an evolving roadmap to automatically determine when to stop (or perhaps change) the planner. This is implemented by iteratively building the roadmap until it satisfies a set of evaluation criteria (see Figure 1). The main difference from the traditional two-phase PRM method [13] is that we partition the roadmap construction into several iterations (expansion steps), each of which adds samples and connections to the evolving roadmap, and we add a new phase called "roadmap evaluation" which tests if the roadmap satisfies some evaluation criteria (stopping condition). If the roadmap passes the stopping condition, then roadmap construction finishes. Otherwise, another iteration is performed to expand the roadmap by adding additional samples and connections. The framework can accept a broad range of stopping criteria, which can be customized for particular applications or user preferences. For example, the criteria can be as simple as satisfying a specified set of queries, or more complicated such as monitoring graph topology.

IMG has several important features, including:

- *Automatic determination of roadmap size.* The most important feature of IMG is that it provides a mechanism to incrementally construct roadmaps and to automatically determine when construction should be halted.
- *Evaluation criteria.* A key requirement for IMG is effective evaluation criteria that can be efficiently tested during roadmap construction. A contribution of this work is to propose evaluation criteria for measuring roadmap quality (e.g., coverage and connectivity) that do not require prior knowledge about

the solution (as do, e.g., test queries) and that do not rely on C-space discretization (so can be efficiently applied to high dof problems).

- *Compatibility with existing sampling-based planners.* IMG is *not* a new sampling method; instead, it is a general strategy that can be used with any sampling-based planner and, moreover, it provides a natural mechanism for adaptive planning. For example, each IMG iteration can use any of the existing adaptive strategies that utilize multiple planners (e.g., [11, 17]) or different strategies could be chosen for different IMG iterations.

2 Related Work

The general PRM methodology [13] consists of a preprocessing phase and a query phase. Preprocessing, which is done once for a given environment, first samples points 'randomly' from the robot's C-space, retaining those that satisfy certain feasibility requirements. Then, these points are connected to form a graph, or roadmap, containing representative paths in the free C-space. The query phase then connects any given start and goal to the same connected component of the roadmap, and if successful, returns a path connecting them.

The probability of failing to find a path in a probabilistic roadmap, when one exists in C-free, decreases exponentially as the number of samples in the roadmap increases [12]. However, it is difficult to decide beforehand the roadmap size required in practice.

The *coverage* and *connectivity* of an ideal PRM roadmap should match that of its underlying C-space. In [7], coverage and maximal connectivity achieved by different sampling methods was compared to that of the C-space being modeled. Coverage indicates how each query can be connected to the roadmap. If there exists a path in the free C-space between two query configurations, maximal connectivity ensures that a path between them can be found in the roadmap. The authors evaluate the time needed to adequately cover and connect the free C-space for various techniques. This work relies on a discretization of C-space and so cannot be applied to high dof problems.

Since there is no principled mechanism to determine when to stop roadmap construction, a commonly used evaluation criterion is to predefine a set of relevant queries in each environment and continue building the roadmap until the query configurations can be connected to the same connected component. This is helpful in environments where the user knows beforehand such a representative query. However, in many situations defining such a query can be problematic, e.g., in cluttered environments or in higher dof problems such as the protein folding applications. The stopping criteria we propose can be applied when it is hard or even impossible to define a representative query for a given problem. In the case when such a query is easy to define or when solving a particular query is the user's objective, then IMG can easily use it as a stopping criterion, i.e., IMG also supports the traditional query-based criterion.

A set of metrics are proposed in [18] to estimate how each new sample improves, or not, the representation of C-space achieved by the planner. With these metrics, the authors identify three phases common to all sampling-based planners: quick learning (a coarse roadmap is constructed), model enhancement (the roadmap is refined), and learning decay (most new samples do not provide additional information). They also demonstrate that the traditional scheme of testing a set of witness queries, which is commonly used in practice as a stopping criterion, can be misleading.

Adaptive hybrid PRM sampling [11] proposes using a mixture of samplers. They adapt the mixture of strategies based on each strategy's past success. In this work, we incorporate hybrid PRM sampling and apply our IMG framework to hybrid PRM to decide when to stop building the roadmap.

3 Incremental Map Generation (IMG)

We propose a new PRM-based framework called Incremental Map Generation (IMG) in which we iteratively build a roadmap until it satisfies a set of evaluation criteria (see Figure 1). Most importantly, this framework provides a systematic way to automatically decide when to stop roadmap construction. Algorithm 3.1 describes IMG. This framework is simple and general. It can be customized for a particular application domain or problem by simply varying the node generation and connection strategies used and the evaluation criteria. In the following sections we discuss two main aspects of our framework: incremental roadmap construction and roadmap evaluation.

Algorithm 3.1. Incremental Map Generation.

Input. An existing roadmap R, a roadmap evaluator E, the size of a node set, n.
Output. A roadmap R that meets the criteria indicated by E.
1: **repeat**
2: *Initialization.* Set parameters for this iteration.
3: *Sampling.* Generate the new node set (n nodes) and add them to roadmap R.
4: *Connection.* Perform connection.
5: **until** R meets criteria in E

3.1 Incremental Roadmap Construction

To build the roadmap incrementally, we first divide roadmap construction into "sets" of size n; the size, or target number of nodes for each set, is specified by the user. Then, for each iteration, IMG performs the following steps.

Initialization. In line 2, Algorithm 3.1, in order to ensure the independence of each set, we seed the random number generator. The seed s is a polynomial function of the *base seed* of the program (e.g., the time execution starts), the type of node generation method used, and the number of sets completed by that node generation method so far. Calculating the seed in a deterministic way based on a (possibly random) base seed supports reproducibility given the same base seed.

Sampling. In line 3, Algorithm 3.1, the sampling strategy selected for that iteration is applied. Recall that IMG is *not* a new sampling method, but rather is a general strategy that can be applied to any sampling-based planner.

In addition, IMG provides a natural mechanism for adaptive planning. For example, each iteration of IMG could select a different sampling strategy or it might use one of the recently proposed adaptive strategies that utilize multiple planners (e.g., [11,17]). To illustrate this feature of IMG, we incorporated hybrid PRM [11] in our current implementation. In hybrid PRM [11], the performance of component samplers is evaluated and the methods with good performance are chosen to run more frequently. We incorporated hybrid PRM in two different ways: (1) it is simply used as described in [11] as the sampling method in IMG, and (2) in each IMG iteration, an initial phase uses the hybrid PRM strategy to select a planner to use for the remainder of that iteration.

Connection. In line 4, Algorithm 3.1, the connection strategy chosen by the user is applied to connect the new set of nodes to the existing roadmap. In the results presented here, we use a variant of the commonly used K-closest connection strategy. K-closest attempts to connect each node to its k "nearby" neighbors, but it does not distinguish successful attempts from failed attempts. Nevertheless, identifying successes and failures in connection attempts provides some information about the complexity of the local area. When a node can be connected to most of its neighbors, it indicates that this node is in an easy to connect area and we probably do not need to try many connection attempts; on the other hand, if a node fails to be connected to most of its neighbors, it indicates that this node is in a difficult local area and it could be useful to try to connect it to more neighbors. In order to adjust the connection effort based on a node's local environment, we use a modified version of K-Closest connection method called *L-Success-M-Failure.* In L-Success-M-Failure, the local planner attempts to connect each node to its $l+m$ "nearby" neighbors, stopping as soon as it has achieved l successful attempts or m failed attempts.

3.2 Roadmap Evaluation

The other key component enabling automatic determination of roadmap size is the stopping or evaluation criteria. In this paper, we propose two classes of evaluation methods: *roadmap progress evaluation* and *application-specific evaluation.* The following sections give examples of evaluators for both classes.

Roadmap Progress Evaluation

Our roadmap progress evaluators are based on metrics for evaluating roadmap *coverage* and *connectivity,* which have been noted as important properties by many researchers working with sampling-based planners (see, e.g., [7]). Here, we are interested in monitoring the contribution of new samples to the coverage and connectivity modeled by the roadmap. Classification of new samples provides a mechanism to perform this evaluation as shown in [18]. In [18], every node is classified as it is inserted into the roadmap (see Figure 2). A node is classified as *cc-create* if it cannot be connected to any existing roadmap component. A node is classified as *cc-merge* if it connects to more than one connected component (CC) in the roadmap. A node is classified as *cc-expand* if it connects to exactly one component in the roadmap and satisfies an expanding criterion. A node is classified as *cc-oversample* if it does not fall in any of the previous categories. In previous work [20, 11], cc-expand and cc-oversample nodes were not always distinguished, in some cases because it was considered too expensive to classify a node as cc-expand.

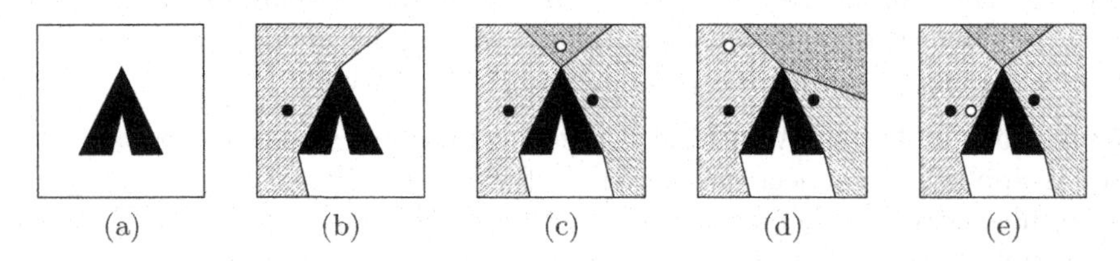

Fig. 2. (a) A 2D C-space. Classification of samples as (b) cc-create, (c) cc-merge, (d) cc-expand, and (e) cc-oversample.

In this work, we use the diameter of the connected components as a measurement of component expansion. The *diameter* of a CC is the length of the longest shortest-path in the CC. The diameter is an interesting metric because its changes correlate with changes in cc-create, cc-merge or cc-expand nodes. Also, the diameter of a graph can be approximated and is independent of the distance metric [22]. We let *max-diameter* be the maximum diameter of all the CCs and *sum-diameter* be the sum of the diameters of all CCs. Note that max-diameter is an approximation of the coverage of the largest connected component. Similarly, sum-diameter is an approximation of the coverage and connectivity of all the components in the roadmap. Then we use the rates of change of max-diameter and sum-diameter to approximate the planner's effectiveness in mapping C-space.

In this evaluation method, we stop building a roadmap when the rates of change of max-diameter and sum-diameter over a certain period of time, e.g., k sets of nodes, is smaller than a user-defined threshold, τ, which is used to define the desired variability in coverage and connectivity (as indicated by the components' diameters). We compute the max-diameter among all CCs and the sum-diameter of all the CCs at the end of each node set.

The percentage of change of the max-diameter ($PCMAX_i$) in the ith set over its k previous sets is computed as:

$$PCMAX_i = \sum_{j=0}^{k-1} \frac{|MD_{i-j} - MD_{i-j-1}|}{MD_{i-j-1}},$$

where MD_{i-j} is the max-diameter in the $(i-j)^{th}$ node set. We define the percentage change of the sum-diameter $(PCSUM_i)$ over all the components in a similar way.

Application-Specific Evaluation

The IMG framework can accept a broad range of stopping or evaluation criteria customized for particular applications or user preferences. In this section, we give two examples of application-specific evaluation methods.

Query Evaluation. This evaluator simply determines whether a roadmap can solve a set of user specified queries. For each query, it attempts to connect the start and goal to the roadmap and returns successful if they are connected to the same connected component. The evaluator returns success when all queries are solved. This type of evaluator is useful when the user wants to solve a particular set of test problems or for a single query application.

Max-flow Evaluation. Some applications require many paths between two configurations. For example, motion planning has been recently applied to study problems in computational biology such as protein folding and transitions [3, 23]. To study how a protein changes between two configurations, we can examine the probable paths between them in the roadmap. We can define this as a maximum flow problem on a network. If a roadmap edge weight, $w(e)$, reflects the likelihood that the protein will move from one configuration to the next, then we can define edge capacity $c(e)$ as $1/w(e)$. The evaluator returns success if the max-flow between the two configurations is above some user specified threshold f.

4 Experiments

IMG is *not* a new sampling method, instead it is a general strategy that can be applied to any sampling-based planner. We investigate how IMG automatically builds roadmaps with an appropriate number of samples using different evaluation criteria. Our experiments use the following sampling methods:

- *Uniform random sampling*: samples are created by picking random values for each dof.
- *Gaussian-biased sampling* [4]: sets of two samples are created, one uniformly at random and the other a distance d away, where d has a Gaussian distribution. A collision-free sample is added to the roadmap when one is collision-free and the other is not.
- *Bridge-test sampling* [9]: similar to Gaussian sampling, it takes two random samples a distance d apart, where d has a Gaussian distribution, until both samples are in collision and their midpoint is not. The collision-free sample is added to the roadmap.

- *Obstacle-based sampling* (OBPRM) [1]: samples are generated near C-obstacle surfaces by first generating a random colliding (resp., collision-free) sample and searching along a random direction until the sample becomes collision-free (resp., in collision).

We implemented all planners with the Parasol Lab motion planning library developed at Texas A&M University and performed collision detection with RAPID [8]. For each problem, we built two types of roadmaps: a tree and a graph. We use the L-Success-M-Failure connection strategy introduced in Section 3. In particular, we apply a 10-Success-20-Failure connection strategy for building trees and 5-Success-20-Failure for building graphs. For rigid-body motion planning, we use two local planners: straight-line and rotate at 0.5 [2], which translates from the start to the midpoint, rotates to the orientation of the goal configuration and then translates to the goal configuration. For articulated linkage motion planning, we only use the straight-line local planner. All results were run on 700MHz Intel PIII Xeon processors.

In the following sections, we discuss the performance of IMG's roadmap progress evaluator, the overhead of the IMG framework, how IMG and hybrid PRM may be combined, and how IMG can be tailored to specific applications such as protein folding.

4.1 Automatically Stopping Roadmap Construction

Here we investigate the performance of the roadmap progress evaluator (see Section 3.2) in the four different environments shown in Figure 3. For these experiments, the node set size is 50 samples. After each set, we compute $PCMAX_i$ and $PCSUM_i$. Roadmap construction stops when both $PCMAX_i$ and $PCSUM_i$ are below a threshold τ. τ represents the desired roadmap improvement over a period of time, i.e., k sample sets. Note that in the beginning of roadmap construction (during the quick learning stage), there will be large changes in $PCMAX$ and $PCSUM$. These changes will drop when the enhancement stage begins.

We studied the impact of τ and k on IMG's performance for each sampling method in each environment by varying k with constant τ and alternatively by varying τ with constant k. Due to space limitations, we only show a subset of these results. A complete set of results can be found in [26].

A Case Study: Varying k with Constant τ

Figure 4 shows IMG's performance at building both trees and graphs for each planner in the hook environment. Here we vary k (the number of sample sets over which the percentage change in diameter is computed) while keeping τ constant at 0.0125. In each plot, the upper two curves show the sum-diameter and max-diameter as a function of roadmap size for a tree, and the lower two curves for a graph. The circles indicate where IMG would stop roadmap construction for various k. For example, the circle labeled $k_1, 1250$ in Figure 4(a) shows that with $k = k_1 = 5$, IMG would stop construction of the tree after 1250 samples. Similarly,

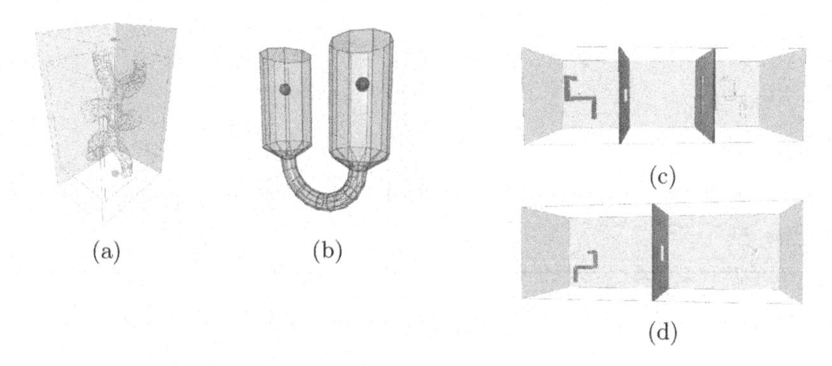

Fig. 3. Problems studied. (a) Maze environment (wire frame): rigid body robot must navigate the maze. (b) U shape environment: rigid body robot must navigate from one chamber to the other. (c) Hook environment: rigid body robot must rotate to move from one side of the walls to the other. (d) Hook manipulator environment: articulated linkage (10 dof) must move from one end to the other.

the circle labeled $k_1, 2150$ indicates IMG would stop construction of the *graph* after 2150 samples with the same k value. All plots use the same random seed.

From the evolution of max-diameter and sum-diameter, it is clear that the roadmap grows rapidly in the beginning and then experiences a long period of refinement until both stabilize. As expected, the diameter in the tree roadmap is larger than the diameter in the graph roadmap. This corresponds to the graph roadmap having shorter and smoother paths. An interesting observation from the graph roadmap is that the "path refinement" stage is clearly shown as the diameters drop.

Overall, we see that for a fixed τ, increasing k causes the planner to stop later because larger k values allow IMG to capture changes over longer periods. This trend appears in all experiments we ran. This means that we can decide how long we want to refine the roadmap by what value we choose for k. It is also clear that for a given k value, different planners stop at different points. In particular, BasicPRM (Figure 4(a)) stops the earliest. The intuition behind this is that BasicPRM is the slowest to progress in terms of samples, and thus needs larger values of k to capture similar changes.

Finally, we defined a witness query from the first chamber to the last chamber. We use this query in the query evaluation method as described in Section 3.2. BasicPRM is unable to solve the query after 15000 samples, while Bridge test solved it after 550 samples, Gauss after 3600 samples, and OBPRM after 500 samples[1]. It is clear from Figure 4 that the max-diameter and sum-diameter still experience large changes after solving the witness query. This confirms the observation in [18] that solving queries is not enough by itself to evaluate whether the planner is still making progress in mapping the space.

[1] The tree and graph roadmaps solve the query at the same point since we used the same random seed.

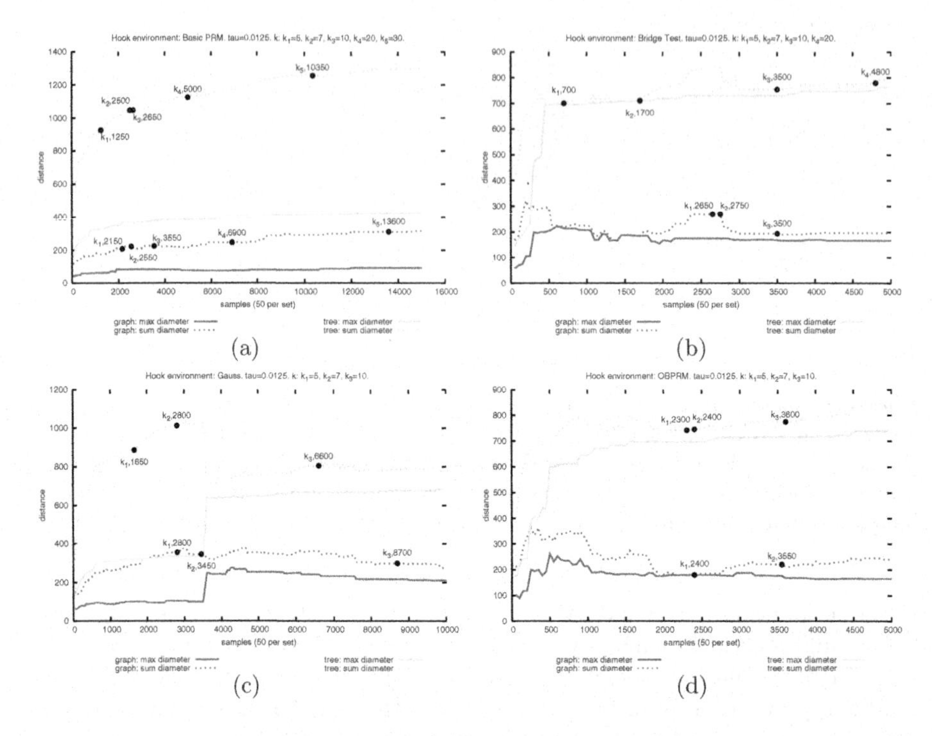

Fig. 4. Performance of IMG for the hook environment for various planners and k values with $\tau = 0.0125$. (a) BasicPRM; unable to solve the witness query with 15000 samples. (b) Bridge test; witness query solved at 550 samples. (c) Gauss; witness query solved at 3600 samples. (d) OBPRM; witness query solved at 500 samples.

A Case Study: Varying τ with Constant k

Here we study the effect of τ (a threshold for desired variability in coverage and connectivity) while fixing k at 10 for OBPRM [1] in different environments. Figure 5 shows results for both trees and graphs using the same random seed. As before, the circles indicate where IMG would stop roadmap construction for the various τ.

Overall, decreasing τ requires the planner to run longer because smaller τ values signify a smaller tolerance for diameter variability. Thus, a smaller τ means the planner has to run longer before the learning stabilizes enough to cause the diameter changes to fall below the threshold.

We set witness queries as described in Figure 3. OBPRM is unable to solve the query for the hook manipulator after 20000 samples, while it solved the query for the maze environment after 3750 samples and the U environment after 2850 samples. For the hook manipulator (Figure 5(a)), we observe that $\tau = 0.025$ roughly marks the end of the "quick learning" stage as it transitions into "model enhancement." Note, the planner remains in "model enhancement" for the entire duration. This is reflected by the fact that $\tau = 0.0007$ was never satisfied and

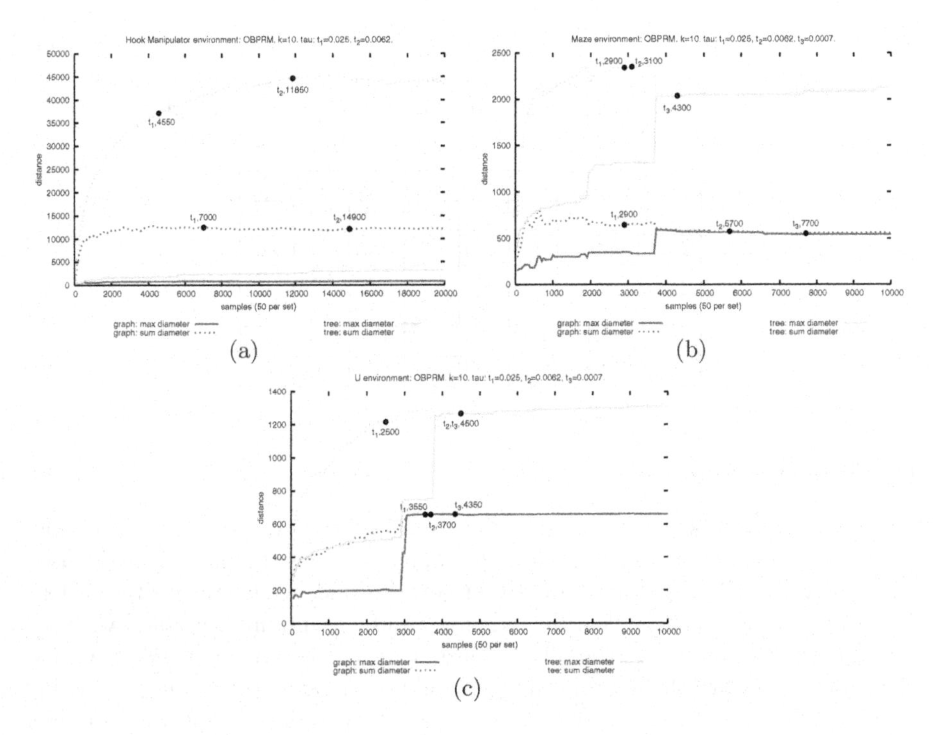

Fig. 5. Performance of IMG with OBPRM for several environments and τ values when $k = 10$. (a) Hook manipulator environment; unable to solve witness query in 20000 samples. (b) Maze environment; witness query solved at 3750 samples. (c) U environment; witness query solved at 2850 samples.

the witness query was never solved. In Figures 5(b) and 5(c), "learning decay" is clearly marked with $\tau = 0.0007$.

Overhead

The overhead incurred by the calculation of roadmap diameter as an evaluation of roadmap progress is affordable. We show in Table 1 the percentage of total running time spent in the diameter computation for the hook environment for the case when the roadmap can contain cycles (a graph). The diameter computation is performed after every 50 samples and for the tree roadmap case is exactly computed by a Dijkstra search. In the cyclic graph case, the diameter is approximated using two Dijkstra searches, with the second search starting from the furthest node found during the first search. While more accurate approximations of cyclic graph diameters exist [6], this was sufficient for our experiments.

Table 1. Diameter computation as a percentage of total running time in the Hook environment. Diameter computation was performed after every 50 samples. For all methods, the overhead for IMG is small, even for large numbers of samples.

Sampling Method	Number of Samples					
	100	500	1000	2000	4000	8000
BasicPRM	0.20	1.04	1.86	3.16	4.75	6.50
OBPRM	0.00	0.44	0.93	1.70	2.84	4.41
Gauss	0.09	0.48	0.95	1.78	3.13	5.39
Bridge test	0.02	0.09	0.19	0.45	0.99	1.93
Hybrid PRM	0.08	0.34	0.71	1.47	2.70	4.64

Combining IMG and Hybrid PRM

The IMG framework can be used with any sampling strategy. In this section, we incorporated hybrid PRM [11] in two different ways. First, we simply used hybrid PRM as the sampling strategy in IMG. Second, we partitioned each IMG iteration into two parts: a "learning window" and a "sampling window," with the learning window at 20% of the total set size. During the learning window, we use hybrid PRM to learn the appropriate probabilities of using each sampler, starting with a uniform distribution. We then fix this distribution during the sampling window. We experimented with several different ways of learning during the learning window and all variants displayed comparable behavior. In the results shown here, we use a fixed uniform probability distribution during the learning window, but learn the probability distribution for the sampling window just as with hybrid PRM sampling. Because the learning window is relatively small, this allows the learner to observe all the samplers. For all experiments shown here, we use the cost-based version of hybrid PRM described in [11] and five component samplers: BasicPRM, two versions of the Bridge test, and two versions of Gauss. The reward mechanism of the original hybrid PRM only rewards a planner when it generates *cc-create* and *cc-merge* samples. However, a sample that expands a roadmap is also important. Therefore, when a planner generates a *cc-expand* sample we reward it equal to the complement of the percentage of successful connections from that sample.

In Figure 6, we show hybrid PRM in the IMG framework. For clarity, we only show BasicPRM, and the version of Bridge test and Gauss that performed best. Figure 6(a) shows where IMG would stop roadmap construction with pure hybrid PRM sampling for various k and τ. This shows similar trends when varying k and τ as seen previously. Figure 6(b) shows the relative number of samples created by each sampler. Figure 6(c) shows the probability of being selected and the percentage of cc-oversample nodes for each sampler. Our results confirm the findings in [11]: BasicPRM is selected early on because it is relatively inexpensive but dies out quickly as other, more powerful and expensive samplers are selected.

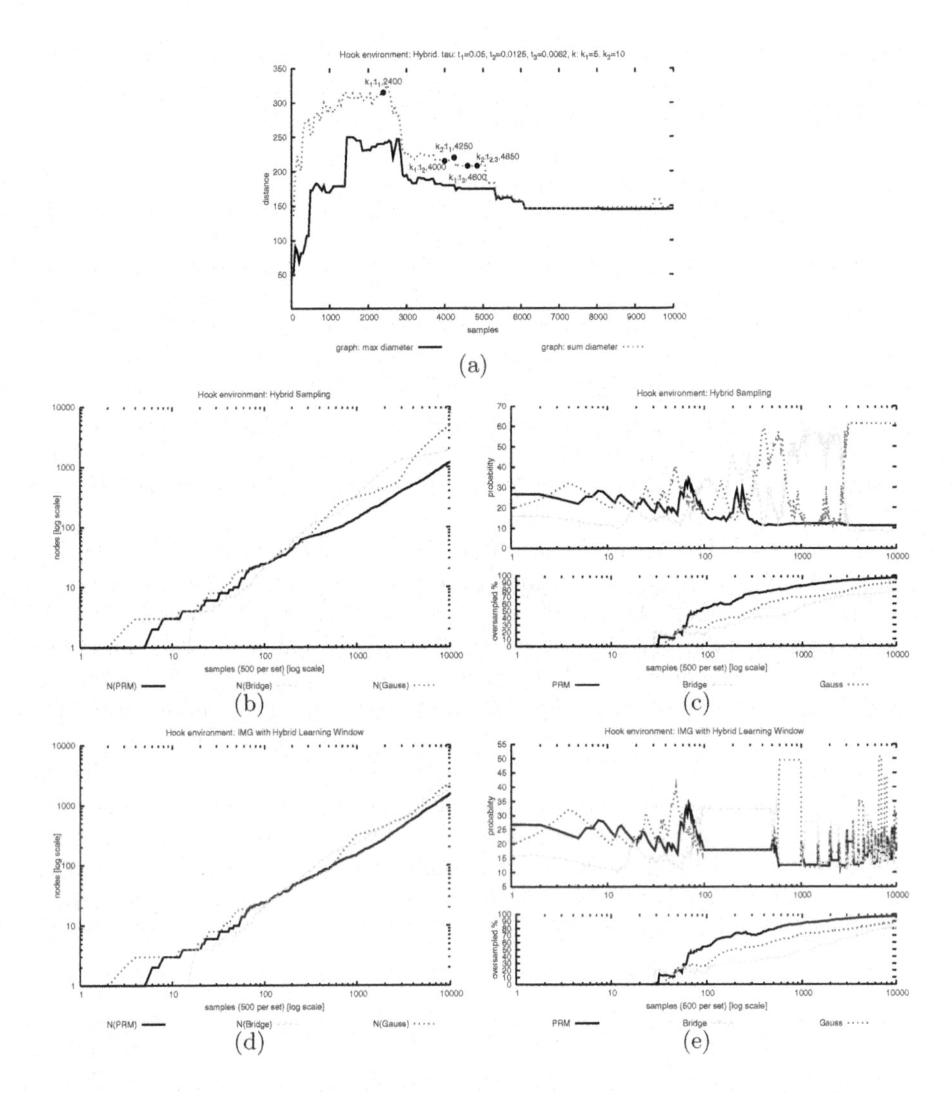

Fig. 6. Applying Hybrid PRM with IMG. Hybrid PRM using IMG: (a) stopping criteria, (b) number of samples per sampler, and (c) sampler probability and oversample %. Hybrid PRM Learning Window: (d) number of samples per sampler and (e) sampler probability and oversample %. The witness query solved after 1440 samples.

In the end, after the witness query is solved, hybrid PRM vacillates between a version of Gauss and a version of Bridge test.

Figure 6(d) and 6(e) show similar plots as 6(b) and 6(c), respectively, for IMG with a hybrid learning window. Unlike the previous plots (b and c), this version does not select a dominant sampler towards the end of roadmap construction, after the witness query is solved. We believe in fact that this is a more accurate evaluation because at this stage all samplers are equally "bad," i.e., none are able to generate useful samples and should not necessarily be distinguished. In particular, note that more than 80% of the nodes created by all samplers are cc-oversample nodes in the later stages of roadmap construction.

4.2 Application-Specific Stopping Criteria

As discussed in Section 3.2, the IMG framework can accept a broad range of stopping criteria that can be customized for particular applications or user preferences. We can apply our framework to study computational biology problems such as protein folding and protein structure transitions.

Here, we compare our general roadmap progress evaluation to an application specific one that measures when the secondary structure formation order (i.e., the high-level order in which the protein folds) has stabilized in the roadmap (see [24] for details). We specifically study the folding of proteins G, L, and mutants of protein G, NuG1 and NuG2 [19] because they are known to fold differently despite having similar structure.

Figure 7 shows how each metric varies during roadmap construction for protein G and NuG1: roadmap progress evaluation indicated by black circles at various k and τ and stable secondary structure formation order indicated by red circles. (Similar plots for all proteins studied can be found in [26].) The plots also show the percentage of folding pathways in the roadmap that follow the same order as seen experimentally. The application specific stopping criteria is noisy in the early phases of roadmap construction and then stabilizes after 4000 nodes. For this application, $PCMAX$ and $PCSUM$ are similar since there is typically one large connected component with the remaining nodes as singletons. Even so, roadmap progress evaluation is able to identify different stopping points for different τ and k values. In fact, for both proteins shown here, $\tau = 0.0062$ and $k = 5$ corresponds to when the percentage of folding pathways matching experimental data stabilizes.

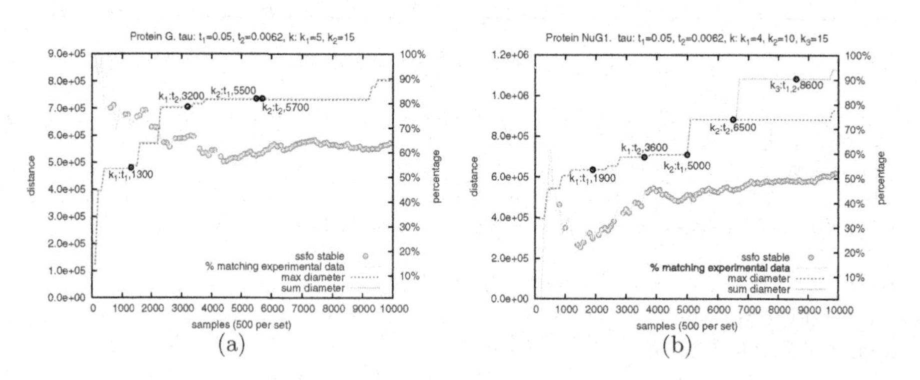

Fig. 7. Comparison of IMG roadmap progress evaluation to application specific evaluation for (a) protein G (112 dof) and (b) a mutant of protein G (114 dof) which are experimentally known to fold differently despite structural similarity.

5 Conclusion

Here, we proposed a framework to automatically determine how many samples a planner needs for a given motion planning problem. This framework can accept a broad range of evaluation criteria which can be customized for particular applications. We provide easy to define parameters that allow users to stop roadmap construction by satisfying criteria based on the quality of the roadmap. This has many potential applications that we plan to study. There are also several other areas that we would like to investigate further. For example, we would like to expand our list of node generation methods to include other types of random sampling and grid-based techniques.

References

1. Amato, N.M., Bayazit, O.B., Dale, L.K., Jones, C.V., Vallejo, D.: OBPRM: An obstacle-based PRM for 3D workspaces. In: Proc. Third Workshop on Algorithmic Foundations of Robotics (WAFR), Houston, TX, Natick, MA, pp. 155–168. A.K. Peters (1998)
2. Amato, N.M., Bayazit, O.B., Dale, L.K., Jones, C.V., Vallejo, D.: Choosing good distance metrics and local planners for probabilistic roadmap methods. IEEE Trans. Robot. Automat. 16(4), 442–447 (2000)
3. Apaydin, M., Singh, A., Brutlag, D., Latombe, J.-C.: Capturing molecular energy landscapes with probabilistic conformational roadmaps. In: Proc. IEEE Int. Conf. Robot. Autom. (ICRA), pp. 932–939 (2001)
4. Boor, V., Overmars, M.H., van der Stappen, A.F.: The Gaussian sampling strategy for probabilistic roadmap planners. In: Proc. IEEE Int. Conf. Robot. Autom. (ICRA), vol. 2, pp. 1018–1023 (1999)
5. Burns, B., Brock, O.: Sampling-based motion planning using predictive models. In: Proc. IEEE Int. Conf. Robot. Autom. (ICRA) (2005)
6. Corneil, D.G., Dragan, F.F., Köhler, E.: On the power of bfs to determine a graph's diameter. Networks 42(4), 209–222 (2003)
7. Geraerts, R., Overmars, M.H.: Reachablility analysis of sampling based planners. In: Proc. IEEE Int. Conf. Robot. Autom. (ICRA), pp. 406–412 (2005)
8. Gottschalk, S., Lin, M.C., Manocha, D.: OBB-tree: A hierarchical structure for rapid interference detection. Comput. Graph. 30, 171–180 (1996); Proc. SIGGRAPH 1996
9. Hsu, D., Jiang, T., Reif, J., Sun, Z.: Bridge test for sampling narrow passages with proabilistic roadmap planners. In: Proc. IEEE Int. Conf. Robot. Autom. (ICRA), pp. 4420–4426 (2003)
10. Hsu, D., Latombe, J.-C., Motwani, R.: Path planning in expansive configuration spaces. In: Proc. IEEE Int. Conf. Robot. Autom. (ICRA), pp. 2719–2726 (1997)
11. Hsu, D., Sánchez-Ante, G., Sun, Z.: Hybrid PRM sampling with a cost-sensitive adaptive strategy. In: Proc. IEEE Int. Conf. Robot. Autom. (ICRA), pp. 3885–3891 (2005)
12. Kavraki, L.E., Latombe, J.-C., Motwani, R., Raghavan, P.: Randomized query processing in robot path planning. In: Proc. ACM Symp. Theory of Computing (STOC), pp. 353–362 (May 1995)

13. Kavraki, L.E., Svestka, P., Latombe, J.C., Overmars, M.H.: Probabilistic roadmaps for path planning in high-dimensional configuration spaces. IEEE Trans. Robot. Automat. 12(4), 566–580 (1996)
14. Latombe, J.-C.: Robot Motion Planning. Kluwer Academic Publishers, Boston (1991)
15. LaValle, S.M., Kuffner, J.J.: Randomized kinodynamic planning. In: Proc. IEEE Int. Conf. Robot. Autom. (ICRA), pp. 473–479 (1999)
16. Mazer, E., Ahuactzin, J.M., Bessiere, P.: The Ariadne's clew algorithm. Journal of Artificial Robotics Research (JAIR) 9, 295–316 (1998)
17. Morales, M., Tapia, L., Pearce, R., Rodriguez, S., Amato, N.M.: A machine learning approach for feature-sensitive motion planning. In: Proc. Int. Workshop on Algorithmic Foundations of Robotics (WAFR), Utrecht/Zeist, The Netherlands, pp. 361–376 (July 2004)
18. Morales, M.A., Pearce, A.R., Amato, N.M.: Metrics for comparing C-Space roadmaps. In: Proc. IEEE Int. Conf. Robot. Autom. (ICRA) (2006)
19. Nauli, S., Kuhlman, B., Baker, D.: Computer-based redesign of a protein folding pathway. Nature Struct. Biol. 8(7), 602–605 (2001)
20. Nissoux, C., Simeon, T., Laumond, J.-P.: Visibility based probabilistic roadmaps. In: Proc. IEEE Int. Conf. Intel. Rob. Syst (IROS), pp. 1316–1321 (1999)
21. Reif, J.H.: Complexity of the mover's problem and generalizations. In: Proc. IEEE Symp. Foundations of Computer Science (FOCS), San Juan, Puerto Rico, pp. 421–427 (October 1979)
22. Seidel, R.: On the all-pairs-shortest-path problem. In: Proc. 24th Annu. ACM Sympos. Theory Comput., pp. 745–749 (1992)
23. Song, G., Amato, N.M.: Using motion planning to study protein folding pathways. In: Proc. Int. Conf. Comput. Molecular Biology (RECOMB), pp. 287–296 (2001)
24. Thomas, S., Tang, X., Tapia, L., Amato, N.M.: Simulating protein motions with rigidity analysis. In: Proc. Int. Conf. Comput. Molecular Biology (RECOMB) (2006)
25. Wilmarth, S.A., Amato, N.M., Stiller, P.F.: MAPRM: A probabilistic roadmap planner with sampling on the medial axis of the free space. In: Proc. IEEE Int. Conf. Robot. Autom. (ICRA), vol. 2, pp. 1024–1031 (1999)
26. Xie, D., Morales, M.A., Pearce, R., Thomas, S., Lien, J.-M., Amato, N.M.: Incremental map generation (IMG). Technical Report TR06-005, Parasol Lab, Dept. of Computer Science, Texas A&M University (March 2006)

Part II

Planning for Movable and Moving Obstacles

Caging Polygons with Two and Three Fingers

Mostafa Vahedi and A. Frank van der Stappen

Department of Information and Computing Sciences, Utrecht University
{vahedi,frankst}@cs.uu.nl

Abstract. We present algorithms for computing *all* placements of two and three fingers that cage a given polygonal object with n edges in the plane. A polygon is caged when it is impossible to take the polygon to infinity without penetrating one of the fingers. Using a classification into squeezing and stretching cagings, we provide an algorithm that reports all caging placements of two disc fingers in $O(n^2 \log n)$ time. Our result extends and improves a recent solution for point fingers. In addition, we construct a data structure requiring $O(n^2)$ storage that can answer in $O(\log n)$ whether two fingers in a query placement cage the polygon. We also study caging with three point fingers. Given the placements of two so-called base fingers, we report all placements of the third finger so that the three fingers jointly cage the polygon. Using the fact that the boundary of the set of placements for the third finger consists of equilibrium grasps, we present an algorithm that reports all placements of the third finger in $O(n^6 \log^2 n)$ deterministic time and $O(n^6 \log n (\log \log n)^3)$ expected time. Our results extend previous solutions that only apply to convex polygons.

1 Introduction

The caging problem was posed by Kuperberg in [7] as a problem of designing an algorithm for finding a set of points that prevents a polygon from moving arbitrarily far from a position. In other words, a polygon is caged when it is impossible to take it to infinity without penetrating a finger. However, to solve the problem it is easier to keep the polygon fixed and move the fingers instead, keeping their mutual distances fixed.

Caging is related to the notions of form (and force) closure grasps (see e.g. Mason's text book [9]), and immobilizing and equilibrium grasps [12]. Rimon et al. [11] introduced the notion of a caging set (or capture region) as the set of placements of fingers that may not immobilize the object but may prevent it from escaping to infinity. A comprehensive review on caging and related problems can be found in [1]. Caging sets have been applied to several problems in manipulation, such as grasping and in-hand manipulation, mobile robot motion planning, parts feeding, and stable pose computation (see [4] for the references).

Using stratified Morse theory, Rimon and Blake [11] showed that in a two-fingered one-parameter gripping system, the hand's configuration at which the cage is broken corresponds to a frictionless equilibrium grasp. These results are extended by Davidson and Blake [2] to a three-fingered one-parameter gripper.

S. Akella et al. (Eds.): Algorithmic Foundation of Robotics VII, STAR 47, pp. 71–86, 2008.
springerlink.com

In the problem of caging a polygon with two disc fingers, it is required to compute all placements of two disc fingers that cage a polygonal object. This problem was first tackled by Sudsang and Luewirawong [14] by computing an acceptable distance for every pair of immobilizing vertices independent of the entire body of the polygon. As a result, the algorithm is not complete as it reports only a subset of all caging placements of two disc fingers. Independently Pipattanasomporn and Sudsang [10] have recently also solved the problem for point fingers in $O(n^2 \log n)$, and also provided a data structure capable of answering queries in $O(\log n)$. However the disc-finger problem has not been analyzed in their paper.

In the problem of caging a polygon with three fingers, the placements of two fingers, called the base fingers, are given. It is required to find all placements of the third finger, such that the resulting fingers cage the polygon. Sudsang [13] stated a sufficient (but not necessary) condition for caging a convex object in the plane with more than two fingers using the width of the object. For a non-convex object it was proposed to divide the object into convex parts and to consider the maximum width sub-part. The concept of the caging set was used by Sudsang and Ponce [15] as a basis for computing a plan for manipulating polygonal objects using three discs. The resulting caging set was very small and hence not complete, as the computation only takes three edges into account. Erickson et al. [4] provided the first complete algorithm for three-finger cagings of convex polygons. Two fingers are placed along the boundary of the polygon and then a region —the caging set— is computed for the third finger. An exact algorithm is provided that determines this region in $O(n^6)$ time, and also an approximation algorithm is provided with pre-specified accuracy. However, the problem of computing the set of all caging placements for non-convex polygons remained open, and is tackled in this paper.

In the first part of this paper (in Section 3) a solution for computing all caging placements of two disc fingers is presented. In addition a data structure is presented that requires $O(n^2)$ storage and is capable of answering in $O(\log n)$ whether a given placement of two fingers is caging. A given placement of two fingers is *squeezing* (*stretching*) caging, if it is caging and anyhow closing (opening) the fingers without penetrating the polygon the fingers remain caging until they reach a minimum (maximum) in which both fingers are on the boundary. It can be proven that any caging is *squeezing* or *stretching*. Using this fact, our work for two fingers extends and improves the result in [10] by providing an algorithm for computing all caging placements of two disc fingers that runs in $O(n^2 \log n)$. To solve the problem we have employed pseudo triangulation, cell decomposition, connectivity graphs, and an event processing technique.

In the second part of this paper (in Section 4), the solution for computing all caging placements of three fingers with a fixed pair of base fingers is presented. It is shown that the boundary of the caging regions made by the third finger corresponds to equilibrium grasps. Our work on three-finger caging uses this fact to extend the result in [4] by providing the first complete algorithm for computing the set of all caging placements for non-convex polygons. To solve this problem we have used similar techniques as in our two fingers solution. The running time of the proposed three point-finger caging algorithm is $O(n^6 \log^2 n)$ deterministic time, and $O(n^6 \log n (\log\log n)^3)$ expected time.

2 Definitions and Assumptions

The given simple closed polygon P has no holes and is bounded by n edges. Let P_d be the Minkowski sum outer-face of P and a closed disc of radius d. More formally $P_d = P \oplus \Delta_d$[1], where Δ_d is the disc of radius d centered at the origin. Placing disc fingers of radius d around P is equivalent to placing point fingers around the generalized polygon P_d. A generalized polygon is a shape bounded by straight segments and circular arcs. As the holes of P_d correspond to placements of a single finger caging P, we discard these holes in our computations of two and three finger cagings.

Without loss of generality we can assume that P_d is enclosed in a sufficiently large rectangle B. Let $F \subset \mathbb{R}^2$ be $F = B \setminus \text{int}(P_d)$. Clearly F is the set of all possible placements for a finger. Throughout the paper we assume that the fingers are points, P is P_d, and we refer to F^2 $(= F \times F)$ and F^3 $(= F \times F \times F)$ as the admissible space for two and three fingers.

To solve the caging problem with two disc fingers, we decompose F into pseudo triangles. A pseudo triangle here is defined as a triangle that has at most one concave circular arc of radius d, and arc angle less than π. To obtain a pseudo-triangulation, new vertices may be added, but no vertex of a pseudo triangle should lie inside the edge or arc of one of its neighbor triangles.

A force vector with its application point fixed uniquely determines a wrench. Our model for wrenches induced by a given polygon differs from that of Markenscoff et al. [8] in the case of a point contact and a convex vertex. Consider a point on a convex vertex of a polygon. In our model, the possible wrenches for this point, similar to the concave case, contains the set of all convex combinations of the two unit normal wrenches to both incident edges which makes a cone. The intuition behind this model is that an ϵ-radius disc-finger on a convex vertex can apply any wrench being a combination of the two normal wrenches with ϵ adjustment.

Using Corollary 4.1 from [12], the grasp made by fingers on edges is an equilibrium grasp if and only if the wrench vectors meet in a common point and the angle between two consecutive wrenches is not more than π. When some of the fingers are at vertices, the grasp is an equilibrium grasp provided that there is a point that lies on the intersection of cones and wrenches and satisfies the angle condition. The number of fingers that make an equilibrium grasp is the number of fingers that exert a non-zero wrench on the object.

3 Two Fingers Caging

In this section the solution for the caging problem of a polygon P with two disc fingers with the same fixed radius is presented. In this problem all placements

[1] $A \oplus B = \{a + b | a \in A, b \in B\}$.

of two fingers that cage the polygon is computed. It is also required to answer quickly whether a given placement of two fingers is caging.

Let C_P be the set of all two-finger caging placements of a polygon P and let $C_{\delta,P}$ be the set of all caging placements for which the fingers are δ apart. Clearly $C_P = \bigcup_{0<\delta\in R} C_{\delta,P}$. Increasing or decreasing δ, the topology of $C_{\delta,P}$ changes at certain critical δ's. We use this fact to construct the set of all cagings.

Let (p_1, q_1) and $(p_2, q_2) \in F^2$ be two placements of a two-finger hand. These placements are *δ-reachable* if $|p_1q_1| = |p_2q_2| = \delta$ and both of them lie in the same connected component of the admissible space of finger placements that are δ apart. They are *δ-max-reachable* (*δ-min-reachable*) if $|p_1q_1|, |p_2q_2| \leq \delta$ ($|p_1q_1|, |p_2q_2| \geq \delta$) and both of them lie in the same connected component of the admissible space of finger placements that are at most (at least) δ apart. When two placements are *δ-reachable*, *δ-max-reachable*, or *δ-min-reachable*, it is possible to move the two-finger hand between the placements keeping the distance of the fingers fixed, at most δ, or at least δ respectively. Note that when two placements are *δ-max-reachable* (*δ-min-reachable*) they are *δ'-max-reachable* (*δ'-min-reachable*) for any $\delta' \geq \delta$ ($\delta' \leq \delta$). Hence, if a δ-apart placement is not *δ-max-reachable* (*δ-min-reachable*) to a δ-apart placement being remote from the polygon, it is squeezing (stretching) caging. If the reachability type is clear from the context or if we want to define something for all types of reachability we will use just the word *reachable*. Because of the lack of space we just mention the following important fact that provides the basis for our approach outlined in Subsection 3.2.

Lemma 3.1. *Given one obstacle in the plane, if two placements (p_1, q_1) and (p_2, q_2) of a two-finger hand satisfying $|p_1q_1| = |p_2q_2| = \delta$ are both* δ-max-reachable *and* δ-min-reachable, *then they are* δ-reachable.

The direct result of Lemma 3.1 is that, if a placement is caging, then it is squeezing caging, stretching caging, or both. In Figure 1(a) a shaded polygon and four δ apart placements (p_1, q_1), (p_2, q_2), (p_3, q_3) and (p_4, q_4) are shown. No two placements are δ-reachable, but (p_1, q_1) and (p_2, q_2) are δ-max-reachable while (p_1, q_1) and (p_3, q_3) are δ-min-reachable. Moreover (p_1, q_1) is not caging, (p_2, q_2) is stretching caging, (p_3, q_3) is squeezing caging, and (p_4, q_4) is both stretching and squeezing caging.

Based on above reachability notions and a pseudo triangulation of the set F, the space F^2 is decomposed into constant-complexity 4D cells for every δ, such that all placements inside each cell are *reachable* from each other. The required property of the pseudo triangulation is stated in the following lemma. The construction is relatively easy and therefore we confine ourselves to mentioning the result.

Lemma 3.2. *It is possible to decompose F in $O(n \log n)$ time in $O(n)$ pseudo triangles such that every pseudo triangle has a constant number of neighbors.*

Based on δ and the cell decomposition of F^2, a *connectivity graph* is defined in Subsection 3.1, with cells as the nodes, and two neighbor nodes are connected by an edge if there are *reachable* placements inside the corresponding cells. Note that all the corresponding placements of one node are either all caging or all noncaging. Hence we associate with a node the caging status (caging or noncaging) of all its placements. Since all noncaging placements are *reachable* from each other, the noncaging nodes form a connected component in the connectivity graph. Therefore, all components in the graph except the one containing noncaging placements represent a set of caging placements.

To compute all caging placements of two fingers, it is possible to start from zero and increase (or equivalently start from a largely enough distance and decrease) the distance of the fingers. Meanwhile, there are *critical distances* at which the connectivity graph changes. The idea is to compute all critical distances and sort them increasingly. Clearly between two consecutive critical distances, the connectivity graph does not change. Therefore it is possible to compute all possible connectivity graphs for all distances by considering the critical distances one by one, and updating the connectivity graph accordingly in a reasonable time (instead of computing the whole connectivity graph from scratch every time). When a *caging* cell merges to or becomes disconnected from the *noncaging* cell, equivalently a connected component of the graph respectively merges to or becomes disconnected from the component of the graph that represents the noncaging placements. To update the graph for every merging or splitting of cells some edges respectively should be added to or deleted from the graph as the update operation. If the update operation includes deletion of edges the components of the graphs should be maintained during the process and it is not possible to do this operation in constant time (at least easily). But if it consists of just addition of edges the components will just merge or emerge and therefore it is possible to maintain the two types of components in constant amortized time. By using the squeezing/stretching fact, defined formally in Subsection 3.1, and increasing/decreasing the distance the cells just merge or emerge and therefore the update operation just include addition of nodes and edges in constant time. Whenever a caging node is going to join a noncaging node by a path, the corresponding 4D cell of the caging node is reported as a set of caging placements; any placement inside this cell is caging. The complete algorithm and the running time analysis is explained in Subsection 3.2.

3.1 Two Fingers Squeezing and Stretching Caging

Let s and s' be two closed subsets of F. The set of admissible placements induced by a pair of subsets of F for two fingers with distance δ is the set

$$R_\delta(s, s') = \{(p, q) \in s \times s' \mid |pq| = \delta\}.$$

The set $R_\delta(s, s')$ consists of a number of 4D connected components. Every connected component corresponds to a set of *δ-reachable* placements. Let $R_\delta^M(s, s')$ be the set of connected components of $R_\delta(s, s')$. Therefore every member of $R_\delta^M(s, s')$ is a subset of F^2 and is called a *cell*. If s and s' have constant complexity, the number of cells in $R_\delta^M(s, s')$ and their complexity will be constant too. Figure 1(b) shows a shaded polygon, its Minkowski-sum outer-face with a

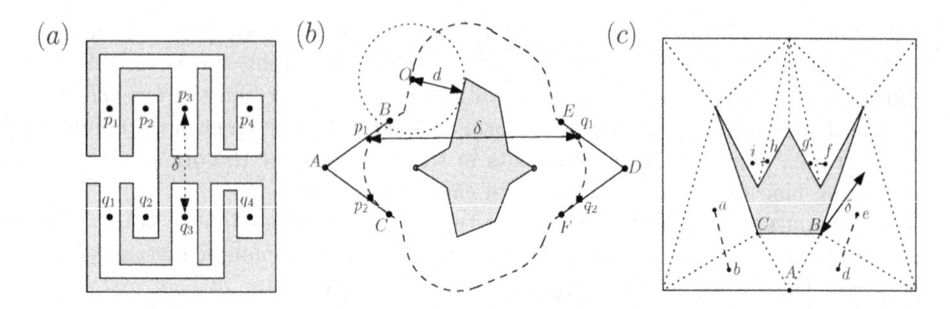

Fig. 1. (a) Reachability notions and caging types, (b) $R_\delta^M(ABC, DEF)$ has one cell, (c) connectivity subgraph of distance δ when fingers are points and the first finger is inside ABC

disc of radius d (displayed dashed), and two pseudo-triangles ABC and DEF. Here $R_\delta^M(ABC, DEF)$ has one cell. It seems that (p_2, q_2) is not *δ-reachable* from (p_1, q_1) using the placements inside the two pseudo-triangles; but the placement is reachable from (p_1, q_1) by moving p_1 toward B and moving q_1 toward F and then moving p_1 toward C.

Let T be a *suitable* pseudo triangulation of F (i.e. satisfies Lemma 3.2). The connectivity graph $CG_{\delta,T}(V, E)$ for T and distance δ is defined by

$$\begin{cases} V = \{r \subset F^2 \mid \exists\, t, t' \in T : r \in R_\delta^M(t, t')\}, \\ \\ E = \{\ (r_1, r_2) \in V^2 \mid \exists\, t_1, t_1', t_2, t_2' \in T : r_1 \in R_\delta^M(t_1, t_1'), \\ r_2 \in R_\delta^M(t_2, t_2') \wedge \exists r \in R_\delta^M(t_1 \cup t_2, t_1' \cup t_2') : r = r_1 \cup r_2\ \}. \end{cases}$$

By definition every cell is assigned a unique node and therefore every admissible placement of fingers is assigned a node in the graph; there is no edge between the cells of a set $R_\delta^M(t, t')$. There is an edge between two cells in $R_\delta^M(t_1, t_1')$ and $R_\delta^M(t_2, t_2')$ when there is a cell in $R_\delta^M(t_1 \cup t_2, t_1' \cup t_2')$ that contains the two cells. In other words there is an edge between two nodes when their corresponding pairs of pseudo triangles are neighbor and their corresponding placements are *reachable*. As every pseudo triangle has a constant number of neighbors in T, the total number of edges is linear in the total number of nodes. Therefore if there are $O(n)$ pseudo triangles in T, there will be $O(n^2)$ nodes and edges in $CG_{\delta,T}(V, E)$.

In Figure 1(c) a shaded polygon bounded in a rectangle, the polygon exterior triangulated (displayed dotted), and one of the triangles ABC are shown. Since it is not easy to draw the whole connectivity graph for distance δ, we show a subset of the graph for point fingers while the first finger is inside ABC. For any triangle if there are δ apart points inside that triangle and ABC, then consider a node in the graph (the nodes are displayed with small letters). Since there are no two points of distance δ inside ABC no node is considered for it. Note that here since no $R_\delta^M(ABC, t)$ has more than one cell, every pair of (ABC, t) has at most one node in the graph. When it is possible to move the second finger from one triangle to its neighbor while keeping the first finger inside ABC and the distance δ, then connect the two corresponding nodes by an edge (displayed with dashed segments).

Lemma 3.3. *Let* $(p_1, q_1) \in r_1 \in R^M_\delta(t_1, t'_1)$ and $(p_2, q_2) \in r_2 \in R^M_\delta(t_2, t'_2)$ *be two placements.* (p_1, q_1) *and* (p_2, q_2) *are* δ-reachable, *if and only if there is a path between* r_1 *and* r_2 *in* $CG_{\delta,T}(V, E)$.

Let v_δ be a noncaging node for which $v_\delta \in R^M_\delta(t_\delta, t'_\delta)$, and $t_\delta, t'_\delta \in T$. Without loss of generality, we can assume that it is possible to compute v_δ based on B, T, and δ such that is not caging (clearly v_δ may change when δ changes). The points on the boundary of B at distance δ e.g. do not cage the polygon being remote from the polygon. Using the fact that all noncaging nodes form a connected component in the graph, a given placement of fingers is caging if and only if there is no path between the corresponding node and v_δ in the graph. Let the set of caging nodes for the polygon P and distance δ be the set

$$CN_{\delta,P} = \{v \in V(CG_{\delta,T}) \mid \text{There is no path in } CG_{\delta,T} \text{ between } v \text{ and } v_\delta\}.$$

and let the set of caging placements obtained by CG graphs be the set

$$CN_P = \{(p, q) \in F^2 \mid \exists v \in CN_{|pq|,P} : (p, q) \in v\}.$$

Consider the following definitions of $R'_\delta(s, s')$ and $R''_\delta(s, s')$ that correspond to the *δ-max-reachable* and *δ-min-reachable* set of placements induced by s and s':

$$R'_\delta(s, s') = \{(p, q) \in s \times s' \mid |pq| \leq \delta\},$$

$$R''_\delta(s, s') = \{(p, q) \in s \times s' \mid |pq| \geq \delta\}.$$

Replacing $R_\delta(s, s')$ with $R'_\delta(s, s')$ and $R''_\delta(s, s')$, and *δ-reachable* with *δ-max-reachable* and *δ-min-reachable* in above definitions results in new definitions of $R'^M_\delta(s, s')$ and $R''^M_\delta(s, s')$, $CG'_{\delta,T}$ and $CG''_{\delta,T}$, $CN'_{\delta,P}$ and $CN''_{\delta,P}$, and CN'_P and CN''_P respectively in order. The adjusted Lemma 3.3 still holds for $CG'_{\delta,T}$ and $CG''_{\delta,T}$. We refer to $CG'_{\delta,T}$ and $CG''_{\delta,T}$ as the max and min connectivity graph and to CN'_P and CN''_P as the set of all squeezing and stretching caging placements respectively. Because of the lack of space we just mention the following important facts. Lemma 3.5 is the direct result of Lemma 3.1.

Lemma 3.4. *Given a polygon* P *and a distance* δ, *it is possible to compute* $CG_{\delta,T}$, $CG'_{\delta,T}$, $CG''_{\delta,T}$, $CN_{\delta,P}$, $CN'_{\delta,P}$, *and* $CN''_{\delta,P}$ *in* $O(n^2)$. *Then using* T *it is possible to determine in* $O(\log n)$ *time whether a given placement of two disc fingers that are* δ *apart cages* P.

Lemma 3.5. $C_P = CN_P = CN'_P \cup CN''_P$.

As it was mentioned in Section 3, it is possible to maintain the graph components in computing the squeezing and stretching caging placements in constant amortized time. Therefore, based on Lemma 3.5 that states the relation between the common notion of caging on one hand and squeezing and stretching caging at the other hand, we compute all caging placements by computing CN'_P and CN''_P separately in Subsection 3.2.

3.2 Two Disc Fingers Caging Algorithm

In this section we present our approach to solving the problem of finding all caging placements of two disc fingers of equal radius. To report all cagings the algorithm uses Lemma 3.5 and reports the two sets CN'_P and CN''_P instead, which both consist of 4D cells corresponding to squeezing and stretching cagings respectively. Each point inside every cell corresponds to a placement of two disc fingers on the plane that cages P. The algorithm consists of three steps for both types of cagings. Since these steps are similar for both types, we focus on the computation of the set CN'_P of squeezing cagings:

1. find and sort the critical distances (see below) induced by all pairs of pseudo-triangles in T (a *suitable* pseudo-triangulation of F),
2. compute $CG'_{\delta,T}$ for all δ by processing the critical distances and updating $CG'_{\delta,T}$ accordingly, meanwhile reporting the possible squeezing caging-placements,
3. report the remaining squeezing caging-placements.

Hereafter we use pseudo triangle and triangle interchangeably. The first step is based on the fact that the structure of $CG'_{\delta,T}$ only changes at particular values of δ, to which we shall refer as *critical distances*. Increasing δ from zero, we distinguish three types of critical distances induced by a single pair (t, t'):

1. $|R'^M_\delta(t, t')|$ increases,
2. $|R'^M_\delta(t, t')|$ decreases,
3. for neighbor pairs of (t_1, t'_1) and (t_2, t'_2) in $R'^M_\delta(t_1 \cup t_2, t'_1 \cup t'_2)$, a member of $R'^M_\delta(t_1, t'_1)$ merges with a member of $R'^M_\delta(t_2, t'_2)$.

The cells of a pair of triangles can only merge and not split, because when two placements become *δ-max-reachable* they remain so for any bigger δ. Therefore, the first type only occurs when a cell emerges, and the second only occurs when two cells merge together. Since the first two types of critical distances depend on t and t' only, the number of such distances is constant for a given t and t'. From the fact that all pseudo-triangles in T have a constant number of neighbors and also $R'^M_\delta(t, t')$ has constant cardinality, it follows that the number of critical distances of the latter type is constant as well. As a result, we can accomplish the computation and sorting of all $O(n^2)$ critical distances in $O(n^2 \log n)$ time.

In the last two steps we use a graph-based data structure to keep track of the changes in $CG'_{\delta,T}$ while increasing δ. When $R'^M_\delta(t, t')$ changes topologically for a critical distance δ, a new cell emerges or two cells merge into a single cell containing the original ones. For each newly emerging or merging cell there is a corresponding node in the graph and all nodes are included in the graph from the start. With every node in the graph we associate a caging status, *caging* or *noncaging*. Initially all nodes are *caging*. Every node also has a *critical distance* that will be determined later; initially it is set to zero. Every pair of triangles has a corresponding set of nodes in the graph and a set that specifies the current nodes in the graph for the current value of δ.

Starting from zero with $CG'_{0,T}$, which is built from scratch, the second step processes the critical distances in order to update the connectivity graph. Since

P does not contain holes, there is no caging node in $CG'_{0,T}$. At each critical distance some actions are taken to determine $CG'_{\delta,T}$ from $CG'_{\delta',T}$, in which δ and δ' are the current and previous critical distances respectively. We recall that between two consecutive critical distances the graph does not change. The actions taken to update the graph depend on the type of critical distance. The first two sets of actions are the same and they are not repeated. They follow in order:

1,2. The current set of nodes for the corresponding pair of triangles is updated. The edges for the new node are computed. If there is no edge or the new edges only connect to caging nodes, the caging status of the node is 'caging'. Otherwise the caging status is 'noncaging'. If the corresponding node is a bridge between a caging node and a noncaging node, a maximal report (see below) is performed; otherwise the *critical distances* of the old nodes (if existing) are set to the current value of δ.

3. An edge is added between the corresponding nodes. If the nodes had a different caging status, a maximal report is performed.

A *maximal report* is done, when the caging status of a node changes from caging to noncaging. This happens for squeezing caging-placements for which δ reaches the critical maximum distance, at which any distance larger than that distance allows the fingers to escape. Look at Figure 2(a) for some of the critical maximum distances that lead to *maximal report*. In this operation, the corresponding 4D cells of all the nodes in the graph that are in the same connected component of the changing node are reported, and their associated *critical distances* are set to the current value of δ and their associated caging statuses are set to 'noncaging'. If a node becomes 'noncaging', it can never become 'caging' again. Hence every node is reported at most once, and the time devoted to reporting the caging cells is linear in the number of nodes and therefore is $O(n^2)$.

Every update operation takes constant time. Clearly every change is local to a node and its neighbors. Since no node is added to the graph, the addition of edges is the only performed operation; the number of neighbor nodes and the number of edges for each node is constant. Therefore, the updates induced by a single critical distance take constant time in total.

Since some of the squeezing cagings may have no critical maximum distance (e.g. (p_4, q_4) in Figure 1(a)), all the remaining squeezing cagings are reported in a separate step at the end. In the third step the final connectivity graph is traversed for nodes that are 'caging' but their *critical distance* fields are still zero. For every such node, the field is set to infinity and its 4D cells is reported. The following theorem follows from the preceding discussion.

Theorem 3.6. *Given a polygon with n edges and two disc fingers of equal radius, it is possible to report all the caging placements in $O(n^2 \log n)$ time.*

Theorem 3.7. *It is possible to compute a data structure with $O(n^2)$ space and time complexity, with which it is possible to answer in $O(\log n)$ whether a given placement of fingers is caging.*

Proof. To answer the caging query using the two final (squeezing and stretching) computed data structures, the corresponding pseudo triangles are located in $O(\log n)$ [3]. Then for the given distance of the fingers, the corresponding two nodes in the graphs can be determined in constant time. There are two cases; If the caging status of any node is caging, the answer is caging. Otherwise, the critical distance field of every node is compared with the query distance. For squeezing/stretching caging, if the query is smaller/larger the answer is caging; otherwise the answer is noncaging. Therefore the total time to answer a query is $O(\log n)$. Clearly the space needed to store the data structure is $O(n^2)$.

4 Three Fingers Caging

In this section, the caging problem for three point fingers is presented. In this problem the placement of two fingers, called the base fingers, is given. It is required to find all placements of the third finger, such that the fingers cage the polygon. The caging placements form some regions on the plane of which the boundaries should be reported. We assume that the base fingers do not cage the polygon without the third finger.

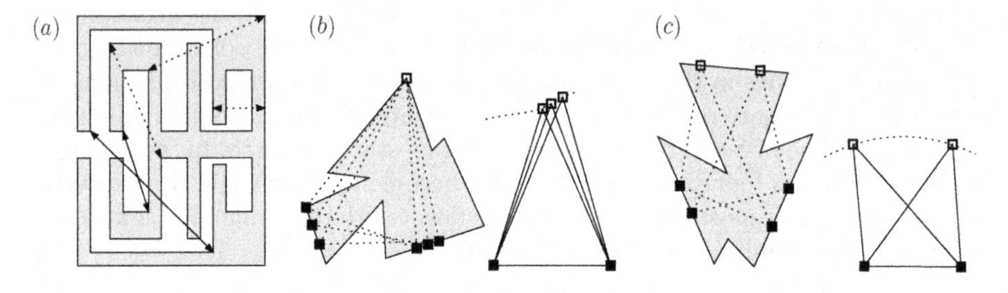

Fig. 2. (a) Three critical maximum distances displayed with dotted arrows and two critical minimum distances displayed with solid arrows. (*b* and *c*) Two loci are displayed for two polygons at the right side in which the filled boxes represent the base fingers and dotted triangles represent equilibrium grasps.

Similar to the solution of the two fingers case, F is triangulated and the *connectivity graph* for F^3 is defined for a given triangulation T and a given vector of distances of three fingers δ. Therefore $R_\delta(s, s', s'')$, $R_\delta^M(s, s', s'')$, $CG_{\delta,T}$, and $CN_{\delta,P}$ are defined similarly. Because of the similarity we have not repeated the definitions and the lemmas, except for the following important lemma.

Lemma 4.1. *Given T and δ, it is possible to compute $CG_{\delta,T}$ and $CN_{\delta,P}$ in $O(n^3)$ and to answer queries about caging status of given fingers placements with δ distances in $O(\log n)$.*

In Subsection 4.2 it is shown that the third finger placed on a point on the caging boundary jointly with the given placements of the base fingers correspond to some equilibrium grasps. It does not mean that the fingers necessarily make an equilibrium grasp at that placement, but there is a corresponding placement, reachable from that placement, at which the fingers make an equilibrium grasp. Every curve on the caging boundaries corresponds to a set of equilibrium grasps that induced by the same pair or triple of features of the polygon. Therefore, it is possible to compute all curves on the caging boundaries by considering every pair or triple of features and computing the possible equilibrium grasps. The resulting grasps when moved to the fixed placement of the base fingers define some 2D curves which we call *curves of equilibrium grasps*; only some parts of these curves constitute the caging boundaries. Based on these facts, the idea is to compute all possible caging intervals on every curve, and then to compute the caging boundaries using the caging intervals. In Subsection 4.1 all equilibrium grasps involving the two base fingers are computed. Note that the base fingers have a fixed distance.

To compute the caging intervals on each *curve of equilibrium grasps* the vector of distances, δ, is altered by changing the position of the third finger along that curve. Therefore for every point on the curve, $CG_{\delta,T}$ is computed accordingly. Similar to the two fingers case, there are critical points on the curve at which $CG_{\delta,T}$ changes and it does not change between two consecutive critical points. The same event processing approach is employed here to compute all possible $CG_{\delta,T}$ when the third finger moves along the curve. It is shown that the total number of possible nodes is $O(n^3)$ and all of them are included in the graph from the start; so there is no need to add or remove nodes from the graph. Here in contrast to the two fingers case, the update operation may require deletion of edges beside addition because an existing cell may split. Therefore, we have to use a special data structure called fully dynamic graph to efficiently query the caging property each time. The complete algorithm and the running time analysis is explained in Subsection 4.2.

4.1 Locus of Three Fingers Equilibrium Grasps

Consider all possible equilibrium grasps involving three fingers, two of which — referred to as base fingers— have a fixed distance, and the triangles defined by the fingers for every such grasp. Now consider two fixed points in the plane with distance equal to that between the base fingers, and draw the triangles such that the fixed points are on the base fingers. It is required to find the locus of the third finger in the plane. There are two general cases depending on the number of fingers that make the equilibrium grasp:

1. two fingers on
   a) two edges: the two edges should be parallel, hence it is possible to move the fingers together in one direction along the edges: the locus of all placements of the third finger describes a circular arc;
   b) an edge and a vertex, or two vertices: the locus of all placements of the third finger describes a circular arc;

2. three fingers with
   a) a base finger at a vertex: the locus of all placements of the third finger describes a line segment;
   b) the third finger at a vertex: the locus of all placements of the third finger describes a limaçon of Pascal or a line segment;
   c) all fingers on edges: the locus of all placements of the third finger describes a circular arc or a line segment;
   d) a base finger and the third finger at vertices: the locus of all placements of the third finger describes a finite number of points.

In Figure 2(*b* and *c*) two loci are displayed for two polygons. The filled boxes represent the base fingers and the empty boxes represent the third finger. Each dotted triangle represents an equilibrium grasp and is redisplayed at the right side with solid lines. For the polygon (*b*), the third finger is at a vertex and the base fingers are on two edges for which the locus is a limaçon of Pascal arc. For the polygon (*c*), all three fingers are on edges for which the locus is a circular arc. The loci are displayed with dotted curves at right side above the triangles for each polygon.

Since the boundary of the caging regions consists of continues curves, the points of case 2.d are not relevant and can be discarded. Since the number of features is at most three, there are $O(n^3)$ curves.

Theorem 4.2. *The locus of all equilibrium grasps made by three fingers of which the distance of base fingers is fixed, when moved to a fixed placement of the base fingers, defines $O(n^3)$ constant complexity 2D curves.*

4.2 Three Fingers Caging Algorithm

We report all placements of a point finger such that it cages P together with the two given base fingers. The output of the algorithm is a set of regions. Each point inside every region corresponds to a placement of the third finger that cages the polygon jointly with the base fingers.

Before explaining the algorithm, it should be shown that the boundary of the caging regions correspond to the boundary of the polygon or to sets of equilibrium grasps. Consider an intersection point of the caging boundary (not on the polygon boundary) and an arbitrary line. The intersection point is a puncture point (see [11] for the definition), because moving the third finger along the line, the caging status changes at that point. Considering the set of fingers consisting of the base fingers and the third finger moving on the line, Proposition 3.3 of [11] states that the corresponding placement corresponds to an equilibrium grasp. Therefore the caging boundaries correspond to the boundary of the polygon or to sets of equilibrium grasps.

Lemma 4.3. *The caging regions of a polygon are bounded by curves of equilibrium grasps or the polygon boundary.*

The algorithm consists of four steps:

1. Compute the locus of the third finger in all possible equilibrium grasps made with three fingers,

2. determine critical points related to a triple of triangles and a curve, and sort all of the critical points for every curve on that curve,
3. determine the caging intervals by computing all possible $CG_{\delta,T}$ for every curve by processing the sorted critical points,
4. report the caging boundaries using the computed caging intervals.

In the first step, all the equilibrium grasps induced by three fingers are computed as a set of curves, *curves of equilibrium grasps.* To ease the computation of caging intervals on the caging boundaries (in the last step), the polygon edges are added to the set of curves.

To explain the second step, the notion of a critical point should be defined first. Consider a curve E of equilibrium grasps and a point p on E at which the third finger is placed. It is possible to build a connectivity graph for p and the base fingers. Moving the third finger along E, there are two groups of critical points on E related to a triple of triangles (t, t', t''):

1. $|R_\delta^M(t, t', t'')|$ changes,
2. (t_2, t_2', t_2'') and (t_1, t_1', t_1'') are neighbors and a member of $R_\delta^M(t_1, t_1', t_1'')$ merges with or becomes disconnected from a member of $R_\delta^M(t_2, t_2', t_2'')$ inside $R_\delta^M(t_1 \cup t_2, t_1' \cup t_2', t_1'' \cup t_2'')$.

In the second step, all critical points related to a triple of triangles and a curve are calculated, and then all of the critical points for every curve are sorted along that curve. Since both the complexity of the triangles and their neighbors and the complexity of each curve are constant, there are constant number of critical points for every triple of triangles. Considering all possible ordered triples of triangles, there are $O(n^3)$ critical points for each curve, including the intersection points of that curve with the polygon and other curves of equilibrium grasps.

In the third step, all possible $CG_{\delta,T}$ are computed for every curve by taking the critical points in order and updating $CG_{\delta,T}$ for that curve; meanwhile the caging intervals are calculated. The approach for every curve E is as follows. One of the critical points p on E is taken as the starting point. The same data structure that was used in the two fingers case is used, without critical distance field. Initially $CG_{\delta,T}$ is built from scratch for p and the base fingers. Changing the position of p on E according to sorted critical points, one of the following actions is taken to update the current $CG_{\delta,T}$, depending on the type of the critical point:

1. the current set of nodes for the corresponding ordered triple of triangles is updated and the edges for the new set of nodes are computed, or
2. an edge is added between or removed from the two corresponding nodes.

Similar to the two fingers case, the update operation on the graph can be done in constant time for every critical point. In addition, however, we need to know the caging status of the current placement. Therefore, it is required to maintain a special data structure to quickly answer whether the current placement is caging.

A placement is caging, if there is no path in the current $CG_{\delta,T}$ between the corresponding node and v_δ, a *noncaging* node (here too, the v_δ may change when δ changes). Using the fully dynamic graph data structure [6, 5], it is possible to query for the connectivity of two nodes in the graph in $O(\log n/\log\log n)$ time and to update the mentioned data structure in $O(\log^2 n)$ deterministic amortized time and in $O(\log n(\log\log n)^3)$ expected amortized time. To find v_δ, choose a placement remote from the polygon in F, and find the corresponding node in the graph by locating the containing ordered triple of triangles in $O(\log n)$ [3].

To properly compute the caging status on the boundary of caging regions we use the trapezoidal map of the arrangement of the curves of equilibrium grasps. Since all the points inside a trapezoid have similar caging status, instead of choosing points on curves, we choose points exactly inside the trapezoids.

In the fourth step, the caging boundaries are reported from the previously computed caging intervals. To report the caging boundaries, first a curve is found on the boundary of each caging region. To do this, every caging interval of every curve is taken and one of its starting points is considered. Clearly, the starting point is on the caging boundary. Considering the caging intervals that include this point, there is a caging interval such that all the other ones lie on one side of it. This interval is on the boundary of a caging region, and the other intervals are inside this caging region. Walking along the corresponding curve such that the interior of the caging region is on the left, the next intersection point on the curve is considered. On every intersection point, the caging intervals are considered that include the intersection point and the rightmost curve is selected. The next intersection point along the selected curve is considered and the same steps are repeated till the same starting point is reached. The same is done for every unvisited caging interval. Recalling that our algorithm computes all three-finger caging regions, we get the following final result.

Theorem 4.4. *Given a polygon with n edges and given placements of base fingers, it is possible to report all placements of the third finger such that the three fingers jointly cage the polygon in $O(n^6 \log^2 n)$ deterministic time, and in $O(n^6 \log n(\log\log n)^3)$ expected time.*

Proof. F can be triangulated in $O(n\log n)$ time [3], and the number of triangles is linear. Since the total number of locus curves is $O(n^3)$ and computing every one takes a constant time, the first step can be done in $O(n^3)$ time (Theorem 4.2). Since every ordered triple of triangles has constant number of critical points, the total number of critical points on each curve is $O(n^3)$. Therefore the second step takes $O(n^6\log n)$. In the third step, for each critical point, adding and removing edges takes constant time, but testing the connectivity takes $O(\log^2 n)$ deterministic and $O(\log n(\log\log n)^3)$ expected amortized time. Since there are $O(n^6)$ critical points, the third step takes $O(n^6\log^2 n)$ deterministic time and $O(n^6\log n(\log\log n)^3)$ expected time. Since every intersection point is visited at most once the fourth step takes $O(n^6)$ time. Hence, the algorithm takes totally $O(n^6 \log^2 n)$ deterministic time and $O(n^6 \log n (\log\log n)^3)$ expected time.

5 Conclusion

In this paper we have presented algorithms for computing all possible caging placements of two disc fingers of equal radius, and three point fingers of which the placements of two base fingers are given. In the case of three fingers, extending the results to disc-shaped fingers is straightforward. Although the curves of equilibrium grasps become more complicated, their degrees remain constant. We intend to implement the algorithms to gain more insight into the shapes of caging regions and their combinatorial complexities with the purpose of improving the worst-case running time of our algorithm. In addition we would like to consider the three-finger caging query as well. Finally, extending the results to 3D seems challenging, because of the problem of decomposing a polyhedron into few simple cells. Hence we will look for alternative ways to tackle the caging problems.

References

1. Bicchi, A., Kumar, V.: Robotic grasping and contact: A review. In: ICRA, pp. 348–353. IEEE, Los Alamitos (2000)
2. Davidson, C., Blake, A.: Caging planar objects with a three-finger one-parameter gripper. In: ICRA, pp. 2722–2727. IEEE, Los Alamitos (1998)
3. de Berg, M., van Kreveld, M., Overmars, M., Schwartzkopf, O.: Computational Geometry: Algorithms and Applications. Springer, Heidelberg (1997)
4. Erickson, J., Thite, S., Rothganger, F., Ponce, J.: Capturing a convex object with three discs. In: ICRA, pp. 2242–2247. IEEE, Los Alamitos (2003)
5. Henzinger, M.R., King, V.: Randomized fully dynamic graph algorithms with polylogarithmic time per operation. J. ACM 46(4), 502–516 (1999)
6. Holm, J., de Lichtenberg, K., Thorup, M.: Poly-logarithmic deterministic fully-dynamic algorithms for connectivity, minimum spanning tree, 2-edge, and biconnectivity. In: STOC 1998: Proceedings of the 30th annual ACM symposium on Theory of computing, pp. 79–89. ACM Press, New York (1998)
7. Kuperberg, W.: Problems on polytopes and convex sets. In: DIMACS Workshop on Polytopes, pp. 584–589 (1990)
8. Markenscoff, X., Ni, L., Papadimitriou, C.H.: The geometry of grasping. Int. J. Robotics Res. 9(1), 61–74 (1990)
9. Mason, M.: Mechanics of Robotic Manipulation. Intelligent Robotics and Autonomous Agents Series. MIT Press, Cambridge (2001)
10. Pipattanasomporn, P., Sudsang, A.: Two-finger caging of concave polygon. In: ICRA, pp. 2137–2142. IEEE, Los Alamitos (2006)
11. Rimon, E., Blake, A.: Caging 2d bodies by 1-parameter two-fingered gripping systems. In: ICRA, vol. 2, pp. 75–91. IEEE, Los Alamitos (1995)
12. Rimon, E., Burdick, J.W.: Mobility of bodies in contact—part i: A 2nd-order mobility index for multiple–finger grasps. IEEE Tr. on Robotics and Automation 14(5), 696–717 (1998)

13. Sudsang, A.: A sufficient condition for capturing an object in the plane with disc-shaped robots. In: ICRA, pp. 682–687. IEEE, Los Alamitos (2002)
14. Sudsang, A., Luewirawong, T.: Capturing a concave polygon with two disc-shaped fingers. In: ICRA, pp. 1121–1126. IEEE, Los Alamitos (2003)
15. Sudsang, A., Ponce, J.: A new approach to motion planning for disc-shaped robots manipulating a polygonal object in the plane. In: ICRA, pp. 1068–1075. IEEE, Los Alamitos (2000)

An Effective Framework for Path Planning Amidst Movable Obstacles

Dennis Nieuwenhuisen, A. Frank van der Stappen, and Mark H. Overmars

Institute of Information and Computing Sciences,
Utrecht University, The Netherlands
{dennis,frankst,markov}@cs.uu.nl

Abstract. This paper addresses the problem of navigating an autonomous moving entity in an environment with both stationary and movable obstacles. If a movable obstacle blocks the path of the entity attempting to reach its goal configuration, the entity is allowed to alter the placement of the obstacle by manipulation (e.g. pushing or pulling), to clear its path. This paper presents a probabilistically complete framework for solving path planning problems among movable obstacles. Heuristics are presented to provide efficient solutions for problems in environments encountered in practical situations.

1 Introduction

Motion planning [8] has been an active area of research for three decades. Over the past years the motivation has gradually extended from the traditional robotics context toward applications in computer-assisted training and advanced games. These applications feature planning problems of huge complexity, as they involve many (often human) entities with large numbers of degrees of freedom. The entities move in environments that are not necessarily fixed but can or even *must* (in the light of the objective of the training or game) be modified by the entities. As complete solutions (that guarantee a solution provided one exists) to motion planning are only feasible for problems involving a few degrees of freedom, research has led to approaches that provide a weaker form of completeness. A particularly successful approach that is suited for complex problems is the probabilistic roadmap method [6] in which a roadmap of the free space is built incrementally. After creation, the roadmap can be queried for a collision-free path for the moving entity. Over the past decade, many planners based on this principle have been proposed.

We explore the relatively unaddressed problem of planning the motions of a moving entity in an environment inhabited by both stationary and movable obstacles. Our motivation comes from an ultimate wish to automatically generate visually-convincing motions for computer-controlled entities in virtual environments and games. A rigorous way to plan motions among stationary and moving obstacles would be to consider the problem in the high-dimensional composite

S. Akella et al. (Eds.): Algorithmic Foundation of Robotics VII, STAR 47, pp. 87–102, 2008.
springerlink.com

configuration space of the moving entity and the movable obstacles. Unfortunately, in all but the simplest instances the complexity is too high to efficiently find a solution.

However, with our motivation in mind, we believe that the above costly approach is not required. A human entity moving in a realistic (e.g. office) environment will plan his or her motions on the basis of knowledge about the layout of the stationary features of the environment. While executing the path, the entity may encounter movable obstacles such as a chair or a trolley with supplies standing in the way, or a door that is closed. If the entity encounters such movable obstacles it will try to move them out of the way by manipulating them or by getting around them, so that it will be able to continue its predetermined path. Only if the required manipulations get truly complicated or require a lot of effort, the entity may start to explore alternative paths toward the goal. Since it is our aim to provide convincing motions of entities, we want the planner to follow a similar strategy.

The difference between our approach and that in the composite configuration space resembles the difference between *decoupled* [5] and *centralized* [10] planning for multiple robots. As in decoupled planning we approach the problem in a lower-dimensional configuration space taking into account stationary features only. The resulting path is subsequently adjusted to resolve collisions with non-stationary features.

Rearrangement planning [2, 1] is a problem closely related to motion planning among movable obstacles. Here, an entity also navigates in an environment among movable obstacles, but the goal is not defined in terms of a configuration for the entity, but rather in terms of configurations for the movable obstacles. Even though rearrangement planning has had considerable attention over the years, motion planning among movable obstacles has not.

Wilfong [13] has shown that motion planning among movable obstacles is NP-hard. Chen and Hwang [4] created a grid based planner that heuristically tries to minimize the cost to move obstacles out of the way. To reduce the cost, they only consider a very limited number of different states. With their planner they are able to solve some simple but realistic problems.

Another planner is the one developed by Stilman and Kuffner [12]. Their global approach uses the fact that the free space of the entity consists of multiple connected components. If start and goal are not in the same connected component then the entity uses manipulation to move obstacles to try to join connected components. To detect if a manipulation action has succeeded, a grid based approach is used. To manipulate an obstacle, contact points are sampled and a set of primitive actions is applied to those points. The candidate obstacles for manipulation are found by using an A^* search on a grid from the current position of the entity to the goal. Obstacles encountered during this search are the candidates. In case of failure, backtracking is used. The authors prove that their planner is resolution complete for a class of problems they call LP_1, analogous to the LP_1 class in rearrangement planning [2]. The LP_1 class contains problems in which disjoint components of the free space can be merged by moving

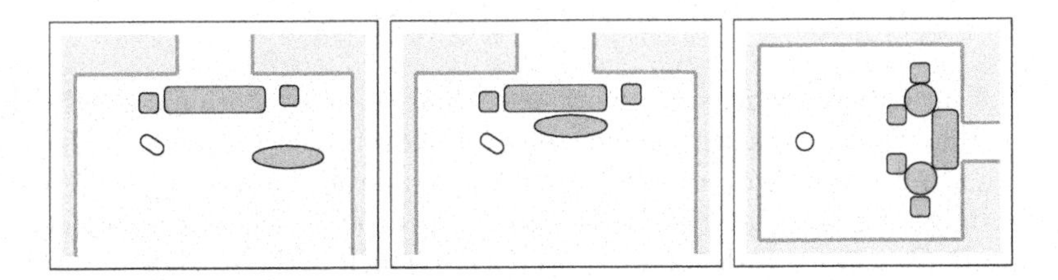

Fig. 1. Rooms consisting of couches, tables and chairs. The entity is represented as the light colored object. The goal for the entity is to leave the room. (a) Moving only the couch is enough to open the path to another room (LP_1). (b) First the table has to be moved before the couch can be moved (LP_2). (c) The couch is blocked by the round tables which on their turn are blocked by the chairs (LP_3).

a single obstacle. Stated differently, only if an obstacle blocks the path of the entity directly, it will be manipulated. Problems that require the manipulation of an obstacle that blocks another obstacle will not be solved. An example of an LP_1 problem is shown as Figure 1a.

Solving a motion planning problem among stationary and movable obstacles can be regarded as finding an alternating sequence of motions toward a movable obstacle and manipulation of these obstacles until the goal configuration is reachable without further manipulation. In this paper a novel framework is presented based on the expansion of a so-called action tree. This action tree represents the complete set of motions toward movable obstacles and manipulations of movable obstacles in every order. Finding a motion plan is equivalent to finding a path in the action tree. If the problem gets more complicated (i.e. more movable obstacles are involved), the construction of the complete action tree may become infeasible. Therefore, we present heuristics inspired by problems encountered in practical applications that guide the construction process of the action tree in favor of promising tree nodes. The class of problems that can be solved using our planner is LP [2], i.e., the set of problems that can be solved by a sequence of manipulations (Figures 1b+c). In contrast to the LP_1 class, the LP class contains problems for which movable obstacles have to be manipulated that do not directly block the path of the entity. For example, the manipulation of a movable obstacle can be blocked by another movable obstacle for which the manipulation is blocked by yet another movable obstacle etc. During the manipulation of a movable obstacle, the other movable obstacles are assumed to be stationary.

This paper is arranged as follows, first in Section 2, we will describe the problem in more detail and state some properties for the solution. Next, in Sections 3 and 4 the action tree and its properties are presented. The heuristics to guide the search process are described in Section 5. Finally, results of our experiments in are presented in Section 6.

2 Problem Statement and Preliminaries

Let E be an entity defined in a workspace that, besides stationary obstacles, contains k movable obstacles (or movables for short) $M = \{M_1, M_2, ...M_k\}$. All movables are assumed to be closed sets. A movable M_i cannot move by itself, but can only be moved by E if M_i is first grasped by E. The distance between two objects O_i and O_j is denoted by $d(O_i, O_j)$, which is the Euclidean distance between the two points on the boundary of O_i and O_j that are closest together. If the Euclidean distance between points p and q is denoted by $e(p, q)$ and $\text{INT}(O)$ denotes the interior of object O, then:

$$d(O_i, O_j) = \begin{cases} \min_{p \in O_i, q \in O_j} e(p, q) & \text{IF INT}(O_i) \cap \text{INT}(O_j) = \emptyset \\ \infty & \text{IF INT}(O_i) \cap \text{INT}(O_j) \neq \emptyset \end{cases}$$

M_i is only said to be grasped by E if $d(E, M_i) = 0$. If M_i is grasped by E, the combination of E and M_i is denoted by M_i^E. A physical model is used to define the set of possible motions of M_i^E, depending on the forces that E is able to apply to M_i. This model can also describe the potential positions on the boundary of M_i where grasps are allowed. Since the physical model is highly dependent on the specific capabilities of E and the properties of the environment, we use a simplified model in which E is capable to push/pull a movable in any direction and grasp it whenever $d(E, M_i) = 0$. Other models can be used without affecting the algorithm. We do not allow E to manipulate two movables at the same time.

Our goal is to create a motion plan for E from a given start to a given goal configuration. The behavior of E should be convincing compared to the behavior of its real (e.g. human) counterpart. For example, if a human has the choice between moving multiple obstacles that block a door and taking a small detour, it will most likely do the latter. Also most problems will be solved locally, i.e. a blocking movable will not have to be pushed a long distance before E will be able to move around it.

In this paper we introduce the concept of an action tree. The action tree is a general framework for solving motion planning problems among movable obstacles. In contrast to a roadmap graph, which represents the free configuration space, the action tree represents the different actions E can perform given the configurations of the movables. Using this framework, a planner is presented that uses heuristics to guide the search process through the action tree in order to efficiently find a solution.

3 Action Tree

In the basic motion planning problem all obstacles are stationary. Here, the movables can also change position during the execution of the planning algorithm. Therefore we introduce the notion of a *worldstate* that encodes the placement of all non-stationary obstacles, i.e. entity E and movables M.

Definition 1 (Worldstate). *A* worldstate $W = (w_e, w_1, ..., w_i, ..., w_k)$ *describes the configuration of* E *and the configurations of all movables in* M.

A worldstate is essentially a point in the composite configuration space of E and M. As stated before however, we will not solve the planning problem in this composite configuration space.

We will define two basic actions that are used to transform one worldstate into another. The first action is GRASP(M_i). A successful call to GRASP(M_i) transforms worldstate $W = (w_e, w_1, ..., w_i, ..., w_k)$ into worldstate $W' = (w'_e, w_1, ..., w_i, ..., w_k)$ satisfying $d(E[w'_e], M_i[w_i]) = 0$, else it reports failure. GRASP(M_i), moves E from its current configuration w_e to a randomly selected configuration w'_e on the boundary of M_i. If M_i cannot be grasped (because E is not able to reach the boundary of M_i) the action reports failure. Note that if $d(E[w_e], M_i[w_i]) = 0$, the GRASP(M_i) action has the effect of re-grasping M_i at another location. A re-grasp can be useful if the current grasp does not suffice, for example if the room for M_i^E to maneuver is limited.

The second action is MANIP(M_i^E), which tries to manipulate the currently grasped movable. A successful call to MANIP(M_i^E) transforms worldstate $W = (w_e, w_1, ..., w_i, ..., w_k)$ with $d(E[w_e], M_i[w_i]) = 0$ into worldstate $W' = (w'_e, w_1, ..., w'_i, ..., w_k)$ with $d(E[w'_e], M_i[w'_i]) = 0$. MANIP($M_i^E$) results in a joint motion of M_i and E satisfying the constraints imposed by the physical model from their current configuration to a randomly selected configuration. If MANIP(M_i^E) does not succeed, for example because the physical model forbids the manipulation or a collision occurs, MANIP(M_i^E) reports failure.

We now define the *action tree* T_A that represents all valid actions in all possible orders.

Definition 2 (Action Tree). *The* action tree *is a description of the space of all valid actions. Every node of the tree corresponds to a worldstate. The edges between the nodes represent the action* (GRASP() or MANIP()) *that results in the transformation from one worldstate to another worldstate.*

At every node n of T_A, a worldstate $W(n)$ is associated. T_A consists of two types of nodes: *manipulation nodes* are the result of a call to MANIP(), *grasp nodes* are the result of a call to GRASP(). A manipulation node will never have a child node that is also a manipulation node. This can be seen as follows. Suppose manipulation node n has a child node m that is also a manipulation node. Then node m could also have been a direct child of the parent of n (e.g. a sibling of n). The same holds for grasp nodes, a grasp node will never have another grasp node as a child because the second movable should have been grasped at once without first grasping the first one. Therefore a grasp node will always have a manipulation node as a parent and vice versa. We do not allow E to grasp two movables at the same time.

After initializing the algorithm by associating the initial worldstate to the root node n_s of T_A, the algorithm constructs T_A by *expanding* nodes. The expansion of a manipulation node adds a child node in which a movable is grasped at a random configuration; this may also be a re-grasp of the currently grasped movable. The expansion of a grasp node adds a child node that manipulates the currently grasped movable.

After the addition of a manipulation node n to T_A, E may be able to reach its destination. If a grasp node is added, this is not possible since no movables have changed their configuration. After manipulating a movable though, a new path may have emerged that brings E to its goal. Therefore after the addition of a manipulation node to T_A, the algorithm checks whether E can reach its goal configuration. If this succeeds, the algorithm terminates, if not, the construction of T_A continues. An example of the construction of an action tree is shown in Figure 2.

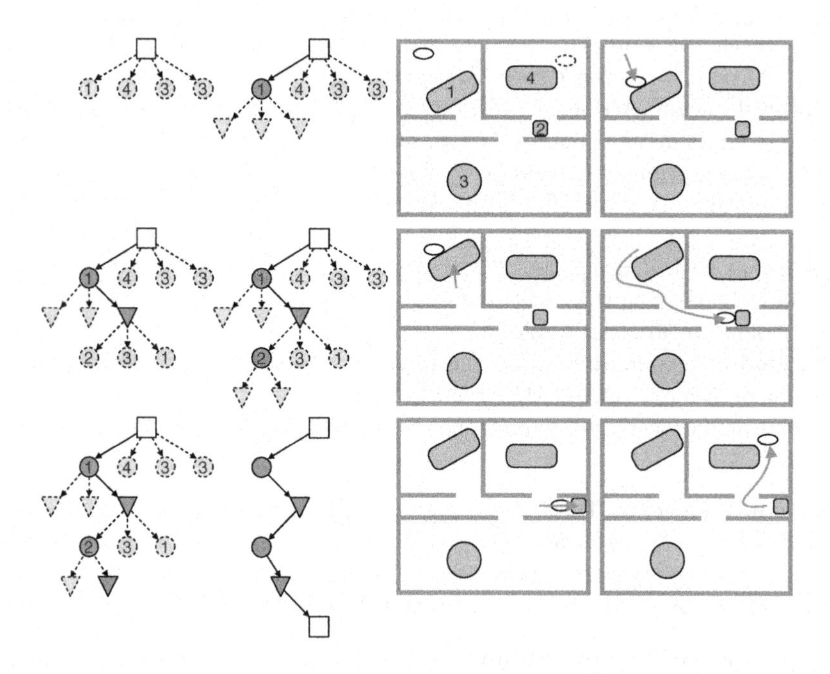

Fig. 2. Example of the construction of T_A. Left the the action tree T_A is shown, at the right the corresponding workspaces. The circle nodes in T_A are the grasp nodes, the triangles are the manipulation nodes. The square nodes are the start and goal configurations. The dashed nodes are omitted parts of T_A that did not contribute to the final solution or were unfeasible. In the first workspace, the start position of E is shown in the top left, the goal in the top right.

A motion plan is a finite alternating sequence of grasp and manipulation actions. In the action tree such a plan is represented by a path from the root to another node. If the physical model is such that during a manipulation M_i^E behaves as a rigid body and the physical model imposes no nonholonomic constraints on the motion, then, eventually every possible grasp and manipulation action will be represented in T_A. Also if nodes in T_A are selected randomly for expansion, every possible order of such actions will be represented by a path in T_A from the root node to another node. Therefore the framework of the action tree is probabilistically complete.

4 Realization of the Action Tree

In this section we will describe how to implement the necessary building blocks for the action tree. In scenes encountered in practical settings, we have observed that after a few expansions of a node, further expansion rarely led to more progress. For that reason we only expand nodes once (i.e. only leaves in T_A are expanded) but add multiple child nodes at once. If a manipulation node is selected for expansion, k children are added all grasping a different movable (including the current one). The expansion of grasp node adds a number of child nodes that manipulate the currently grasped movable.

To grasp a movable or to check whether the goal configuration can be reached, a graph G is used. G represents the free space for E with respect to the stationary obstacles. It should preferably contain cycles to provide alternative routes. Any path planning technique that results in a graph for E that represents feasible paths can be used to create G. A well known example of such a technique is the probabilistic roadmap method [6]. Without loss of generality, we will assume that the start and goal configurations for E are configurations in G.

4.1 Checking If the Goal Can Be Reached

After the addition of a manipulation node n, E may be able to reach its destination. Since G is collision free for the stationary obstacles only, it needs to be updated according to $W(n)$. Edges in G that collide with one of the movables have to be invalidated such that queries are guaranteed to be collision free.

A manipulation transforms worldstate $W(p(n)) = (w_e, w_1, ..., w_i, ..., w_k)$ to worldstate $W'(n) = (w'_e, w_1, ..., w'_i, ..., w_k)$, where $p(n)$ is the parent of node n. To be able to quickly update G for a given worldstate, at every manipulation node, a list of invalidated edges of G is stored. This list is equivalent to the list of $p(n)$ except for edges that intersect with either $M_i[w_i]$ or $M_i[w'_i]$. Only these edges have to be collision checked against M_i. To find these efficiently, the endpoints of the edges are stored in a Kd-tree [3]. A Kd-tree allows for quickly identifying the edges that are close to w_i so that they can be checked for collision. When the root node is added to T_A, G needs to be checked once against all movables in order to create the initial list of invalidated edges. Grasp nodes simply copy the list from their parent. After updating G, w'_e is connected to G using standard procedures from motion planning. Finally, a query is performed to see if E can reach its destination.

4.2 Grasping a Movable Obstacle

The GRASP$(M_i[w_i])$ action creates a path for E from its current configuration w_e to w'_e where $d(E[w'_e], M_i[w_i]) = 0$. Here, w'_e is a randomly chosen grasp on the boundary of M_i. Depending on the application, the selection of grasp configurations can be customized. As described in the previous section, G is updated w.r.t. the current worldstate. Next, w_e and w'_e are attempted to be connected to G. If this succeeds, a query can be executed between w_e and w'_e. If the query is successful, a new node is added to T_A.

4.3 Manipulating a Movable Obstacle

After E has grasped M_i, resulting in a grasp node, it will try to manipulate M_i by executing the MANIP(M_i^E) action. To move M_i^E we will use an approach based on the Rapidly-exploring Random Trees (RRT) algorithm [9]. An RRT is aimed at growing a tree from a given start configuration in an attempt to cover the free space. Here, the RRT operates in the configuration space of M_i^E where all movables M_j with $i \neq j$ are considered stationary. An RRT quickly generates many different paths away from the start configuration. This property is very useful in our situation because our target is to move M_i^E away from its current configuration to many different goal configurations.

The vertices[1] in the RRT represent configurations for M_i^E. The edges represent paths between them. The RRT algorithm works as follows. First the start configuration is added to the RRT as a vertex. Next, a random (not necessarily collision free) configuration $c_r = (w_e, w_i)$ with $d(E[w_e], M_i[w_i]) = 0$ for M_i^E is generated. The nearest configuration $c_n = (w'_e, w'_i)$ in the RRT to c_r is found (not necessarily a vertex of the RRT) and a path is tested for collision moving from c_n to c_r. If c_r is reached, it is added as a vertex to the RRT together with the edge (c_n, c_r). If a collision occurs before c_r is reached, the last collision free configuration $c_s = (w''_e, w''_i)$ is added to the RRT together with the edge (c_n, c_s). This process is repeated until some stop criterion is met.

The RRT algorithm needs to check whether a path exists between c_n and c_r. To verify the existence of such a path a *local planner* is used in many sampling based motion planning techniques. Given two configurations the local planner checks whether the path between them is feasible. Because of the many collision checks involved, a call to a local planner may be relatively expensive. Therefore, usually the local planner only checks whether the straight line connection between two configurations is feasible. Rotational parameters are often interpolated. In our algorithm, the generated path also needs to comply with the physical model for the specific type of manipulation (e.g. pushing/pulling). The local planner is allowed to use any physical model as long as it is capable of deciding whether a manipulation path exists between two configurations or, in case of a collision, what the closest reachable configuration is to c_r.

Since the RRT algorithm works incrementally by nature, it is easy to implement our MANIP() action using an RRT. Every grasp node n contains an RRT. The child nodes of n (which are all manipulation nodes) represent the vertices of that RRT.

---

[1] Note the difference between the action tree that contains nodes, and the RRT that contains vertices.

We have now described all the building blocks necessary for expanding a node. Algorithm 1 shows how a node is expanded. Calling this function with n_s (that contains the initial worldstate), initiates the creation of T_A. Nodes are expanded in a breadth first manner to provide an optimal solution. In the next section, we will describe heuristics to tailor this concept to problem settings encountered in practical settings.

Algorithm 1. EXPANDNODE (n_s, G, W_g)

```
1: if n_s.type = MANIPULATIONNODE then
2:    for all M_i ∈ M do
3:       W' ← GRASP (M_i, W(n_s)) {grasp M_i at a random position}
4:       if W' ≠ NULL then
5:          n_s.AddChild(W') {add a grasp node}
6: else {n_s is a grasp node}
7:    n_s.InitRRT() {n_s is the container of the RRT}
8:    for i = 0 to MAXRRTVERTICES do
9:       W' ← EXTENDRRT (W(n_s))
10:      if W' ≠ NULL then
11:         n_s.AddChild(W') {add a manipulation node}
12:         UPDATEENTITYGRAPH(G, W') {update G for W'}
13:         if PATHEXISTS(G, W', W_g) {is there a path to W_g?} then
14:            return PATHFOUND {the goal can be reached}
15: for i = 1 to n_s.nrChildren do
16:    EXPANDNODE (n_s.Child[i], G, W_g)
```

5 Planner

Although breadth-first expansion of T_A guarantees an exhaustive exploration of all possible sequences of manipulations, this strategy becomes computationally expensive or even infeasible when the number of movables is large. In problems encountered in practical settings however, only a small subset of the movables are involved in the final motion plan of E. In that respect, many nodes of T_A will most likely not contribute to the final solution. For example, if a motion plan needs to be created that moves E from room A to B, then often the movables present in room C will not be part of the final motion plan. On the other hand, a movable that is not directly impeding the path of E, may very well be blocking the manipulation of another movable and because of that be part of a feasible motion plan. Because of the above considerations, expanding nodes in a breadth first search manner is not the most efficient way to find a motion plan. In this section we will describe how to focus the expansion process toward promising nodes, such that a solution can be found rapidly without affecting the probability of finding a solution.

5.1 Choosing a Path through the Action Tree

Instead of expanding T_A in a breadth first manner, we will use heuristics to guide the expansion process. To be able to do this, every node n is assigned a probability $q(n)$. The following holds:

$$\sum_{m \in \text{children}(n)} q(m) = 1 \tag{1}$$

Nodes are now selected for expansion by creating a path from the root node to a not yet expanded node (a leaf). Starting at the root (having probability 1), a child node is selected randomly based on its probability. This process is repeated until a leaf is reached. Then that leaf is selected for expansion.

After expanding a node n, all its children are initially assigned equal probabilities. Later on in the process we will increase probabilities for a node depending on the progress that is made. For example, if n is successful in getting closer to the goal, $q(n)$ is increased. Increasing $q(n)$ should not violate Equation 1. In addition we must assure that $q(n)$ never reaches 0 as this would exclude its selection. If $q(n)$ is already very high (e.g. $q(n) = 0.95$), there is little reason to increase it more. Using the above observations, we use the following procedure to adapt the probabilities: a fraction $f \in [0, 1]$ is used to increase $q(n)$ of a node n. We denote the updated value of the probability of n by $q'(n)$. The siblings of node n are denoted by the set $S(n)$.

$$q'(n) = (1 - q(n))f + q(n) \tag{2}$$

$$\forall_{m \in S(n)} : q'(m) = q(m)(1 - f) \tag{3}$$

Similarly if, after the expansion of a node, no or little progress is made, the probabilities of selecting that node should be decreased. The following procedure is used to decrease the probabilities of node n if n has at least one sibling:

$$q'(n) = q(n) - f \cdot q(n) \tag{4}$$

$$\forall_{m \in S(n)} : q'(m) = \frac{1 - q(m)}{|S(n)| - 1 + q(n)} f \cdot q(n) + q(m) \tag{5}$$

Equations 2-5 lead to the following lemma, for which the proof is omitted:

Lemma 1. *The updates given by Equations 2 and 3 and by 4 and 5 maintain Equation 1.*

The probability that n will be selected for expansion is: $q(n) \cdot \prod_{m \in \text{ancestors}(n)} q(m)$.
If $q(n)$ is increased (at the expense of its siblings), the probability of n being selected will increase only by a small amount (especially when n is not close to the root). To solve this issue, the increased probability of n is *propagated* along the path from n to the root, i.e. the probabilities of all ancestors of n are increased as well. We must make sure however that after a few updates, the probabilities

of nodes higher in the tree do not become too high. If we consider the path from n to the root, the further away a node n' is from n, the less similarity between $W(n)$ and $W(n')$. Therefore f is lowered during the propagation. The procedures to increase and decrease the probabilities of nodes are shown as Algorithms 2a and 2b.

Algorithm 2. (a) Increasing and (b) decreasing probabilities

INCREASEPROBABILITY (n, f)	DECREASEPROBABILITY (n, f)
1: $q'(n) = (1 - q(n)) \cdot f + q(n)$	1: **if** $\lvert S(n)\rvert > 1$ **then**
2: **for all** $m \in S(n)$ **do**	2: $q'(n) = q(n) - f \cdot q(n)$
3: $q'(m) = q(m) * (1 - f)$	3: **for all** $m \in S(n)$ **do**
4: $f = factor * f$ {lower f by a factor}	4: $q'(m) = \frac{1-q(m)}{\lvert S(n)\rvert - 1 + q(n)} + q(m)$
5: INCREASEPROBABILITY $(p(n), f)$	5: $f = factor * f$ {lower f by a factor}
	6: DECREASEPROBABILITY $(p(n), f)$

An example of increasing the probability of a node is shown in Figure 3. Since only the nodes (and their children) on the path from n to the root are updated, the cost of the propagation is $O(r)$, where r is the rank of n.

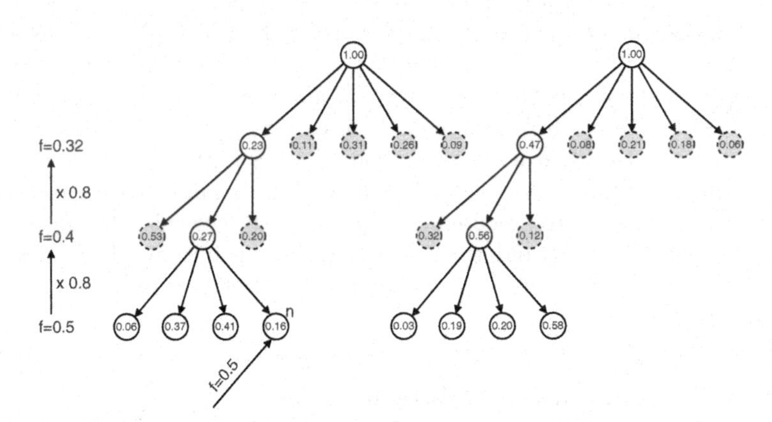

Fig. 3. Increasing the probability of a node. The values in the nodes are the probabilities. (a) The probability of node n is increased by a fraction of 0.5. If f is propagated to the parent node, it is multiplied by a factor of 0.8 (shown left). (b) The resulting probabilities after propagation using Algorithm 2a.

The value of the fraction f determines the global behavior of the algorithm. If f is small, the differences between the probabilities of the nodes will not be large, resulting in a breadth first type of expansion. If f is high however, a small number of nodes will receive a high probability, resulting in a more depth first type of expansion.

5.2 Adapting Probabilities

As stated before, often only a small subset of M will be blocking the path of E. Therefore we will increase the probability of manipulating movables that actually block the path of E. A movable M_i can block the path of E either *directly* or *indirectly*. Directly blocking means that the path of E actually collides with M_i, indirectly blocking means that some M_j, $i \neq j$ blocks the manipulation of M_i.

Directly Blocking Movable Obstacles

After the expansion of a manipulation node n, it is checked whether E can reach its destination (Section 4). If no path to the goal is found, it is necessary to manipulate movables. Selecting a good movable candidate for manipulation involves taking into account our target of creating convincing paths for E. One of the properties of such a path is that a small detour is favorable over manipulating many movables. Therefore a second version of G is created in which no edges are invalidated but rather get a penalty when colliding with one of the movables. This graph is denoted by G^p. The cost of traversing an edge in G^p is a function of its length in C-space and the penalty. Using G, G^p can be constructed by assigning a penalty to edges instead of invalidating them. No additional collision checks are necessary to construct G^p.

After reconnecting E to the graph, a shortest path query on G^p, using for example A^* or D^* Lite ([11], [7]) yields a path that prefers to avoid movables. By using the result of the query, the *first colliding movable* M_i on the path to the goal in G^p can be easily determined. M_i is called a *directly blocking movable.* The probability of the child node of n grasping M_i is increased.

The above procedure is repeated a few times to make sure that the expansion process does not become too focused toward one path. For this, the edges in G^p that collide with M_i are invalidated and a new query is initialized. If successful, again the first colliding movable is determined and the corresponding node's probability is increased.

Indirectly Blocking Movable Obstacles

Movables that block the path of E directly, can be determined by a shortest path query. However, a movable can also block the manipulation action of another movable, thus blocking the path of E indirectly (see Figure 4a for an example). Using the properties of the RRT algorithm, these can be determined easily. Recall that an RRT is extended by trying to create a path for M_i^E from the RRT to a randomly chosen configuration. If this random configuration is not reached, M_i^E collides with either a stationary obstacle or another movable M_j (Figure 4b). In the latter case, M_j blocks the manipulation path of M_i^E and M_j is identified as an *indirectly blocking movable.*

During the expansion of a grasp node n in which M_i is grasped (creating child nodes that are vertices of the RRT), if a blocking movable M_j is encountered, the probability of the sibling of n that grasps M_j is increased using Algorithm

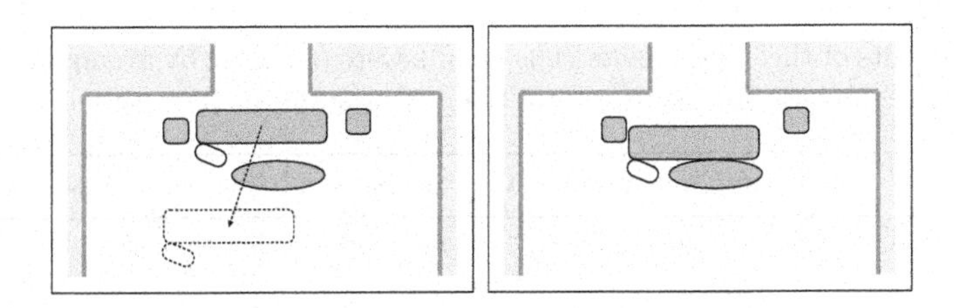

Fig. 4. E tries to pull the couch to the dotted target position. (a) The start of the manipulation action. (b) The coffee table blocks the path. This configurations is added to the RRT and the probability of the node in which the coffee table is grasped is increased.

2a. The more often M_j acts as indirectly blocking movable, the more likely it is that M_j is selected for manipulation.

Decreasing Probabilities

If a node n in T_A has many successful descendants, $q(n)$ will increase because of the propagation algorithm. However, if its descendants cease to make progress, its probability should be lowered. For this, we need a method to measure progress. A cheap measure of progress is the graph distance of E to the destination in G. The current position of E is connected to G and using a shortest path query to the destination provides an estimate of the current distance to the goal. The distance estimate is saved in the node such that the progress between a node and its parent can easily be determined. If no or little progress is made, the probability of the node is decreased by a small fraction using Algorithm 2b.

5.3 Lazy Expansion

Expanding a node (Algorithm 1) can be a costly operation. Luckily many operations can be postponed until the moment a node is actually selected for expansion. Manipulation nodes contain child nodes that grasp a movable. Checking the feasibility of the grasp (i.e. can the grasp configuration be reached by E from its current configuration) is not necessary until the moment the child node is selected for expansion. So if a manipulation node is chosen for expansion, it adds child nodes that represent grasps of movables without verifying that such grasps actually exists.

If a child node is added to a grasp node, the entity graphs (G and G^p) need to be updated. This update involves collision checks and is thus relatively expensive. Only when the child node is selected for expansion this update is necessary. Therefore the calculation of the update of G and G^p can be postponed until the node is actually chosen for expansion.

If lazy expansion is used then it is uncertain if n is feasible at the moment it is added to T_A. Only if n is selected by the random process for expansion, its feasibility is checked. If it is not feasible, n is declared a *dead end* and $q(n)$ is set

Table 1. Results of the experiments. The results were obtained by averaging 100 runs. The average tree size is the average total number of nodes in the tree.

Scene	Avg. running time	Avg. tree size	Avg. rank of solution
1	1.06s	81	14.4
2	8.1s	167	43.8
3	29.0s	506	13.5
4	22.0s	400	23.3

to 0. The probabilities of $S(n)$ are updated such that Equation 1 is maintained. Now another random leaf node is selected for expansion.

6 Experiments

We implemented our algorithm in C++ and conducted experiments with three scenes on a Pentium 2.40GHz with 1GB of memory. In all scenes the heuristics as described in Section 5 were used. The different values for f were determined experimentally and the same values turned out to be useful in all experiments. A node grasping a blocking movable received an increase of its probability using $f = 0.8$. The probabilities of nodes containing an indirectly blocking movable were increased using $f = 0.05$ for every collision. If a new node did not result in progress toward the goal, the probability of the corresponding node was decreased using $f = 0.2$. The number of child nodes added to a grasp node was at most 5, the edge penalty used in G^p was 100 times its length.

For the first experiment the LP_3 scene of Figure 1c was used. Next, for the second experiment, to verify that our method also scales to multiple indirectly blocking movables, we took three versions of the scene of Figure 1c and connected them together, effectively creating a problem involving a series of three LP_3 problems. The third scene is shown in Figure 5. The pitfall in this scene is that the shortest path (shown as the dotted arrow) is blocked by an immovable obstacle close to the goal. Thus, to reach a solution, the algorithm probably needs backtracking (depending on the random choices). The fourth experiment (Figure 6) has two indirectly blocking movables. For all scenes, we conducted 100 experiments and averaged the results, shown in Table 1.

7 Discussion

In this paper we have presented a probabilistically complete framework based on an action tree to solve motion planning queries among movable obstacles. The nodes of the tree represent worldstates, the edges the transitions between the worldstates. During those transitions, the entity either manipulates one of the movables or (re)grasps a movable. A path in the action tree represents a motion plan. Constructing the complete action tree (in a breadth first search manner) can be infeasible because of the huge number of nodes.

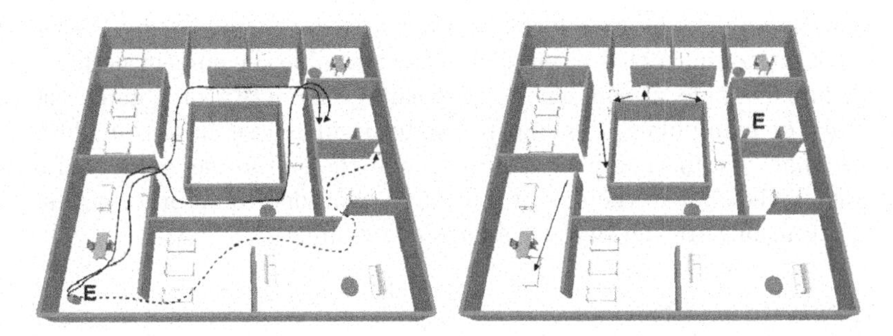

Fig. 5. The third scene. (a) The entity E is represented by a cylinder. The two solid arrows show feasible paths, the dotted path results in a dead end (behind the first movable is another one that blocks the manipulation of the first). (b) The situation after the entity has reached its destination.

Fig. 6. The fourth scene. The entity needs to move from the top left to the right bottom. (a) The two encircled movables are indirectly blocking movables. (b) The situation after the entity has reached its destination.

In environments encountered in practical settings often only a small subset of the motion plans in the action tree are useful. Therefore we presented heuristics that focus the node expansion process toward these motion plans. This process is solely guided by adapting the probabilities of selecting nodes. By continuously adapting these probabilities, using information gathered during the process, an efficient algorithm is obtained that is capable of solving realistic problems in reasonable running times.

An issue we did not address in this paper is smoothing. To smooth the path of the entity during a grasp action, standard smoothing techniques can be used. Also, since the final path is the result of a probabilistic process, some nodes from the action tree in the solution may be redundant. A smoothing procedure may be able to by-pass them.

Even though the heuristics are often successful in guiding the probabilistic process in realistic problems, in certain situations the process may become slow. This happens if many movables are close together and need to be manipulated in a certain order. If n movables are involved, there are $n!$ sequences of manipulating the movables. Since none of these result in overall progress, many nodes get comparable probabilities resulting in breadth first search behavior. Experiments have also shown that sometimes computation time is spent on parts of the action tree that are quite similar (e.g. they only slightly differ in the configuration of a movable).

A solution could be to use information gathered in the process not only locally in the action tree but rather in all nodes that resemble the current node in some aspects. For example, if a movable is impeding the entity in a node, then its probability of manipulation should not only be increased in that node but in all nodes that contain a similar subproblem. If the subproblem is solved in one node, then a shortcut in the action tree could be added to all other nodes. This type of extensions are the subject of current research.

Acknowledgments

Part of this research has been funded by the Dutch BSIK/BRICKS project. The authors would like to thank Jur van den Berg for fruitful discussions and suggestions.

References

1. Alami, R., Laumond, J.P., Siméon, T.: Two Manipulation Planning Algorithms. In: Workshop on Algorithmic Foundations of Robotics, pp. 109–125 (1994)
2. Ben-Shahar, O., Rivlin, E.: Practical Pushing Planning for Rearrangement Tasks. IEEE Trans. on Robotics and Automation 14(4), 549–565 (1998)
3. Bentley, J.L.: Multidimensional Binary Search Trees used for Associative Searching. Commun. ACM 18, 509–517 (1975)
4. Chen, P.C., Hwang, Y.K.: Practical Path Planning among Movable Obstacles. In: Proc. IEEE International Conference on Robotics and Automation, pp. 444–449 (1991)
5. Erdmann, M., Lozano-Pérez, T.: On multiple moving objects. Algorithmica 2, 477–521 (1987)
6. Kavraki, L., Švestka, P., Latombe, J.-C., Overmars, M.H.: Probabilistic Roadmaps for Path Planning in High-Dimensional Configuration Spaces. IEEE Trans. on Robotics and Automation 12, 566–580 (1996)
7. Koenig, S., Likhachev, M.: Improved Fast Replanning for Robot Navigation in Unknown Terrain. In: Proc. of the 2002 IEEE International Conference on Robotics and Automation, pp. 968–975 (2002)
8. LaValle, S.M.: Planning Algorithms. Cambridge University Press, Cambridge (2006), `http://msl.cs.uiuc.edu/planning/`
9. LaValle, S.M., Kuffner, J.J.: Rapidly-exploring random trees: Progress and prospects. In: Donald, B.R., Lynch, K.M., Rus, D. (eds.) Algorithmic and Computational Robotics: New Directions, A.K. Peters, Wellesley, MA, pp. 293–308 (2001)
10. Schwartz, J.T., Sharir, M.: On the "piano movers" problem III: coordinating the motion of several independent bodies: the special case of circular bodies moving amidst polygonal barriers. Int. Journal of Robotics Research 2(3), 46–75 (1983)
11. Stenz, A.: The Focused D^* Algorithm for Real-Time Replanning. In: Proc. of the International Joint Conference on Artificial Intelligence, pp. 1652–1659 (1995)
12. Stilman, M., Kuffner, J.: Navigation Among Movable Obstacles: Real-time Reasoning in Complex Environments. Int. Journal of Humanoid Robotics 2(4), 479–504 (2005)
13. Wilfong, G.: Motion Planning in the Presence of Movable Obstacles. In: Proc. of the 4th Annual Symposium on Computational Geometry, pp. 279–288 (1988)

Planning the Shortest Safe Path Amidst Unpredictably Moving Obstacles*

Jur van den Berg and Mark Overmars

Department of Information and Computing Sciences,
Universiteit Utrecht, The Netherland
{berg,markov}@cs.uu.nl

Abstract. In this paper we discuss the problem of planning safe paths amidst unpredictably moving obstacles in the plane. Given the initial positions and the maximal velocities of the moving obstacles, the regions that are possibly not collision-free are modeled by discs that grow over time. We present an approach to compute the *shortest path* between two points in the plane that avoids these growing discs. The generated paths are thus guaranteed to be collision-free with respect to the moving obstacles while being executed. We created a fast implementation that is capable of planning paths amidst many growing discs within milliseconds.

1 Introduction

An important challenge in robotics is motion planning in dynamic environments. That is, planning a path for a robot from a start location to a goal location that avoids collisions with the moving obstacles. In many cases the motions of the moving obstacles are not known beforehand, so often their future trajectories are estimated by extrapolating current velocities (acquired by sensors) in order to plan a path [2, 5, 10]. This path may become invalid when some obstacle changes its velocity (say at time t), so then a new path should be planned. However, there is actually no time for planning; as the world is continuously changing, the computation would already be outdated even before it is finished.

To overcome this problem, often a fixed amount of time, say τ, is reserved for planning [6, 9]. The planner then takes the expected situation of the world at time $t + \tau$ as initial world state, and the plan is executed when the time $t + \tau$ has come. This scheme carries two problems:

- The predicted situation of the world at time $t + \tau$ may differ from the actual situation when some obstacles change their velocities again during planning. This may result in invalid paths.
- The path the robot will follow between time t and time $t + \tau$ is not guaranteed to be collision-free, because this path was computed based on the old velocities of the obstacles.

* This research was supported by the IST Programme of the EU as a Shared-cost RTD (FET Open) Project under Contract No IST-2001-39250 (MOVIE - Motion Planning in Virtual Environments).

S. Akella et al. (Eds.): Algorithmic Foundation of Robotics VII, STAR 47, pp. 103–118, 2008.
springerlink.com

In this paper we take a first step to overcome these problems. We present an approach to compute a path from a start location to a goal location that is guaranteed to be collision-free, no matter how often the obstacles change their velocities in the future. Replanning might still be necessary from time to time, to generate trajectories with more appealing global characteristics, but the two problems identified above do not occur in our case. The first problem is solved by incorporating all the possible situations of the world at time $t+\tau$ in the world model. The second problem is solved as the computed paths are guaranteed to be collision-free regardless of what the moving obstacles do.

We assume that all obstacles and the robot are modeled as discs in the plane, and that the robot and each of the obstacles have a (known) maximum velocity. The maximum velocity of the obstacles should not exceed the maximum velocity of the robot. The problem is solved in the configuration space, that is, the radius of the robot is added to the radii of the obstacles, so that we can treat the robot as a point.

Given the initial positions of each of the obstacles, the regions of the space that are possibly not collision-free are modeled by discs that grow over time with rates corresponding to the maximal velocities of the obstacles. Our goal is to compute a *shortest path* (a minimum time path) from a start to a goal configuration that avoids these growing discs (see Fig. 1).

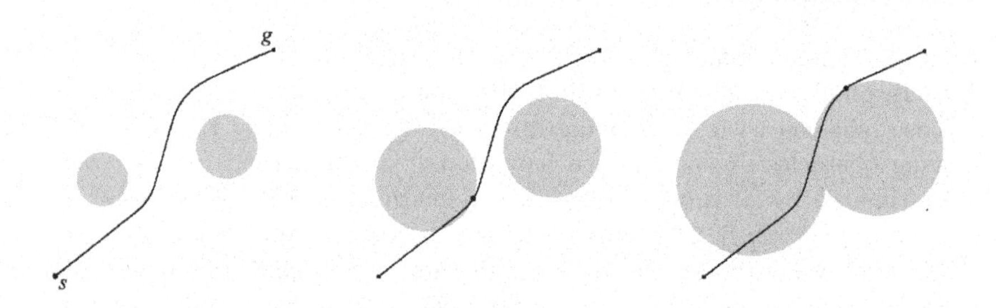

Fig. 1. An environment with two moving obstacles and a shortest path. The pictures show the growing discs at $t = 0$, $t = 1$ and $t = 2$, respectively. A small dot indicates the position along the path.

Although computing shortest paths is a well studied topic in computational geometry (see [8] for a survey), the problem we study in this paper is new. In fact, it is a three-dimensional shortest path problem, as the time accounts for an additional dimension. Such problems are NP-hard in general, yet we present an $O(n^3 \log n)$ algorithm (n being the number of discs) for our problem in the restricted case that all discs have the same growth rate.[1] In case the growth rates are different, we cannot give a time bound expressed in n. Instead, we

[1] Note that the special case of discs with zero growth rate gives a two-dimensional shortest path problem, which can be solved in $O(n^2 \log n)$ time (see e.g. [4]).

implemented a practical algorithm for this general case, that appears to work well: Experimental results show that we are able to generate shortest paths amidst many growing discs within only milliseconds of computation time.

The rest of the paper is organized as follows. We formally define the problem in Section 2. In Section 3 we examine the structure of shortest paths amidst growing discs. We sketch our global approach in Section 4, and in Section 5 we present efficient algorithms for the restricted and general case. Experimental results are given in Section 6, and Section 7 concludes the paper.

2 Problem Definition

The problem is formally defined as follows. Given are n moving obstacles $O_1, \ldots, O_n$ which are discs in the plane. The centers of the discs (i.e. the positions of the obstacles) at time $t = 0$ are given by the coordinates $p_1, \ldots, p_n \in \mathbb{R}^2$, and the radii of the discs by $r_1, \ldots, r_n \in \mathbb{R}^+$. All of the obstacles have a maximal velocity, given by $v_1, \ldots, v_n \in \mathbb{R}^+$. The robot is a point (if it is a disc, it can be treated as a point when its radius is added to the radii of the obstacles), for which a path should be found between a start configuration $s \in \mathbb{R}^2$ and a goal configuration $g \in \mathbb{R}^2$. The robot has a maximal velocity $V \in \mathbb{R}^+$ which should be larger than each of the maximal velocities of the obstacles, i.e. $(\forall i :: V > v_i)$.

As we do not assume any knowledge of the velocities and directions of motion of the moving obstacles, other than that they have a maximal velocity, the region that is guaranteed to contain all the moving obstacles at some point in time t is bounded by $\bigcup_i B(p_i, r_i + v_i t)$, where $B(p, r) \subset \mathbb{R}^2$ is an open disc centered at p with radius r. In other words, each of the moving obstacles is conservatively modeled by a disc that grows over time with a rate corresponding to its maximal velocity (see Fig. 1 for an example environment).

Definition 1. *A point $p \in \mathbb{R}^2$ is* collision-free *at time $t \in \mathbb{R}^+$ if $p \notin \bigcup_i B(p_i, r_i + v_i t)$.*

The goal is to compute the shortest possible path $\pi : [0, t_g] \to \mathbb{R}^2$ between s and g (i.e. a minimal time path with minimal t_g where $\pi(0) = s$ and $\pi(t_g) = g$) that is collision-free with respect to the growing discs for all $t \in [0, t_g]$.

3 Properties of Shortest Paths

In this section we deduce some elementary properties of shortest paths amidst growing discs. We first show that we are actually dealing with a three-dimensional path planning problem: As the discs grow over time, we can see the obstacles as cones in a three-dimensional space (see Fig. 2), where the third dimension represents the time. Each obstacle O_i transforms into a cone C_i, whose central axis is parallel to the time-axis of the coordinate frame, and intersects the xy-plane at point p_i. The maximal velocity v_i determines the opening angle of the cone, and together with the initial radius r_i, it determines the (negative) time-coordinate of the apex. The equation of cone C_i is given by:

$$C_i : (x - p_{ix})^2 + (y - p_{iy})^2 = (v_i t + r_i)^2. \tag{1}$$

The goal configuration g is transformed into a line parallel to the time-axis, where we want to arrive as soon as possible (i.e. for the lowest value of t). In the three-dimensional space it is easier to reason about the properties of shortest paths.

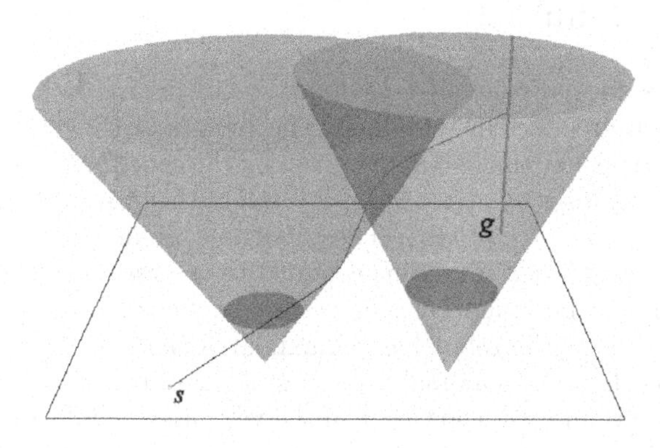

Fig. 2. The three-dimensional space of the same environment as Fig. 1

3.1 Maximal Velocity

We will first show that a shortest path is always traversed at the maximal velocity V, and hence a shortest path makes a constant angle $\arctan(1/V)$ with the xy-plane.

Lemma 1. *A point $p \in \mathbb{R}^2$ that is collision-free at time $t = t'$, is collision-free for all $t :: 0 \leq t \leq t'$.*

Proof. If $t_1 \leq t_2$, we know that $\bigcup_i B(p_i, r_i + v_i t_1) \subseteq \bigcup_i B(p_i, r_i + v_i t_2)$. Thus if a point p is collision-free at time t_2, i.e. $p \notin \bigcup_i B(p_i, r_i + v_i t_2)$, it is certainly not in $\bigcup_i B(p_i, r_i + v_i t_1)$. Hence point p is collision-free at time t_1 as well. □

Theorem 1. *The velocity $\frac{\|(\delta x, \delta y)\|}{\delta t}$ of a shortest path is constant and equal to the maximal velocity V.*

Proof. Suppose π is a path to g, of which a sub-path has a velocity smaller than V. Then this sub-path could have been traversed at maximal velocity, so that points further along the path would be reached at an earlier time. Lemma 1 proves that these points are then collision-free as well, so also g could have been reached sooner, and hence π is not a shortest path. □

3.2 Straight-Line Segments and Spiral Segments

Next, we prove that that a shortest path can only consist of straight-line motions, and motions that stay in contact with the growing discs. These latter motions follow curves 'winding' around a disc while it grows. They lie on the surface of a cone, when viewed in the three-dimensional space.

Theorem 2. *A shortest path solely consists of straight-line segments, and segments on the boundary of a growing disc.*

Proof. Theorem 1 implies that the time it takes to traverse a path is proportional to its length. Hence, parts of the path in 'open' space can always be shortcut by a straight-line segment. Only when the path stays in contact with a growing disc, it is not possible to shortcut. □

We next show that in fact, as both the velocity of the path and the growth rate of the discs are constant, the segments on the boundary of a disc are supported by a *logarithmic spiral.*

Without loss of generality, we assume that the disc has radius 0 at $t = 0$, that the disc is centered at the origin, and that the disc grows with velocity 1 (other discs can be transformed such that these conditions hold). Let the velocity of the path be V. We express the equations of the path curve in polar coordinates $(r(t), \theta(t))$, parametrized by the time t. The radius $r(t)$ of the curve at time t is equal to the radius of the disc at time t, thus:

$$r(t) = t. \tag{2}$$

The angle $\theta(t)$ is not trivially deduced, but we know that

$$\sqrt{x'(t)^2 + y'(t)^2} = V, \tag{3}$$

as the velocity along the path is constantly equal to V. From this equation, we deduce a closed form for $\theta(t)$:

$$\begin{aligned}
\{x(t) &= r(t)\cos\theta(t),\\
y(t) &= r(t)\sin\theta(t)\},\\
\{x'(t) &= r'(t)\cos\theta(t) - r(t)\theta'(t)\sin\theta(t),\\
y'(t) &= r'(t)\sin\theta(t) + r(t)\theta'(t)\cos\theta(t)\},\\
x'(t)^2 + y'(t)^2 &= r'(t)^2 + r(t)^2\theta'(t)^2 = 1 + t^2\theta'(t)^2.
\end{aligned} \tag{4}$$

Combining Equations (4) and (3), and solving for $\theta(t)$ gives:

$$\begin{aligned}
\sqrt{1 + t^2\theta'(t)^2} &= V,\\
1 + t^2\theta'(t)^2 &= V^2,\\
\theta'(t) &= \pm\frac{\sqrt{V^2 - 1}}{t},\\
\theta(t) &= \pm\sqrt{V^2 - 1}\log t + \theta_0.
\end{aligned} \tag{5}$$

Equations (2) and (5) together define a curve which is well known as the *logarithmic spiral* [11]. The $\pm$ indicates whether the spiral revolves counterclockwise (+), or clockwise (−) about the growing disc. The term θ_0 gives the starting angle of the spiral.

3.3 Path Smoothness

Theorem 3. *A shortest path is C^1-smooth.*

Proof. Suppose path π is not C^1-smooth and contains sharp turns. Then these turns could be shortcut by a straight-line segment. Hence π is not a shortest path. □

This theorem implies that in a (general) shortest path the straight-line segments and spiral segments alternate each other, and that the straight-line segments must be *tangent* to the supporting spirals of the spiral segments. In terms of the three-dimensional space this means that the straight-line segments (which "connect" two spiral segments), are *bitangent* to the cones on which the spirals lie.

3.4 Departure Curves

There are four ways in which a straight-line segment can be bitangent to a pair of cones (say C_i and C_j): left-left, right-right, left-right and right-left. In each of these cases, there is an infinite number of possible segments (whose slope corresponds to the maximal velocity V) that are tangent to both C_i and C_j. However, the possible tangency points at the surface of C_i form a continuous curve on that surface. We call such curves *departure curves.* They play a major role in our algorithm to compute a shortest path.

Definition 2. *For two cones C_i and C_j, the set $DC(C_i, C_j)$ is defined as the collection of points on the surface of C_i, for which the straight-line of slope $1/V$ that is tangent to C_i in that point is also tangent to C_j. The set $DC(C_i, C_j)$ consists of four continuous curves, each associated with one of the tangency cases. We call them* departure curves. *They are denoted $DC_{ll}(C_i, C_j)$, $DC_{lr}(C_i, C_j)$, $DC_{rl}(C_i, C_j)$ and $DC_{rr}(C_i, C_j)$, respectively.*

The set $DC(C_i, g)$ to the goal configuration g is defined similar, but then the tangent line segment should go through the goal configuration g. In this case the departure curves $DC_r(C_i, g)$ and $DC_l(C_i, g)$ are distinguished.

We now show how we can deduce equations for the departure curves on the surface of a cone C. Again, without loss of generality, we assume that the disc associated with the cone has radius 0 at $t = 0$, that the disc is centered at the origin, and that the disc grows with velocity 1. Let the velocity of the path be V. The surface of C can be parametrized by two variables, time T and angle Θ:

$$C : (T, \Theta) \rightarrow \{T \cos \Theta, T \sin \Theta, T\}\,.$$

Let us consider the counterclockwise spirals about this cone. Each of them is uniquely defined by the initial angle θ_0 (see Equation (5)). Each point (T, Θ) on the surface of the cone has a unique spiral that goes through that point. This spiral can be found by solving $\theta(T) = \Theta$ for θ_0:

$$\theta_0 = -\sqrt{V^2 - 1}\log T + \Theta. \tag{6}$$

Hence, the spiral though (T, Θ) is described in Euclidean coordinates as:

$$\begin{aligned}\{x(t) &= t\cos\left(\sqrt{V^2-1}\log t - \sqrt{V^2-1}\log T + \Theta\right),\\ y(t) &= t\sin\left(\sqrt{V^2-1}\log t - \sqrt{V^2-1}\log T + \Theta\right)\}.\end{aligned}$$

If we walk along this spiral, we can depart for another cone if the straight-line segment tangent to the spiral is tangent to another cone as well. The straight-line segment ℓ tangent to the spiral at point (T, Θ) is represented by:

$$\begin{aligned}\ell(t) &= \{x(T) + (t-T)x'(T),\ y(T) + (t-T)y'(T)\} = \\ &= \{t\cos\Theta - (t-T)\sqrt{V^2-1}\sin\Theta,\ t\sin\Theta + (t-T)\sqrt{V^2-1}\cos\Theta\}.\end{aligned} \tag{7}$$

This segment must be tangent to another cone, say C_i with position p_i, initial radius r_i and velocity v_i, in order for point (T, Θ) to be on a departure curve of $DC(C, C_i)$. The surface of C_i is given by Equation (1). If we fill in line ℓ in (1), by substituting $x = \ell_x(t)$ and $y = \ell_y(t)$, and solve for t, we get a solution of the following form:

$$t_{1,2} = A(T, \Theta) \pm \sqrt{D(T, \Theta)}. \tag{8}$$

Here, $D(T, \Theta)$ is the discriminant whose sign indicates whether or not line ℓ intersects C_i. When $D(T, \Theta) = 0$, ℓ is tangent to C_i, hence $D(T, \Theta) = 0$ is an implicit equation for the set $DC_r(C, C_i)$. We can make this explicit by solving $D(T, \Theta) = 0$ for T. In Fig. 3 this function is plotted for various values of v_i (note that the function has a period of 2π). In each of these cases we see two sine-like curves (for $v_i = 1$, it is degenerate). They correspond with $DC_{rl}(C, C_i)$ and $DC_{rr}(C, C_i)$, respectively. The other departure curves $DC_{lr}(C, C_i)$ and $DC_{ll}(C, C_i)$ can be found when considering clockwise spirals.

Given a position (T, Θ) on the surface of cone C for which $D(T, \Theta) = 0$, the arrival time of the straight-line segment at cone C_i is given by $A(T, \Theta)$. The departure time of the segment is given by T. For some points along the departure curve $A(T, \Theta)$ is smaller than T. They correspond with bitangencies in the negative direction, i.e. the arrival time on C_i is smaller than the departure time at C. In the plots this is indicated by dashed curves. In the remainder of this paper these improper curves are ignored when we refer to departure curves.

We also have to take into account departure curves of $DC(C, g)$ associated with segments tangent to C and leading to the goal configuration g. In this case, we have to solve the system of equations $[\ell(t) = g]$ for T, to get a closed form for the departure curve.

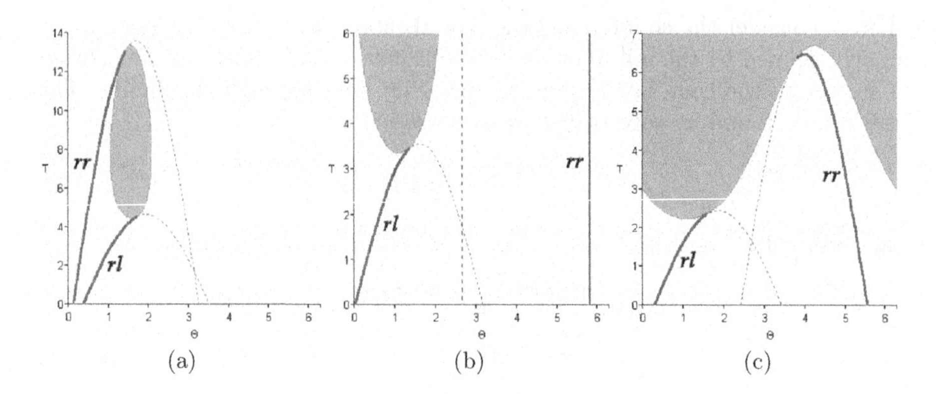

Fig. 3. Departure curves $DC_r(C, C_i)$ on the surface of C parametrized by angle Θ and time T for different values of v_i. (a) $v_i < 1$. (b) $v_i = 1$. (c) $v_i > 1$. The dashed curves are improper departure curves. The gray area, given by the inequality $(T \cos \Theta - p_{ix})^2 + (T \sin \Theta - p_{iy})^2 < (r_i + v_i T)^2$, is the region on the surface of C that is penetrated by C_i, i.e. these points are not collision-free.

4 A Naive Algorithm

With the notions introduced so far, we can devise a first, rather naive algorithm to find a shortest path amidst growing discs from some start configuration s to some goal configuration g. Our approach grows a tree of possible shortest paths that is rooted in the start configuration at time $t = 0$. A leaf is *expanded* if the length of its path from the start configuration is minimal among all leafs of the tree. To this end, each leaf is maintained in a *priority queue*, with a *key value* equal to its time coordinate (which equals the length of its path from s). The priority queue is initialized with the initial motions from the start configuration s that possibly belong to a shortest path. These are straight-line segments with slope $1/V$ leading either directly to the goal configuration, or to a tangency point on the surface of one of the cones. Some of these segments may intersect other cones, which would make them invalid, so only the collision-free segments are considered. The endpoint of each valid segment is put into the priority queue with a key corresponding to its t-value.

Now, the algorithm proceeds by handling the point with the lowest t-value in the queue (the front element of the queue). This point is either the goal configuration, in which case the shortest path has been found, or a point on the surface of a cone. In this latter, more general case we proceed by walking along a spiral about the cone. This spiral *either* runs into an obstacle (another cone), in which case there is no valid continuation of the path, *or* it encounters a departure curve on the surface of the cone. In this case there are two outgoing branches: (1) continuing along the spiral on the surface of the cone to find a next departure curve, and (2) departing for the other cone by a straight-line segment. If this latter segment is collision-free, its endpoint is inserted into the queue. Also for the first option an entry is enqueued.

This procedure is repeated until the goal configuration is popped from the priority queue. In this case the shortest path has been found, and can be read out if backpointers have been maintained during the algorithm. If the priority queue becomes empty, or if the front element of the queue has a time-value for which the goal configuration is not collision-free anymore (it is occupied by one of the growing discs), no valid path exists. In Algorithm 1, the algorithm is given in pseudocode.

Algorithm 1. SHORTESTPATHNAIVE(s, g)

```
1: Initialize priority queue Q with endpoints of all valid outgoing segments from s.
2: while Q is not empty do
3:    Pop the front element ⟨q, t⟩ from the queue.
4:    if the goal configuration is not collision-free anymore at time t then
5:       Path does not exist. Terminate.
6:    else if q = g then
7:       Shortest path found! Terminate.
8:    else
9:       q is on the surface of a cone, say C_i, so proceed along the spiral about C_i until
         it runs into another cone, or encounters a departure curve.
10:      if the spiral encounters a departure curve, say DC(C_i, C_j), then
11:         ⟨q', t'⟩ ← the intersection point of the spiral and the departure curve.
12:         ⟨q'', t''⟩ ← arrival point of the bitangent segment on the surface of C_j.
13:         Insert ⟨q', t'⟩ into Q.
14:         if segment ⟨q', t'⟩, ⟨q'', t''⟩ is collision-free then
15:            Insert ⟨q'', t''⟩ into Q.
16: Path does not exist.
```

In the above algorithm, we have to identify the spiral we are on (let us assume that it is a counterclockwise spiral), given a point on the surface of the cone (line 9). Let q be a point on the surface of some cone, say C_i, given in Euclidean coordinates (x, y, t). Then the corresponding coordinates (T, Θ) on the surface of C_i are given by $(T, \Theta) = (t, \arctan \frac{y - p_{iy}}{x - p_{ix}})$.

The spiral on the surface of C_i going through (T, Θ) is given by θ_0 as computed in Equation (6). Equation (5) then gives a function for the angle $\theta(t)$ along the spiral through (T, Θ). In line 10 of Algorithm 1, we wish to know whether the spiral encounters any departure curves. To this end, we should find the intersections of the spiral and the departure curves on the surface of C_i. Recall that we can deduce an implicit equation $D(T, \Theta) = 0$ for the departure curves of any pair of cones (see Equation (8)). The intersections are thus found by solving $D(t, \theta(t)) = 0$ for t.

When we have found an intersection for some value $t = T$ of the spiral and a departure curve of, say, $DC(C_i, C_j)$, we wish to know what kind of departure curve we have encountered. The arrival time at cone C_j when departed from time T is $A(T, \theta(T))$ (see Equation (8)). If this arrival time is smaller than T, the intersection can be ignored. If it is larger, we like to know whether the tangent straight-line segment arrives on the left side of C_j (and should be succeeded by a clockwise spiral on C_j), or on the right side of C_j (and should be succeeded by a counterclockwise spiral). This is determined by the derivative of $D(t, \theta(t))$ to t. If this derivative is negative at point T, we have arrived on the left side. If it is positive, we have arrived on the right side. The exact arrival location on the surface of cone C_j is given by $\ell(A(T, \theta(T))$ (see Equation (7)). From this information we can deduce the parameters defining the spiral on C_j on which we have arrived.

5 An Efficient Algorithm

The algorithm described above will indeed find a shortest path to the goal within a finite amount of time. However, in order to have a bound on the running time we must define *nodes* that can provably be visited only once in a shortest path, such that we can do *relaxation* on them as in Dijkstra's algorithm [7]. We will show that this is easy to achieve in the restricted case where all discs have equal growth rates, and present an $O(n^3 \log n)$ algorithm (n being the number of discs). For the general case this problem is left open, but we will present an algorithm that is very fast in practice, by pruning large parts of the search tree.

5.1 Discs with Equal Growth Rates

If a point $q = (T, \Theta)$ on the surface of a cone has been visited during the search for a shortest path to the goal, all points on the cone that are reachable from q by following some collision-free path with a velocity less than the maximal velocity V (i.e., $\frac{\|(\delta x, \delta y)\|}{\delta t} < V$), can never lie on a shortest path from the start to the goal (this follows directly from Theorem 1). These points are contained in the wedge formed by the clockwise and counterclockwise spiral going through q (see Fig. 4; a spiral appears as an exponential function in the ΘT-coordinate frame), as we know that on the spirals $\frac{\|(\delta x, \delta y)\|}{\delta t} = V$. We call this region the *wedge region* of q.

Let us consider the *arrangement* [1] on the surface of a cone containing all (proper) departure curves and all obstacle regions (other cones penetrating the surface) on that cone (see Fig. 5 for an impression). Note that the departure curves may be subdivided into a number of *collision-free intervals* by the obstacle regions. In case all discs have the same growth rate, say $v_i = v < V$ for all i, these intervals satisfy an interesting property (see Fig. 3(b)): let (T, Θ) be a point on some departure curve interval, then all points (T', Θ') on the same interval for which $T' > T$ are within the wedge region of (T, Θ). To prove this, we must show that for the departure curves hold that $\frac{\|(\delta x, \delta y)\|}{\delta t} \leq V$. As the proof is rather technical, we omit it here.

This means that these departure curve intervals can serve as *nodes* in our Dijkstra-algorithm. Only the path arriving earliest in an interval can contribute to a shortest path. Paths arriving later in the interval cannot be part of the shortest path, because the path arriving earliest in the interval can be extended with a traversal along the interval to end up at the same position (and time) as the path arriving later in the interval.

Each node (an interval) has two outgoing *edges*. Let the interval be a segment of a departure curve of $DC(C_i, C_j)$, then the first edge is a spiral segment to the next departure curve on the surface of C_i, and the second edge consists of a bitangent straight-line segment and a spiral segment and arrives in the first departure curve encountered on the surface of C_j. For the first edge, which stays on the cone, we have to determine the next departure curve that is encountered if we proceed by moving along the spiral about the cone. This can be done efficiently using the arrangement, if we have computed its *trapezoidal map* [1], where the sides of the trapezoids are spiral segments.

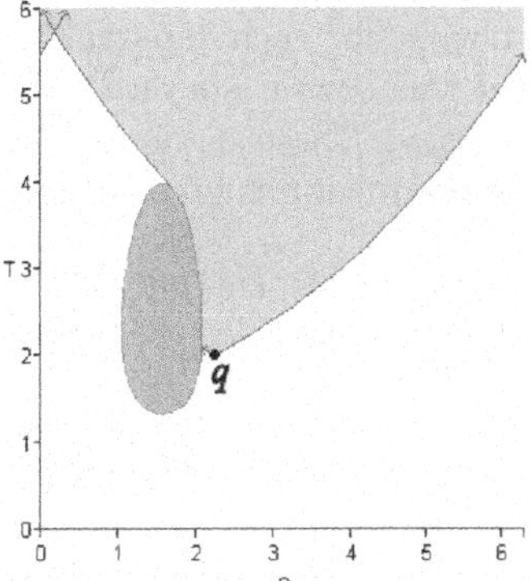

Fig. 4. The region (light grey) on the surface of a cone that is reachable from point q by paths with $\frac{\|(\delta x, \delta y)\|}{\delta t} \leq V$. The dark grey area is an obstacle.

For the second edge, which traverses to another cone, we have to determine what the first departure curve is we will encounter there. This can be done efficiently using the arrangement we have computed on that cone. Using a point-location query, we can determine in what cell of the arrangement the straight-line segment has arrived, and using the trapezoidal map we know what the first departure curve is we will encounter if we proceed from there.

Finally, we must ascertain that each edge is collision-free with respect to the other cones. Spiral segments may collide with other cones if these penetrate the

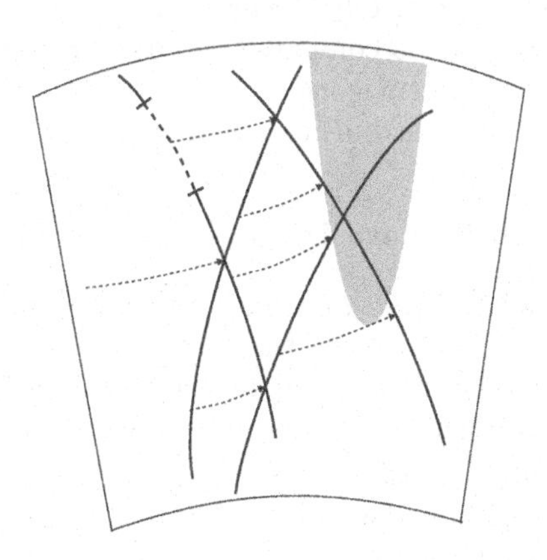

Fig. 5. An impression of an arrangement on the surface of the cone. The thick lines are the departure curves, of which one has a shadow interval (dashed). The thin dashed lines are spiral segments that delimit trapezoidal regions that have the same next departure curve or collision (only the counter-clockwise spirals are shown). The gray area depicts an obstacle area of another cone penetrating the surface, and cutting several departure curves into two intervals.

spiral's cone surface. Since obstacle areas are incorporated into the arrangement, such collisions are easily detected. Straight-line segments may collide with any cone, so for each departure curve and each cone, we calculate the *shadow interval* this cone casts on the departure curve, in which a departure will result in collision. These shadow intervals are stored in the arrangement as well. In Fig. 5, an impression is given of how such an arrangement might look.

Theorem 4. *The algorithm to compute a shortest path amidst n growing discs with equal growth rates runs in* $O(n^3 \log n)$ *time.*

Proof. For each pair of cones there are $O(1)$ departure curves. Since there are $O(n^2)$ pairs of cones, there are $O(n^2)$ departure curves in total. Each of the departure curves can be segmented into at most $O(n)$ intervals, as there are at most $O(n)$ cones intersecting the departure curve (each cone can split the departure curve into at most two segments). Hence, there are $O(n^3)$ departure curve intervals. Each departure curve interval has $O(1)$ outgoing edges, making a total of $O(n^3)$ edges.

The complexity of Dijkstra's algorithm is known to be $O(N \log N + E)$ where N is the number of nodes, and E the number of edges. Each edge requires some additional work. Firstly, we have to find the departure curve interval in which it will arrive, by doing a point-location query in the trapezoidal map of one of the arrangements. This takes $O(\log n)$ time. Further, we must determine whether an edge is collision-free. Using the shadow intervals stored at the departure curves, this can be done in $O(\log n)$ time as well. Thus, as both N and E are $O(n^3)$, Dijkstra's algorithm will run in $O(n^3 \log n)$ time in total.

Computing the arrangements and their trapezoidal maps takes $O(n^2)$ time per cone, as there are $O(n)$ departure curves on each cone, and $O(n)$ intersection areas of other cones. As there are $O(n)$ cones, this step takes $O(n^3)$ time in total. All the shadow intervals can be computed in $O(n^3)$ time as well, as there are $O(n^2)$ departure curves and $O(n)$ cones.

Overall, we can conclude that our algorithm runs in $O(n^3 \log n)$ time. □

5.2 General Case: Discs Have Different Growth Rates

In the general case, where the discs may have different growth rates, the problem becomes much harder. We can follow the same approach as above, but let us look at what happens to the slope of the departure curves in this case (see Figs. 3(a) and (c)). In the case where the arrival cone has a slower growth rate, the departure curves (provably) satisfy $\frac{\|(\delta x, \delta y)\|}{\delta t} \leq V$ (see Fig. 3(a)). However, in the case where the arrival cone has a faster growth rate (Fig. 3(c)), it is clear that this is not the case. The departure curve DC_{rr} is horizontal at some point, meaning that $\frac{\|(\delta x, \delta y)\|}{\delta t} = \infty$. Hence, we cannot define intervals on these departure curves that serve as nodes in the search process.

We can still use Algorithm 1 for the general case, but a problem is that this algorithm considers many branches in the search tree of which we know that they will not lead to a shortest path. For instance, it lets the spirals wind around the cones forever, thereby encountering many departure curves, which in turn generate other spirals on other cones. Hence, it lets the size of the search tree blow up quickly.

In order to have an algorithm that runs fast in practice, we need to prune these useless branches of the search tree. The key observation we use for this is that a point (T, Θ) on the surface of cone C_i cannot be part of shortest path if we have visited (T', Θ) already (where $T' < T$) *and* the vertical line segment on the surface of the cone between (T', Θ) and (T, Θ) is collision-free. This is because (T, Θ) is then in the wedge region of (T', Θ) (note that the velocity $\frac{\|(\delta x, \delta y)\|}{\delta t}$ along the vertical line segment equals $v_i < V$). Hence a spiral encountering (T, Θ) need not be expanded any further.

To implement this practically, we only do this test for a constant number of Θ's. To this end, we augment Algorithm 1 by choosing a small constant ε, and drawing $\frac{2\pi}{\varepsilon}$ evenly distributed vertical lines on the surface of each cone. These vertical lines are segmented into collision-free intervals by obstacle regions on the surface. Now, these intervals will serve as *nodes* in our practical algorithm on which we perform *relaxation.*

This means that if we walk along a spiral on the surface of a cone, and the spiral crosses a vertical line, we have to check whether this spiral is the first to arrive in the particular interval. If not, this spiral can never be part of a shortest path, for the same reasons as above. Thus, this branch of the search tree can be pruned.

The smaller ε is chosen, the sooner the spirals can be pruned, and hence the smaller the size of the search tree will be. On the other hand, a smaller ε also causes the algorithm to perform more (costly) relaxation checks, with diminishing returns. So ε should not be chosen too small. Even though we are unable to bound the running time of this algorithm in terms of the number of discs (n) or the value of ε, it turns out to be very fast in practice, as we will see next.

5.3 Implementation Details

We created a fast implementation of Algorithm 1, augmented with the pruning heuristic presented above. We did not create an arrangement of all vertical lines and all obstacle regions on each cone. This would take too much time. Instead, we maintain for each vertical line m one time-value at which it was last visited, say t_m. Given the order in which the points are considered in the priority queue, we know that when a point q is popped from the queue, it has a higher time value than any point previously considered. So, if point q lies on vertical line m, its time value q_t is larger than the time-value t_m of the point previously considered on that line. If the line segment between t_m and q_t on m is collision-free, q is in a previously visited interval, and hence this point is not expanded. However, if the vertical line segment between these two points is not collision-free, point q is the first to arrive in a new interval, and its outgoing edges must be inserted into the priority queue.

From this moment on, q_t is set as the time value attached to the vertical line m, as we know that no point below q_t will be considered anymore.

Outgoing edges of a point q on a vertical line segment are a spiral segment to the next vertical line, and –in case this spiral segment crosses one or more departure curves– segments to vertical lines on other cones. In our implementation, the intersection between spiral segments and departure curves is found using a combination of two approximate root-finding algorithms [3].

Collision-checking straight-line segments is done by testing them for intersections with all cones, except the ones they are tangent to. We approximate a spiral segment between two consecutive vertical lines by one or more small straight-line segments, and collision-check them in the same way (in our implementation, we use a single straight-line segment, as the radial distance ε between two consecutive vertical lines is small).

Finally, the Dijkstra paradigm was replaced by an equally suited A*-method [7], that is faster in practice as it focusses the search to the goal. It adds a lower bound estimate of the distance to the goal to the key-value of each point in the priority queue. In our implementation, the lower bound estimate is simply the Euclidean distance divided by the maximal velocity.

6 Experimental Results

We created an interactive application for planning paths amidst growing discs. The properties of the growing discs (position, size, growth rate) can be changed by the user, and on-the-fly a new path is computed. From this application we report results. Experiments were run on a Pentium IV 3.0GHz with 1 GByte of memory. The value of ε was optimized and fixed at $\frac{2\pi}{40}$.

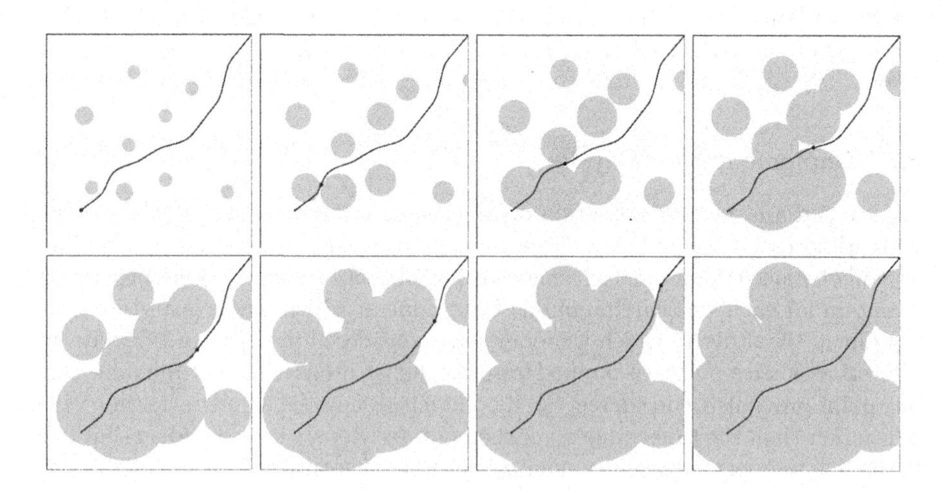

Fig. 6. A shortest path amidst 10 growing discs. A small dot indicates the position along the path at $t = 0, 1, \ldots, 7$. The pictures were generated by our application.

We report the running times of the algorithm for a varying number of discs. As the running time of the algorithm does not only depend on the number of obstacles, but also on the exact configuration of the discs, and how well the A* method manages to focus the search, etc., we averaged the running times over various positions of the start configuration for each experiment. In Fig. 7 the results are given.

What first of all can be seen from the results is that our implementation is very fast. Even for 15 growing discs, the running time is only 0.0042 seconds, well within real-time requirements. We did not show results for more than 15 discs, as it appeared to be difficult to find sensible setups with this many discs that still contain a valid path to the goal. From the figure it seems that the running time is more or less quadratically related to the number of discs. This is what we expected based on the implementation. In Fig. 6, snapshots are shown of a shortest path amidst 10 growing discs.

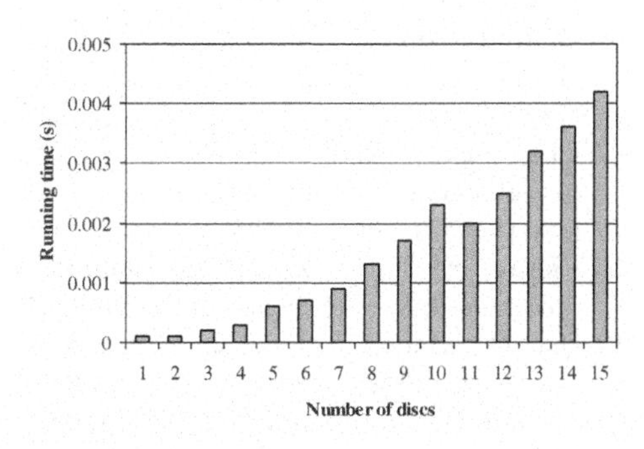

Fig. 7. Results of our experiments

7 Conclusion

In this paper we presented an algorithm for computing shortest paths (minimum time paths) amidst discs that grow over time. A growing disc can model the region that is guaranteed to contain a moving obstacle of which the maximal velocity is given. Hence, using our algorithm, paths can be found that are guaranteed to be collision-free in the future, regardless of the behavior of the moving obstacles. As the regions grow fast over time, a new path should be planned from time to time –based on newly acquired sensor data– to generate paths with more appealing global characteristics. Our implementation shows that such paths can be generated very quickly. A great advantage over other methods is that this replanning can be done safely. The old path that is still used *during* replanning is guaranteed to be collision-free. A requirement though, is that the robot has a higher maximal velocity than any of the moving obstacles.

A drawback of the method we presented is that a path to the goal often does not exist. This occurs when the goal is covered by a growing disc before it can be reached. A solution to this problem would be to find the path that comes closest to the goal. It seems that this can easily be incorporated into our algorithm. Other possible extensions include allowing obstacles with different shapes (other than discs), and fixed obstacles in the environment, but they are still subject of ongoing research.

References

1. de Berg, M., van Kreveld, M., Overmars, M., Schwarzkopf, O.: Computational Geometry, Algorithms and Applications, ch. 6 and 8, 2nd edn. Springer, Berlin, Heidelberg (2000)
2. van den Berg, J., Ferguson, D., Kuffner, J.: Anytime path planning and replanning in dynamic environments. In: Proc. IEEE Int. Conf. on Robotics and Automation (ICRA) (2006)
3. Burden, R.L., Faires, J.D.: Numerical analysis, ch.2, 7th edn. Brooks/Cole, Pacific Grove (2001)
4. Chang, E.C., Choi, S.W., Kwon, D.Y., Park, H., Yap, C.K.: Shortest path amidst disc obstacles is computable. In: Proc. Ann. Symposium on Computational Geometry (SoCG), pp. 116–125 (2005)
5. Fiorini, P., Shiller, Z.: Motion planning in dynamic environments using velocity obstacles. Int. J. of Robotics Research 17(7), 760–772 (1998)
6. Hsu, D., Kindel, R., Latombe, J., Rock, S.: Randomized kinodynamic motion planning with moving obstacles. Int. J. of Robotics Research 21(3), 233–255 (2002)
7. LaValle, S.M.: Planning Algorithms, ch. 2. Cambridge University Press, New York (2006)
8. Mitchell, J.S.B.: Geometric shortest paths and network optimization. In: Handbook of Computational Geometry, pp. 633–701. Elsevier Science Publishers, Amsterdam (2000)
9. Petty, S., Fraichard, T.: Safe motion planning in dynamic environments. In: Proc. IEEE Int. Conf. on Intelligent Robots and Systems (IROS), pp. 3726–3731 (2005)
10. Vasquez, D., Large, F., Fraichard, T., Laugier, C.: High-speed autonomous navigation with motion prediction for unknown moving obstacles. In: Proc. IEEE Int. Conf. on Intelligent Robots and Systems (IROS), pp. 82–87 (2004)
11. Weisstein, E.W.: Logarithmic Spiral. In MathWorld – a Wolfram web resource, `http://mathworld.wolfram.com/LogarithmicSpiral.html`

Planning Among Movable Obstacles with Artificial Constraints

Mike Stilman and James J. Kuffner

Carnegie Mellon University
5000 Forbes Ave.
Pittsburgh, PA 15213 USA
{robot,jkuffner}@cmu.edu

Summary. This paper presents artificial constraints as a method for guiding heuristic search in the computationally challenging domain of motion planning among movable obstacles. The robot is permitted to manipulate unspecified obstacles in order to create space for a path. A plan is an ordered sequence of paths for robot motion and object manipulation. We show that under monotone assumptions, anticipating future manipulation paths results in constraints on both the choice of objects and their placements at earlier stages in the plan. We present an algorithm that uses this observation to incrementally reduce the search space and quickly find solutions to previously unsolved classes of movable obstacle problems. Our planner is developed for arbitrary robot geometry and kinematics. It is presented with an implementation for the domain of navigation among movable obstacles.

1 Introduction

A robot that can move obstacles out of its way is capable of more autonomous tasks. For example, in Figure 1, the robot cannot directly plan a path to the goal. By manipulating four objects, the robot changes its configuration space and opens free space for a path. This capacity comes at the cost of computational complexity. We explore a method for allowing robots to constrain their action space and create computationally manageable search spaces.

A simple path planning task in the movable obstacle domain becomes a complex manipulation planning problem with a partially specified goal. The robot can change its own configuration and the configurations of other objects. Each of these changes alters the workspace of the robot by increasing or decreasing the free space for future motions. The size of the search space is exponential in the number of movable objects. Furthermore, the branching factor of forward search is linear in the number of all possible world interactions [1]. A simplified variant of this domain involving only one movable obstacle is NP-hard.[2] More recent results demonstrated NP-hardness results for trivial problems where square blocks are pushed in block-size increments on a planar grid.[3]

In this paper we show that allowing one interaction with each object and reverse planning let the robot constrain its initial search space. We do not escape the curse of dimensionality. The proposed method of *artificial constraints* enables fast heuristic search in a domain where standard proximity heuristics

S. Akella et al. (Eds.): Algorithmic Foundation of Robotics VII, STAR 47, pp. 119–135, 2008.
springerlink.com

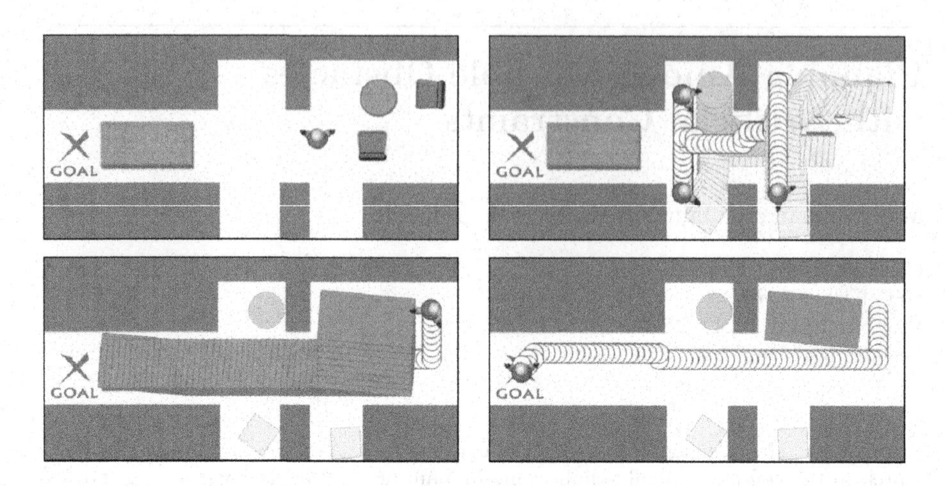

Fig. 1. A simulated solution to a problem of Navigation Among Movable Obstacles. The robot is instructed to reach the goal. After constructing a plan, it first moves the three smaller objects to the niches. The robot uses the new free space to move the table. Finally it clears a path and navigates to the goal.

provide little or no insight. We demonstrate that our method is directly applicable to robot tasks in a simulated domain. Furthermore, we introduce a problem formulation and runtime analysis that form a basis for future work.

2 Related Work

Obstacles moving along specified trajectories is a problem addressed by bounding the velocities of the obstacles and augmenting the configuration space with time.[4] A point in the free space ensures that a configuration is valid at the given time in which it takes place. This approach has been extended to kinodynamic domains, [5] as well as real-time deformable plans.[6] These algorithms do not allow the robot to affect the world.

Initial work in coordinating robot motions can be found in [7, 8, 9]. Most recent research that deals with robots repositioning *multiple objects* is in assembly planning. Assembly planners focus on separating a collection of parts and typically ignore the robot/manipulator. Domain operators also allow unassembled parts to be removed to "infinity."[10, 11]

In the *movable obstacle* domain, objects cannot move unless manipulated by the robot. The motion of the objects is constrained to the workspace of the robot, while the robot is constrained to move along collision-free paths. *Rearrangement planning* is the domain that is most closely related. [12, 13, 14] The final configurations of all objects are specified, and the robot must find coordinated transport paths. For instance, when a manipulation path to the goal collides with other objects, [14] heuristically selects intermediate configurations for interfering obstacles.

In our domain, final configurations for objects are unspecified. Hence the robot must decide not only where to move objects but which objects must be moved. [15] searched a graph of robot paths, allowing objects to be pushed away from the robot trajectory. This method is effective on small problems, but easily encounters local minima. [1] and [16] propose to consider joining regions of robot free space and constructing graphs of interfering obstacles respectively. Neither planner handles objects that interfere with the motion of other objects. [17] ignores the robot, but offers some insight into graph-based chronological and spatial coordination of movable objects.

[15, 1, 16, 14] were developed for mobile robots. Our work addresses the problem generally for any kinematic structure of the robot. This is important when considering manipulation problems where robot geometry varies significantly for different portions of the workspace. We will base our domain on configuration space operators first described in [12]. Our constraint approach is related to [8], however we do not assume a priority on object motions, rather we must search the space of object choices and orders.

3 Movable Obstacle Domain

In this section, we develop a geometric model for movable obstacles. Our choice of space and operators make the presented approach general for any robot kinematics in the framework of rigid body motion and prehensile manipulation. Section 7 gives an example of how the tools developed in this framework can be applied to a specific robot problem.

Consider a path planning problem in a 2D or 3D Euclidian space that contains a set of fixed objects $\mathbf{O}_F = \{F_1, \ldots, F_f\}$ and a set of movable objects $\mathbf{O}_M = \{O_1, \ldots, O_m\}$. The space also contains an n degree of freedom robot, R. While paths are not explicitly parameterized by time, we will use the variable t to refer to a chronological ordering on states and operations. A world state at time-step t is the tuple consisting of t, the robot configuration r^t and the configuration q_i^t of each movable object: $W^t = (t, r^t, q_1^t, q_2^t, \ldots, q_m^t)$.

Let $\mathcal{C}_W$ be the space of all possible W^t. We permit the robot to move one object at a time. Consequently, we are interested in subspaces or *slices* of $\mathcal{C}_W$:

$\mathcal{C}_R(W^t) = (\{r\}, q_1^t, q_2^t, \ldots, q_m^t)$ - the slice of robot configurations, and
$\mathcal{C}_{O_i}(W^t) = (r^t, q_1^t, q_2^t, \ldots, \{q_i\}, \ldots, q_m^t)$ - configurations of object O_i.

Observe that any slice is parameterized by the positions of other objects. Following [18] we define free space to be the subspace of collision free configurations. First consider the configuration space obstacles (CO). Let $A(q) = \{x \in \mathbb{R}^k | x \text{ is a point of object } A \text{ in configuration } q\}$. For any set of points S in $\mathbb{R}^k$, a configuration space obstacle in $\mathcal{C}_B$ is the set: $CO_B(S) = \{p \in \mathcal{C}_B | B(p) \cap S \neq \emptyset\}$. Let q be a configuration of object A and p be a configuration of object B. Since no two objects can occupy the same space in $\mathbb{R}^n$, CO is symmetric:

$$p \in CO_B(A(q)) \Rightarrow B(p) \cap A(q) \neq \emptyset \Rightarrow q \in CO_A(B(p)) \tag{1}$$

To simplify notation, let $CO_R^{q_i} = CO_R(O_i(q_i))$ and $CO_{O_j}^{q_i} = CO_{O_j}(O_i(q_i))$ represent obstacles due to O_i in $\mathcal{C}_R$ and $\mathcal{C}_{O_j}$ respectively.

Let $\overline{CO_A(B)}$ be the complement of $CO_A(B)$ in $\mathcal{C}_A$. The free space of a movable object, $\mathcal{C}_A^{free}(W^t)$, is the set of configurations where the object is not in collision with fixed or movable obstacles.

$$\mathcal{C}_{O_i}^{free}(W^t) = \bigcap_k \overline{CO_{O_i}(F_k)} \bigcap_{O_j \neq O_i} \overline{CO_{O_i}^{q_j^t}} \tag{2}$$

Collisions between a moving object and the robot are treated separately from Eq. 2 since the motion of an object also implies the motion of the robot. $\mathcal{C}_R^{free}(W^t)$ is expressed analogously in terms of CO_R.

In spaces with external forces, such as gravity, objects will not remain static in arbitrary configurations. Manipulated objects must be released in *placement* configurations $\mathcal{C}_{O_i}^{place}(W^t) \subseteq \mathcal{C}_{O_i}^{free}(W^t)$. When solving three dimensional problems, we propose form closure to develop this set. In our two dimensional examples, we assume gravity is orthogonal to the object plane and hence $\mathcal{C}_{O_i}^{place}(W^t) = \mathcal{C}_{O_i}^{free}(W^t)$.

Having defined the sets of free configurations for the robot and movable objects, we now address the allowable interactions between the robot and the environment. Following [12], we define two parameterized operators on the $\mathcal{C}_{space}$: *Transit* and *Transfer*. *Transit* creates a path for the robot. *Transfer* represents the motion of the robot and a single movable object.

TRANSIT: We first define a continuous path τ in the configuration space of the robot: $\tau : [0,1] \rightarrow r$ for $r \in \mathcal{C}_R$. $\tau(r_i, r_j)$ will shorten the notation for a path where $\tau[0] = r_i$ and $\tau[1] = r_j$. The $Transit$ operator is a function that maps a world state and path to another world state.

$$Transit : (W^t, \tau(r^t, r^{t+1})) \rightarrow W^{t+1} \tag{3}$$

This operator is valid if and only if the following condition holds:

$$\tau(s) \in \mathcal{C}_R^{free}(W^t) \qquad \forall s \in [0,1] \tag{4}$$

TRANSFER: When an object is rigidly grasped, its configuration is fully determined by a transformation of the generalized pose of the robot end effector. $K : \mathcal{C}_R \rightarrow x$ $(x \in \mathbb{R}^n)$ is the kinematic mapping of robot configurations to end effector positions/orientations. We will consider a discrete or sampled set of grasps for each movable object: $GS(O_i) = \{\mathbf{G}_{O_i}\}$. Each $\mathbf{G}_{O_i}$ is a rigid transform from the robot pose to a configuration of O_i. $\mathbf{G}_{O_i}(K(r)) = q$ states that the robot configuration r grasps O_i in configuration q.

For any grasp $\mathbf{G}_{O_i}^k \in GS(O_i)$, the $Transfer$ operator maps a world state and a path in $\mathcal{C}_R$ to a new state where the robot and an object are displaced:

$$Transfer : (W^t, O_i, \mathbf{G}_{O_i}^k, \tau(r^t, r^{t+1})) \rightarrow W^{t+1} \tag{5}$$

Notice that for any robot path τ we can compute τ_{O_i} for the object as the path $\tau_{O_i} = \mathbf{G}_{O_i}(K(\tau))$. A valid $Transfer$ operator must satisfy:

$$\tau(s) \in \mathcal{C}_R^{free}(W^t) \cup CO_R^{q_i^t} \quad \tau_{O_i}(s) \in \mathcal{C}_{O_i}^{free}(W^t) \qquad \forall s \in [0,1] \tag{6}$$

$$\tau_{O_i}(0) = q_i^t \tag{7}$$

$$\tau_{O_i}(1) \in \mathcal{C}_{O_i}^{place}(W^t) \tag{8}$$

$$R(\tau(s)) \cap O_i(\tau_{O_i}(s)) = \emptyset \qquad \forall s \in [0,1] \tag{9}$$

Eq. 8 requires that the final configuration of the object be statically stable. Eq. 9 ensures that the robot does not collide with obstacle O_i.

4 Motions of Multiple Objects

The problems we are interested in are realistic domains with numerous movable objects. Due to the dimension of these spaces, finding meaningful sub-domains and heuristics takes precedence over completeness. In earlier work [1] we observed that $\mathcal{C}_R^{free}$ can be partitioned into disjoint subsets $\{C_1, C_2, \ldots, C_d\}$ such that a robot in configuration $r_i \in C_i$ can access any configuration in C_i via a *Transit* operation but no configuration in $C_j \neq C_i$.

Our planner detected objects that could be moved in order to give the robot access to other components of $\mathcal{C}_R^{free}$. For two subsets $C_i, C_j \in \mathcal{C}_R^{free}(W^t)$ and a border obstacle O_l we pursued a k-length sequence of *Transit* and *Transfer* operations that yield a merged component $C_{mrg} \subset \mathcal{C}_R^{free}(W^{t+k})$:

$$C_{mrg} = (C_i \cup C_j \bigcup CO_R^{q_l^t}) \bigcap \overline{CO_R^{q_l^{t+k}}} \tag{10}$$

$$\forall r_i, r_j \in C_{mrg} \text{ there exists } \tau(r_i, r_j) \text{ s.t. } \forall s(\tau[s] \in C_{mrg}) \tag{11}$$

Based on the concept of connecting free space components, we defined the class of *linear problems* (*LP*). A problem has a linear solution when there exists a sequence of free space components $\{C_1, C_2, \ldots, C_n\}$ such that merging adjacent components C_i and C_j does not constrain the C_{space} required to merge adjacent C_k and C_l where $(i < j \leq k < l)$. [1] presented a resolution complete algorithm for problems in L_1, where only one object must be displaced to merge two components.

Extending [1] to L_k problems where up to k objects may be moved to connect free space components is challenging even for $k = 2$. In the best case, every robot path between two components would pass through two objects, O_1 and O_2, allowing the planner to locally search the joint motion space of size $|O_1| \times |O_2|$. However, as seen in Figure 1, the path between C_i, $C_j \in \mathcal{C}_R^{free}$ might only pass through one object (the table). A complete L_2 planner must consider all possibilities for the choice of second object. In general, for L_k problems, we may need to enumerate 2^{k-1} possible sets of objects that do not directly interfere with a path to the goal.

4.1 Proposed Hierarchy

In order to manage the increased complexity when local search requires the motion of multiple objects, we propose further classification of the movable obstacle domain to *monotone* plans. In assembly planning, monotone plans refer to plans where each application of an operator yields a subassembly that is part of the final assembly [11]. We do not enforce a final assembly and define *monotone plans* as those in which a transferred object cannot be moved again:

$$W^{t+1} = Transfer(W^t, O_i, \mathbf{G}^k_{O_i}, \tau) \quad \Rightarrow \quad q_i^T = \tau_{O_i}(1) \quad (T > t) \tag{12}$$

Monotone search decouples the joint motion space of objects into individual path plans. The search must decide which objects to displace, the $Transfer$ paths for each object and the ordering of object motion.

Notice that any plan can be expressed as a sequence of monotone plans:

$$Plan_{NM} = \ldots, \tau_1, (O_i, \tau_2), \tau_3, (O_j, \tau_4), \tau_5, (O_i, \tau_6), \tau_7, \ldots \equiv$$
$$Plan_{M_1} = \ldots, \tau_1, (O_i, \tau_2), \tau_3, (O_j, \tau_4), \tau_5 \text{ and } Plan_{M_2} = (O_i, \tau_6), \tau_7, \ldots \tag{13}$$

Let W^6 be the world state after the operation $Transit(W^5, \tau_5)$, prior to the second displacement of O_i. We refer to W^6 as an intermediate world state. A problem can be characterized in its *non-monotone degree* by the number of intermediate states necessary to construct a sequence of monotone plans.

We propose the following classes of problems:

- L_k Linear problems where components of C_R^{free} can be connected independently. k is the maximum number of objects that must be displaced to connect two components.
- NL Non-linear problems that require the planner to consider interactions between keyholes.
- M Monotone problems where each object needs to only be displaced once throughout the plan.
- NM_i Non-monotone problems that can be expressed as i monotone problems with intermediate states.

A planner can operate in the space of one or two of these classes. For instance a planner in L_3NM_6 would seek linear solutions that require manipulating at most three objects and using six intermediate states to merge two free space components. Our proposed algorithm operates in L_kM.

5 Artificial Constraints

The monotone class of problems helps organize the study of movable objects. It still preserves a number of the computational challenges of our domain. The planner determines a subset $\{O_1, \ldots, O'_m\} = \mathbf{O}'_M \subset \mathbf{O}_M$ of movable objects to displace. It constructs a valid set of paths $\{\tau_{O_1}, \ldots, \tau_{O'_m}\}$ for displacing the objects and $\{\tau_1, \ldots, \tau_{m+1}\}$ $Transit$ operations between grasps. Additionally, the planner decides an ordering for object motion. This section will analyze the retained problem complexity and present our solution.

5.1 Complexity of Forward Search

Suppose we were to perform a standard forward search of obstacle motion. In the monotone case, we do not need to consider all possible $Transit$ and $Transfer$ paths. At each time-step t we select an object O_i for motion and a goal configuration $q_i^{t+2} \in \mathcal{C}_{O_i}^{place}(W^{t+2})$. We verify that there exists a robot configuration $r^{t+1} \in \mathcal{C}_R^{free}(W^t)$ that satisfies $q_i^t = \mathbf{G}_{O_i}^k(K(r^{t+1}))$ for some k. Additionally, we check the existence of valid paths:

$$\begin{aligned} & Transit(W^t, \tau_1(r^t, r^{t+1})) \text{ and} \\ & Transfer(W^{t+1}, O_i, \mathbf{G}_{O_i}^k, \tau_2(r^{t+1}, r^{t+2})) \\ \text{such that} \quad & q_i^{t+2} = \mathbf{G}_{O_i}^k(K(r^{t+2})) \end{aligned} \tag{14}$$

Assume that verification could be performed in constant time, and that the number of placements is $O(d^n)$, where d is the resolution of each of the n dimensions of $\mathcal{C}_{O_i}$. Typically, $n = 3$ or 6. At $t = 0$, this algorithm would select from m objects and d^n configurations for each object: $O(md^n)$. Expanding the search to depth 2, there are now $m - 1$ objects and d^n placements for each object: $O(md^n \times (m-1)d^n)$. This algorithm has an asymptotic runtime of $O(md^n \times (m-1)d^n \times ... \times 2d^n \times d^n) = O(m!d^{nm})$.

The difficulty lies in finding an informed heuristic for the exponentially large space of object placements. A *good* placement for the object is one that respects the motion of subsequent obstacles. Since the motion of future objects is postponed in the search, good placements are unknown.

5.2 Reverse Search

Reverse planning is common in *assembly* problems. However, the implementation and motivation of reverse planning is different in our domain. Assembly planners have fixed goal configurations for all objects in which the motion of the objects is typically highly constrained. Consequently, the reverse search space has a much smaller branching factor due to *actual constraints.*

In the domain of movable obstacles, the final configuration is not predetermined, hence object motion must be planned from the initial state. Search reversal is performed in regard to the ordering of object motions (i.e. a $Transfer$ of the last object is performed as the first step of the search). At the start of search, the branching factor is large due to the non-existence of goal configurations. However, *artificial constraints* yield significant space reduction when searching prior motions.

Artificial Constraints

Let W^0 be the initial world state. Assume that at some future time step $t > 0$, the robot will perform a $Transit(W^t, \tau^t(r^t, r^{t+1}))$ operation. This operation is valid only when $\tau^t(s) \in \mathcal{C}_R^{free}(W^t)$ (Eq. 4). Let q_j^t be the configuration of obstacle O_j at time t. By the definition of free configuration space (Eq. 2):

$$\tau^t(s) \notin CO_R(O_j(q_j^t)) \tag{15}$$

Due to the symmetry of CO (Eq. 1), we can invert this relationship.

$$q_j^t \notin CO_{O_j}(R(\tau^t(s))) \qquad \forall s \in [0,1] \tag{16}$$

The robot motion along τ^t defines a *swept volume* in $\mathbb{R}^n$. Let $V(\tau^t)$ be the volume of points occupied by the robot during its traversal of τ^t:

$$Transit(W^t, \tau^t(r^t, r^{t+1})) \rightarrow V(\tau^t) = \bigcup_{s\in[0,1]} R(\tau^t(s)) \tag{17}$$

$$q_j^t \notin CO_{O_j}(V(\tau^t)) \tag{18}$$

Analogously, if we assume a valid $Transfer(W^t, O_i, \mathbf{G}_{O_i}^k, \tau(r^t, r^{t+1}))$ at step t ($t > 0$), we would define $V(\tau^t, O_i, \mathbf{G}_{O_i}^k)$ as the volume of points occupied by the robot and the object during their joint motion:

$$Transfer(W^t, O_i, \mathbf{G}_{O_i}^k, \tau^t(r^t, r^{t+1})) \rightarrow$$
$$V(\tau^t, O_i, \mathbf{G}_{O_i}^k) = \bigcup_{s\in[0,1]} [R(\tau^t(s)) \cup O_i(\tau_{O_i}^t(s))] \tag{19}$$

From Eq. 6 we find

$$\tau^t(s) \notin CO_R(O_j(q_j^t)) \qquad \tau_{O_i}^t(s) \notin CO_{O_i}(O_j(q_j^t))) \qquad (j \neq i). \tag{20}$$

Due to the symmetry of CO:

$$q_j^t \notin CO_{O_j}[R(\tau^t(s)) \cup O_i(\tau_{O_i}^t(s))] \qquad \forall s \in [0,1] \tag{21}$$

$$q_j^t \notin CO_{O_j}(V(\tau^t, O_i, \mathbf{G}_{O_i}^k)) \tag{22}$$

Eq. 18 and 22 indicate that the swept volume of any $Transit$ or $Transfer$ operation in W^t places a constraint on the configurations of movable objects: $V^t = V(\tau^t)$ or $V(\tau^t, O_i, \mathbf{G}_{O_i}^k)$ respectively. Since objects remain fixed unless moved by $Transfer$, then for some final time T:

$$q_j^0 \notin CO_{O_j}(V^T) \text{ or there exists a time } t (0 \leq t < T) \text{ such that}$$
$$Transfer(W^t, O_j, \mathbf{G}_{O_j}^k \tau^t(r^t, r^{t+1})) \text{ and } \tau_{O_j}^t(1) \notin CO_{O_j}(V^T) \tag{23}$$

Due to our assumption of monotone plans, if the initial configuration of an obstacle conflicts with V^T, there is exactly one $Transfer$ operator that displaces it to a non-conflicting configuration at some time-step t ($t < T$).

6 Algorithm

In order to apply the method of artificial constraints, our planner consists of two modules: *obstacle identification* and *constraint resolution*. Obstacle identification decides the last object that will be manipulated prior to reaching the goal or a subgoal. Constraint resolution plans a $Transfer$ path for this object and the following $Transit$ to the goal. The two paths form artificial constraints. We detect objects that violate the constraints in W^0 and recursively plan corresponding $Transfer$ and $Transit$ operations. The first grasping configuration identified by a successful resolution step is used as the preceding subgoal for obstacle identification. Both modules backtrack on their choices when the algorithm fails to resolve the constraints.

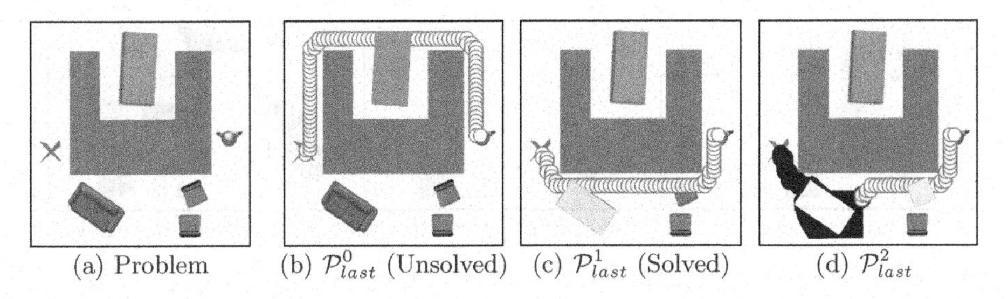

(a) Problem (b) $\mathcal{P}^0_{last}$ (Unsolved) (c) $\mathcal{P}^1_{last}$ (Solved) (d) $\mathcal{P}^2_{last}$

Fig. 2. $\mathcal{P}_{last}$ selects the last object for manipulation by the robot. The planner is called for alternative selections (c), and for preceding subgoals (d).

6.1 Obstacle Identification

The search is initialized by a constrained relaxed planner $\mathcal{P}_{last}$. [1] $O_L \leftarrow \mathcal{P}_{last}$ operates in $\mathcal{C}_R$. It is permitted to pass through movable configuration space obstacles with a heuristic one-time cost for entering any object. $\mathcal{P}_{last}$ finds a path to the goal and selects the last colliding obstacle, O_L, to schedule for manipulation. In Figure 2(b), $\mathcal{P}^0_{last}$ selects the table.

Constraint resolution, described in Section 6.2, validates the heuristic selection with a sequence of $Transit$ and $Transfer$ operations. If no such sequence is possible, $\mathcal{P}_{last}$ is called again, prohibiting any transition into $CO_R^{q_i^0}$. Since constraint resolution fails on the table, $\mathcal{P}^1_{last}$ selects the couch for motion in Figure 2(c). We ensure completeness over the selection of final objects by aggregating $\mathbf{O}_{avoid}$, a set of prohibited transitions for $\mathcal{P}_{last}$.[1]

When resolution is successful, $\mathcal{P}_{last}$ is called with the goal of reaching the initial grasping configuration identified by constraint resolution. Figure 2(d) shows that after successfully scheduling the manipulation of the couch, $\mathcal{P}^2_{last}$ selects the chair for motion. Finally, when $\mathcal{P}_{last}$ finds a collision-free path to the subgoal, the algorithm terminates successfully.

6.2 Constraint Resolution

Let T index the final time step of the plan and t be the current time step. We will maintain the following sets:

$\mathbf{O}_f^t$ - the set of objects O_i scheduled for manipulation after time t.
$\mathbf{O}_c^t$ - the set of objects scheduled for motion prior to time t.
$\mathbb{V}^t$ - the union of all artificial constraints $V^{t'}(t < t' < T)$.

$\mathbb{V}^T$ is initialized as an empty volume of space. $\mathbf{O}_f^T$ and $\mathbf{O}_c^T$ are empty sets.

We begin by applying $O_L \leftarrow \mathcal{P}_{last}$ and adding O_L to $\mathbf{O}_c^T$. Constraint resolution attempts to move all objects from $\mathbf{O}_c$ to $\mathbf{O}_f$. Objects may be added to $\mathbf{O}_c$ when they interfere with manipulation. The following three procedures are performed recursively. Each iteration of recursion will plan from state W^t, such that operations that follow time step t are assumed to be known.

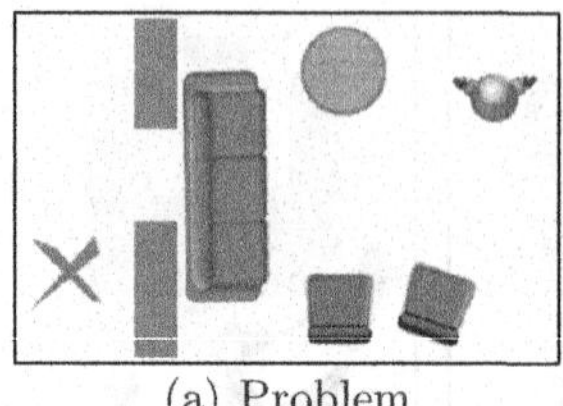
(a) Problem

(b) Manipulation Plan

(c) Constraint Resolution

Fig. 3. (a) To free a goal path in W^T, $\mathcal{P}$ chooses to manipulate the couch in W^{T-2}. (b) The planner selects the manipulation of the couch that minimizes collision. (c) In W^{T-4} our planner manipulates the table to clear space for transferring the couch.

(1) Choosing an Obstacle and Grasp

First, we select an obstacle $O_d \in \mathbf{O}_c^t$. We then choose a grasp, $\mathbf{G}_{O_d}^k$ from a predefined set $\{\mathbf{G}^1, ..., \mathbf{G}^n\}$. Each grasp corresponds to a robot configuration r_{gi}. If the robot is redundant the space of inverse kinematic solutions is sampled, yielding a set of robot configurations $\{r_{g1}, r_{g2}, ..., r_{gn}\}$.

From the set of grasping configurations we select r^{t-2} such that for some k: $\mathbf{G}_{O_d}^k(r^{t-2}) = q_d^{t-2} = q_d^0$. The grasp transform specifies that the robot configuration r^{t-2} is grasping object O_d in the objects initial configuration.

We plan a path $\tau_g(r_0, r^{t-2})$ to verify that the grasp configuration can be reached by the robot without passing through previously scheduled obstacles in their initial configurations:

$$\tau_g(s) \notin (\bigcup_{O_i \in \mathbf{O}_f^t} CO_R^{q_i^0}) \cup CO_R^{q_d^0} \qquad \forall s \in [0,1] \tag{24}$$

If such a path does not exist, the planner searches over alternative grasps.

(2) Dual Planning for Transfer and Transit

The *Transit* operation to the subsequent grasp, or goal, occurs after the *Transfer* of an object. Chronologically it should be planned first. However, we have not yet determined the initial configuration for *Transit* since it is equivalent to the final configuration of the *Transfer* task. We propose assembling the *Transit* path from two segments: τ_1 is a path from the initial grasp of the object to r^T and τ_2 is the *Transfer* path of the object. The robot returns to its initial grasping configuration, r^{t-1}, during transit.

τ_1) Plan a partial path τ_1 from r^{t-2} to r^t. The path must not pass through any future scheduled obstacle:

$$\tau_1(s) \notin \bigcup_{O_i \in \mathbf{O}_f^t} CO_R^{q_i^0} \qquad \forall s \in [0,1] \tag{25}$$

Notice that taken alone this path is not intended for a *Transit* operation. In the world state W^{t-2}, object O_d may still block this path. We choose this path heuristically to pass through the least number of objects in their initial configuration and minimize euclidian path length. If no such path is possible, a different grasp, r^{t-2}, is selected.

τ_2) Plan $Transfer(W^{t-2}, O_d, \mathbf{G}^k_{O_d}, \tau_2^{t-2})$. The robot configuration at the start of the plan is $\tau_2(0) = r^{t-2}$. The final configuration of the robot must be selected by the planner. Given that τ_2 maps to the object path $\mathbf{G}^k_{O_d}(\tau_2) \rightarrow \tau_{2O_d}$, we require the paths to adhere to the following constraints:

$$\tau_2(s) \notin \bigcup_{O_i \in \mathbf{O}^t_f} CO^{q_i^0}_R \quad \tau_{2O_d}(s) \notin \bigcup_{O_i \in \mathbf{O}^t_f} CO^{q_i^0}_{O_d} \quad \forall s \in [0,1] \tag{26}$$

$$\tau_{2O_d}(1) \notin CO_{O_d}[R(\tau_1(s)) \cup R(\tau_2(s))] \qquad \forall s \in [0,1] \tag{27}$$

$$\tau_{2O_d}(1) \notin CO_{O_d}(\mathbb{V}^t) \tag{28}$$

Eq. 26 states that the object and the robot may not pass through the configuration space obstacles of future scheduled objects. Eq. 27 states that the final configuration of O_d may not interfere with neither path segment τ_1 nor τ_2. Eq. 28 requires the final configuration of O_d to be consistent with the artificial constraints imposed by future motion.

Merging τ_1 and τ_2 into a single τ, we can define the operation $Transit(W^{t-1}, \tau)$. The transit is valid after the obstacle has been displaced.

(3) Composing Artificial Constraints

Having selected $Transfer$ and $Transit$ operations in W^t, we can advance the search to W^{t-2}. To do so, we will update the three sets described earlier:

$$\mathbf{O}^{t-2}_f \leftarrow \mathbf{O}^t_f \cup \{O_d\} \tag{29}$$

$$\begin{aligned} \mathbb{V}^{t-2} &\leftarrow \mathbb{V}^t \cup V(\tau_1) \cup V(\tau_2, O_d, \mathbf{G}^k_{O_d}) \\ &= \mathbb{V}^t \cup R(\tau_1(s)) \cup R(\tau_2(s)) \cup O_d(\tau_{2O_d}(s)) \quad \forall s \in [0,1] \end{aligned} \tag{30}$$

$$\mathbf{O}^{t-2}_c \leftarrow \{O_i \mid O_i \notin \mathbf{O}^{t-2}_f \wedge q_i^0 \in CO_{O_i}(\mathbb{V}^{t-2})\} \tag{31}$$

Eq. 29 fixes the configuration of O_d to the initial configuration and marks it as resolved in future states. Eq. 30 updates the artificial constraint to include the $Transfer$ and $Transit$ in W^{t-2} and W^{t-1} respectively. Eq. 31 updates the set of conflicting objects that must be moved earlier than W^{t-2} to resolve the constraints.

6.3 Depth First Search

Section 6.1 and 6.2 detailed the components of our planner. We now introduce pseudo-code that reflects the structure of the search. IDENTIFY-OBSTACLE is called to initialize the plan. The algorithm is implemented as depth first search to conserve space required for planning and help with the interpretability. □ indicates a successful base case while (NIL) reflects backtracking.

```
IDENTIFY-OBSTACLE(r^t, (V^t, O_f, O_c))
 1  O_avoid ← ∅
 2  while O_L ← P_last(W^0, r^t, O_avoid) ≠ NIL
 3  do
 4      if O_L = none return □
 5      O_c ← {O_L}
 6      (Plan, r^{t-n}, (V^{t-n}, O_f^{t-n})) ← RESOLVE-CONSTRAINTS(r^t, (V^t, O_f, O_c))
 7      if Plan ≠ NIL
 8         then PastPlan ← IDENTIFY-OBSTACLE(r^{t-n}, (V^{t-n}, O_f^{t-n}, O_c^{t-n}))
 9              if PastPlan ≠ NIL
10                 then return (PastPlan append Plan)
11      O_avoid ← O_avoid ∪ {O_L}
12  return NIL
```

```
RESOLVE-CONSTRAINTS(r^t, V^t, O_f^t, O_c^t)
 1  if O_c^t = ∅ return □
 2  for each O_d ∈ O_c
 3  do
 4      Choose r^{t-2} : G^k_{O_d}(r^{t-2}) = q_d^0
 5         s.t. exists τ_g(r_0, r^{t-2}) satisfying Eq. 24
 6      Choose τ_1(r^{t-2}, r^t)
 7         Satisfying Eq. 25
 8      Choose τ_2(r^{t-2}, r^{t-1})
 9         Satisfying Eq. 26 – 28
10      if no valid choices
11         then return NIL
12      determine (O_f^{t-2}, V^{t-2}, O_c^{t-2}) by Eq. 29 – 31
13      (Plan, r^{t-n}, (V^{t-n}, O_f^{t-n}, O_c^{t-n})) ←
14                   RESOLVE-CONSTRAINTS(r^{t-2}, (V^{t-2}, O_f^{t-2}, O_c^{t-2}))
15      if Plan ≠ NIL
16         then Plan append Transfer(W^{t-2}, O_d, G^k_{O_d} τ_2(r^{t-2}, r^{t-1}))
17              Plan append Transit(W^{t-1}, τ_2 + τ_1)
18              return (Plan, r^{t-n}, (V^{t-n}, O_f^{t-n}, O_c^{t-n}))
19  return NIL
```

7 Implementation

The algorithm described in this paper is entirely general for two and three dimensional spaces with arbitrary configuration spaces for the manipulator. In this section we will discuss our implementation of the algorithm in the domain of *Navigation Among Movable Obstacles* (NAMO) [1].

NAMO is two dimensional domain where obstacles are represented by polygons. The robot is a circular disc that can grasp objects when the center of the disc is at a given distance from pre-defined contact points. The domain is selected due to its interpretability and the simple property of object placement:

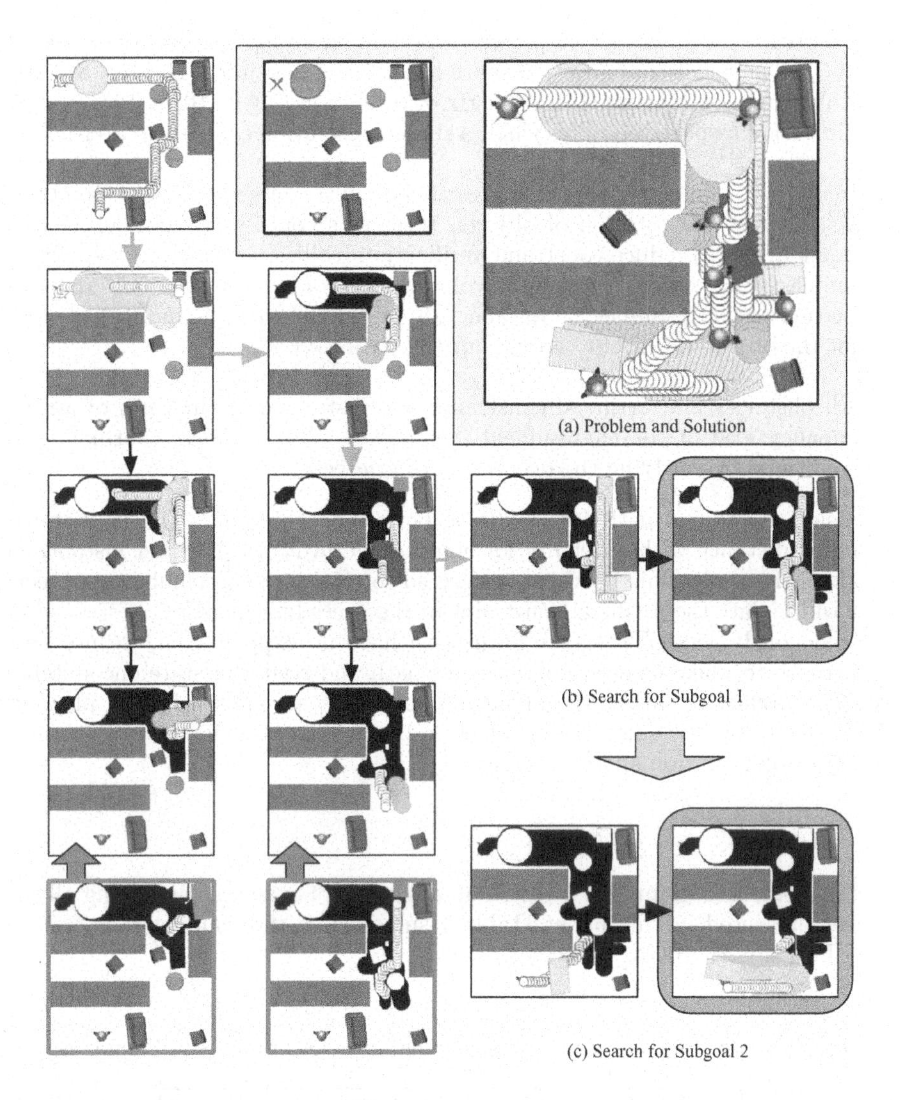

Fig. 4. A search tree for the given example. Large upward arrows indicate backtracking when an object cannot be resolved.

$\mathcal{C}_{O_i}^{place}(W^t) \subset \mathcal{C}_{O_i}(W^t)$. The figures in this paper are constructed by the implemented planner in our NAMO simulation environment.

7.1 Planning Details

When constructing a plan for the NAMO domain, we directly apply the algorithm in Section 6 by selecting a computational representation of paths and artificial constraints:

- For paths, we choose a grid planner based on an evenly spaced discretization of $\mathcal{C}_R$. The robot configuration space has three dimensions: $(\mathbb{R} \times \mathbb{R} \times SO(2))$. Robot paths are planned in a matrix of resolution $(10cm, 10cm, 10^o)$ in each dimension respectively. This yields a simple, resolution complete search space.

- In the two dimensional domain, artificial constraints are sets of points in $\mathbb{R}^2$. Due to the rotation of objects, these sets could have complex curved boundaries. To reduce constraint verification (collision detection) to polygon intersection, we construct swept volumes using a local convex hull approximation method similar to [19] and [20]. We create local bounding polygons for the object and robot throughout the path.

- All obstacles and artificial constraints are rasterized in the form of an occupancy grid of the environment. Set membership in world coordinates is confirmed by verifying the occupancy of grid cells.

Choosing a $Transit$ path (τ_1) in $\mathcal{C}_R$ is performed using A^*. The heuristic is euclidian distance with a penalty for entering $CO_R(O_i)$ for the first instance of O_i along the path. This heuristic is selected to minimize the number of objects that will violate the artificial constraint in the preceding plan.

Analogously, since $Transfer$ paths (τ_2) have no explicit goal, we use best first search to make a selection. The first path and resulting state encountered by the search that satisfy the artificial constraints are chosen by the planner. Heuristically, we penalize states where robot or the transferred object enter movable configuration space obstacles.

7.2 Results

The implemented planner was tested on a number of examples, including all the figures presented in this paper. Table 1 summarizes the running times on an Intel Pentium M 1.6Ghz processor.

Table 1. Quantities of objects and running times for examples in Figures 1-4

	Fig.1	Fig.2	Fig.3	Fig.4
# Objects	4	4	4	9
# Transferred	4	2	2	6
Planning Time	0.77s	0.05s	0.10s	2.08s

Of the presented examples, Fig. 1, 2 and 4 cannot be solved by existing planners [15, 1, 16]. In Fig. 3, the proposed method is asymptotically faster than [1] due to the early selection of $Transit$ paths as constraints in contrast to path validation during $Transfer$ search. However, this choice precludes completeness in the proposed implementation. In L_1 problems, [1] will discover remote $Transit$ paths that are not considered by the proposed implementation.

We find these results encouraging towards the implementation of this planner on a real robot system. Since the planner searches locally in the configuration space of the robot, the same algorithm can be applied directly to very high dimensional configuration spaces by replacing grid search methods with sampling-based alternatives.

7.3 Complexity

Since IDENTIFY-OBSTACLE never considers an obstacle more than once at any level of the search tree, it can generate at most $m!$ sequences. Each sequence can contain m objects to be resolved by RESOLVE-CONSTRAINTS. A breadth first search of $\mathcal{C}_R$ of resolution d in n dimensions has runtime $O(d^n)$. The overall algorithm is asymptotically $O(m!d^n)$. This is a vast overestimate. In most cases only a few sequences will satisfy the conditions of the planner.

Notice, however, that each of three "Choose" statements in RESOLVE-CONSTRAINTS is an opportunity for backtracking (Lines 4, 6 and 8). Selecting a different simple path for $Transfer$ or $Transit$ will yield distinct artificial constraints for the remainder of the search. While enumerating all possible simple paths for robot motion and manipulation is computationally expensive, selecting a subset of these paths may prove to be valuable.

8 Conclusion and Future Work

In this paper, we have presented a general planner for movable obstacles in arbitrary configuration spaces. The heuristic methods of artificial constraints have proven to be fast and effective in resolving complex examples from the sample domain of Navigation Among Movable Obstacles.

Future work will consider the possibility of reducing the number of object orderings and examining alternative object paths. Some likely classes of heuristics are the following:

Accessibility Constraints - Currently we search through all orderings of objects that violate an artificial constraint. However, clearly some objects cannot be reached by the robot before others are moved. These objects must be moved at a later time-step.

Path Heuristics - Reverse search carries significant advantages to forward search in selecting alternative paths. Simply by finding paths that explore distinct, or distant, portions of space we would change the topology of artificial constraints and therefore open distinct possibilities for prior object placements.

In addition to the investigation of heuristics, it will be interesting to study the potential for using artificial constraints to determine the necessity of intermediate states. Doing so will enable planners to address the greater challenges of non-monotone problems.

Acknowledgements

We are grateful to C. G. Atkeson for his input and support in the development of this project. We thank the anonymous reviewers for their thorough reading and insightful comments. This research was partially supported by NSF grants DGE-0333420, ECS-0325383, ECS-0326095, and ANI-0224419.

References

1. Stilman, M., Kuffner, J.J.: Navigation among movable obstacles: Real-time reasoning in complex environments. In: IEEE/RAS Int. Conf. on Humanoid Robotics, pp. 322–341. IEEE, Los Alamitos (2004), http://www.golems.org/NAMO
2. Wilfong, G.: Motion panning in the presence of movable obstacles. In: ACM Symp. Computat. Geometry, pp. 279–288. ACM Press, New York (1988)
3. Demaine, E., Demaine, M., O'Rourke, J.: Pushpush and push-1 are np-hard in 2d. In: 12th Canadian Conference on Computational Geometry, Fredericton, New Brunswick, Canada, August 16-19, pp. 211–219 (2000)
4. Reif, J.H., Sharir, M.: Motion planning in the presence of moving obstacles. In: Proc. 26th Annual Symposium on Foundations of Computer Science, Portland, Oregon, pp. 144–154. IEEE Computer Society Press, Los Alamitos (1985)
5. Hsu, D., Kindel, R., Latombe, J.C., Rock, S.: Randomized kinodynamic motion planning with moving obstacles. The Int. J. of Robotics Research 21(3), 233–255 (2002)
6. Brock, O., Khatib, O.: Elastic strips: Real-time path modification for mobile manipulation. In: International Symposium of Robotics Research, Hayama, Japan, pp. 117–122. Springer, Heidelberg (1997)
7. Fortune, S., Wilfong, G., Yap, C.: Coordinated motion of two robot arms. In: IEEE Int. Conf. on Robotics and Automation, pp. 1216–1223 (April 1986)
8. Erdmann, M., Lozano-Perez, T.: On multiple moving objects. In: IEEE Int. Conf. on Robotics and Automation, April 7-10, pp. 1419–1424 (1986)
9. Schwartz, J., Sharir, M.: On the piano movers' problem. iii: Coordinating the motion of several independent bodies. the special case of circular bodies moving amidst polygonal barriers. Int. J. Robotics Research 2(3), 46–74 (1983)
10. Goldwasser, M.H., Motwani, R.: Complexity measures for assembly sequences. The International Journal of Computational Geometry and Applications 9(4 & 5), 371–418 (1999)
11. Randall, H.: Wilson. On Geometric Assembly Planning. PhD thesis, Department of Computer Science, Stanford University, Stanford, CA, USA, UMI Order Number: UMI Order No. GAX92-21686 (1992)
12. Alami, R., Laumond, J.P., Sim'eon, T.: Two manipulation planning algorithms. In: Goldberg, K., Halperin, D., Latombe, J., Wilson, R. (eds.) Workshop on the Algorithmic Foundations of Robotics, San Francisco, California, United States, pp. 109–125. A. K. Peters, Ltd., Natick (1994)
13. Ben-Shahar, O., Rivlin, E.: Practical pushing planning for rearrangement tasks. IEEE Trans. on Robotics and Automation 14(4), 549–565 (1998)
14. Ota, J.: Rearrangement of multiple movable objects. In: IEEE Int. Conf. Robotics and Automation (ICRA), New Orleans, LA, vol. 2, pp. 1962–1967. IEEE, Los Alamitos (2004)
15. Chen, P.C., Hwang, Y.K.: Pracitcal path planning among movable obstacles. In: IEEE Int. Conf. Robotics and Automation (ICRA), Sacramento, CA, pp. 444–449. IEEE, Los Alamitos (1991)

16. Okada, K., Haneda, A., Nakai, H., Inaba, M., Inoue, H.: Environment manipulation planner for humaonid robots using task graph that generates action sequence. In: IEEE/RSJ Int. Conf. on Intelligent Robots and Systems (IROS 2004), Sendai, Japan, pp. 1174–1179. IEEE, Los Alamitos (2004)
17. Chadzelek, T., Eckstein, J., Schömer, E.: Heuristic motion planning with movable obstacles. In: 8th Canadian Conference on Computational Geometry, Ottawa, Canada, August 12-15, pp. 131–136. Carleton University Press (1996)
18. Lozano-Perez, T.: Spatial planning: A configuration space approach. IEEE Trans. Comput. 32(2), 108–120 (1983)
19. Ganter, M.A.: Dynamic Collision Det. using Kinematics and Solid Modeling Techniques(Mechanical Engineering). PhD thesis, University of Wisconsin (1985)
20. Foisy, A., Hayward, V.: A safe swept-volume approach to collision detection. In: Robotics Research, Sixth International Symposium (1994)

Part III

Navigation, SLAM, and Error Models for Filtering/Control

Inferring and Enforcing Relative Constraints in SLAM

Kristopher R. Beevers and Wesley H. Huang

Department of Computer Science, Rensselaer Polytechnic Institute
{beevek,whuang}@cs.rpi.edu

Abstract. Most algorithms for simultaneous localization and mapping (SLAM) do not incorporate prior knowledge of structural or geometrical characteristics of the environment. In some cases, such information is readily available and making some assumptions is reasonable. For example, one can often assume that many walls in an indoor environment are rectilinear. In this paper, we develop a SLAM algorithm that incorporates prior knowledge of relative constraints between landmarks. We describe a "Rao-Blackwellized constraint filter" that infers applicable constraints and efficiently enforces them in a particle filtering framework. We have implemented our approach with rectilinearity constraints. Results from simulated and real-world experiments show the use of constraints leads to consistency improvements and a reduction in the number of particles needed to build maps.

1 Introduction

The simultaneous localization and mapping (SLAM) problem is for a mobile robot to concurrently estimate both a map of its environment and its pose with respect to the map. Most SLAM algorithms make few assumptions about the environment; thus, SLAM does not take advantage of prior information when the environment is known to have specific structural characteristics. For example, a robot designed to operate indoors can often assume its environment is "mostly" rectilinear.

In many cases structural or geometrical assumptions can be represented as information about relative constraints between landmarks in a robot's map, which can be used in inference to determine which landmarks are constrained and the parameters of the constraints. In the rectilinearity example, such a formulation can be used to constrain the walls of a room separately from, say, the boundary of a differently-aligned obstacle in the center of the room.

Given relative constraints between landmarks, they must be enforced. Some previous work has enforced constraints on maps represented using an extended Kalman filter (EKF) [6, 11, 14]. In this paper, we develop techniques to instead enforce constraints in maps represented by a Rao-Blackwellized particle filter (RBPF). The major difficulty is that RBPF SLAM relies on the conditional independence of landmark estimates given a robot's pose history, but relative constraints introduce correlation between landmarks.

S. Akella et al. (Eds.): Algorithmic Foundation of Robotics VII, STAR 47, pp. 139–154, 2008.
springerlink.com

Our approach exploits a property similar to that used in the standard SLAM Rao-Blackwellization: conditioned on values of constrained state variables, unconstrained state variables are independent. We use this fact to incorporate per-particle constraint enforcement into RBPF SLAM. We have also developed a technique to address complications which arise when initializing a constraint between groups of landmarks that are already separately constrained; the technique efficiently recomputes conditional estimates of unconstrained variables when modifying the values of constrained variables.

Incorporating constraints can have a profound effect on the computation required to build maps. A motivating case is the problem of mapping with sparse sensing. In previous work [3], we have shown that particle filtering SLAM is possible with limited sensors such as small arrays of infrared rangefinders, but that many particles are required due to increased measurement uncertainty. By extending sparse sensing SLAM to incorporate constraints, an order-of-magnitude reduction in the number of particles can be achieved.

The paper proceeds as follows. We first discuss previous work on constrained SLAM. Then, in Section 2, we briefly review the general SLAM problem and the ideas behind RBPF, and discuss the assumption of unstructured environments made by most SLAM algorithms. In Section 3 we formalize the idea of SLAM with relative constraints and describe a simple but infeasible approach. We then introduce the Rao-Blackwellized constraint filter: Section 4 describes an RBPF-based algorithm for enforcing constraints, and Section 5 incorporates inference of constraints. Finally, in Section 6 we describe the results of simulated and real-world experiments with a rectilinearity constraint.

1.1 Related Work

Most work on SLAM focuses on building maps using very little prior information about the environment, aside from assumptions made in feature extraction and data association. A thorough coverage of much of the state-of-the art in unconstrained SLAM can be found in, e.g., [8].

The problem of inferring when constraints should be applied to a map is largely unexplored. Rodriguez-Losada *et al.* [11] employ a simple thresholding approach to determine which of several types of constraints should be applied.

On the other hand, several researchers have studied the problem of enforcing *a priori* known constraints in SLAM. In particular, Durrant-Whyte [6] and Wen and Durrant-Whyte [14] have enforced constraints in EKF-based SLAM by treating the constraints as zero-uncertainty measurements. More recently, Csorba, Newman and Durrant-Whyte [4, 10] and Deans and Hebert [5] have built maps where the state consists of relationships between landmarks; they apply constraints on the relationships to enforce map consistency. From a consistent relative map an absolute map can be estimated.

Finally, others have studied general constrained state estimation using the EKF. Simon and Chia [12] derive Kalman updates for linear equality constraints (discussed in detail in Section 3.1) that are equivalent to projecting

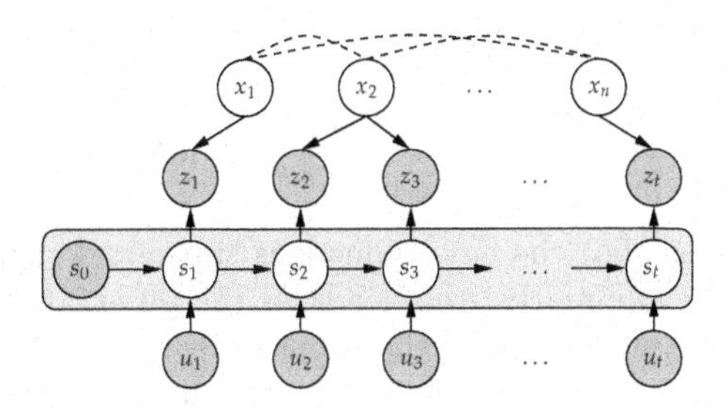

Fig. 1. A Bayes network showing common SLAM model assumptions. Input variables are represented by shaded nodes; the objective of SLAM is to estimate values for the unshaded nodes. Arcs indicate causality or correlation between variables. (Correspondence variables n_t are omitted for clarity — observations are connected directly to the corresponding landmarks.) Correlations between landmarks due to structure in the environment (dashed arcs) are typically ignored in SLAM.

the unconstrained state onto the constraint surface. In [13], Simon and Simon extend this approach to deal with linear inequality constraints.

2 The SLAM Problem

The goal of SLAM is to simultaneously estimate both a map M of the environment and the robot's (time-dependent) pose s_t with respect to the map. A number of map representations exist; we focus on landmark-based mapping with $M = \{x_1, x_2, \ldots, x_n\}$, where each landmark x_i is a parameterized geometric object such as a point or a line. In the basic SLAM process, the robot executes a motion and estimates its new pose using odometry. It then takes a sensor reading and extracts geometric features from the raw sensor data. Data association matches features with landmarks in the map, and the map and pose estimates are updated.

SLAM is often posed in a Bayesian filtering formulation where the goal is to estimate a posterior distribution over poses and maps given all of the measurements z^t, commanded motions u^t, and correspondences n^t between features and landmarks [8]. (The superscript notation indicates a set of values $1 \ldots t$ over all time steps.) A Bayes network depicting the standard SLAM model assumptions is shown in Fig. 1. The filter can be written recursively:

$$p(s_t, M | z^t, u^t, n^t) = \eta p(z_t | s_t, x_{n_t}, n_t) \int p(s_t | s_{t-1}, u_t) p(s_{t-1}, M | z^{t-1}, u^{t-1}, n^{t-1}) \, ds_{t-1} \tag{1}$$

where $p(z_t|s_t, x_{n_t}, n_t)$ is the measurement model, $p(s_t|s_{t-1}, u_t)$ models the robot's motion, and η is a normalization constant. In this approach, SLAM is usually done using the extended Kalman filter (EKF).

An alternative is to filter over the entire robot trajectory s^t, i.e.:

$$p(s^t, M|n^t, z^t, u^t) = \eta p(z_t|s_t, x_{n_t}, n_t)\, p(s_t|s_{t-1}, u_t) p(s^{t-1}, M|n^{t-1}, z^{t-1}, u^{t-1}) \quad (2)$$

Under the assumption that the environment is static and that no direct correlations exist between landmarks, this leads to the observation that landmarks are conditionally independent given the robot's trajectory, since correlation between landmarks arises only through robot pose uncertainty [9]. In Fig. 1, the highlighted variables (the robot's trajectory) *d-separate* the landmark variables. Thus, the posterior over trajectories and maps can be factored:

$$p(s^t, M|n^t, z^t, u^t) = p(s^t|n^t, z^t, u^t) \prod_{i=1}^{n} p(x_i|s^t, n^t, z^t) \quad (3)$$

This factorization is known as Rao-Blackwellization. To perform SLAM based on Eqn. 3, the posterior over trajectories can be represented with a particle filter where each particle samples a single trajectory. Associated with a particle are a number of separate small filters (typically EKFs) to analytically estimate each landmark in the particle's map. This approach is known as Rao-Blackwellized particle filtering (RBPF) and is the basis for the well-known FastSLAM algorithm [8].

2.1 Structured Environments

Typically, SLAM approaches assume the environment is *unstructured*, i.e., that landmarks are randomly and independently distributed in the workspace. Often this is not the case, as in indoor environments where landmarks are placed methodically. Thus, some correlation exists between landmarks, due not to the robot's pose uncertainty, but rather to the *structure* in the environment. (This is represented by the dotted arcs in Fig. 1).

Correlation between landmarks can arise in many ways, making it difficult to include in the SLAM model. In this paper, we assume that structure in the environment takes on one of a few forms — i.e., that the space of possible (structured) landmark relationships is small and discrete. When this is the case, the model shown in Fig. 2 can be used. Here, arcs indicating correlation between landmarks are parameterized. The parameters $c_{i,j}$ indicate the constraints (or lack thereof) between landmarks x_i and x_j. We perform inference on the constraint parameter space, and then enforce the constraints. In this paper we focus on the pairwise case, but more complicated relationships can in principle be exploited.

3 SLAM with Relative Constraints

We begin by addressing the issue of efficiently *enforcing* known relative constraints. Parallel to this is the problem of *merging* constraints when new relationships are found between separate groups of already constrained landmarks.

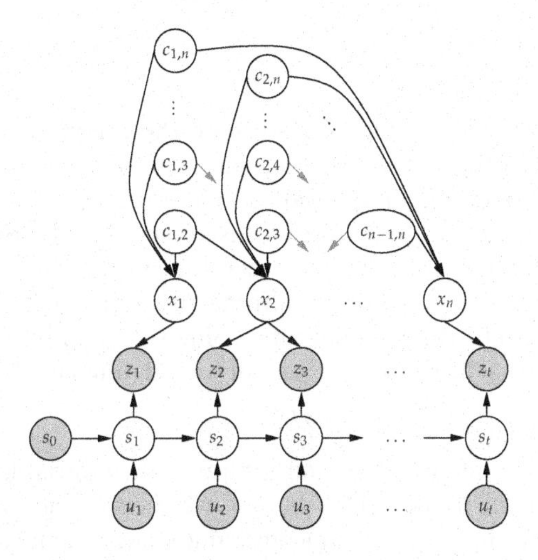

Fig. 2. Bayes network for a SLAM model that incorporates pairwise constraints between landmarks, parameterized by the variables $c_{i,j}$. Inference in the space of relationship parameters can be used to determine correlations between landmark parameters; relative constraints on the landmarks enforce inferred relationships.

Throughout the rest of the paper we omit time indices for clarity. Variables are vectors unless otherwise noted. We use P_i to represent the covariance of the landmark estimate x_i. We assume that measurements of a landmark are in the parameter space of the landmark (i.e., measurements are of the landmark state). Measurements that do not meet this condition can easily be transformed. Finally, while we present our formulation for a single constraint, the approach can be applied in parallel to several types of constraints.

3.1 The Superlandmark Filter

There is an immediate problem with SLAM when the environment is structured: landmark correlations lead to interdependencies that break the factorization utilized in Eqn. 3, which assumes correlation arises only through robot pose uncertainty. We first describe a simple (but ultimately impractical) approach to deal with the correlation, which leads to an improved technique in Section 4. Note that the RBPF factorization still holds for unconstrained landmarks; we rewrite the filter, grouping constrained landmarks. Formally, partition the map into groups:

$$\mathcal{L} = \{\{x_{a_1}, x_{a_2}, \ldots\}, \{x_{b_1}, x_{b_2}, \ldots\}, \{x_c\}, \ldots\} \tag{4}$$

Each group ("superlandmark") $L_i \in \mathcal{L}$ contains landmarks constrained with respect to each other; correlation arises only among landmarks in the same group. We immediately have the following version of the RBPF SLAM filter:

$$p(s^t, M|n^t, z^t, u^t) = p(s^t|n^t, z^t, u^t)\prod_{i=1}^{|\mathcal{L}|} p(L_i|s^t, n^t, z^t) \tag{5}$$

We can still apply a particle filter to estimate the robot's trajectory. Each superlandmark is estimated using an EKF, which accounts for correlation due to constraints since it maintains full covariance information.

There are several ways to enforce constraints on a superlandmark. One approach is to treat the constraints as zero-uncertainty measurements of the constrained landmarks [6, 14, 11]. An alternative is to directly incorporate constrained estimation into the Kalman filter. Simon and Chia [12] have derived a version of the EKF that accounts for equality constraints of the form

$$DL_i = d \tag{6}$$

where L_i represents the superlandmark state with n variables, D is an $s \times n$ constant matrix of full rank, and d is a $s \times 1$ vector; together they encode s constraints. In their approach, the unconstrained EKF estimate is computed and then repaired to account for the constraints. Given the unconstrained state L_i and covariance matrix P_{L_i}, the constrained state and covariance are computed as follows (see [12] for the derivation):

$$\tilde{L}_i \leftarrow L_i - PD^T(DPD^T)^{-1}(DL_i - d) \tag{7}$$

$$\tilde{P}_{L_i} \leftarrow P_{L_i} - P_{L_i}D^T(DP_{L_i}D^T)^{-1}DP_{L_i} \tag{8}$$

i.e., the unconstrained estimate is projected onto the constraint surface.

If a constraint arises between two superlandmarks they are easily merged:

$$L_{ij} \leftarrow \begin{bmatrix} L_i \\ L_j \end{bmatrix}, \quad P_{ij} \leftarrow \begin{bmatrix} P_i & P_i \frac{\partial L_j}{\partial L_i}^T \\ \frac{\partial L_j}{\partial L_i} P_i & P_j \end{bmatrix} \tag{9}$$

Unfortunately, the superlandmark filter is too expensive unless the size of superlandmarks can be bounded by a constant. In the worst case the environment is highly constrained and, in the extreme, the map consists of a single superlandmark. EKF updates for SLAM take at least $O(n^2)$ time and constraint enforcement using Eqns. 7 and 8 requires $O(n^3)$ time for a superlandmark of size n. If the particle filter has N particles, the superlandmark filter requires $O(Nn^3)$ time for a single update. We thus require a better solution.

3.2 Reduced State Formulation

A simple improvement can be obtained by noting that maintaining the full state and covariance for each landmark in a superlandmark is unnecessary. Constrained state variables are redundant: given the value of the variables from one "representative" landmark, the values for the remaining landmarks in a superlandmark are determined. In the rectilinearity example, with landmarks represented as lines parameterized by distance r and angle θ to the origin, a full superlandmark state vector has the form: $[r_1\ \theta_1\ r_2\ \theta_2\ \ldots\ r_n\ \theta_n]^T$. If the $\{\theta_i\}$ are constrained the state can be rewritten as: $[r_1\ \theta_1\ r_2\ g_2(c_{1,2};\theta_1)\ \ldots\ r_n\ g_n(c_{1,n};\theta_1)]^T$.

Thus, filtering of the superlandmark need only be done over the reduced state: $[r_1\, r_2\, \ldots\, r_n\, \theta_1]^T$. The function $g_i(c_{j,i}; x_{j,\rho})$ with parameters $c_{j,i}$ maps the constrained variables $x_{j,\rho}$ of the representative landmark x_j to values for $x_{i,\rho}$; in the rectilinearity case, $c_{j,i} \in \{0, 90, 180, 270\}$ and $g_i(c_{j,i}; \theta_j) = \theta_j - c_{j,i}$. We assume the constraints are invertible: the function $h_i(c_{j,i}; x_{i,\rho})$ represents the reverse mapping, e.g., $h_i(c_{j,i}; \theta_i) = \theta_i + c_{j,i}$. We sometimes refer to the unconstrained variables of landmark x_i as $x_{i,\overline{\rho}}$.

4 Rao-Blackwellized Constraint Filter

From the reduced state formulation we see it is easy to separate the map state into constrained variables $M^c = \{x_{1,\rho}, \ldots, x_{n,\rho}\}$, and unconstrained variables $M^f = \{x_{1,\overline{\rho}}, \ldots, x_{n,\overline{\rho}}\}$. By the same reasoning behind Eqn. 3, we factor the SLAM filter as follows:

$$p(s^t, M|n^t, z^t, u^t) = p(s^t, M^c|n^t, z^t, u^t) \prod_{i=1}^{|M^f|} p(x_{i,\overline{\rho}}|s^t, M^c, n^t, z^t) \qquad (10)$$

In other words, conditioned on *both* the robot's trajectory and the values of all constrained variables, free variables of separate landmarks are independent.

Eqn. 10 suggests that we can use a particle filter to estimate both the robot trajectory and the values of the constrained variables. We can then use separate small filters to estimate the unconstrained variables conditioned on sampled values of the constrained variables. The estimation of representative values for the constrained variables is accounted for in the particle filter resampling process, where particles are weighted by data association likelihood.

4.1 Particlization of Landmark Variables

We first discuss initialization of constraints between previously unconstrained landmarks. Given a set $\mathcal{R} = \{x_1, x_2, \ldots, x_n\}$ of landmarks to be constrained, along with constraint parameters $c_{1,i}$ for each $x_i \in \mathcal{R}, i = 2 \ldots n$ (i.e., with x_1 as the "representative" landmark — see Section 3.2), we form a superlandmark from $\mathcal{R}$. Then, we perform a *particlization procedure*, sampling the constrained variables from the reduced state of the superlandmark. Conditioning of the unconstrained variables of every landmark in the superlandmark is performed using the sampled values. We are left with an EKF for each landmark that estimates only the values of unconstrained state variables.

In selecting values of the constrained variables on which to condition, we should take into account all available information, i.e., the estimates of the constrained variables from each landmark. We compute the maximum likelihood estimate of the constrained variables:

$$P_{\hat{\rho}} \leftarrow \left(\sum_{x_j \in \mathcal{R}} P_{j,\rho}^{-1} \right)^{-1}, \quad \hat{\rho} \leftarrow P_{\hat{\rho}}^{-1} \left(\sum_{x_j \in \mathcal{R}} h_j(c_{1,j}; x_{j,\rho}) P_{j,\rho}^{-1} \right) \qquad (11)$$

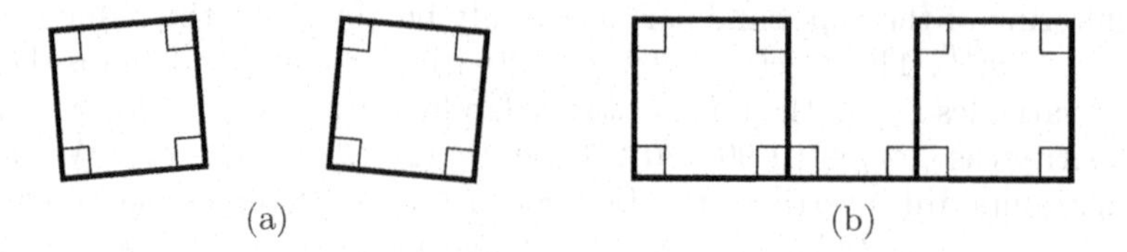

Fig. 3. Merging groups of constrained landmarks. (a) Two constrained groups of landmarks. (b) After finding a new landmark constrained with respect to both groups, all landmarks are constrained together.

To choose values for ρ, we can either sample, e.g., according to $\mathcal{N}(\hat{\rho}, P_{\hat{\rho}})$; or we can simply pick $\hat{\rho}$, which is the approach we take in our implementation.

Once values of constrained variables are selected, we condition the unconstrained variables on the selected values. To condition x_i with covariance P_i on values for $x_{i,\rho}$, we first partition the state and covariance:

$$x_i = [x_{i,\overline{\rho}} \;\; x_{i,\rho}]^T \quad , \quad P_i = \begin{bmatrix} P_{i,\overline{\rho}} & P_{i,\overline{\rho}\rho} \\ P_{i,\rho\overline{\rho}} & P_{i,\rho} \end{bmatrix} \tag{12}$$

Then given $x_{i,\rho} = g_i(c_{1,i}; \hat{\rho})$ and since landmark state is estimated by an EKF, the standard procedure for conditioning the Normal distribution yields:

$$\tilde{x}_{i,\overline{\rho}} \leftarrow x_{i,\overline{\rho}} + P_{i,\overline{\rho}\rho} P_{i,\rho}^{-1} (g_i(c_{1,i}; \hat{\rho}) - x_{i,\rho}) \tag{13}$$

$$\tilde{P}_{i,\overline{\rho}} \leftarrow P_{i,\overline{\rho}} - P_{i,\overline{\rho}\rho} P_{i,\rho}^{-1} P_{i,\overline{\rho}\rho}^T \tag{14}$$

For purposes of data association it is convenient to retain the full state and covariance, in which case $\tilde{x}_{i,\rho} = g_i(c_{1,i}; \hat{\rho})$ and $\tilde{P}_{i,\rho} = \tilde{P}_{i,\overline{\rho}\rho} = \tilde{P}_{i,\rho\overline{\rho}} = [\mathbf{0}]$.

4.2 Reconditioning

Particlization is straightforward if none of the landmarks is already constrained. This is not the case when a new landmark is added to a superlandmark or when merging several constrained superlandmarks. Since the values of unconstrained state variables are already conditioned on values of the constrained variables, we cannot change constrained variables without invalidating the conditioning. Such a situation is depicted in Fig. 3.

One solution is to "rewind" the process to the point when the landmarks were first constrained and then "replay" all of the measurements of the landmarks, conditioning on the new values of the constrained variables. This is clearly infeasible. However, we can achieve an equivalent result efficiently because the order in which measurements are applied is irrelevant. Applying k measurements to the landmark state is equivalent to merging $k+1$ Gaussians. Thus, we can "accumulate" all of the measurements in a single Gaussian and apply this instead, in unit time.

From this, we obtain the following reconditioning approach:

1. Upon first constraining a landmark x_i, store its pre-particlization unconstrained state $\beta_i = x_i, \Lambda_i = P_i$, initialize the "measurement accumulator" $\mathcal{Z}_i = [\mathbf{0}], \mathcal{Q}_i = [\infty]$, and particlize the landmark.
2. For a measurement z with covariance R of the constrained landmark update both the conditional state and the measurement accumulator:

$$x_i \leftarrow x_i + P_i(P_i + R)^{-1}(z - x_i) \tag{15}$$
$$P_i \leftarrow P_i - P_i(P_i + R)^{-1}P_i^T \tag{16}$$
$$\mathcal{Z}_i \leftarrow \mathcal{Z}_i + \mathcal{Q}_i(\mathcal{Q}_i + R)^{-1}(z - \mathcal{Z}_i) \tag{17}$$
$$\mathcal{Q}_i \leftarrow \mathcal{Q}_i - \mathcal{Q}_i(\mathcal{Q}_i + R)^{-1}\mathcal{Q}_i^T \tag{18}$$

3. When instantiating a new constraint on x_i, recondition x_i on the new constrained variable values by rewinding the landmark state ($x_i = \beta_i$, $P_i = \Lambda_i$), computing the conditional distribution $\tilde{x}_i, \tilde{P}_i$ of the state (Eqns. 13-14), and replaying the measurements since particlization with:

$$x_i \leftarrow \tilde{x}_i + \tilde{P}_i(\tilde{P}_i + \mathcal{Q}_i)^{-1}(\mathcal{Z}_i - \tilde{x}_i) \tag{19}$$
$$P_i \leftarrow \tilde{P}_i - \tilde{P}_i(\tilde{P}_i + \mathcal{Q}_i)^{-1}\tilde{P}_i^T \tag{20}$$

The reconditioning technique can be extended to handle multiple types of constraints simultaneously by separately storing the pre-particlization state and accumulated measurements for each constraint. Only completely unconstrained state variables should be stored at constraint initialization, and only the measurements of those variables need be accumulated.

4.3 Discussion

A potential issue with our approach is that reconditioning neither re-evaluates data associations nor modifies the trajectory of a particle. In practice we have observed that the effect on map estimation is negligible.

Computationally, the constrained RBPF approach is a significant improvement over the superlandmark filter, requiring only $O(Nn)$ time per update.[1] At first it appears that more particles may be necessary since representative values of constrained variables are now estimated by the particle filter. However, incorporating constraints often leads to a significant reduction in required particles by reducing the degrees of freedom in the map. In a highly constrained environment, particles only need to filter a few constrained variables using the reduced state, and the EKFs for unconstrained variables are smaller since they filter only over the unconstrained state. By applying strong constraint priors where appropriate, the number of particles required to build maps is often reduced by an order of magnitude, as can be seen in Section 6.

[1] We note that while the data structures that enable $O(N \log n)$ time updates for FastSLAM [8] can still be applied, they do not improve the complexity of constrained RBPF since the reconditioning step is worst-case linear in n.

4.4 Inequality Constraints

So far we have only considered equality constraints, whereas many useful constraints are inequalities. For example, we might specify a prior on corridor width: two parallel walls should be at least a certain distance apart. In [13], the authors apply inequality constraints to an EKF using an active set approach. At each time step, the applicable constraints are tested. If a required inequality is violated, an equality constraint is applied, projecting the unconstrained state onto the boundary of the constraint region.

While this approach appears to have some potential problems (e.g., it ignores the landmark PDF over the unconstrained half-hyperplane in parameter space), a similar technique can be incorporated into the Rao-Blackwellized constraint filter. After updating a landmark, applicable inequality constraints are tested. Constraints that are violated are enforced using the techniques described in Section 4. The unconstrained state is accessible via the measurement accumulator, so if the inequality is later satisfied, the parameters can be "de-particlized" by switching back to the unconstrained estimate.

5 Inference of Constraints

We now address the problem of deducing the relationships between landmarks, i.e., deciding when a constraint should be applied. A simple approach is to just examine the unconstrained landmark estimates. In the rectilinearity case, we can easily compute the estimated angle between two landmarks. If this angle is "close enough" to one of $0°, 90°, 180°$, or $270°$, the constraint is applied to the landmarks. (A similar approach is used by Rodriguez-Losada *et al.* [11].) However, this technique ignores the confidence in the landmark estimates.

We instead compute a PMF over the space $\mathcal{C}$ of pairwise constraint parameters; the PMF incorporates the landmark PDFs. In the rectilinearity example, $\mathcal{C} = \{0, 90, 180, 270, \star\}$, where $\star$ is used to indicate that landmarks are unconstrained. Given a PMF over $\mathcal{C}$, we sample constraint parameters for each particle to do inference of constraints. Particles with incorrectly constrained landmarks will yield poor data associations and be resampled.

We compute the PMF of the "relationship" of landmarks x_i and x_j using:

$$p(c_{i,j}) = \int p(x_{i,\rho}) \int_{h_j(c_{i,j};x_{j,\rho})-\delta}^{h_j(c_{i,j};x_{j,\rho})+\delta} p(x_{j,\rho})\, dx_{j,\rho}\, dx_{i,\rho} \tag{21}$$

for all $c_{i,j} \in \mathcal{C} \setminus \star$. Then, $p(\star) = 1 - \sum_{c_{i,j} \in \mathcal{C} \setminus \star} p(c_{i,j})$. The parameter δ encodes "prior information" about the environment: the larger the value of δ, the more liberally we apply constraints. A benefit of this approach is that the integrals can be computed efficiently from standard approximations to the Normal CDF since the landmarks are estimated by EKFs.

In the rectilinearity case, given orientation estimates described by the PDFs $p(\theta_i)$ and $p(\theta_j)$, for $c_{i,j} \in \{0, 90, 180, 270\}$, we have:

Algorithm 1. INITIALIZE-LANDMARK$(x_{n+1}, P_{n+1}, \mathcal{L})$

1: $\beta_{n+1} \leftarrow x_{n+1};\ \Lambda_{n+1} = P_{n+1}$ *// initialize backup state*
2: $\mathcal{Z}_{n+1} \leftarrow [\mathbf{0}];\ \mathcal{Q}_{n+1} \leftarrow [\infty]$ *// initialize measurement accumulator*
3: $\mathcal{R} \leftarrow \{\}$ *// initialize constraint set*
4: **for all** $L_i \in \mathcal{L}$ **do** *// previously constrained groups*
5: $\quad c_{n+1,j} \sim p(c_{n+1,j}),\ \forall x_j \in L_i$ *// draw constraint parameters*
6: $\quad$ **if** $\exists x_j \in L_i$ such that $c_{n+1,j} \neq \star$ **then** *// constrained?*
7: $\quad\quad$ **for all** $x_j \in L_i$ **do**
8: $\quad\quad\quad \mathcal{R} \leftarrow \mathcal{R} \cup \{x_j\}$ *// add x_j to constraint set*
9: $\quad\quad \mathcal{L} \leftarrow \mathcal{L} \setminus L_i$ *// remove old superlandmark*
10: **if** $\mathcal{R} = \emptyset$ **then**
11: $\quad$ **return** *// no constraints on x_{n+1}*
12: $\mathcal{R} \leftarrow \mathcal{R} \cup \{x_{n+1}\}$ *// add new landmark to constraint set*
13: $\mathcal{L} \leftarrow \mathcal{L} \cup \{\mathcal{R}\}$ *// add new superlandmark*
14: **for all** $x_j \in \mathcal{R}$ **do** *// for all constrained landmarks*
15: $\quad \hat{x}_j \leftarrow \beta_j + \Lambda_j \mathcal{Q}_j^{-1}(\mathcal{Z}_j - \beta_j)$ *// compute unconstrained state estimate*
16: $\quad \hat{P}_j \leftarrow \Lambda_j - \Lambda_j \mathcal{Q}_j^{-1} \Lambda_j^T$ *// compute unconstrained covariance*
17: $P_{\hat{\rho}} \leftarrow \left(\sum_{x_j \in \mathcal{R}} P_{j,\rho}^{-1}\right)^{-1}$ *// covariance of ML estimate of ρ*
18: $\hat{\rho} \leftarrow P_{\hat{\rho}}^{-1}\left(\sum_{x_j \in \mathcal{R}} h_j(c_{n+1,j}; x_{j,\rho}) P_{j,\rho}^{-1}\right)$ *// ML estimate of ρ*
19: **for all** $x_j \in \mathcal{R}$ **do** *// for all constrained landmarks*
20: $\quad x_j \leftarrow \beta_j;\ P_j \leftarrow \Lambda_j$ *// "rewind" state to pre-particlized version*
21: $\quad x_{j,\overline{\rho}} \leftarrow x_{j,\overline{\rho}} + P_{j,\overline{\rho}\rho} P_{j,\rho}^{-1}\left(g_j(c_{n+1,j}; \hat{\rho}) - x_{j,\rho}\right)$ *// conditional mean given ρ*
22: $\quad P_{j,\overline{\rho}} \leftarrow P_{j,\overline{\rho}} - P_{j,\overline{\rho}\rho} P_{j,\rho}^{-1} P_{j,\overline{\rho}\rho}^T$ *// conditional covariance*
23: $\quad x_{j,\rho} \leftarrow g_j(c_{n+1,j}; \hat{\rho});\ P_{j,\rho} \leftarrow [\mathbf{0}];\ P_{j,\overline{\rho}\rho} \leftarrow [\mathbf{0}]$ *// fix constrained variables*
24: $\quad x_j \leftarrow x_j + P_j(P_j + \mathcal{Q}_j)^{-1}(\mathcal{Z}_j - x_j)$ *// "replay" meas. since particlization*
25: $\quad P_j \leftarrow P_j - P_j(P_j + \mathcal{Q}_j)^{-1} P_j^T$

Algorithm 2. UPDATE-LANDMARK(x_j, P_j, z, R)

1: $x_j \leftarrow x_j + P_j(P_j + R)^{-1}(z - x_j)$ *// update state*
2: $P_j \leftarrow P_j - P_j(P_j + R)^{-1} P_j^T$ *// update covariance*
3: **if** $\exists L \in \mathcal{L}, x_k \in L$ such that $x_j \in L$ and $x_j \neq x_k$ **then** *// is x_j constrained?*
4: $\quad \mathcal{Z}_j \leftarrow \mathcal{Z}_j + \mathcal{Q}_j(\mathcal{Q}_j + R)^{-1}(z - \mathcal{Z}_j)$ *// update measurement accumulator*
5: $\quad \mathcal{Q}_j \leftarrow \mathcal{Q}_j - \mathcal{Q}_j(\mathcal{Q}_j + R)^{-1} \mathcal{Q}_j^T$ *// update accumulator covariance*
6: **else** *// not constrained*
7: $\quad \beta_j \leftarrow x_j;\ \Lambda_j \leftarrow P_j$ *// update backup state/covariance*

$$p(c_{i,j}) = \int_{-\infty}^{\infty} p(\theta_i) \int_{\theta_i + c_{i,j} - \delta}^{\theta_i + c_{i,j} + \delta} p(\theta_j)\, d\theta_j\, d\theta_i \tag{22}$$

which gives a valid PMF as long as $\delta \leq 45°$.

6 Results

We have now described the complete approach for implementing constrained RBPF SLAM. Algorithm 1 gives pseudocode for initializing a landmark x_{n+1} given the current set of superlandmarks $\mathcal{L}$. Algorithm 2 shows how to update a (possibly constrained) landmark given a measurement of its state. The algorithms simply collect the steps described in detail in Sections 4 and 5.

We have implemented the Rao-Blackwellized constraint filter for the rectilinearity constraint described earlier, on top of our algorithm for RBPF SLAM with sparse sensing [3], which extracts features using data from multiple poses. Because of the sparseness of the sensor data, unconstrained SLAM typically requires many particles to deal with high uncertainty. We performed several experiments, using both simulated and real data, which show that incorporating prior knowledge and enforcing constraints leads to a significant improvement in the resulting maps and a reduction in estimation error.

6.1 Simulated Data

We first used a simple kinematic simulator based on an RWI MagellanPro robot to collect data from a small simulated environment with two groups of rectilinear features. The goal was to test the algorithm's capability to infer the existence of constraints between landmarks. Only the five range sensors at $0°, 45°, 90°, 135°$, and $180°$ were used (i.e.,). Noise was introduced by perturbing measurements and motions in proportion to their magnitude. For a laser measurement of range r, $\sigma_r = 0.01r$; for a motion consisting of a translation d and rotation ϕ, the robot's orientation was perturbed with $\sigma_\theta = 0.03d + 0.08\phi$, and its position with $\sigma_x = \sigma_y = 0.05d$.

Fig. 4 shows the results of RBPF SLAM with a rectilinearity prior (as described in Section 5, with $\delta = \frac{\pi}{10}$). The filter contained 20 particles and recovered the correct relative constraints. The edges of the the inner "box" were constrained, and the edges of the boundary were separately constrained.

A separate experiment compared the consistency of the rectilinearity-constrained filter and the unconstrained filter (all other filter parameters were kept identical, including number of particles). A filter is inconsistent if it significantly underestimates its own error. It has been shown that RBPF SLAM is generally inconsistent [1]; our experiments indicate that using prior knowledge and enforcing constraints improves (but does not guarantee) consistency.

Fig. 5 depicts the consistency analysis. The ground truth trajectory from the simulation was used to compute the normalized estimation error squared (NEES) [2, 1] of the robot's trajectory estimate. For ground truth pose s_t and estimate $\hat{s}_t$ with covariance $\hat{P}_{s_t}$ (estimated from the weighted particles assuming they are approximately normally distributed), the NEES is $(s_t - \hat{s}_t)\hat{P}_{s_t}^{-1}(s_t - \hat{s}_t)^T$. For more details of how NEES can be used to examine SLAM filter consistency, see [1]. The experiment used 200 particles for each of 50 Monte Carlo trials, with a robot model similar to the previous simulation.

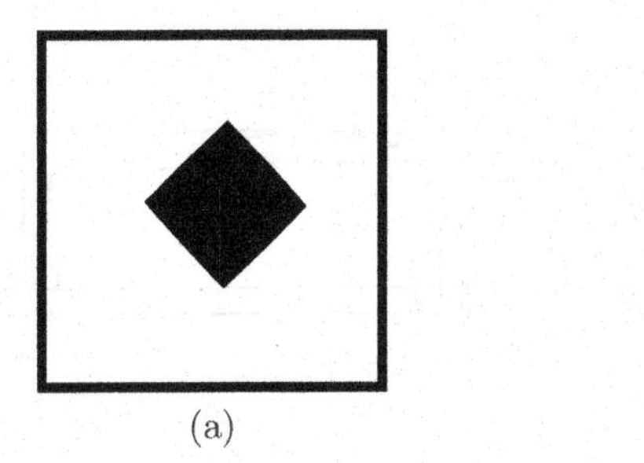

(a)

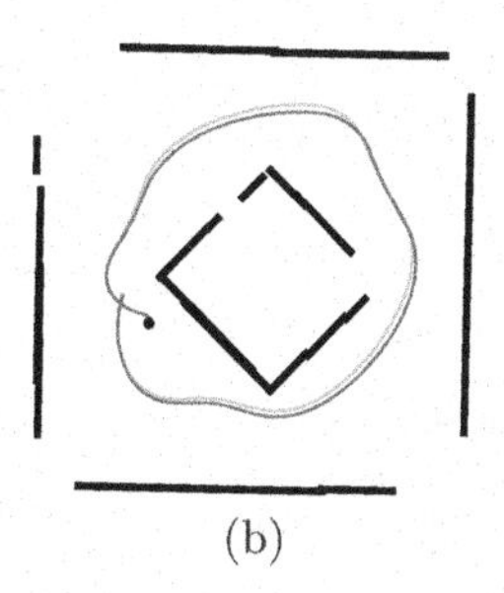

(b)

Fig. 4. (a) Simulated environment (ground truth). (b) Results of applying constrained SLAM. The dark curved line is the trajectory estimate, the light curved line is the ground truth trajectory, and the dot is the starting pose. The landmarks on the boundary form one constrained group; those in the interior form the other.

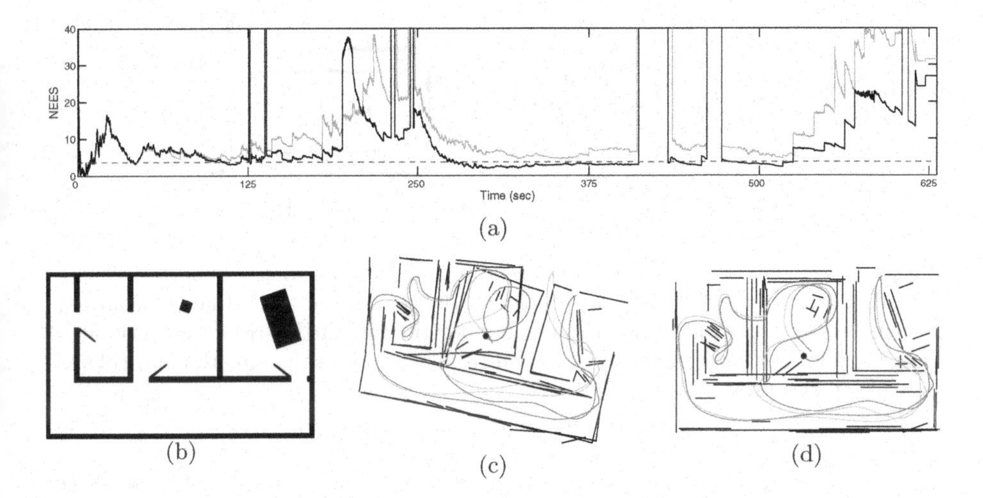

Fig. 5. (a) Normalized estimation error squared (NEES) of the robot's estimated pose with respect to the ground truth, computed over 50 Monte Carlo trials for the environment in (b). The gray plot is the error for standard (unconstrained) RBPF SLAM. The black plot is the error for our algorithm with rectilinearity constraints. Error significantly above the dashed line indicates an optimistic (inconsistent) filter. Our approach is less optimistic. (Sharp spikes correspond to degeneracies due to resampling upon loop closure.) (c) A typical map produced by unconstrained sparse sensing SLAM. (d) A typical rectilinearity-constrained map.

6.2 Real-World Data

Our real-world experiments used data from Radish [7], an online repository of SLAM datasets. Most of the datasets use scanning laser rangefinders. Since our goal is to enable SLAM with limited sensing, we simply discarded most of the data in each scan, keeping only the five range measurements at $0°, 45°, 90°, 135°$,

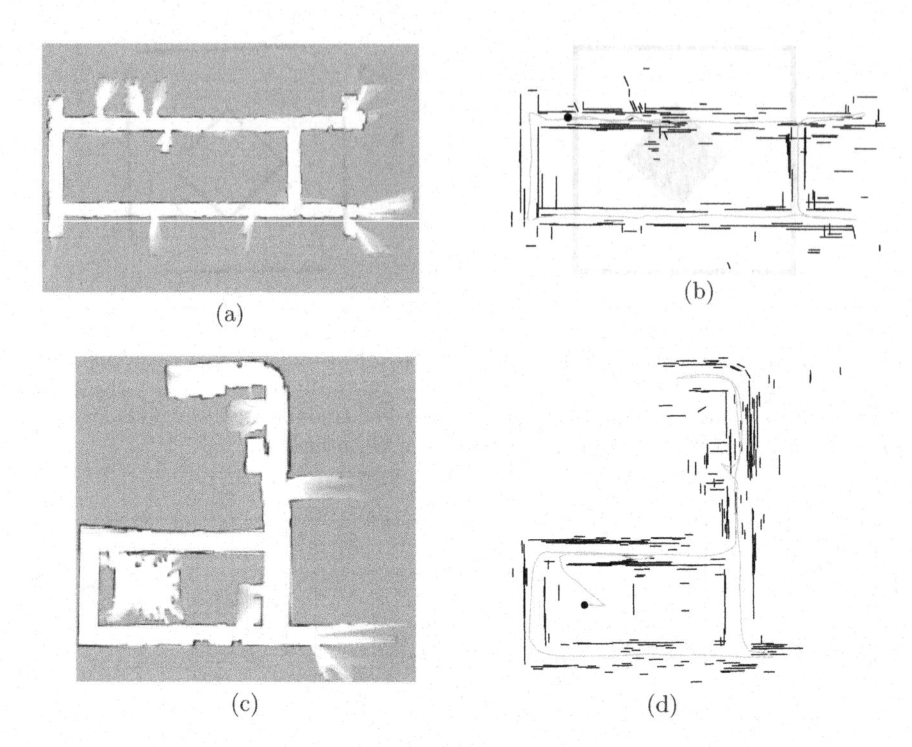

Fig. 6. (a) and (b) show the USC SAL Building, second floor (dataset courtesy of Andrew Howard). (c) and (d) show Newell-Simon Hall Level A at CMU (dataset courtesy of Nicholas Roy). (a) and (c) Occupancy data for the corrected trajectories (generated using the full laser data for clarity). (b) and (d) The estimated landmark maps (black) and trajectories (gray).

and 180°. We also restricted the sensor range (see Table 1). We used the same rectilinearity prior as for the simulated examples ($\delta = \frac{\pi}{10}$).

Fig. 6 shows the results of our algorithm for two datasets. The *USC SAL* dataset consists of a primary loop and several small excursions. Most landmarks are constrained, in three separate groups. For the *CMU NSH* experiment, the maximum sensing range was restricted to 3 m, so the large initial loop (bottom) could not be closed until the robot finished exploring the upper hallway. Aside from several landmarks in the curved portion of the upper hallway, most landmarks are constrained.

Table 1 gives mapping statistics. Also included is the number of particles required to successfully build an unconstrained map, along with running times for comparison. (The complete results for unconstrained sparse sensing SLAM can be found in [3].) All tests were performed on a P4-1.7 GHz computer with 1 GB RAM. Incorporating constraints enables mapping with many fewer particles — about the same number as needed by many unconstrained SLAM

Table 1. Experiment statistics

	USC SAL	CMU NSH
Dimensions	39×20 m^2	25×25 m^2
Particles (constrained)	20	40
Particles (unconstrained)	100	600
Avg. Runtime (constrained, 30 runs)	11.24 s	34.77 s
Avg. Runtime (unconstrained, 30 runs)	32.02 s	268.44 s
Sensing range	5 m	3 m
Path length	122 m	114 m
Num. landmarks	162	219
Constrained groups	3	3

algorithms that use full laser rangefinder information. This leads to significant computational performance increases when constraints are applicable.

One caveat is that the conditioning process is sensitive to the landmark cross-covariance estimates. (The cross-covariances are used in Eqns. 13-14 to compute a "gain" indicating how to change unconstrained variables when conditioning on constrained variables.) Because we use sensors that give very little data for feature extraction, the cross-covariance of $[r \ \ \theta]^T$ features is only approximately estimated. This leads to landmark drift in highly constrained environments since landmarks are frequently reconditioned, as can be seen in, e.g., the upper right corner of the NSH map in Fig. 6(d). Future research will examine alternative feature estimators and map representations (e.g., relative maps [10, 5]) that may alleviate this issue.

7 Conclusions

In this paper we have described a Rao-Blackwellized particle filter for SLAM that exploits prior knowledge of structural or geometrical relationships between landmarks. Relative constraints between landmarks in the map of each particle are automatically inferred based on the estimated landmark state. By partitioning the state into constrained and unconstrained variables, the constrained variables can be sampled by a particle filter. Conditioned on these samples, unconstrained variables are independent and can be estimated by EKFs on a per-particle basis.

We have implemented our approach with rectilinearity constraints and performed experiments on simulated and real-world data. For SLAM with sparse (low spatial resolution) sensing, incorporating constraints significantly reduced the number of particles required for map estimation.

Most of this work has focused on linear equality constraints. While we have described a way to extend the approach to inequality constraints, this remains an area for future work. Also, while constraints clearly help in mapping with limited sensing, they do not significantly improve data association inaccuracies related to sparse sensing, another potential avenue for improvement.

References

1. Bailey, T., Nieto, J., Nebot, E.: Consistency of the FastSLAM algorithm. In: IEEE Intl. Conf. on Robotics and Automation, pp. 424–427 (2006)
2. Bar-Shalom, Y., Li, X.R., Kirubarajan, T.: Estimation with applications to tracking and navigation. Wiley, New York (2001)
3. Beevers, K.R., Huang, W.H.: SLAM with sparse sensing. In: IEEE Intl. Conf. on Robotics and Automation, pp. 2285–2290 (2006)
4. Csorba, M., Durrant-Whyte, H.: New approach to map building using relative position estimates. SPIE Navigation and Control Technologies for Unmanned Systems II 3087(1), 115–125 (1997)
5. Deans, M., Hebert, M.: Invariant filtering for simultaneous localization and mapping. In: IEEE Intl. Conf. on Robotics and Automation, pp. 1042–1047 (2000)
6. Durrant-Whyte, H.: Uncertain geometry in robotics. IEEE Journal of Robotics and Automation 4(1), 23–31 (1988)
7. Howard, A., Roy, N.: The Robotics Data Set Repository (Radish) (2003)
8. Montemerlo, M.: FastSLAM: a factored solution to the simultaneous localization and mapping problem with unknown data association. PhD thesis, Carnegie Mellon University, Pittsburgh, PA (2003)
9. Murphy, K.: Bayesian map learning in dynamic environments. In: Advances in Neural Information Processing Systems, vol. 12, pp. 1015–1021. MIT Press, Cambridge (2000)
10. Newman, P.: On the structure and solution of the simultaneous localization and mapping problem. PhD thesis, University of Sydney, Australia (1999)
11. Rodriguez-Losada, D., Matia, F., Jimenez, A., Galan, R.: Consistency improvement for SLAM – EKF for indoor environments. In: IEEE Intl. Conf. on Robotics and Automation, pp. 418–423 (2006)
12. Simon, D., Chia, T.: Kalman filtering with state equality constraints. IEEE Transactions on Aerospace and Electronic Systems 39, 128–136 (2002)
13. Simon, D., Simon, D.: Aircraft turbofan engine health estimation using constrained Kalman filtering. In: ASME Turbo Expo. (2003)
14. Wen, W., Durrant-Whyte, H.: Model-based multi-sensor data fusion. In: IEEE Intl. Conf. on Robotics and Automation, pp. 1720–1726 (1992)

Second-Order Theory of Error Propagation on Motion Groups

Yunfeng Wang[1] and Gregory S. Chirikjian[2]

[1] Department of Mechanical Engineering, The College of New Jersey, Ewing, NJ 08628
jwang@tcnj.edu
[2] Department of Mechanical Engineering, Johns Hopkins University, Baltimore, MD 21218
gregc@jhu.edu

Summary. Error propagation on the Euclidean motion group arises in a number of areas such as and in dead reckoning errors in mobile robot navigation and joint errors that accumulate from the base to the distal end of manipulators. We address error propagation in rigid-body poses in a coordinate-free way. In this paper we show how errors propagated by convolution on the Euclidean motion group, $SE(3)$, can be approximated to second order using the theory of Lie algebras and Lie groups. We then show how errors that are small (but not so small that linearization is valid) can be propagated by a recursive formula derived here. This formula takes into account errors to second-order, whereas prior efforts only considered the first-order case [8, 9].

Keywords: Recursive error propagation, Euclidean group, spatial uncertainty.

1 Introduction

In this section we review the literature on error propagation, and review the terminology and notation used throughout the paper.

1.1 Literature Review

Murray, Li and Sastry [3], and Selig [4] presented Lie-group-theoretic notation and terminology to the robotics community, which has now become standard vocabulary. Chirikjian and Kyatkin [1] showed that many problems in robot kinematics and motion planning can be formulated as the convolution of functions on the Euclidean group. The representation and estimation of spatial uncertainty has also received attention in the robotics and vision literature. Two classic works in this area are due to Smith and Cheeseman [6] and Su and Lee [7]. Recent work on error propagation by Smith, Drummond and Roussopoulos [5] describes the concatenation of random variables on groups and applies this formalism to mobile robot navigation. In all three of these works, errors are assumed small enough that covariances can be propagated by the formula [8, 9]

$$\Sigma_{1*2} = Ad(g_2^{-1})\Sigma_1 Ad^T(g_2^{-1}) + \Sigma_2, \tag{1}$$

S. Akella et al. (Eds.): Algorithmic Foundation of Robotics VII, STAR 47, pp. 155–168, 2008.
springerlink.com

where Ad is the adjoint operator for $SE(3)$. This equation essentially says that given two 'noisy' frames of reference $g_1, g_2 \in SE(3)$, each of which is a Gaussian random variable with 6×6 covariance matrices[1] Σ_1 and Σ_2, respectively, the covariance of $g_1 \circ g_2$ will be Σ_{1*2}. This approximation is very good when errors are very small. We extend this linearized approximation to the quadratic terms in the expansion of the matrix exponential parametrization of $SE(3)$. Results for $SO(3)$ are generated in the process.

1.2 Review of Rigid-Body Motions

The Euclidean motion group, $SE(3)$, is the semi direct product of $\mathbb{R}^3$ with the special orthogonal group, $SO(3)$. We denote elements of $SE(3)$ as $g = (\mathbf{a}, A) \in SE(3)$ where $A \in SO(3)$ and $\mathbf{a} \in \mathbb{R}^3$. For any $g = (\mathbf{a}, A)$ and $h(\mathbf{r}, R) \in SE(3)$, the group law is written as $g \circ h = (\mathbf{a} + A\mathbf{r}, AR)$, and $g^{-1} = (-A^T\mathbf{a}, A^T)$. Alternately, one may represent any element of $SE(3)$ as a 4×4 homogeneous transformation matrix of the form

$$H(g) = \begin{pmatrix} A & \mathbf{a} \\ \mathbf{0}^T & 1 \end{pmatrix},$$

in which case the group law is matrix multiplication.

For small translational (rotational) displacements from the identity along (about) the ith coordinate axis, the homogeneous transforms representing infinitesimal motions look like

$$H_i(\epsilon) \stackrel{\triangle}{=} \exp(\epsilon \tilde{E}_i) \approx I_{4\times 4} + \epsilon \tilde{E}_i$$

where

$$\tilde{E}_1 = \begin{pmatrix} 0 & 0 & 0 & 0 \\ 0 & 0 & -1 & 0 \\ 0 & 1 & 0 & 0 \\ 0 & 0 & 0 & 0 \end{pmatrix}; \quad \tilde{E}_2 = \begin{pmatrix} 0 & 0 & 1 & 0 \\ 0 & 0 & 0 & 0 \\ -1 & 0 & 0 & 0 \\ 0 & 0 & 0 & 0 \end{pmatrix}; \quad \tilde{E}_3 = \begin{pmatrix} 0 & -1 & 0 & 0 \\ 1 & 0 & 0 & 0 \\ 0 & 0 & 0 & 0 \\ 0 & 0 & 0 & 0 \end{pmatrix};$$

$$\tilde{E}_4 = \begin{pmatrix} 0 & 0 & 0 & 1 \\ 0 & 0 & 0 & 0 \\ 0 & 0 & 0 & 0 \\ 0 & 0 & 0 & 0 \end{pmatrix}; \quad \tilde{E}_5 = \begin{pmatrix} 0 & 0 & 0 & 0 \\ 0 & 0 & 0 & 1 \\ 0 & 0 & 0 & 0 \\ 0 & 0 & 0 & 0 \end{pmatrix}; \quad \tilde{E}_6 = \begin{pmatrix} 0 & 0 & 0 & 0 \\ 0 & 0 & 0 & 0 \\ 0 & 0 & 0 & 1 \\ 0 & 0 & 0 & 0 \end{pmatrix}.$$

These are related to the basis elements $\{E_i\}$ for $so(3)$ as

$$\tilde{E}_i = \begin{pmatrix} E_i & \mathbf{0} \\ \mathbf{0}^T & 0 \end{pmatrix}$$

when $i = 1, 2, 3$.

[1] Exactly what is meant by a covariance for a Lie group is quantified later in the paper.

Large motions are also obtained by exponentiating these matrices. For example,

$$\exp(t\tilde{E}_3) = \begin{pmatrix} \cos t & -\sin t & 0 & 0 \\ \sin t & \cos t & 0 & 0 \\ 0 & 0 & 1 & 0 \\ 0 & 0 & 0 & 1 \end{pmatrix} \quad \text{and} \quad \exp(t\tilde{E}_6) = \begin{pmatrix} 1 & 0 & 0 & 0 \\ 0 & 1 & 0 & 0 \\ 0 & 0 & 1 & t \\ 0 & 0 & 0 & 1 \end{pmatrix}.$$

More generally, it can be shown that every element of a matrix Lie group G can be described with the exponential parametrization

$$g = g(x_1, x_2, ..., x_N) = \exp\left(\sum_{i=1}^{N} x_i \tilde{E}_i\right). \tag{2}$$

This kind of relationship is common in the study of Lie groups and algebras.

One defines the 'vee' operator, $\vee$, such that

$$\left(\sum_{i=1}^{N} x_i \tilde{E}_i\right)^{\vee} = \begin{pmatrix} x_1 \\ x_2 \\ x_3 \\ \vdots \\ x_N \end{pmatrix}$$

The vector, $\mathbf{x} \in \mathbb{R}^N$, can be obtained from $g \in G$ from the formula

$$\mathbf{x} = (\log g)^{\vee}. \tag{3}$$

When integrating a function over a group in a neighborhood of the identity, a weight $w(\mathbf{x})$ is defined as

$$\int_G f(g)dg = \int_{\mathbb{R}^N} f(g(\mathbf{x}))w(\mathbf{x})d\mathbf{x}.$$

It may be shown that due to the nature of the exponential parameterization, $w(\mathbf{x}) = 1 + O(\|\mathbf{x}\|^2)$ near the identity, and so the approximation $w(\mathbf{x}) = 1$ can be used in the first order theory. However, in the current presentation we retain $w(\mathbf{x})$ for higher order errors.

We calculate

$$w(\mathbf{x}) = \det\left[\left(g^{-1}\frac{\partial g}{\partial x_1}\right)^{\vee}, \cdots, \left(g^{-1}\frac{\partial g}{\partial x_N}\right)^{\vee}\right]. \tag{4}$$

If the approximation $g = I + X + X^2/2 + X^3/6$ is used, then to second order we can write

$$w(\mathbf{x}) = 1 - \frac{1}{2}\mathbf{x}^T K \mathbf{x}$$

for some matrix K that depends on the structure of the group. K is computed for $SO(3)$ and $SE(3)$ in the Appendix.

2 Nonparametric Second-Order Theory

Let $g_1, g_2 \in SE(3)$ be two precise reference frames. Then $g_1 \circ g_2$ is the frame resulting from stacking one relative to the other. Now suppose that each has some uncertainty. Let $\{h_i\}$ and $\{k_j\}$ be two sets of frames of reference that are distributed around the identity. Let the first have N_1 elements, and the second have N_2. What will the covariance of the set of $N_1 \cdot N_2$ frames $\{(g_1 \circ g_2)^{-1} \circ g_1 \circ h_i \circ g_2 \circ k_j\}$ (which are assumed to be distributed around the identity) look like ?

A pdf, ρ, on a Lie group G is said to have mean at the identity if the function

$$C(g) = \int_G \|[\log(g^{-1} \circ h)]^\vee\|^2 \rho(h) dh$$

is minimized at $g = e$. For this kind of pdf, the covariance is defined as

$$\Sigma = \int_G \log(g)^\vee [\log(g)^\vee]^T \rho(g) dg. \tag{5}$$

A similar expression can be defined for discrete cloud of frames, which is equivalent to replacing $\rho(g)$ with a weighted sum of Dirac delta functions.

Let $\rho_i(g)$ be a unimodal pdf with mean at the identity and which has a preponderance of its mass concentrated in a unit ball around the identity (where distance from the identity is measured as $\|(\log g)^\vee\|$). Then $\rho_i(g_i^{-1} \circ g)$ will be a distribution with the same shape centered at g_i. In general, the convolution of two pdfs is defined as

$$(f_1 * f_2)(g) = \int_G f_1(h) f_2(h^{-1} \circ g) dh,$$

and in particular if we make the change of variables $k = g_1^{-1} \circ h$, then

$$\rho_1(g_1^{-1} \circ g) * \rho_2(g_2^{-1} \circ g) = \int_G \rho_1(k) \rho_2(g_2^{-1} \circ k^{-1} \circ g_1^{-1} \circ g) dk.$$

Making the change of variables $g = g_1 \circ g_2 \circ q$, where q is a relatively small displacement measured from the identity, the above can be written as

$$\rho_{1*2}(g_1 \circ g_2 \circ q) = \int_G \rho_1(k) \rho_2(g_2^{-1} \circ k^{-1} \circ g_2 \circ q) dk. \tag{6}$$

The essence of this paper is the efficient approximation of covariances associated with (6) when those of ρ_1 and ρ_2 are known. This problem reduces to the efficient approximation of

$$\Sigma_{1*2} = \int_G \int_G \log(q)^\vee [\log(q)^\vee]^T \rho_1(k) \rho_2(g_2^{-1} \circ k^{-1} \circ g_2 \circ q) dk dq. \tag{7}$$

In many practical situations, discrete data are sampled from ρ_1 and ρ_2 rather than having complete knowledge of the distributions themselves. Therefore, sampled covariances can be computed by making the following substitutions:

$$\rho_1(g) = \sum_{i=1}^{N_1} \alpha_i \Delta(h_i^{-1} \circ g) \tag{8}$$

and

$$\rho_2(g) = \sum_{j=1}^{N_2} \beta_j \Delta(k_j^{-1} \circ g) \tag{9}$$

where

$$\sum_{i=1}^{N_1} \alpha_i = \sum_{j=1}^{N_2} \beta_j = 1.$$

Here $\Delta(g)$ is the Dirac delta function for the group G, which has the properties

$$\int_G f(g)\Delta(h^{-1} \circ g)dg = f(h) \quad \text{and} \quad \Delta(h^{-1} \circ g) = \Delta(g^{-1} \circ h).$$

Using these properties, if we substitute (8) into (5), the result is

$$\Sigma_1 = \int_G \log(g)^\vee [\log(g)^\vee]^T \sum_{i=1}^{N_1} \alpha_i \Delta(h_i^{-1} \circ g)dg = \sum_{i=1}^{N_1} \alpha_i \log(h_i)^\vee [\log(h_i)^\vee]^T \tag{10}$$

Substitution of the sampled ρ_1 into (7) yields

$$\Sigma_{1*2} = \sum_{i=1}^{N_1} \alpha_i \int_G \log(q)^\vee [\log(q)^\vee]^T \rho_2(g_2^{-1} \circ h_i^{-1} \circ g_2 \circ q)dq. \tag{11}$$

Similarly, substitution of the sampled ρ_2 into the above equation kills the integral and substitutes values of q for which $g_2^{-1} \circ h_i^{-1} \circ g_2 \circ q = k_j$. This yields

$$\Sigma_{1*2} = \sum_{i=1}^{N_1} \sum_{j=1}^{N_2} \alpha_i \beta_j \log(g_2^{-1} \circ h_i \circ g_2 \circ k_j)^\vee [\log(g_2^{-1} \circ h_i \circ g_2 \circ k_j)^\vee]^T. \tag{12}$$

While this equation is exact, it has the drawback of requiring $O(N_1 \cdot N_2)$ arithmetic operations. In the first-order theory of error propagation, we made the approximation

$$\log(k^{-1} \circ q) = X - Y,$$

or equivalently

$$[\log(k^{-1} \circ q)]^\vee = \mathbf{x} - \mathbf{y},$$

where $k = \exp Y$ and $q = \exp X$ are elements of the Lie group $SE(3)$. This decouples the summations and makes the computation $O(N_1 + N_2)$. However, the first-order theory breaks down for large errors. Therefore, we explore here a second-order theory that has the benefits of greater accuracy, while retaining good computational performance.

In the second-order theory of error propagation on Lie groups (and $SE(3)$ in particular), we now make the approximation

$$\log(k^{-1} \circ q) = X - Y + \frac{1}{2}[X, Y],$$

or equivalently

$$[\log(k^{-1} \circ q)]^{\vee} = \mathbf{x} - \mathbf{y} + \frac{1}{2} ad(\mathbf{x})\mathbf{y}. \tag{13}$$

Interestingly, terms such as X^2 and Y^2 do not appear in the approximation (13).

Here $[\cdot, \cdot]$ denotes the Lie bracket, and

$$[X, Y] = \sum_{i,j,k} C_{ij}^{k} x_i y_j \tilde{E}_k,$$

which means that the kth component of $[X, Y]^{\vee}$ will be of the form $\sum_{i,j} C_{ij}^{k} x_i y_j$ which is a weighted product of elements from $\mathbf{x}$ and $\mathbf{y}$. We therefore write

$$[X, Y]^{\vee} = \mathbf{x} \wedge \mathbf{y} = ad(X)\mathbf{y}.$$

$ad(X)$ should not be confused with $Ad(g)$, which is defined by $Ad(g)\mathbf{x} = (gXg^{-1})^{\vee}$. The relationship between these two is $Ad(\exp X) = \exp(ad(X))$. See the Appendix for a more complete review.

In addition to the approximation in (13) we use two additional properties of the log function:

$$[\log(k^{-1})]^{\vee} = -[\log(k)]^{\vee} \tag{14}$$

and

$$[\log(g \circ h \circ g^{-1})]^{\vee} = Ad(g)[\log(h)]^{\vee}. \tag{15}$$

Using (13), (14) and (15), then to second order,

$$\log(g_2^{-1} \circ h_i \circ g_2 \circ k_j)^{\vee} = Ad(g_2^{-1})\mathbf{y}_i + B_i \mathbf{z}_j$$

where $\mathbf{y}_i = (\log h_i)^{\vee}$, $\mathbf{z}_j = (\log k_j)^{\vee}$, and $B_i = B(Ad(g_2^{-1})\mathbf{y}_i)$ where

$$B(\mathbf{x}) = I + \frac{1}{2} ad(\mathbf{x}).$$

Note also that $B(\mathbf{y})\mathbf{y} = \mathbf{y}$ because $[Y, Y] = 0$, and therefore $[B(\mathbf{y})]^{-1}\mathbf{y} = \mathbf{y}$ as well. Substitution into the formula (12) for Σ_{1*2} then yields

$$\Sigma_{1*2} = \sum_{i=1}^{N_1} \sum_{j=1}^{N_2} \alpha_i \beta_j (B_i \mathbf{z}_j + A\mathbf{y}_i)(B_i \mathbf{z}_j + A\mathbf{y}_i)^T$$

where $A = Ad(g_2^{-1})$.

Assuming that the sampled distributions are centered around the identity (so that cross terms sum to zero), allows the summations over i and j to decouple. The result is written as

$$\Sigma_{1*2} = A\Sigma_1 A^T + \sum_{i=1}^{N_1} \alpha_i B_i \Sigma_2 B_i^T \tag{16}$$

In practice, $\alpha_i = 1/N_1$ and $\beta_j = 1/N_2$.

Note, in the first order theory we approximated $B_i = I$, and the above reduced to

$$\Sigma_{1*2} = Ad_{g_2^{-1}} \Sigma_1 Ad^T_{g_2^{-1}} + \Sigma_2.$$

3 Numerical Examples

Evaluating the robustness of the first-order (1) and the second-order (16) covariance propagation formula over a wide range of kinematic errors is essential to understand effectiveness of these formulas. In this section, we test these two covariance propagation formulas with concrete numerical examples.

Consider a spatial serial manipulator, PUMA 560. The link-frame assignments of PUMA 560 for D-H parameters is the same as those given [2]. Table 1 lists the D-H parameters of PUMA 560, where $a_2 = 431.8$ mm, $a_3 = 20.32$ mm, $d_3 = 124.46$ mm, and $d_4 = 431.8$ mm. The solution of forward kinematics is the homogeneous transformations of the relative displacements from one D-H frame to another multiplied sequentially.

In order to test these covariance propagation formulas, we first need to create some kinematic errors. Since joint angles are the only variables of the PUMA 560, we assume that errors exist only in these joint angles. We generated errors by deviating each joint angle from its ideal value with uniform random absolute errors of $\pm\epsilon$. Therefore, each joint angle was sampled at three values: $\theta_i - \epsilon$, θ_i, $\theta_i + \epsilon$. This generates $n = 3^6$ different frames of references $\{g^i_{ee}\}$ that are clustered around desired g_{ee}. Here g_{ee} denotes the position and orientation of the distal end of the manipulator relative to the base in the form of homogeneous transformation matrix.

Three different methods for computing the same error covariances for the whole manipulator are computed. The first is to apply brute force enumeration, which gives the actual covariance of the whole manipulator:

$$\Sigma = \frac{1}{n} \sum_{i=1}^{n} \mathbf{x}_i \mathbf{x}_i^T \tag{17}$$

where $\mathbf{x}_i = [\log(g^{-1} \circ g_i)]^\vee$, and the formula (17) is used to all the 3^6 different frames of references $\{g^i_{ee}\}$. The second method is to apply the first-order propagation formula (1). The third is to apply the second-order propagation formula (16). For the covariance propagation methods, we only need to find the covariance of each individual link. Then the covariance of the whole manipulator can be recursively calculated using the corresponding propagation formula. In our case, all the individual links have the same covariance since we assumed the same kinematic errors at each joint angle.

In order to quantify the robustness of the two covariance approximation methods, we define a measure of deviation of results between the first/second order formula and the actual covariance using the Hilbert-Schmidt (Frobenius) norm as

Table 1. DH PARAMETERS of PUMA 560

i	α_{i-1}	a_{i-1}	d_i	θ_i
1	0	0	0	θ_1
2	-90°	0	0	θ_2
3	0	a_2	d_3	θ_3
4	-90°	a_3	d_4	θ_4
5	90°	0	0	θ_5
6	-90°	0	0	θ_6

$$deviation = \frac{\|\Sigma_{prop} - \Sigma_{actual}\|}{\|\Sigma_{actual}\|}, \tag{18}$$

where Σ_{prop} is the covariance of the whole manipulator calculated using either the first-order (1) or the second-order (16) propagation formula, Σ_{actual} is the actual covariance of the whole manipulator calculated using (17), and $\|\cdot\|$ denotes the Hilbert-Schmidt (Frobenius) norm.

With all the above information, we now can conduct the specific computation and analysis. Our numerical simulations have showed that different configurations of the manipulator will not influence the end-effector covariances too much. Here the ideal joint angles from θ_1 to θ_6 were taken as $[0, \pi/2, -\pi/2, 0, 0, \pi/2]$. The joint angle errors ϵ were taken from 0.1 rad to 0.6 rad. The covariances of the whole manipulator corresponding to these kinematic errors were then calculated through the three aforementioned methods. The results of the first-order and second-order propagation formula were graphed in Fig. 1 in terms of deviation defined by Eq. (18). It was shown that the second-order propagation formula makes significant improvements in terms of accuracy than that of the first-order formula. The second-order propagation theory is much more robust than the first-order formula over a wide range of kinematic errors. These two methods both work well for small errors, and deviate from the actual value more and more as the errors become large. However, the deviation of the first-order formula grows rapidly and breaks down while the second-order propagation method still retains a reasonable value.

To give the readers a sense of what these covariances look like, we listed the values of the covariance of the whole manipulator for the joint angle error $\epsilon = 0.3$ rad below.

The ideal pose of the end effector can be found easily via forward kinematics to be

$$g_{ee} = \begin{pmatrix} 0.0000 & -1.0000 & 0 & 0.0203 \\ -1.0000 & -0.0000 & 0 & 0.1245 \\ 0 & 0 & -1.0000 & -0.8636 \\ 0 & 0 & 0 & 1.0000 \end{pmatrix}.$$

The actual covariance of the whole manipulator calculated using equations (17) is

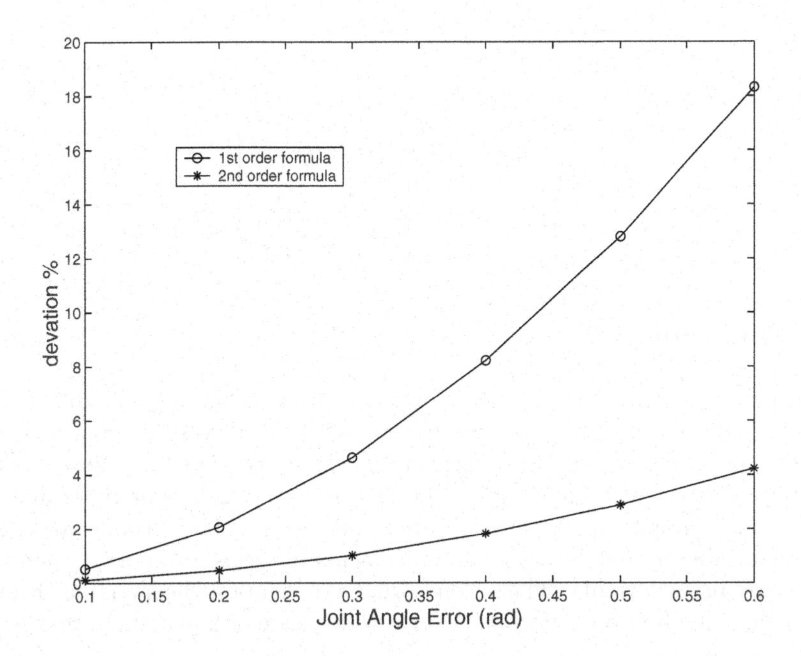

Fig. 1. The Deviation of the First and Second-order Propagation Methods

$$\Sigma_{actual} = \begin{pmatrix} 0.1748 & 0.0000 & 0.0000 & 0.0000 & -0.0755 & -0.0024 \\ 0.0000 & 0.0078 & 0.0000 & 0.0034 & -0.0000 & 0.0003 \\ 0.0000 & 0.0000 & 0.1747 & 0.0012 & -0.0072 & -0.0000 \\ 0.0000 & 0.0034 & 0.0012 & 0.0025 & -0.0001 & 0.0001 \\ -0.0755 & -0.0000 & -0.0072 & -0.0001 & 0.0546 & 0.0015 \\ -0.0024 & 0.0003 & -0.0000 & 0.0001 & 0.0015 & 0.0011 \end{pmatrix},$$

the covariance using the first-order propagation formula (1) is

$$\Sigma_{prop,\,1st} = \begin{pmatrix} 0.1800 & 0.0000 & 0 & 0.0000 & -0.0777 & -0.0024 \\ 0.0000 & 0.0000 & 0 & 0.0000 & -0.0000 & -0.0000 \\ 0 & 0 & 0.1800 & 0.0012 & -0.0075 & 0 \\ 0.0000 & 0.0000 & 0.0012 & 0.0000 & -0.0002 & -0.0000 \\ -0.0777 & -0.0000 & -0.0075 & -0.0002 & 0.0569 & 0.0016 \\ -0.0024 & -0.0000 & 0 & -0.0000 & 0.0016 & 0.0000 \end{pmatrix},$$

and the covariance using the second-order propagation formula (16) is

$$\Sigma_{prop,\,2nd} = \begin{pmatrix} 0.1765 & 0.0000 & 0.0000 & 0.0000 & -0.0762 & -0.0024 \\ 0.0000 & 0.0079 & 0.0000 & 0.0034 & -0.0000 & 0.0003 \\ 0.0000 & 0.0000 & 0.1765 & 0.0012 & -0.0072 & 0.0000 \\ 0.0000 & 0.0034 & 0.0012 & 0.0025 & -0.0001 & 0.0001 \\ -0.0762 & -0.0000 & -0.0072 & -0.0001 & 0.0551 & 0.0015 \\ -0.0024 & 0.0003 & 0.0000 & 0.0001 & 0.0015 & 0.0011 \end{pmatrix},$$

where the covariance of one link is

$$\Sigma_{one-link} = \begin{pmatrix} 0 & 0 & 0 & 0 & 0 & 0 \\ 0 & 0 & 0 & 0 & 0 & 0 \\ 0 & 0 & 0.0600 & 0 & 0 & 0 \\ 0 & 0 & 0 & 0 & 0 & 0 \\ 0 & 0 & 0 & 0 & 0 & 0 \\ 0 & 0 & 0 & 0 & 0 & 0 \end{pmatrix}.$$

4 Conclusions

In this paper, first-order kinematic error propagation formulas are modified to include second-order effects. This extends the usefulness of these formulas to errors that are not necessarily small. In fact, in the example to which the methodology is applied, errors in orientation can be as large as a radian or more and the second-order formula appears to capture the error well. The second-order propagation formula makes significant improvements in terms of accuracy than that of the first-order formula. The second-order propagation theory is much more robust than the first-order formula over a wide range of kinematic errors.

References

1. Chirikjian, G.S., Kyatkin, A.B.: Engineering Applications of Noncommutative Harmonic Analysis. CRC Press, Boca Raton (2001)
2. Craig, J.J.: Introduction to Robotics Mechanics and Control, 3rd edn. Prentice Hall, Englewood Cliffs (2005)
3. Murray, R.M., Li, Z., Sastry, S.S.: A Mathematical Introduction to Robotic Manipulation. CRC Press, Ann Arbor MI (1994)
4. Selig, J.M.: Geometrical Methods in Robotics. Springer, New York (1996)
5. Smith, P., Drummond, T., Roussopoulos, K.: Computing MAP trajectories by representing, propagating and combining PDFs over groups. In: Proceedings of the 9th IEEE International Conference on Computer Vision, vol. II, pp. 1275–1282, Nice (2003)
6. Smith, R.C., Cheeseman, P.: On the Representation and Estimation of Spatial Uncertainty. The International Journal of Robotics Research 5(4), 56–68 (1986)
7. Su, S., Lee, C.S.G.: Manipulation and Propagation of Uncertainty and Verification of Applicability of Actions in assembly Tasks. IEEE Transactions on Systems, Man, and Cybernetics 22(6), 1376–1389 (1992)
8. Wang, Y., Chirikjian, G.S.: Error Propagation on the Euclidean Group with Applications to Manipulator Kinematics. IEEE Transactions on Robotics (in press)
9. Wang, Y., Chirikjian, G.S.: Error Propagation in Hybrid Serial-Parallel Manipulators. In: IEEE International Conference on Robotics and Automation, Orlando Florida, pp. 1848–1853 (May 2006)

A Appendix

A.1 Background

Given $g \in SE(3)$ of the form,

$$g = \begin{pmatrix} R & \mathbf{t} \\ \mathbf{0}^T & 1 \end{pmatrix}$$

and $X \in se(3)$ of the form

$$X = \begin{pmatrix} \Omega & \mathbf{v} \\ \mathbf{0}^T & 0 \end{pmatrix},$$

if

$$\mathbf{x} = (X)^\vee = \begin{pmatrix} \boldsymbol{\omega} \\ \mathbf{v} \end{pmatrix},$$

then $Ad(g)$ is defined by the expression

$$(gXg^{-1})^\vee = Ad(g)\mathbf{x}$$

and explicitly

$$Ad(g) = \begin{pmatrix} R & 0 \\ TR & R \end{pmatrix}. \tag{19}$$

The matrix T is skew-symmetric, and $\text{vect}(T) = \mathbf{t}$.

Similarly, $ad(X)$ (which can also be written as $ad(\mathbf{x})$), is defined by

$$[X, Y]^\vee = ad(X)\mathbf{y},$$

where $[X, Y] = XY - YX$ is the Lie bracket. Explicitly,

$$ad(X) = \begin{pmatrix} \Omega & 0 \\ V & \Omega \end{pmatrix} \tag{20}$$

where the matrix V is skew-symmetric, and $\text{vect}(V) = \mathbf{v}$.

A.2 Second Order Approximation of Volume Weighting Function for $SO(3)$ and $SE(3)$

Let g be an element of the matrix Lie group G, and X be an arbitrary element of the associated Lie algebra, $\mathcal{G}$. Let $g = \exp X$. Then we can truncate the Taylor series expansion for g and g^{-1} as:

$$g = I + X + X^2/2 + X^3/6 + O(X^4) \quad \text{and} \quad g^{-1} = I - X + X^2/2 - X^3/6 + O(X^4).$$

If $X = \sum_i x_i E_i$, this means that

$$\frac{\partial g}{\partial x_i} = E_i + \frac{1}{2}(E_i X + X E_i) + \frac{1}{6}(E_i X^2 + X E_i X + X^2 E_i) + O(X^3).$$

Therefore,

$$g^{-1}\frac{\partial g}{\partial x_i} = E_i + \frac{1}{2}[E_i, X] - \frac{1}{3}XE_iX + \frac{1}{6}(E_iX^2 + X^2E_i) + O(X^3).$$

Taking the $\vee$ of both sides yields the columns of the Jacobian matrix, the determinant of which provides the desired weighting function. Note that $(E_i)^\vee = \mathbf{e}_i$ and $([E_i, X])^\vee = -([X, E_i])^\vee = -ad(X)\mathbf{e}_i$, and so we can write the ith column as:

$$\left(g^{-1}\frac{\partial g}{\partial x_i}\right)^\vee = \mathbf{e}_i - \frac{1}{2}ad(X)\mathbf{e}_i - \frac{1}{3}(XE_iX)^\vee + \frac{1}{6}(E_iX^2 + X^2E_i)^\vee. \quad (21)$$

If we define,

$$J_1(\mathbf{x}) = [(XE_1X)^\vee, (XE_2X)^\vee, (XE_3X)^\vee]$$

and

$$J_2(\mathbf{x}) = [(E_1X^2 + X^2E_1)^\vee, (E_2X^2 + X^2E_2)^\vee, (E_3X^2 + X^2E_3)^\vee],$$

then to second order,

$$J(\mathbf{x}) = I - \frac{1}{2}ad(X) - \frac{1}{3}J_1(\mathbf{x}) + \frac{1}{6}J_2(\mathbf{x}). \quad (22)$$

Details for *SO*(3)

In the case of $SO(3)$,

$$X = \begin{pmatrix} 0 & -x_3 & x_2 \\ x_3 & 0 & -x_1 \\ -x_2 & x_1 & 0 \end{pmatrix}$$

and $ad(X) = X$. Direct calculation shows that the matrix $J_1^{so(3)}$ can be written as

$$J_1^{so(3)}(\mathbf{x}) = -\begin{pmatrix} x_1^2 & x_1x_2 & x_1x_3 \\ x_1x_2 & x_2^2 & x_2x_3 \\ x_1x_3 & x_2x_3 & x_3^2 \end{pmatrix}$$

Similarly, for $J_2^{so(3)}$ one finds

$$J_2^{so(3)}(\mathbf{x}) = -\begin{pmatrix} 2x_1^2 + x_2^2 + x_3^2 & x_1x_2 & x_1x_3 \\ x_1x_2 & x_1^2 + 2x_2^2 + x_3^2 & x_2x_3 \\ x_1x_3 & x_2x_3 & x_1^2 + x_2^2 + 2x_3^2 \end{pmatrix}$$

Now, to second order, the full Jacobian is

$$J^{so(3)}(\mathbf{x}) = I - \frac{1}{2}X - \frac{1}{3}J_1^{so(3)}(\mathbf{x}) + \frac{1}{6}J_2^{so(3)}(\mathbf{x}) = I - \frac{1}{2}X + \frac{1}{6}X^2,$$

where of course

$$X^2 = \begin{pmatrix} -x_2^2 - x_3^2 & x_1x_2 & x_1x_3 \\ x_1x_2 & -x_1^2 - x_3^2 & x_2x_3 \\ x_1x_3 & x_2x_3 & -x_1^2 - x_2^2 \end{pmatrix}$$

The second-order approximation of the determinant of $J^{so(3)}(\mathbf{x})$ is then

$$w^{so(3)}(\mathbf{x}) = \det J^{so(3)}(\mathbf{x}) \approx \det(I - \frac{1}{2}X) \cdot \det(I + \frac{1}{6}X^2) \approx (1 + \frac{1}{4}\|\mathbf{x}\|^2)(1 - \frac{1}{3}\|\mathbf{x}\|^2).$$

The reason why this is justified is that all terms in the cofactor expansion of the det depend on X^2 will be of higher than second order, except those on the diagonal. This is due to the fact that second-order terms here will multiply the diagonal entries of the identity matrix yielding second-order terms.

Finally, this means that to second order,

$$w^{so(3)}(\mathbf{x}) = 1 - \frac{1}{2}\mathbf{x}^T K \mathbf{x} \quad \text{where} \quad K = \frac{1}{6}I \tag{23}$$

Details for $SE(3)$

It is convenient to write an arbitrary element of $se(3)$ as

$$X = \begin{pmatrix} \Omega & \mathbf{v} \\ \mathbf{0}^T & 0 \end{pmatrix}$$

where Ω is an arbitrary element of $so(3)$ and $\mathbf{v}$ is an arbitrary element of $\mathbb{R}^3$.

In this case,

$$ad(X) = \begin{pmatrix} \Omega & 0 \\ V & \Omega \end{pmatrix}$$

where $(V)^\vee = \mathbf{v}$.

Referring back to (21), we can compute each term directly to find:

$$(XE_iX)^\vee = \begin{pmatrix} (\Omega E_i \Omega)^\vee \\ \Omega E_i \mathbf{v} \end{pmatrix}$$

for rotational components ($i = 1, 2, 3$) and

$$(XE_iX)^\vee = \mathbf{0}$$

for transitional components ($i = 4, 5, 6$). This means that

$$J_1^{se(3)}(\omega, \mathbf{v}) = \begin{pmatrix} J_1^{so(3)}(\omega) & 0 \\ \omega \wedge (I \wedge \mathbf{v}) & 0 \end{pmatrix}.$$

Similarly,

$$(E_iX^2 + X^2E_i)^\vee = \begin{pmatrix} (E_i\Omega^2 + \Omega^2 E_i)^\vee \\ E_i \Omega \mathbf{v} \end{pmatrix}.$$

for rotational components ($i = 1, 2, 3$) and

$$(E_i X^2 + X^2 E_i)^\vee = \begin{pmatrix} \mathbf{0} \\ \Omega^2 \mathbf{e}_i \end{pmatrix}.$$

for transitional components ($i = 4, 5, 6$) and

$$J_2^{se(3)}(\boldsymbol{\omega}, \mathbf{v}) = \begin{pmatrix} J_2^{so(3)}(\boldsymbol{\omega}) & 0 \\ (\Omega \mathbf{v})^\wedge & \Omega^2 \end{pmatrix}.$$

Substituting into (22) and taking the determinant,

$$\det J^{se(3)}(\boldsymbol{\omega}, \mathbf{v}) = |\det J^{so(3)}(\boldsymbol{\omega})|^2$$

This means that to second order,

$$w^{se(3)}(\mathbf{x}) = 1 - \frac{1}{2}\mathbf{x}^T K \mathbf{x} \quad \text{where} \quad K = \begin{pmatrix} \frac{1}{3}I & 0 \\ 0 & 0 \end{pmatrix}. \tag{24}$$

Extensive Representations and Algorithms for Nonlinear Filtering and Estimation

Ethan Stump[1], Ben Grocholsky[2], and Vijay Kumar[1]

[1] GRASP Laboratory, University of Pennsylvania
{estump,kumar}@seas.upenn.edu
[2] Robotics Institute, Carnegie Mellon University
grocholsky@ri.cmu.edu

Summary. Most estimation problems in robotics are difficult because of (a) the nonlinearity in observation models; and (b) the lack of suitable probabilistic models for the process and observation noise. In this paper we develop a set-valued approach to estimation that overcomes both these limitations and illustrates the application to localization of multiple, mobile sensor platforms with range sensors.

1 Introduction

Practical estimation tasks require us to deal with nonlinearities that are inherent in process dynamics and observation models. Common solutions to deal with such nonlinear state transition and measurement models require linearization of at least some portion of the problem. The concern is that linearization can lead to inconsistent error handling, removes the ability to directly represent ambiguous confidence sets, and can not be applied when the underlying state is unobservable.

Many techniques have been proposed to either avoid or delay linearization. Some of the most popular in recent years are sampling-based approaches such as Monte Carlo Localization, introduced by Fox *et al.* [9]. They rely on estimating a probability distribution for the system state, but instead of maintaining a simple parameterized distribution, which may require linearization, the distribution is discretely sampled to allow for arbitrary densities. Further refinements have allowed the fusion of sampled representations with standard parameterized ones to solve more challenging problems such as simultaneous localization and mapping tasks [10].

The opposite approach is to delay linearization by simply storing all measurement and state updates until a later time when the larger data base allows for the use of consistency to improve the estimates. The GraphSLAM algorithm presented by Thrun, Burgard, and Fox [4] is an example of such a technique; after several measurements have been taken, an estimate is formed by iteratively linearizing the state propagation and measurement equations and solving a least squares problem to maximize agreement with measurements and problem dynamics. Several different variations on this theme have been used by others such as Folkesson and Christensen [11] and Konolige [12].

S. Akella et al. (Eds.): Algorithmic Foundation of Robotics VII, STAR 47, pp. 169–184, 2008.
springerlink.com

The disadvantage of such delayed measurement integration is that best estimates are not available during data acquisition, making it impossible to knowledgeably improve the collection process. As a compromise, Thrun *et al.* [4] propose the Sparse Extended Information Filter to proactively incorporate each measurement as it is taken while being able to explicitly manage the information links between different entities formed through measurements and motion. The final picture allows for intuitive identification of how features are related but still requires linearization.

For many applications, such linearization may prove to be acceptable, but not for our current application of localization using range-only measurements. In the most general form of this problem, there is no sense of direction and so any attempt to linearize a range measurement will likely result in crippling inconsistency after further measurements and motion. Compelling sampling-based estimation approaches to this problem have been demonstrated by Djugash, Singh and Corke [5], but such implementations may require large numbers of samples making them computationally unattractive.

An important evolution in the methodology of Information-type filters was presented by Hanebeck [6]. The central idea is a nonlinear embedding, sometimes referred to as an over-parameterization, that maps the system states into a extended state space in a way such that the measurement equations become linear. The resulting framework lends itself to the application of the methodology introduced by Schweppe in the field of dynamic estimation under bounded noise [3]. Successful application of such set-based estimation techniques to static localization tasks involving range-only measurements and relative bearing measurements are demonstrated in [7]. While set-based techniques have been investigated by other researchers [14,15,16,17], Hanebeck's representation allows exact representations in the extended state space.

We build on Hanebeck's work and address the range-only localization problem that is frequently encountered in robotics. First, we present techniques that allow estimates of the actual state from the extended state representation. In addition, we present the first incorporation of dynamics into the framework, bringing the system closer to useful implementation on mobile robotics. The resulting filter is simple, robust, recursive, and avoids linearization. The forms are Information-like and so the filter behaves much like the SEIF [4] when it comes to identifying information links between different entities and watching how these links change through motion.

In Section 2, we present notation and equations for a mobile robot system using range-only measurements in Section 2. This work makes extensive use of ellipsoidal calculus, which we introduce in Section 3. The nonlinear transformation framework is presented and demonstrated in Section 4 and we present new techniques for approximate inversion of the filtered sets in Section 5. Finally, we introduce motion in Section 6 and discuss in Section 7.

2 Problem Formulation

We consider a mobile sensor network equipped with relative-range measurement capabilities with some capability of local sensing such as odometry or inertial measurements. This network consists of n standard nodes that have either unknown or only partially known positions and m anchor nodes that have fully known positions with respect to some global reference frame. For convenience, we will assume these m nodes are stationary.

Expressed in the global frame, the position of the ith standard node is a variable, $\bar{x}_i = [\, x_i \; y_i \,]^T$, and the position of the lth anchor node is a constant, $\bar{a}_l$. The total state of the network is $\tilde{x} = [\, \bar{x}_1^T \, \dots \, \bar{x}_n^T \,]^T$, and belongs to the space $\mathcal{S} = \Re^{2n}$. A measurement between standard node i and standard node j has the form:

$$z_{ij} = h_{ij}(\tilde{x}) + e = \|\bar{x}_i - \bar{x}_j\| + e \tag{1}$$

while a measurement between standard node i and anchor node l has the form:

$$z_i^l = h_i^l(\tilde{x}) + e = \|\bar{x}_i - \bar{a}_l\| + e \tag{2}$$

These measurements have noise e; for the purposes of this paper we assume that this noise is bounded with constant bound ϵ. Thus $e \in [-\epsilon, \epsilon]$. These assumptions could be relaxed to include other models of bounded noise.

In a mobile sensor network, the n standard nodes can be considered to be attached to mobile robots. We adopt, for simplicity, a point model:

$$\bar{x}_{i,k+1} = \bar{x}_{i,k} + \bar{u}_{i,k}, \quad i = 1,\, 2,\, \dots\, n \tag{3}$$

where $\bar{u}_i$ is the control input for the ith mobile node at time k. The state of the system evolves discretely, with the dynamic transition from step k to $k+1$ given by (3), and a set of inter-node measurements taken at each step k:

$$\tilde{z}_k = \tilde{h}_k(\tilde{x}_k) + \tilde{e}_k$$

where $\tilde{h}_k$ is a combination of the measurement types expressed in (1) and (2).

3 Ellipsoids

Because we rely extensively on the results in [2], we now summarize their notation and definitions. We will use x, x_0 to denote the state and z to denote observations without worrying about the notation $(\bar{\cdot})$ or $(\tilde{\cdot})$ in this section.

An ellipsoid can be defined by two quantities: a vector specifying the position of its center, and a symmetric positive semi-definite matrix that encodes the directions and lengths of its semi-axes as the eigenvectors and eigenvalues respectively. Given $x_0 \in \Re^n$ and $E \in \mathcal{S}_+^n$, an n-dimensional ellipsoid is defined by the set:

$$\varepsilon_n(x_0, E) = \left\{ \, x \;\middle|\; (x - x_0)^T E (x - x_0) \le 1 \, \right\} \tag{4}$$

If E is singular, then the resulting ellipsoid is degenerate and possesses directions, corresponding to the eigenvectors of the zero eigenvalues, where x is unconstrained. The center in this case is actually only a single representative point of the affine set at the center of the ellipsoid.

3.1 Fusion

Analogous to the fusion operation in sensor fusion, we define fusion for set valued estimates to be the operation that takes two ellipsoids and finds an ellipsoid that tightly bounds their intersection. The minimum-volume bounding ellipsoid can be found using iterative algorithms such as that of [13], but, as noted there, the complexity of this procedure is an open problem. However, the suboptimal approach taken in [2], repeated here, involves the minimization of a convex function over a bounded interval and so is simple and fast. Given two n-dimensional ellipsoids, $B_1 = \varepsilon_n(x_1, E_1)$ and $B_2 = \varepsilon_n(x_2, E_2)$, a one-parameter family of fusing ellipsoids is $\varepsilon_n^\lambda(x_0, E)$, $\lambda \in [0, 1]$, defined by:

$$\begin{aligned} X &= \lambda E_1 + (1 - \lambda)E_2 \\ k &= 1 - \lambda(1 - \lambda)(x_2 - x_1)^T E_2 X^{-1} E_1 (x_2 - x_1) \\ x_0 &= X^{-1}(\lambda E_1 x_1 + (1 - \lambda)E_2 x_2) \\ E &= \frac{1}{k} X \end{aligned} \tag{5}$$

and the fused ellipsoid is taken as the ε_n^λ that has minimum volume. This amounts to either solving a bounded minimization problem over λ using the above, or finding the zero of the derivative of the volume as in Theorem 3 of [2]. This approximate intersection is denoted by $\tilde{\cap}$:

$$\varepsilon_n(x_0, E) \leftarrow \varepsilon_n(x_1, E_1) \,\tilde{\cap}\, \varepsilon_n(x_2, E_2)$$

3.2 Propagations

We have interest in two different ellipsoid propagations, both paralleling the state operations of our system given by (1), (2), and (3). Theorem 1 of [2] provides the general operation that is specialized to these two special cases.

We consider first a 1-dimensional ellipsoid associated with a single measurement (also seen as an interval) $\varepsilon_1(z, 1/\epsilon^2)$. If this measurement is obtained with a linear observation model, $z = H_{1\times n}x$, its pre-image is the n-dimensional degenerate ellipsoid: $\varepsilon_n(H^\dagger z, (1/\epsilon^2)H^T H)$. $H^\dagger z$ can be any solution of the linear map, but we will use $H^\dagger$ as the pseudoinverse of H; note that the ellipsoid is rank 1 due to the form of its matrix. Thus there is an $n - 1$ dimensional affine set of points that are consistent with this one dimensional observation.

Given an n-dimensional ellipsoid $\varepsilon_n(x_0, E)$, its image under the linear map $y = A_{n\times n}x + b_{n\times 1}$ is the n-dimensional ellipsoid: $\varepsilon_n(Ax_0 + b, AEA^T)$. This is the same expression encountered in propagation of Gaussian distributions in linear systems theory.

3.3 Slicing

It can be shown that the intersection of an n-dimensional ellipsoid $\varepsilon_n(x_0, E)$ with an m-dimensional ($m \leq n$) affine set in $\Re^n$, $\mathcal{A} = \{\, y_0 + Yc \mid c \in \Re^m \,\}$, produces an ellipsoid $\varepsilon_m(\eta, F')$, where:

$$
\begin{aligned}
F &= Y^T E Y \\
\eta &= F^{\dagger} Y^T E (x_0 - y_0) \\
k &= 1 - (x_0 - y_0)^T E (x_0 - y_0) + \eta^T F \eta \\
F' &= \frac{1}{k} F
\end{aligned}
\tag{6}
$$

If $k < 0$ then the affine set and the ellipsoid do not intersect.

3.4 Projection

An ellipsoid projection finds the "shadow" of an ellipsoid in some of its components. It can be shown that the projection of the n-dimensional ellipsoid $B = \varepsilon_n(x_0, E)$ onto its first m components is given by:

$$
\mathcal{P}(B) = \varepsilon_m(x_{0,m}, E_{11} - E_{12} E_{22}^{-1} E_{12}^T) \tag{7}
$$

where $x_{0,m}$ are the first m components of x_0, and $E = \begin{bmatrix} E_{11} & E_{12} \\ E_{12}^T & E_{22} \end{bmatrix}$, with E_{11} as an m-dimensional block.

Furthermore, if Y is a basis for any subspace of dimension m and Z is a basis for its null space of dimension $n - m$, then the ellipsoid $B = \varepsilon_n(x_0, E)$ projected onto this subspace is given by $B_Y = \varepsilon_m(Y^T x_0, E_Y)$, with E_Y given by:

$$
E_Y = Y^T E Y - Y^T E Z (Z^T E Z)^{-1} Z^T E Y \tag{8}
$$

The projection operation for ellipsoids is analogous to marginalization of multivariate Gaussians (see [4]).

4 Nonlinear Embedding

Hanebeck [6] introduced a novel framework involving a nonlinear embedding that maps the system states into an extended state space in such a way that the measurement equations become linear. The basic idea is shown in Figure 1.

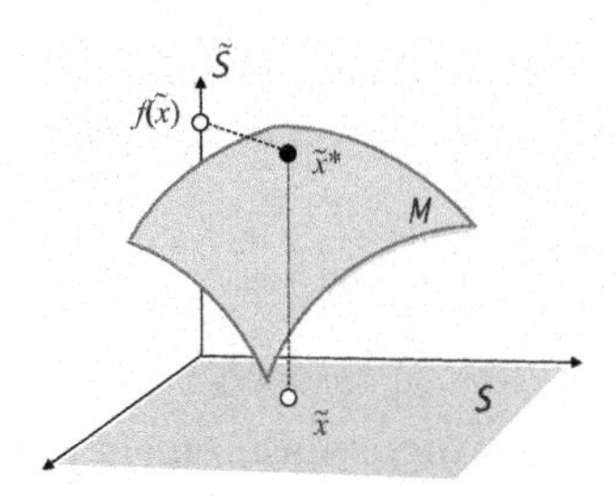

Fig. 1. The state space $\mathcal{S}$ is extended by adding functionally dependent coordinates of $\mathcal{S}$ to create the extended state space $\mathcal{S}^\star = \mathcal{S} \oplus \tilde{\mathcal{S}}$

The *extended space* $\mathcal{S}^{\star}$ is formed by augmenting the *base space* $\mathcal{S}$ with additional dimensions. Define a smooth map $f : \mathcal{S} \to \tilde{\mathcal{S}}$. Recall that $\tilde{x}$ denotes elements of $\mathcal{S}$ and let $\tilde{x}^{\star}$ denote elements of $\mathcal{S}^{\star}$. The map f defines a $2n$-dimensional smooth sub-manifold in $\mathcal{S}^{\star}$ with coordinates $\tilde{x}$:

$$\tilde{x}^{\star} = g(\tilde{x}) = \begin{bmatrix} \tilde{x} \\ f(\tilde{x}) \end{bmatrix}$$

The Jacobian of g,

$$\frac{\partial g}{\partial \tilde{x}} = \begin{bmatrix} I \\ \frac{\partial f}{\partial \tilde{x}} \end{bmatrix}$$

is always full rank and the resulting manifold M is diffeomorphic to $\mathcal{S}$. Our goal is to choose f and transform the system equations in such a way as to make them linear in the p-dimensional extended space $\mathcal{S}^{\star}$ ($p > 2n$). Note that we will ultimately be interested only in those points that lie on the manifold, *i.e.*, $\tilde{x}^{\star} \in M$. The procedure is illustrated with an example next.

It should be noted that we have no automatic procedure for choosing f and instead rely on inspection of the system equations. However, Hanebeck [6] suggests that general polynomial bases such as Bernstein polynomials could be useful in this regard.

4.1 Application to Range Measurements

As an illustration, begin with the range measurement equation between two standard nodes:

$$z_{ij} - e = \|\bar{x}_i - \bar{x}_j\|$$

and square both sides. The left hand side represents the interval $[z_{ij} - \epsilon, z_{ij} + \epsilon]$ which, when squared using interval arithmetic, becomes $[z_{ij}^2 - 2z_{ij}\epsilon + \epsilon^2, z_{ij}^2 + 2z_{ij}\epsilon + \epsilon^2]$. By letting $z_{ij}^{\star} = z_{ij}^2 + \epsilon^2$ and $w \in [-2z_{ij}\epsilon, 2z_{ij}\epsilon]$, the transformed measurement equation is:

$$z_{ij}^{\star} = \bar{x}_i \cdot \bar{x}_i + \bar{x}_j \cdot \bar{x}_j - 2\bar{x}_i \cdot \bar{x}_j + w$$

If a different bounded-noise model is used for e, such a transformation is still possible as long as care is taken to ensure that the modified estimate and bounds are conservative.

Notice that this equation is nonlinear in the system variables $\bar{x}_i$ and $\bar{x}_j$ but is linear in the variables $\bar{x}_i \cdot \bar{x}_i$, $\bar{x}_j \cdot \bar{x}_j$, and $\bar{x}_i \cdot \bar{x}_j$:

$$z_{ij}^{\star} = [1\ 1\ {-2}] \begin{bmatrix} \bar{x}_i \cdot \bar{x}_i \\ \bar{x}_j \cdot \bar{x}_j \\ \bar{x}_i \cdot \bar{x}_j \end{bmatrix} + w$$

Applying this process to the range measurement equation between a standard node and an anchor node leads to:

$$z_i^{l,\star} - \bar{a}_l \cdot \bar{a}_l = [-2\bar{a}_l^T\ 1] \begin{bmatrix} \bar{x}_i \\ \bar{x}_i \cdot \bar{x}_i \end{bmatrix} + w$$

Accordingly, we define f so that:

$$f(\tilde{x}) = [\ldots, \bar{x}_i \cdot \bar{x}_i, \bar{x}_j \cdot \bar{x}_j, \bar{x}_i \cdot \bar{x}_j, \ldots]^T.$$

Thus $\mathcal{S}^\star$ is constructed by adding at most $n + {}^nC_2$ dimensions to $\mathcal{S}$, corresponding to all dot product combinations of the positions of the nodes that appear in the measurement equations. The measurement equations, after suitable modifications to the additive noise, are now linear in $\mathcal{S}^\star$ while having bounded noise.

We are not limited to range-only sensors. Indeed, the measurement equations for bearing-only sensors can also be made linear with a similar embedding (see [7]).

4.2 Recursive Filtering

We saw that an appropriate definition of the map f allows us to write each measurement equation at time step k in the form:

$$z_k^\star - w_k^\star = H_k^\star \tilde{x}_k^\star$$

where $\tilde{x}_k^\star \in \mathcal{S}^\star$. By viewing the interval quantity on the left-hand side as a 1-dimensional ellipsoid, we apply the results of Section 3.2 to define $\mathcal{Z}_k = \varepsilon_P((H_k^\star)^\dagger z_k^\star, (1/(w_k^\star)^2)(H_k^\star)^T H_k^\star)$ as the feasibility ellipsoid in $\mathcal{S}^\star$ consistent with this measurement.

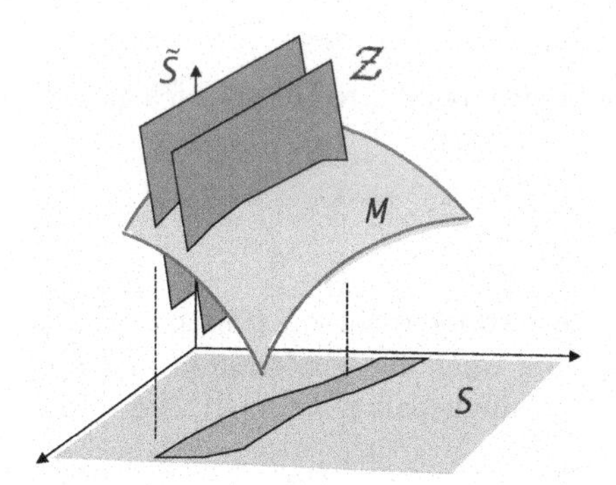

Fig. 2. A measurement $\mathcal{Z}$, transformed into a set bounded by two hyperplanes, defines an interesting set in the base space $\mathcal{S}$ when intersected with M

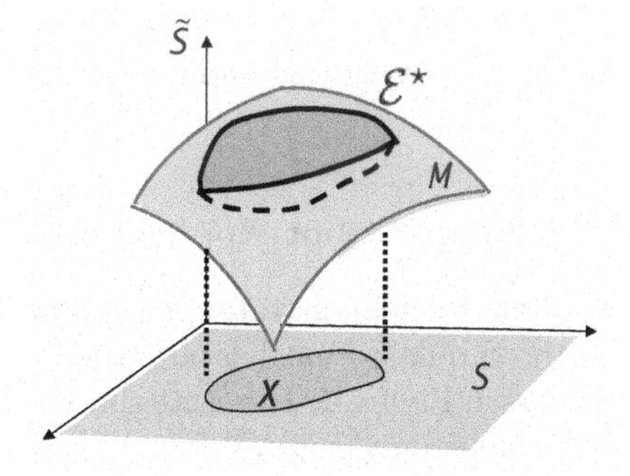

Fig. 3. The feasibility set $\mathcal{X} \subset \mathcal{S}$ is found by intersecting an ellipsoid $\mathcal{E}^\star \subset \mathcal{S}^\star$ with the manifold M

Each $\mathcal{Z}_k$ can be seen as a pair of bounding hyperplanes constraining the possible embedded states. However, since only the values of $\mathcal{S}^\star$ lying in M have meaning, the actual set described by $\mathcal{Z}_k$ can be interesting, as portrayed in Figure 2. As more measurements are included, more bounds need to be incorporated. However, rather than tracking an increasing number of hyperplanes, each new $\mathcal{Z}_k$ is incorporated into an aggregate state estimate ellipsoid $\mathcal{E}_k^\star$ using the fusion introduced in Section 3.1:

$$\mathcal{E}_k^\star \leftarrow \mathcal{E}_k^\star \,\tilde{\cap}\, \mathcal{Z}_k$$

When the filtering is started, $\mathcal{E}_k^\star$ can be initialized with $\varepsilon_p(0, 0)$ to reflect the fact that nothing is known about the state. This is analogous to the initialization of the Information form of the Kalman filter (see [4]). Unlike the case of an extended Kalman filter, problematic estimate initialization procedures are not necessary.

4.3 Set Inversion

After performing filtering steps, the feasibility ellipsoid $\mathcal{E}_k^\star$ contains all $\tilde{x}_k^\star$ consistent with measurements up to step k, but not all of these elements have physical meaning. Only the $\tilde{x}_k^\star \in M$ actually represent the images of states in $\mathcal{S}$. This feasible set, $\mathcal{X}_k \subset \mathcal{S}$ is found by:

$$\mathcal{X}_k = \{\, \tilde{x} \in \mathcal{S} \mid g(\tilde{x}) \in \mathcal{E}_k^\star \,\}$$

The basic idea is shown in Figure 3.

This inversion can be carried out exactly using the implicit form (4) of $\mathcal{E}_k^\star = \varepsilon_p(x_0, E)$ together with g:

$$(g(\tilde{x}_k) - x_0)^T E (g(\tilde{x}_k) - x_0) \leq 1 \qquad (9)$$

Any $\tilde{x}_k$ satisfying this implicit nonlinear inequality belong to the true feasibility set.

4.4 Single Robot Application

In order to demonstrate the operation of this framework, we present results for the simulated localization of one robot using range measurements to known anchors in Figure 4. The state space is given by a single pair $\tilde{x} = [\,x\ y\,]^T \in \mathcal{S} = \Re^2$ and is embedded into $\mathcal{S}^\star$ by

$$\tilde{x}^\star = [\,x,\, y,\, x^2 + y^2\,]^T. \qquad (10)$$

The only measurements are from the single standard node to one of n anchor nodes at positions $\bar{a}_i$, $i \in \{1, \ldots, \text{n}\}$. The transformed measurement equation to anchor i is given by:

$$((z^i)^2 + \epsilon^2 - \bar{a}_i \cdot \bar{a}_i) + w^i = [\,-2\bar{a}_i^T\ 1\,]\, \tilde{x}^\star$$

where z^i is the measured range from the robot to the anchor, ϵ is the symmetric noise bound on this measurement, and $w^i \in [-2\epsilon z^i, 2\epsilon z^i]$.

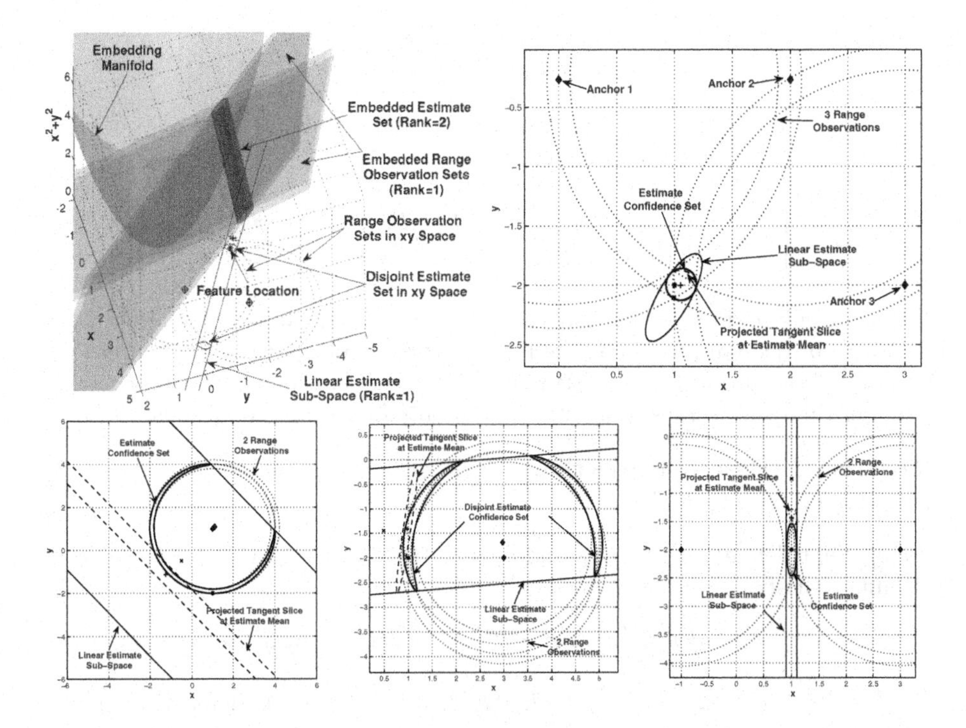

Fig. 4. Visualization of the proposed embedding for a single robot range-only localization problem along with examples of possible estimate uncertainties represented by this formulation. In this case a closed form expression is available for the position estimate set. This set can take the form of an annulus or either a single or two disjoint transformed ellipses. The linear estimate sub-space and projected tangent slice results are discussed in Section 5.

5 Approximate Inversion

Since (9) describes the feasible set in $\mathcal{S}$ as a single nonlinear implicit inequality in $\dim(\mathcal{S})$ variables, it may not be suitable for finding feasible sets of large systems. We propose two methods for approximating this inversion: the first makes use of the projection idea introduced in Section 3.4 and is provably conservative, and the second makes use of the recognition of the embedded manifold M in an intuitive yet not provably conservative result. In order to demonstrate the effectiveness of these approximations, the simulation begun in Section 4.4 is scaled up to include more standard nodes, making it impossible to meaningfully use the exact inversion (9).

5.1 Base Projection

Working within an extended space, an ellipsoid $\varepsilon_p(x_0, E)$ can have its parameters split according to terms belong to the base space and the augmentations:

$$x_0 = \begin{bmatrix} x_{0,L} \\ x_{0,N} \end{bmatrix} \quad E = \begin{bmatrix} E_L & E_C \\ {E_C}^T & E_N \end{bmatrix}$$

with L, N, and C denoting linear, nonlinear, and coupling terms, respectively.

The exact inversion would be, using (9):

$$\mathcal{X}_k = \left\{ \tilde{x} \in \mathcal{S} \;\middle|\; \begin{bmatrix} \tilde{x} - x_{0,L} \\ f(\tilde{x}) - x_{0,N} \end{bmatrix}^T \begin{bmatrix} E_L & E_C \\ {E_C}^T & E_N \end{bmatrix} \begin{bmatrix} \tilde{x} - x_{0,L} \\ f(\tilde{x}) - x_{0,N} \end{bmatrix} \leq 1 \right\}$$

but using ellipsoid projection gives a conservative bound on $\mathcal{X}_k$. We also refer to this base projection process as the linear estimate sub-space in the figures. The validity of the process follows straight from the operation of projection, but a proof of the result serves to identify potentially important conditions:

Theorem 1 (Base Projection). *If* E_N *is invertible, then* $\mathcal{X}_k \subset \varepsilon_{2n}(x_{0,L}, E_L - E_C E_N^{-1} E_C^T)$.

Proof. First define $y \equiv \tilde{x} - x_{0,L}$ and $q(y) \equiv f(y + x_{0,L}) - x_{0,N}$. Then, points in $\mathcal{X}_k$ must satisfy:

$$y^T E_L y \leq 1 - 2y^T E_C q(y) - q(y)^T E_N q(y)$$

The ellipsoid $\varepsilon_N(x_{0,L}, E_L - E_C E_N^{-1} E_C^T)$, once shifted, is defined by the inequality:

$$y^T (E_L - E_C E_N^{-1} E_C^T) y \leq 1$$

and so all points of $\mathcal{X}_k$ belong to this ellipsoid if:

$$1 - 2y^T E_C q(y) - q(y)^T E_N q(y) - y^T E_C E_N^{-1} E_C^T y \leq 1$$

The maximum of the left hand side is found by the program:

$$\max_{y} \quad 1 - \begin{bmatrix} y \\ q(y) \end{bmatrix}^T \begin{bmatrix} E_C E_N^{-1} E_C^T & E_C \\ {E_C}^T & E_N \end{bmatrix} \begin{bmatrix} y \\ q(y) \end{bmatrix} \tag{11}$$

Call this matrix A; its Schur complement is $S = E_C E_N^{-1} E_C^T - E_C E_N^{-1} E_C^T = 0$. Using results from [1], $A \succeq 0$ since $S \succeq 0$, and so the program (11) has a finite maximum at 1. Thus, $\mathcal{X}_k \subset \varepsilon_n(x_{0,L}, E_L - E_C E_N^{-1} E_C^T)$. ◇

The first question to ask is whether it can be expected that E_N will be full rank. Intuition suggests that this will be true after many measurements have been incorporated. Consider the fact that the nonlinear terms from f have only been introduced to correspond with terms in the measurement equations. After a full set of measurements have been taken, each nonlinear term will have shown up at least once in the measurement equations and so information will be known about it.

However, this intuition only makes sense in well-conditioned cases where there are several interconnected measurement that can serve to isolate the contributions of each nonlinear term to the final estimate. If this is not the case, then further reference to [1] tells us that the claim $S \succeq 0 \implies A \succeq 0$ made above will be true if $(I - E_N E_N^\dagger)E_C = 0$.

These conditions only come about because we would like to completely ignore the details of the nonlinear transformation when trying to approximate the true set. If the simple tests fail, then we must take f into account, and the base projection is only conservative if it can be shown that:

$$\inf_y \begin{bmatrix} y \\ q(y) \end{bmatrix}^T \begin{bmatrix} E_C E_N^\dagger E_C^T & E_C \\ E_C{}^T & E_N \end{bmatrix} \begin{bmatrix} y \\ q(y) \end{bmatrix} \geq 0$$

over the domain in question.

If the base projection is indeed conservative, then the coordinates of the true state must lie within it.

5.2 Tangent Slices

We now present an alternate technique that does take the contributions of the nonlinear transformation into account when approximating the true set inversion. By recognizing M as a manifold embedded into $\mathcal{S}^\star$ by g, the tangent space of M at a point $\tilde{x}^\star \in M$ can be found using the Jacobian of g.

Now $\frac{\partial g}{\partial \tilde{x}}$ and $\tilde{x}^\star$ define an affine set in $\mathcal{S}^\star$ with the same dimension as $\mathcal{S}$. By restricting $\mathcal{E}_k^\star$ to this set using the slicing operation of Section 3.3, an approximate representation is found in $\mathcal{S}$.

Use of this operation requires two steps: first finding a single point of M to use, and then calculating the Jacobian at this point and slicing $\mathcal{E}_k^\star$. The point could be chosen by an optimization procedure that sought to find the closest suitable point to the mean of $\mathcal{E}_k^\star = \varepsilon_p(x_0, E)$:

$$\min_{\tilde{x}} \quad \|g(\tilde{x}) - x_0\|_2$$

where x_0 could possibly be an affine set, $x_0 + Null(E)\lambda$, if $\mathcal{E}_k^\star$ has degenerate directions.

This optimization problem has the potential to be very nasty and the effects of choosing a non-optimal point are not known. As a simple heuristic, we have taken $\tilde{x}^\star = g(x_{0,L})$ as the point to linearize about.

The over- or under-estimation of the true set would seem to be intimately related with the curvature of M, but we have no proofs regarding the quality of this approximation, only demonstrations of its use in the following examples. Accordingly, the true state coordinates need not lie within the tangent slice estimate.

5.3 Multiple Robot Application

As a comparison of the set approximation techniques, we present simulation results of an experiment with two standard nodes and three anchor nodes in Figure 5. Both the base projection and tangent slicing approximations are compared against a brute force calculation found by griding the x, y coordinates of both standard nodes and then checking for inclusion in $\mathcal{E}_k^\star$ by using the implicit form (9).

The system state consists of a two position vectors $\tilde{x} = [\bar{x}_1^T \; \bar{x}_2^T]^T = [x_1 \; y_1 \; x_2 \; y_2]^T \in \mathcal{S} = \Re^4$ and the state is transformed into $\mathcal{S}^\star$ according to:

$$\tilde{x}^\star = \left[\bar{x}_1, \bar{x}_2, \bar{x}_1 \cdot \bar{x}_1, \bar{x}_2 \cdot \bar{x}_2, \bar{x}_1 \cdot \bar{x}_2 \right]^T \tag{12}$$

As a representative example, the measurement between standard node 1 and anchor i would take the form:

$$((z_1^i)^2 + \epsilon^2 - \bar{a}_i \cdot \bar{a}_i) + w_1^i = \left[-2\bar{a}_i^T \; 0_{1\times 2} \; 1 \; 0 \; 0 \right] \tilde{x}^\star$$

where z_1^i is the measured range from robot 1 to the anchor, ϵ is the symmetric noise bound on this measurement, and $w_1^i \in [-2\epsilon z_1^i, 2\epsilon z_1^i]$. Measurements to robot 2 take a similar form.

The inter-robot measurement is:

$$((z_{1,2})^2 + \epsilon^2) + w_{1,2} = \left[0_{1\times 2} \; 0_{1\times 2} \; 1 \; 1 \; -2 \right] \tilde{x}^\star$$

where $z_{1,2}$ is the measured range from robot 1 to robot 2, ϵ is the symmetric noise bound on this measurement, and $w_{1,2} \in [-2\epsilon z_{1,2}, 2\epsilon z_{1,2}]$.

When calculating the tangent slice approximation, the Jacobian of this system is:

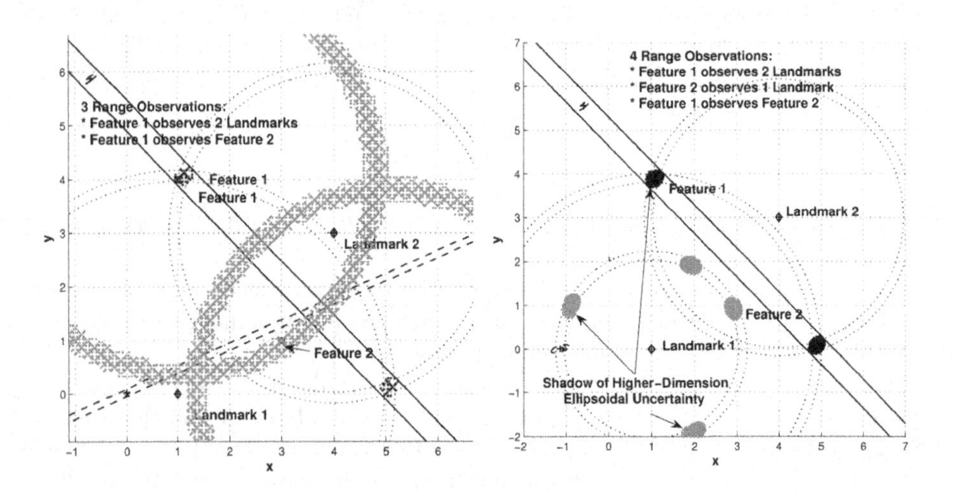

Fig. 5. Two robot localization examples that illustrate the ability of the proposed representation to capture complex structured estimate uncertainty. Note that the base projection of the higher-dimensional uncertainty bounds a smaller estimate recoverable from the intersection of $\mathcal{E}_k^\star$ with M.

$$\frac{\partial T}{\partial \tilde{x}} = \begin{bmatrix} 1 & 0 & 0 & 0 \\ 0 & 1 & 0 & 0 \\ 0 & 0 & 1 & 0 \\ 0 & 0 & 0 & 1 \\ 2x_1 & 2y_1 & 0 & 0 \\ 0 & 0 & 2x_2 & 2y_2 \\ x_2 & y_2 & x_1 & y_1 \end{bmatrix}$$

This example is taken further by including more standard nodes in Figure 6. The brute force representation of the feasible set is computationally daunting, so we present only the approximate representations.

6 Incorporating Motion

The static examples seen so far have shown the power of the representation, but have failed to take advantage of the dynamics of the problem. In applications, the incorporation of motion often transforms a poorly posed problem and makes it possible to estimate the state effectively. In [6], Hanebeck makes no use of dynamics, and further work in [7] and [8] also deals only with static systems. The main problem with incorporating dynamics into this framework is due to the fact that the chosen embedding renders the measurement equations linear but does not necessarily do so for the state update and, even worse, can make linear dynamics become nonlinear.

However, for the problem at hand, we will show that dynamic updates can in fact be incorporated using the fairly restrictive assumptions of a point model and perfect input. Despite these limitations, this inclusion represents a step forward for the theory and relaxations may be possible.

We consider a single robot once more and apply the transformation (10) to the dynamics given by (3). By inserting (3) into (10), we derive an expression relating the state at a given time step to the state at a previous time step:

$$\begin{aligned} D\begin{bmatrix} x \\ y \\ x^2 + y^2 \end{bmatrix} &= \begin{bmatrix} x + u_x \\ y + u_y \\ (x + u_x)^2 + (y + u_y)^2 \end{bmatrix} \\ &= \begin{bmatrix} 1 & 0 & 0 \\ 0 & 1 & 0 \\ 2u_x & 2u_y & 1 \end{bmatrix} \tilde{x}^\star + \begin{bmatrix} u_x \\ u_y \\ u_x^2 + u_y^2 \end{bmatrix} \end{aligned}$$

A similar derivation is possible for the two robot case as well, using (12):

$$D\begin{bmatrix} \bar{x}_1 \\ \bar{x}_2 \\ \bar{x}_1 \cdot \bar{x}_1 \\ \bar{x}_2 \cdot \bar{x}_2 \\ \bar{x}_1 \cdot \bar{x}_2 \end{bmatrix} = \begin{bmatrix} 1 & 0 & 0 & 0 & 0\,0\,0 \\ 0 & 1 & 0 & 0 & 0\,0\,0 \\ 0 & 0 & 1 & 0 & 0\,0\,0 \\ 0 & 0 & 0 & 1 & 0\,0\,0 \\ 2u_{1,x} & 2u_{1,y} & 0 & 0 & 1\,0\,0 \\ 0 & 0 & 2u_{2,x} & 2u_{2,y} & 0\,1\,0 \\ u_{2,x} & u_{2,y} & u_{1,x} & u_{1,y} & 0\,0\,1 \end{bmatrix} \tilde{x}^\star + \begin{bmatrix} u_{1,x} \\ u_{1,y} \\ u_{2,x} \\ u_{2,y} \\ u_{1,x}^2 + u_{1,y}^2 \\ u_{2,x}^2 + u_{2,y}^2 \\ 0 \end{bmatrix}$$

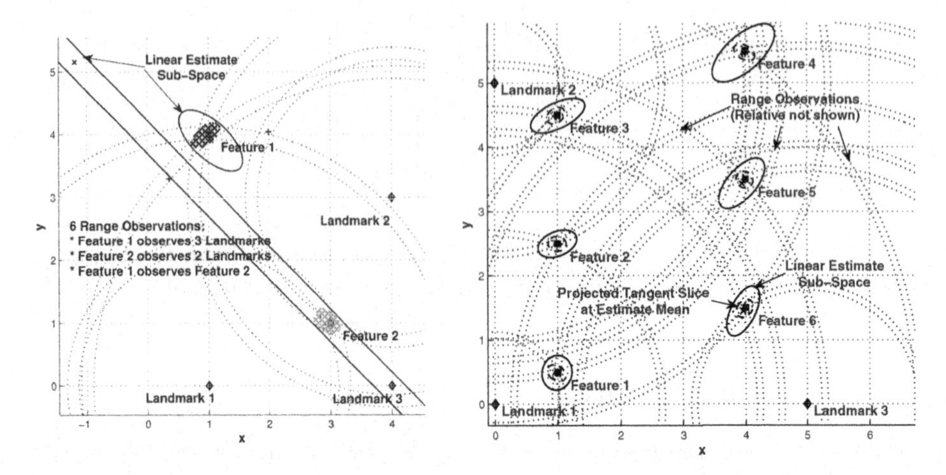

Fig. 5. *(Continued)* Two additional range observations provide a unique estimate solution captured by this representation. In this case the projected linear subspace is unbounded.

Fig. 6. Three landmarks are sufficient to yield a unique bounded estimate from the linear sub-space in this network localization example. The tangent slice yields much tighter estimates.

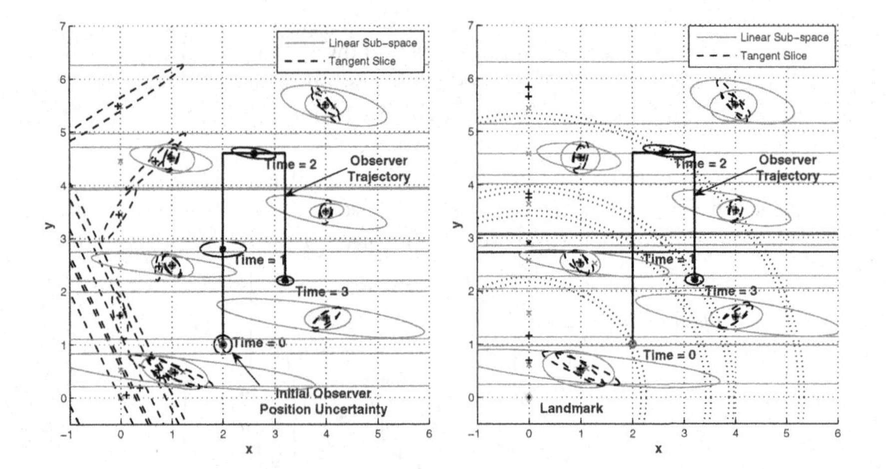

Fig. 7. Two solutions illustrating unique and consistent simultaneous localization and landmark mapping through robot motion: Given an initial robot location estimate *(left)*, and the aid of a known landmark given no prior estimate information *(right)*. A 27 state sparse linear filter fully captures the complex uncertainty structure. The linear sub-space approximation is unbounded when the true uncertainty set is unbounded or non-unique. This approximation becomes bounded in both cases once the robot has performed motion and measurements in two distinct directions.

This demonstrates that the process can extend to any number of point robots.

The resulting update equation now has a nonlinear coupling between the state and input appearing in the transition matrix, making it impossible to allow input disturbances under the current framework. If we allow for perfect input knowledge, then each dynamic update step is performed using the propagation results of Section 3.2 to find $\mathcal{E}^{\star}_{k+1}$. Simulations show that this approach works well for small time steps but may break down over larger intervals – the reasons for this have not been fully explored.

6.1 Application to Landmark Mapping

To demonstrate the effectiveness of incorporating motion into the filtering procedure, we present simulated results of an experiment with a standard node mounted on a mobile robot moving through a system of unlocalized static nodes. Two cases are presented in Figure 7. Global reference is provided by either giving the robot an initial position fix or observing a known landmark. Perfect inputs are applied in accordance with the dynamic update assumptions stated above. After the robot has completed multiple turns, the static nodes have been mapped.

7 Conclusion

We present a novel application of set-based estimation theory that lends itself to simultaneous localization and mapping with multiple mobile sensor platforms. While this paper focussed on range sensors, it is easy to include other types of sensors. The main advantage is our ability to incorporate sensors with nonlinear observation models without any knowledge of the noise.

While our approach shows robustness to modeling uncertainties and to initialization, there are two main limitations that we are currently addressing. First, the approach, as presented, is limited to Euclidean dynamic models. We are exploring alternative representations that will allow us to treat the dynamics and the observation as linear processes with additive noise. Second, while our extensive representation incorporates *all* relevant information in $\mathcal{S}^{\star}$ and makes checking data for consistency and correspondence very easy, it is computationally difficult to translate this information to the base space $\mathcal{S}$. This is because of the complexity of computing the intersection of the feasibility set with the manifold $\mathcal{M}$. However, because the underlying representation is algebraic, it is possible to use symbolic computation software for polynomial algebra to delineate this set and this is an area of ongoing investigation.

Finally, we are also addressing the control of vehicles to actively reduce the volume of the uncertainty with the set-valued representation discussed here. In [18] we present an experimental study with multiple robots localizing static features.

Acknowledgments

This work was in part supported by NSF grants IIS02-22927, IIS-0427313 and ARO MURI Grant DAAD19-02-01-0383.

References

1. Boyd, S., Vandenberghe, L.: Convex optimization. University Press, Cambridge (2004)
2. Ros, L., Sabater, A., Thomas, F.: An ellipsoidal calculus based on propagation and fusion. IEEE Transactions on Systems, Man., and Cybernetics 32(4) (August 2002)
3. Schweppe, F.C.: Recursive state estimation: unknown but bounded errors and system inputs. IEEE Transactions on Automatic Control AC-13(1), 22–28 (1968)
4. Thrun, S., Burgard, W., Fox, D.: Probabilistic Robotics. The MIT Press, Cambridge (2005)
5. Djugash, J., Singh, S., Corke, P.: Further results with localization and mapping using range from radio. In: Proceedings, Fifth Int'l Conf. on Field and Service Robotics, Pt. Douglas, Australia (July 2005)
6. Hanebeck, U.D.: Recursive nonlinear set- theoretic estimation based on pseudo-ellipsoids. In: Int'l Conf. on Multisensor Fusion and Integration for Intelligent Systems, pp. 159–164 (2001)
7. Briechle, K., Hanebeck, U.D.: Localization of a mobile robot using relative bearing measurements. IEEE Transactions on Robotics and Automation 20(1), 36–44 (2004)
8. Horn, J., Hanebeck, U.D., Riegel, K., Heesche, K., Hauptmann, W.: Nonlinear set-theoretic position estimation of cellular phones. In: Proceedings of SPIE. AeroSense Symposium, vol. 5084, pp. 51–58 (2003)
9. Fox, D., Burgard, W., Dellaert, F., Thrun, S.: Monte Carlo localization: Efficient position estimation for mobile robots. In: Proceedings of the National Conference on Artificial Intelligence (AAAI), Orlando (1999)
10. Murphy, K.: Bayesian map learning in dynamic environments. In: Advances in Neural Information Processing Systems (NIPS), MIT Press, Cambridge (2000)
11. Folkesson, J., Christensen, H.I.: Graphical SLAM: A self-correcting map. In: Proceedings of the International Symposium on Autonomous Vehicles, Lisboa, PT (2004)
12. Konolige, K.: Large-scale map-making. In: Proceedings of the AAAI National Conference on Artificial Intelligence, San Jose, CA, pp. 457–463 (2004)
13. Yildirim, E.A.: On the Minimum Volume Covering Ellipsoid of Ellipsoids. Technical Report, Dept. of Applied Mathematics and Statistics, Stony Brook University (2005)
14. Atiya, S., Hager, G.D.: Real-time vision-based robot localization. IEEE Trans. on Robotics and Automation 9(6), 785–800 (1993)
15. Hanebeck, U.D., Schmidt, G.: Set theoretical localization of fast mobile robots using an angle measurement technique. In: Proc. IEEE Int. Conf. on Robotics and Automation, pp. 1387–1394 (1996)
16. Di Marco, M., Garulli, A., Giannitrapani, A., Vicino, A.: A set theoretic aproach to dynamic robot localization and mapping. Autonomous Robots 16, 23–47 (2004)
17. Spletzer, J., Taylor, C.: A bounded uncertainty approach to multi-robot localization. In: Proc. IROS, pp. 1258–1265 (2003)
18. Grocholsky, B., Stump, E., Shiroma, P., Kumar, V.: Control for Localization of Targets Using Range-Only Sensors. In: Proceedings of the International Symposium on Experimental Robotics, Rio de Janeiro, Brazil (2006)

Part IV

Geometric Computations and Applications

An Experimental Study of Weighted k-Link Shortest Path Algorithms

Ovidiu Daescu[1], Joseph S.B. Mitchell[2], Simeon Ntafos[1], James D. Palmer[1], and Chee K. Yap[3]

[1] Department of Computer Science, The University of Texas at Dallas, Richardson, TX 75080, USA
{daescu,ntafos,jdp011100}@utdallas.edu*

[2] Department of Applied Mathematics and Statistics, Stony Brook University, Stony Brook, NY 11794, USA
jsbm@ams.sunysb.edu**

[3] Department of Computer Science, Courant Institute, New York University, New York, NY 10012, USA
yap@cs.nyu.edu***

Abstract. We consider the problem of computing a minimum-weight polygonal path between two points in a weighted polygonal subdivision, subject to the constraint that the path have few segments (*links*). We give an algorithm that generates paths of weighted length at most $(1+\epsilon)$ times the weight of a minimum-cost k-link path, for any fixed $\epsilon > 0$, while using at most $2k - 1$ links. This is an improvement over the previous $(1 + \epsilon)$-approximation algorithm, which used at most $5k - 2$ links. Further, we have implemented our new algorithm and we have conducted a performance study of these algorithms (old and new) on a variety of real-world and synthetic data, comparing not only the efficiency but also the quality of paths generated using these algorithms. We also consider the implications of these results on the practical usage of these algorithms.

1 Introduction

Consider a mobile robot that moves within an environment that is partitioned into regions, each having an associated *weight*, which represents the cost per unit distance for travel in the region. Furthermore, assume that there is a cost associated with each turn that the robot makes. Given an environment specified by a weighted polygonal subdivision, R, and given a positive integer k, our goal is to compute a minimum-cost polygonal path for the robot, from point a to point b, subject to the constraint that the path have at most a specified number, $k - 1$, of turns, and therefore at most k links (edges).

* Daescu's research is supported by NSF grants CCF-0430366 and CCF-0635013.

** J. Mitchell is partially supported by the U.S.-Israel Binational Science Foundation (2000160), NASA (NAG2-1620), NSF (CCR-0098172, ACI-0328930, CCF-0431030), and Metron Aviation.

*** Yap's work is supported by NSF Grant CCF-0430836.

S. Akella et al. (Eds.): Algorithmic Foundation of Robotics VII, STAR 47, pp. 187–202, 2008.
springerlink.com

In a weighted subdivision the distance between two points a and b within the same region $R_i \in R$ is defined as the product of the weight w_i of R_i and the Euclidean length $|ab|$ of the line segment ab. For a path p (not necessarily polygonal), the portion of p that is contained within some region $R_i \in R$ has its length defined as the product of the Euclidean length of $p \cap R_i$ and the weight w_i. The length $||p||$ of the path p is the sum of the lengths over each region it intersects. The weight of a path that follows an edge $e_{i,j}$ that forms the boundary between two regions R_i and R_j is defined as $\min\{w_i, w_j\}$. We assume R is triangulated and has n vertices in general position.

1.1 Related Work

For a survey of optimal path algorithms in geometry, see [14,15].

Minimum-cost paths in weighted subdivisions were first studied algorithmically by Mitchell and Papadimitriou [16], who give a $(1+\epsilon)$-approximation algorithm to compute optimal paths that runs in polynomial time (logarithmic in $1/\epsilon$). Alternative solution methods for the weighted region problem (WRP) are based on discretizing edges of the subdivision, placing Steiner points judiciously, and interconnecting them to form a discretization graph, $G(V, E)$, which is then searched for a shortest path. The first experimental studies of the weighted region problem ([11,13]) were based on implementations of such algorithms. Much of the subsequent WRP work has focused on the placement of Steiner points either on edges [1,2,21] or on face bisectors [3], with clever techniques to speed the computation of shortest paths in the discretization graph [21, 3]. None of these algorithms for the WRP permit one to bound the number of links/turns in the produced path.

The special case of the 1-link minimum-weight path between two regions in a weighted subdivision, often called the optimal "link" or "penetration" problem, was first studied in [5,6]. In [5] the optimal link problem is reduced to $O(n^2)$ subproblems each of which minimizes a two-variable function $f(x, y)$ over a convex domain D, where $f(x, y)$ is given as a sum of $O(n)$ fractional terms. An optimal link between two regions must pass through a vertex in the subdivision [7], a property that has been exploited to create efficient approximation algorithms based on a *prune and search* approach and a *sum of fractionals* approach [9].

The first algorithms to approximate k-link shortest paths in weighted regions, with guaranteed approximation bounds, are proposed in [9]. Two different algorithms for approximating k-link shortest path are proposed based on the computation of *exact* optimal links or *approximate* optimal links. Their solutions produce approximation paths that are within an $(1+\epsilon)$-factor from optimal (ϵ-approximation paths for short) and have $5k-2$ links and $14k$ links, respectively. As in the algorithms in [3, 12, 21], a discretization graph $G(V, E)$ is constructed from Steiner points placed on edges in the subdivision. The construction of G differs in two important ways from discretizations for the shortest path problem. The first difference arises in the placement of Steiner points. A vertex-vicinity (see Section 1.3) is chosen small enough so that no line can intersect three vertex-vicinities. The second difference is in the definition of G. Nodes in V correspond to Steiner edges (not points) and vertices in the subdivision. Edges in E correspond to optimal links between the nodes in V.

The problem of computing optimal k-link paths in weighted regions has been studied in air traffic management applications, where the weights correspond to traversability or risk factors associated with hazardous weather or congestion, and the turn constraint corresponds to the need to bound the human factors complexity of flight trajectories, both from the pilot's perspective and from the air traffic control perspective. The algorithms implemented for this application in Krozel et al. [10] rely on searching grids, using Bellman-Ford methods, but do not provide guarantees of solution quality.

The problem of computing shortest k-link paths in *unweighted* settings (e.g., in simple polygons or polygonal domains) has been studied by Piatko et al. [19, 4, 17].

1.2 Our Contribution

In this paper we describe two new techniques for approximating k-link shortest paths in weighted regions, we implement a solver with a suite of algorithms (new and old), and we perform an experimental investigation:

- We prove that by shrinking the vertex-vicinity by a constant factor of μ, which depends on some parameters of the subdivision, we can find a $(2k-1)$-link path p formed from exact optimal links and turning only on edges, such that $||p|| \leq (1+2\epsilon)||p_k||$ where p_k is an optimal k-link path from s to t.
- We show that using approximate optimal links we can find a $(2k-1)$-link path p formed from approximate optimal links and turning only on edges such that $||p|| \leq (1+7\epsilon)||p_k||$ where p_k is an optimal k-link path from s to t.
- We implement our new methods, as well as those previously proposed. We have built an extensive software system, "k-LinkSolver", which is available for public use.
- We conduct the first experimental investigation of these k-link path solutions. We compare and contrast the run-time performance and solution quality of our new algorithms with those described in [9], and consider the implications of our results on the practical usage of these algorithms. This is the first intensive study on the real-world performance of any of these algorithms.

We refer to a path as being an ϵ-*good approximation* if its weighted length is within a $(1+\epsilon)$-factor of the weight of a corresponding optimal path. That is, p is an ϵ-good approximation of a path p_k if $||p|| \leq (1+\epsilon)||p_k||$. A $C\epsilon$-*good approximation* refers to a path whose weighted length is within a $(1+C\epsilon)$-factor of the weight of an optimal path where C is a constant. By letting $\epsilon' = \epsilon/C$ it is clear that a $C\epsilon$-good approximation forms an ϵ'-good approximation. That is, any $C\epsilon$-good approximation trivially forms an ϵ-good approximation. From this point forward, we will generally refer to paths as ϵ-good disregarding the constant C to mean this and we will state results as being $C\epsilon$-good or within a $(1+C\epsilon)$-factor to acknowledge the specific constant C required. Thus, the two solutions presented here form ϵ-good approximations.

1.3 Definitions and Notations

We begin by describing the terminology and discretization scheme used here. Whenever possible, our definitions and notations are similar to those in [9]. Let E_R be the set of edges that bound the weighted regions of the weighted subdivision R. Let $E_R(v)$ be the set of boundary edges incident to a point v and let $d(v)$ be the minimum Euclidean distance between v and the edges in $E_R \setminus E_R(v)$. For each edge $e \in E_R$, let $d(e) = \sup_{v \in e} d(v)$. Let $v(e)$ be a point on e such that $d(v(e)) = d(e)$. The cell formed by the regions incident to a vertex v is called the *vertex-cell* of v and is denoted by $C(v)$ (C_i for short for a vertex v_i).

For each vertex of R in each possible vertex triplet formed by the vertices in R we compute the minimum distance to the line supporting the opposite edge. Let γ_i be the minimum such distance obtained from a triplet containing the vertex v_i, and let $\gamma = \min\{\gamma_i/2 \mid i = 1, 2, \ldots, n\}$. Using a result in [8], γ can be found in $O(n^2 \log n)$ time.

For each vertex v of R, let $r(v) = \min(\epsilon d(v)/c, \gamma)$ where c is an appropriately chosen constant (see Lemma 2). The disk of radius $r(v)$ centered at a vertex v defines the *vertex-vicinity* $S(v)$ of v (or S_i for short for a vertex v_i). We refer to $r(v)$ as the *vertex-vicinity radius* for v. Note that the choice for $r(v)$ ensures no line can cross more than two vertex-vicinities. This depends on the non-degeneracy assumption that no three vertices of the subdivision are collinear.

We now describe how the Steiner points on an edge $e = \overline{v_1 v_2}$ are chosen. Each vertex v_i, where $i = 1, 2$, has a vertex-vicinity S_i of radius $r(v_i)$ and the Steiner points $v_{i,1}, \ldots, v_{i,j_i}$ are placed on e such that $|v_i v_{i,1}| = r(v_i)$ and $|v_{i,m} v_{i,m+1}| \leq \epsilon d(v_{i,m})$, $m = 1, 2, \ldots, j_i - 1$ (where equality holds in all cases except possibly when $m = j_i - 1$). The value of j_i is such that $v_{i,j_i} = v(e)$. Let δ be the maximum number of Steiner points placed on an edge of the subdivision. The line segment formed by two adjacent Steiner points $v_{i,m}$ and $v_{i,m+1}$ is called a *Steiner edge*. The pairing of any two Steiner edges forms a quadrilateral shape called a *Steiner strip*. The shape could be degenerate if the Steiner edges are on the same boundary edge or share a vertex.

If a path in R is restricted to turn only on edges it is said to be *edge-restricted* otherwise the path is said to be *edge-unrestricted*. If a link that makes up a polygonal path follows an edge of the subdivision (rather than crossing a face) it is said to be *edge-crawling*.

A key subproblem in approximating k-link shortest paths is that of finding "good" 1-link paths (links for short). We refer to two different kinds of good links. An exact optimal link is one that has been solved such that no error exists. An approximate optimal link is within a $(1+\epsilon)$-factor from an optimal link. Finally, for a path p_k, we use $|p_k|$ to denote its Euclidean length and $||p_k||$ to denote its weighted length.

2 Approximating k-Link Shortest Paths

We recall a key theorem from [9]:

Theorem 1. *Given two points s and t of R, there exists a path p between s and t with at most $(2k-1)$-links, that turns only on the edges of the subdivision and such that the weighted length of p is at most that of a k-link shortest path p_k from s to t.*

The edges added to transform an edge-unrestricted path to an edge-restricted path can be thought of as cutting the "corners" off the edge-unrestricted path against the edges of the subdivision. We call such links *corner-cutting* links.

In [9] it is shown how to construct "normalized" paths that either completely or partially avoid vertex-vicinities. The penalty of such an approach is often a higher constant multiplier bounding the number of links. In this section we propose an alternative that decreases the number of links in an approximating path to $(2k-1)$ at the cost of increased complexity in the discretization graph.

We do this by decreasing the size of the vertex-vicinity radius to $r(v) = \min(\mu\epsilon d(v)/c, \gamma)$ where $\mu = w_{min}/w_{max}$ and c is an appropriately chosen constant (see Lemma 2). The discretization graph $G(V, E)$ is defined similarly to [9] but with a few subtle changes. Nodes in V still correspond to Steiner edges and vertices in the subdivision but also include what we will refer to as *interior Steiner edges*. An interior Steiner edge is the edge between a vertex and the first Steiner point on an edge incident to the vertex. Edges in E then correspond to optimal links between pairs of nodes in V.

Lemma 1. *A k-link shortest path, p_k, completely contained in a vertex-vicinity has a maximum weighted length of $2\epsilon w_{min} d(v)/c$.*

Proof. Let l be a link from the start of the path p_k to the end of the path p_k (see Fig. 1). Clearly, if $||p_k|| > ||l||$ then p_k is not an optimal path. Therefore, $||p_k|| \leq ||l||$. The length of l is constrained by the size of the vertex-vicinity and thus, $|l| < 2\epsilon\mu d(v)/c$ or $||l|| < 2\epsilon w_{min} d(v)/c$. And finally, $||p_k|| < 2\epsilon w_{min} d(v)/c$. We may also conclude that only one link is required for a path p'_k to approximate a path p_k such that $||p'_k|| < 2\epsilon w_{min} d(v)/c$.

In this paper, we assume the end points of a path p_k are separated by a distance $\Omega(d(v_i))$ for any vertex-vicinity S_i through which p_k passes. If $|p_k| \ll d(v_i)$, then p_k is a trivial path.

Lemma 2. *A k-link path, p_k, that turns on edges can be approximated by a 2ϵ-good $(2k-1)$-link path made up of k optimal links and $k-1$ connecting links.*

Proof. Let the path p_k be divided into j subpaths, $p_{k_1}, \ldots, p_{k_j}$, as defined below (see Fig. 2). Let $p_{k_0} = \emptyset$. If $i = 1$ then subpath p_{k_i} begins where p_k begins otherwise subpath p_{k_i} begins where $p_{k_{i-1}}$ ends. With $p_{k_{i-1}}$ known, the subpath p_{k_i} is defined as follows. Consider the first vertex-vicinity S_i that p_k crosses after $p_{k_{i-1}}$.

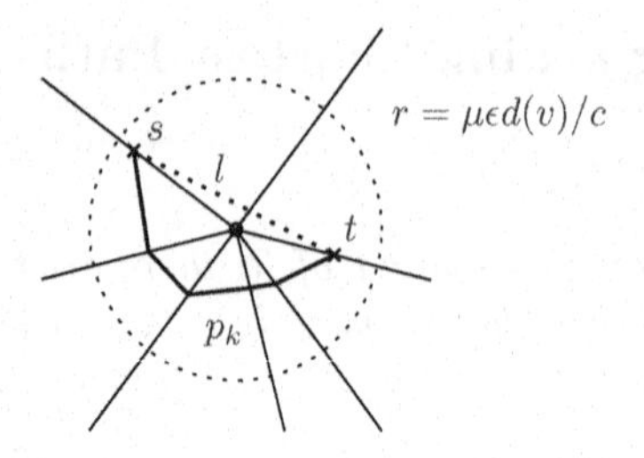

Fig. 1. p_k has a maximum weighted length of $2\epsilon w_{min} d(v)/c$ where c is an appropriately chosen constant

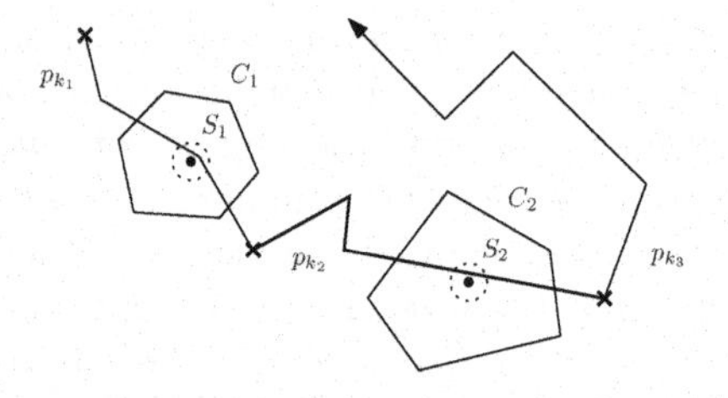

Fig. 2. Three subpaths, p_{k_1}, p_{k_2} and p_{k_3}

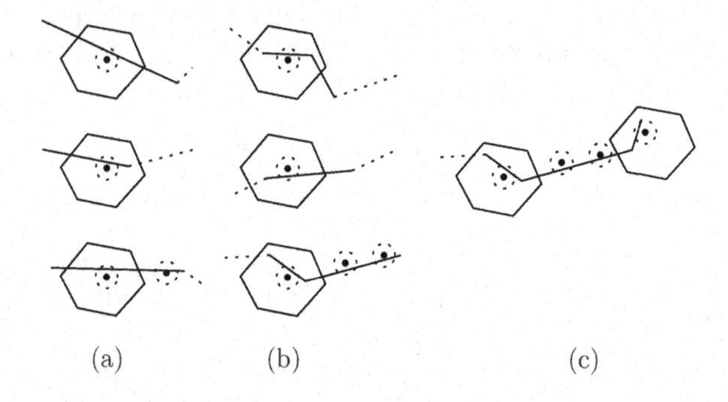

Fig. 3. Several examples of (a) Case 1 and (b) Case 2 described in Lemma 2 and (c) where Case 2 is applied twice on a single subpath

Case 1: If p_k crosses the boundary of the vertex-cell C_i before entering S_i then let the subpath p_{k_i} terminate at the first turn after p_k exits S_i. If no such turn exists, or p_k does not exit S_i, then let p_{k_i} end where p_k ends.

Case 2: If p_k does not cross the boundary of C_i before entering S_i, then let the subpath p_{k_i} terminate at the first turn after p_k exits C_i. If no turn exists after p_k exits C_i then let p_{k_i} end where p_k ends. If p_k does not exit C_i, then we let $p_{k_{i-1}}$ end where p_k ends.

A subpath p_{k_i} so constructed may intersect no more than four vertex-vicinities and $|p_{k_i}| = \Omega(d(v))$ for any vertex-vicinity v crossed by p_{k_i} (see below for a more precise lower bound). Four vertex-vicinities may be crossed if Case 2 is applied once in a situation similar to the last example in Fig. 3 (b) and then again for a small link that crosses a vertex-vicinity as illustrated in Fig. 3 (c).

Note that $(1+\epsilon)||p_k|| = \sum_{i=1}^{j}(1+\epsilon)||p_{k_i}||$. Let p'_k be a path that approximates p_k with k optimal links and $k-1$ connecting links. We prove p'_k can be constructed such that $||p'_k|| < (1+2\epsilon)||p_k||$ by proving $||p'_{k_i}|| < (1+2\epsilon)||p_{k_i}||$.

First consider a subpath π of p_{k_i} that does not cross a vertex-vicinity and its approximating subpath π'. Such a path π is made up of one or more links, $l_1, \ldots, l_{\xi_i}$. Then $||\pi|| = \sum_{j=1}^{\xi_i} ||l_j||$ and by extension, $(1+\epsilon)||\pi|| = \sum_{j=1}^{\xi_i}(1+\epsilon)||l_j||$. Let e_1 and e_2 be the Steiner edges that a link $l_j \in \pi$ originates and terminates at respectively. Let l_j^* be an optimal link between e_1 and e_2. Clearly, $||l_j^*|| \leq ||l_j||$. Consider the contribution of the last region that l_j crosses before connecting to l_{j+1}. Let this region be R_1 with corresponding weight w_1. Let $d_1 = |R_1 \cap l_j|$. Since in general the endpoints of l_j^* and l_{j+1}^* on e_2 are distinct, we need to add a single link, l_{m_j}, to connect l_j^* and l_{j+1}^* on e_2 (l_{m_j} is an edge-crawling connecting link). The link l_{m_j} can have length no greater than that of the Steiner edge on which it lays.

We have $|l_{m_j}| \leq \epsilon d(v_2)$, with v_2 an endpoint of the Steiner edge e_2. By definition, $d(v_2) \leq d_1$. Therefore, $|l_{m_j}| \leq \epsilon d_1$. The contribution l_{m_j} makes to approximate l_j is no more than $\epsilon w_1 d_1$, i.e., $d_1 w_1 + ||l_{m_j}|| \leq (1+\epsilon) w_1 d_1$. Using l_j^* and l_{m_j} to approximate l_j we have $||l_j^*|| + ||l_{m_j}|| \leq (1+\epsilon)||l_j||$, since $||l_j^*|| \leq ||l_j||$. It then follows that $||\pi'|| = \sum_{j=1}^{\xi_i}(||l_j^*|| + ||l_{m_j}||) < \sum_{j=1}^{\xi_i}(1+\epsilon)||l_j|| = (1+\epsilon)||\pi||$. Equivalently, $\sum_{j=1}^{\xi_i} ||l_{m_j}|| < \epsilon||\pi||$.

Next consider what happens when p_{k_i} crosses a vertex-vicinity $S(v)$ of vertex v. From Lemma 1, if p_{k_i} consists of several links contained in $S(v)$, they can be replaced by a single link, of weighted length no more than $2\epsilon w_{min} d(v)$.

Let l_s be a link in the path p_{k_i} that enters $S(v)$ and let l_t be a link that exits $S(v)$ (see Fig. 4). Let l_s^* be an optimal link that begins and ends on the same Steiner edges as l_s. Let l_t^* be an optimal link that begins and ends on the same Steiner edges as l_t. Clearly, $||l_s^*|| \leq ||l_s||$ and $||l_t^*|| \leq ||l_t||$. Let l_m be a connecting link (not necessarily edge-crawling) that connects l_s^* to l_t^*. We need to bound l_m in terms of p_{k_i}. We can find a lower bound for the unweighted length of p_{k_i} by using $d(v)$ and the vertex-vicinity radius of $S(v)$, i.e., $(1-\epsilon\mu/c)d(v) < |p_{k_i}|$

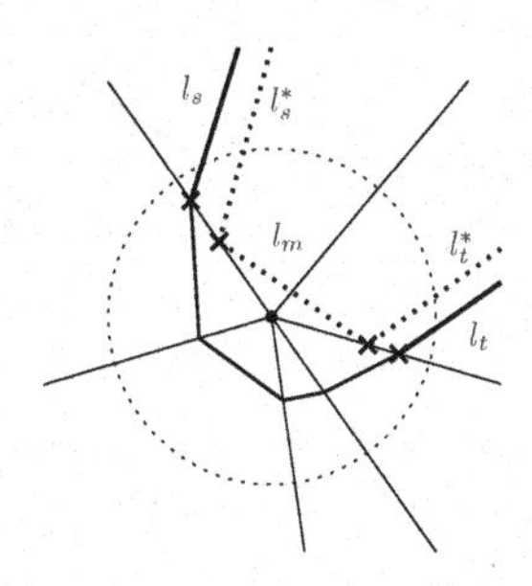

Fig. 4. Approximating path construction inside a vertex-vicinity

and thus $d(v) < c|p_{k_i}|/(c - 1/2)$. The link l_m can achieve an unweighted value of no more than twice the radius of the vertex-vicinity, thus l_m's maximum weighted value is bounded by $||l_m|| \leq 2\epsilon w_{min} d(v)/c < 2\epsilon w_{min}|p_{k_i}|/(c-1/2) \leq 2\epsilon||p_{k_i}||/(c-1/2)$. Also recall that up to four vertex-vicinities may exist in a single subpath p_{k_i}. If we let $c = 17$, then $4||l_m|| < \frac{\epsilon}{2}||p_{k_i}||$.

Combining this result with the result we found for edge-crawling links l_{m_j} not inside the vertex-vicinity, we have a path p'_{k_i} with $||p'_{k_i}|| < (1 + 3\epsilon/2)||p_{k_i}|| < (1 + 2\epsilon)||p_{k_i}||$. That is, a 2ϵ-good subpath p'_{k_i} can be guaranteed. And thus, a 2ϵ-good path p'_k can also be guaranteed.

Theorem 2. *Given two points s and t of R, a k-link shortest path between s and t can be approximated with a 2ϵ-good $(2k-1)$-link path that turns on edges.*

Proof. Consider a link l_i^* on the k-link shortest path and refer to Fig. 5. If l_i^* ends on an edge of the subdivision then Lemma 2 applies and it can be replaced in the approximating path by one optimal link and one connecting link. If l_i^* ends inside a face of R then Theorem 1 applies, adding a corner-cutting link that crosses the face to connect l_i^* to l_{i+1}^*. No additional edge-crawling connecting link is required because the corner-cutting link crosses only one face and therefore its length is adequately captured in the discretization.

This new fact that links which cross a single face do not require additional connecting links was not considered in [9]. Applying this result, we find the $5k - 2$ link and $14k$ link results for optimal links and approximate optimal links are improved to $3k - 2$ links and $8k$ links respectively.

When computing the discretization graph $G(V, E)$, defined earlier, the number of vertices in V is clearly bounded by the number of Steiner points $O(\delta n)$ and the number of edges in E is $O((\delta n)^2)$. Computing a single edge in G corresponds to solving a 1-link shortest path problem for a specific subproblem. Let this time be $T_h(n)$. Thus, the time to compute G is $O((\delta n)^2 T_h(n))$. Once G is constructed, we can use dynamic programming to find a k-link shortest path in G in $O(k(\delta n)^2)$ time.

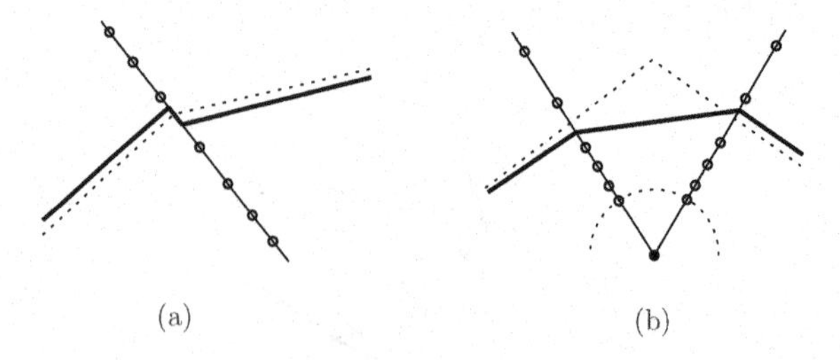

Fig. 5. a) Joining two optimal links with an edge-crawling connecting link (Case 1) and (b) joining two links not constrained to edges of the subdivision with a corner-cutting link (Case 2)

3 Approximate Optimal Links

In practice, optimal links can be time consuming to compute. Approximate optimal links are links that take advantage of the discretization scheme already described to find a single link within a $(1+\epsilon)$ factor of optimal. Such links were introduced in [9] but unfortunately the normalization process required several more links than it did for optimal links. In this section we show that approximate optimal links can be used with our new technique while still maintaining an approximating path with no more than $2k-1$ links.

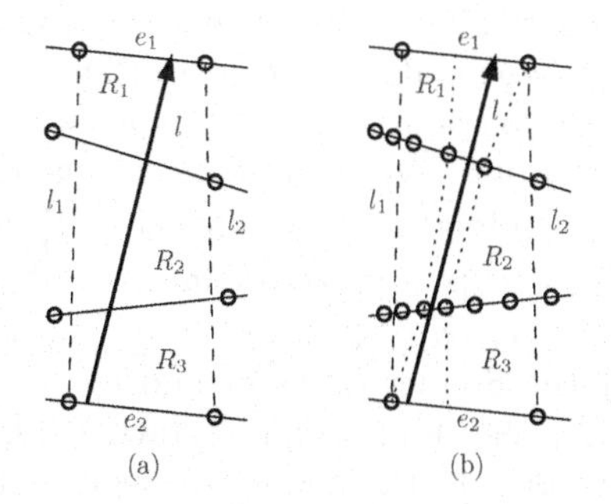

Fig. 6. A Steiner strip formed by lines l_1 and l_2 may intersect edges that are more coarsely (a) or more finely (b) sampled. In (b) the dotted lines describe an hourglass defined by Steiner points.

Consider an optimal link l between two edges e_1 and e_2. Many Steiner points may be captured inside the Steiner strip formed by e_1 and e_2. See Fig. 6. These points define a set of *hourglass* shaped regions, that represent the space over which a link can be translated and rotated without passing a Steiner point. An optimal link clearly must lie in one such hourglass. It has been shown in [9] that an arbitrary link computed for the same hourglass differs from an optimal link by only a factor of $1+2\epsilon$. Here we extend this result to include optimal links that cross or turn inside a vertex-vicinity.

Theorem 3. *Given two points s and t of R, a k-link shortest path between s and t can be approximated with a 7ϵ-good $(2k-1)$-link path made up of k approximate optimal links connected by $k-1$ connecting links.*

Proof. We divide the approximating path p'_k introduced earlier into j subpaths, $p'_{k_1}, \ldots, p'_{k_j}$, as already described in Lemma 2. We seek to bound the error between a subpath p''_{k_i} made of approximate optimal links and a subpath p'_{k_i} made of exact optimal links.

Consider a subpath p'_{k_i}. Such a path is made up of one or more optimal links, $l'_1, \ldots, l'_{\xi_i}$, and one or more small connecting links (but no more than $\xi_i - 1$). The connecting links will remain constrained to the size of the Steiner edge on which they lay or to twice the radius of a vertex-vicinity. Therefore, we are only concerned with how much l'_j and l''_j differ in length. In what follows we show $||l''_j||$ differs from $||l'_j||$ by at most a $(1+\epsilon)$ factor.

The weighted length of a single link, l'_j, intersecting m regions, $R_1, \ldots, R_m$, is given by $\sum_{i=1}^{m} w_i d_i$. We must consider the contribution of each segment to guarantee that the length of an approximating link l''_j is within a $(1+\epsilon)$-factor of the length of l'_j.

First consider a link l'_j that does not cross a vertex-vicinity. Let e_1 and e_2 be the Steiner edges that l'_j originates and terminates at respectively. Let V_S be the set of Steiner points contained in the Steiner strip formed from e_1 and e_2. V_S defines $O((\delta n)^2)$ hourglasses where δ is the number of Steiner points placed on an edge. Let the weighted length of an arbitrary line segment l''_j that passes through a particular hourglass approximate the length of the optimal line segment l'_j that passes through the same hourglass. Clearly each term that makes up the description of l''_j can vary from l'_j by at most the weighted lengths of the Steiner edges s_i and s_{i+1} corresponding to that term, i.e., $||l''_j|| = \sum_{i=1}^{m} w_i * d_i(l''_j) \leq \sum_{i=1}^{m}(w_i * d_i(l'_j) + w_i|s_i| + w_i|s_{i+1}|)$. By definition $|s_i| \leq \epsilon d_i(l')$ and $|s_{i+1}| \leq \epsilon d_i(l')$. Thus, $||l''|| = \sum_{i=1}^{m} w_i * d_i(l'') \leq (1+2\epsilon)\sum_{i=1}^{m} w_i d_i(l') = (1+2\epsilon)||l'||$.

The optimal link for the Steiner edges e_1 and e_2 is captured by one of the $O((\delta n)^2)$ hourglasses. The length of a representative link for one hourglass can be found in time proportional to the number of regions intersected by the link, and thus in $O(n)$ time. By taking the link of smallest weighted length over all hourglasses we have a link which approximates the optimal link for the Steiner strip within a factor of $(1+2\epsilon)$. The overall computation of this link takes $O(n(\delta n)^2)$ time.

Next, assume l'_j crosses a vertex-vicinity $S(v)$. Note that one or more links in p_{k_i} may pass through the vertex-vicinity. Links completely contained in the vertex-vicinity are in fact not captured at all, but we have shown the maximum amount of error induced by such links in Lemma 1. We focus our attention on two links in particular, the link entering and the link exiting the vertex-vicinity (which may be one in the same link).

As a link l'_j enters the vertex-vicinity $S(v)$, it must first pass a Steiner edge before it passes an internal Steiner edge. An arbitrary link l''_j that falls in the same hourglass as l'_j is captured by a Steiner strip with an internal Steiner edge. Clearly the term associated with the description of l'_j in this strip can vary by at most $||s_i|| + \epsilon w_{min} d(v)/c$, where s_i is as above. If $\epsilon w_{min} d(v)/c > \epsilon d_i(l'_j)$ then the $(1+2\epsilon)$ bound we just discovered earlier does not hold. It is necessary to consider the error in our approximation that is induced by using internal Steiner edges not per link but per subpath.

The additional error associated with l''_j entering the vertex vicinity and crossing an internal Steiner edge is $\epsilon w_{min} d(v)/c$. Since the link l''_j may also exit the vertex-vicinity by crossing another Steiner strip formed from one Steiner edge and one internal Steiner edge, l''_j can vary by twice as much. The weighted length inside the vertex vicinity is also captured by $2\epsilon w_{min} d(v)/c$ where c is an appropriately chosen constant and thus crossing a single vertex-vicinity may cost $4\epsilon w_{min} d(v)/c$. Since a subpath p'_{k_i} may cross four vertex-vicinities, the maximum cost per subpath is $16\epsilon w_{min} d(v)/c < 16\epsilon w_{min} |p'_{k_i}|/(c-1/2) < 16\epsilon ||p'_{k_i}||/(c-1/2)$ and choosing $c = 17$ we have that $16\epsilon||p'_{k_i}||/(17-1/2) < \epsilon||p'_{k_i}||$.

That is, p''_{k_i} differs from p'_{k_i} by at most a factor of ϵ due to error from crossing internal Steiner edges. Adding this to the error incurred by crossing arbitrary Steiner edges, p''_{k_i} differs from p'_{k_i} by at most a factor of 3ϵ. Then, the difference between p''_{k_i} and p_{k_i} is $(1+3\epsilon)(1+3\epsilon/2) < 1+7\epsilon$; p''_{k_i} is a 7ϵ-good approximation of p_{k_i} and p''_k is a 7ϵ-good approximation of p_k.

Substituting the link computation time $O(n(\delta n)^2)$ for $T_h(n)$ in the previous section we find the time to compute the discretization graph $G(V, E)$ is $O(n(\delta n)^4)$. As before, once G is constructed, we can use dynamic programming to find a k-link shortest path in G in $O(k(\delta n)^2)$ time.

4 Experiments

We have implemented the algorithms described in this paper as well as the algorithms described in [9] as part of a C++ application we call "k-LinkSolver". We refer to the k-link approximation algorithms in [9] using either optimal links or approximate optimal links as K-LINK-OPT and K-LINK-APPROX, respectively, or K-LINK collectively. We will refer to the new approximations based on either optimal links or approximate optimal links presented in this paper as K-LINK-MU-OPT and K-LINK-MU-APPROX, respectively, or K-LINK-MU collectively.

We also define a new heuristic. Instead of solving for either an approximate or an optimal link between two Steiner edges we simply solve for an arbitrary link within a Steiner strip. When applied to the original discretization scheme we will call this algorithm K-LINK-HEUR and when applied to the discretization scheme used in this paper we call the technique K-LINK-MU-HEUR. This heuristic makes no optimality guarantees but when ϵ is very small we might expect that an arbitrary link between two Steiner edges is very close in value to an optimal link between those two edges.

The k-LinkSolver solves for a k-link path in three stages: (1) discretization, (2) link generation and (3) path finding. The first and second stages do not depend on the start and end points of the path while the third stage does. If one is interested in finding several k-link paths in a single subdivision the first two stages can be thought of as preprocessing steps and the third stage can be executed as many times as required.

We use three different data sets in these experiments. The first is a traced MRI scan from the Visible Human Project [18] which emphasizes structure features of the brain (MRI-DATA). The second is a topological map of Santiago Peak from Big Bend National Park, Texas (TOPO-DATA). The third is a simple subdivision generated using Shewchuk's triangulator [20], Triangle, with specific quality and area guarantees (TRI-DATA). The discretization of MRI-DATA, TOPO-DATA, and TRI-DATA is shown in Fig. 7.

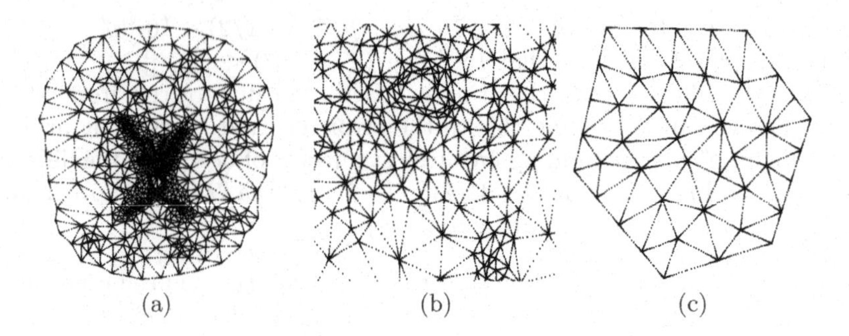

Fig. 7. A discretization of (a) MRI-DATA, (b) TOPO-DATA, and (c) TRI-DATA

4.1 Discretization Performance

The first stage, discretization, is the most efficient. $O(n\delta)$ Steiner points are generated in $O(n\delta)$ time, with $\delta = O(1/\epsilon)$ (the hidden constant in the $O(\cdot)$ depends on some parameters of R, such as the maximum edge length and μ. Each Steiner point can be computed in constant time and each Steiner point becomes part of the discretization graph, making the discretization process inherently output sensitive.

Since Steiner points are associated (through edges) to nodes in the discretization graph, the number of Steiner points generated directly affects the number of links that must be calculated in the link generation phase and the number of links that must be considered in the path finding stage. Ideally we would like to generate as few Steiner points as possible that will make a specific $(1+\epsilon)$ approximation guarantee.

Unlike K-LINK, K-LINK-MU depends on the ratio w_{min}/w_{max}. In our first series of tests, we fix ϵ to 0.5 and then vary μ to gauge μ's effect on the discretization process used in K-LINK-MU. See Fig. 8 (a). Very small values for μ can have a significant impact on Steiner point generation.

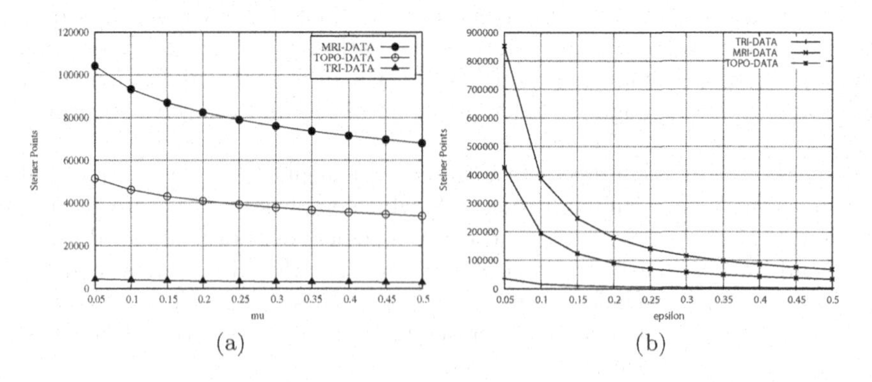

Fig. 8. (a) The effect of fixing ϵ and varying μ. (b) The effect of fixing μ and varying ϵ.

In our second series of tests, we fix μ to 0.5 and then vary ϵ to gauge the effect that ϵ has on Steiner point generation. See Fig. 8 (b). Note that once μ is fixed the resulting vertex-vicinity radii for the K-LINK and K-LINK-MU algorithms differ by only a constant factor. Although we would like to choose a very small ϵ to guarantee the quality of the k-link path we generate, the number of Steiner points grows quickly as ϵ becomes small.

4.2 Link Generation Performance

In the link generation stage $O((n\delta)^2)$ links must be calculated. In our experiments we computed "exact" links using the prune-and-search scheme (see [9]), which returns an optimal link accurate to any user specified precision.

In the link generation stage we compare the running time for all of the approaches mentioned earlier. See Fig. 9. The "exact" solutions are clearly the most expensive while the heuristic solutions are the least expensive. Despite the fact that the heuristic solutions do not offer any quality guarantees, in our experiments they provided paths very close to those generated by the algorithms that rely on computing approximate links. This is an interesting finding given the expensive nature of link generation.

4.3 Path Finding Performance and Quality

Link generation is the true performance bottleneck in finding k-link paths. But if we assume that discretization and link generation are preprocessing steps to potentially many path finding operations, the performance of the path finding step becomes of critical importance.

In our algorithms the path finding step is straightforward and requires simple, $O(k(n\delta)^2)$ computation.

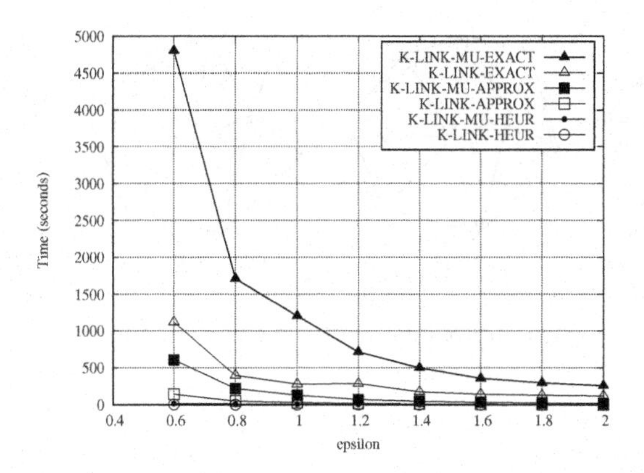

Fig. 9. Timed tests of the link generation stage

An important issue is the quality of the paths being generated. How closely do generated paths follow the optimal path? How closely do generated paths come to the true or optimal k-link path value in practice?

From our experiments we have seen that when ϵ is quite large there is a tendency for the path to change quite a bit as ϵ changes. Making a relatively small change in ϵ can change the path drastically. However, as ϵ becomes smaller there is a tendency for the path to "lock in" and not change drastically. This also seems to have the effect that tightening ϵ to extremely small values often does not yield proportionally better k-link path lengths.

Fig. 10 is one example of the effect of changing ϵ. Notice how little the path changes between Fig. 10 (c) and (d).

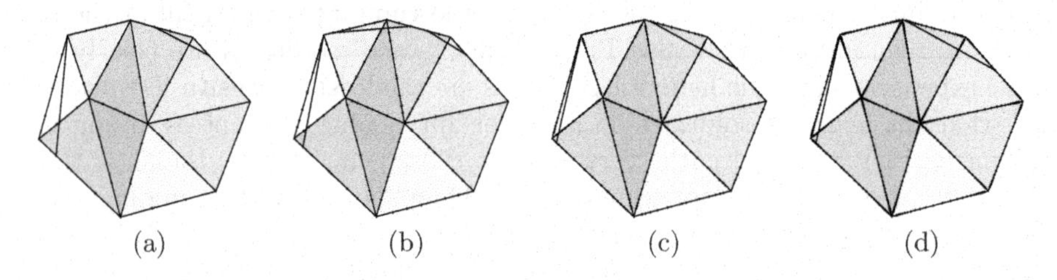

Fig. 10. Four k-link path approximations where $k = 3$ and (a) $\epsilon = 2.0$, (b) $\epsilon = 1.0$, (c) $\epsilon = 0.5$, and (d) $\epsilon = 0.25$

Changing the number of links, k, a path may utilize is usually more profound. Fig. 11 is an example of the effect of changing k with a fixed ϵ. In this example the path for $k = 2$ is actually not a bad approximation of a path with $k = 9$ in that it crosses all of the same faces. This is less likely to be true when considering more complex subdivisions.

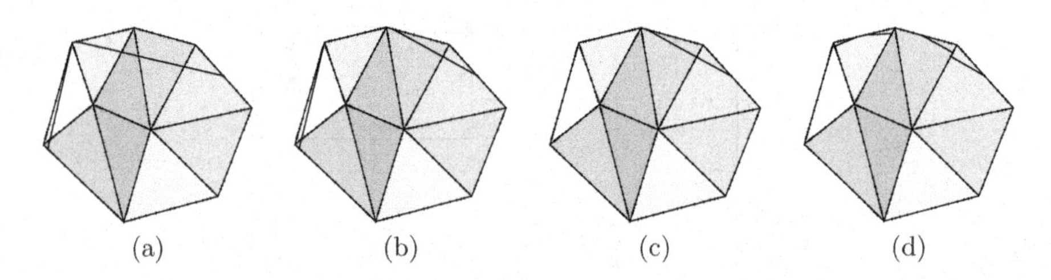

Fig. 11. Four k-link path approximations where $\epsilon = 0.5$ and (a) $k = 2$, (b) $k = 3$, (c) $k = 4$, and (d) $k = 9$

5 Conclusions

In terms of theory, we believe it will be difficult to improve on the $2k - 1$ link result for an approximating k link path that turns only on edges. Future work that seeks to improve this result may need to consider discretization schemes where Steiner points could also be placed inside the faces of R.

Our experiments highlight many of the difficulties and computation intensive facets of this problem. While it may be difficult to improve on the number of links in an approximating path we believe there is potential for improving the running time for finding solutions using new techniques. For example, while the K-Link-Mu result provides a nicer theoretical result by reducing the number of links in an approximating path, the relationship between μ and Steiner point generation often makes K-Link the more desirable algorithm.

References

1. Aleksandrov, L., Lanthier, M., Maheshwari, A., Sack, J.-R.: An ϵ-approximation algorithm for weighted shortest paths on polyhedral surfaces. In: Arnborg, S. (ed.) SWAT 1998. LNCS, vol. 1432, pp. 11–22. Springer, Heidelberg (1998)
2. Aleksandrov, L., Maheshwari, A., Sack, J.-R.: Approximation algorithms for geometric shortest path problems. In: Proc. 32nd ACM Sympos. Theory Computing, pp. 286–295 (2000)
3. Aleksandrov, L., Maheshwari, A., Sack, J.-R.: Determining approximate shortest paths on weighted polyhedral surfaces. Journal of the ACM 52(1), 25–53 (2005)
4. Arkin, E.M., Mitchell, J.S.B., Piatko, C.D.: Bicriteria shortest path problems in the plane. In: Proc. 3rd Canadian Conf. Computational Geometry, pp. 153–156 (1991)
5. Chen, D.Z., Daescu, O., Hu, X., Wu, X., Xu, J.: Determining an optimal penetration among weighted regions in two and three dimensions. Journal of Combinatorial Optimization 5(1), 59–79 (2001)
6. Chen, D.Z., Hu, X., Xu, J.: Optimal beam penetration in two and three dimensions. Journal of Combinatorial Optimization 7(2), 111–136 (2003)
7. Daescu, O.: Improved optimal weighted links algorithms. In: Proc. ICCS 2nd International Workshop on Computational Geometry and Applications, pp. 65–74 (2002)
8. Daescu, O., Luo, J.: Proximity problems on line segments spanned by points. In: Proc. 17th Canadian Conf. Computational Geometry, pp. 224–228 (2005)
9. Daescu, O., Mitchell, J.S.B., Ntafos, S., Palmer, J.D., Yap, C.K.: k-link shortest paths in weighted subdivisions. In: Proc. 9th Workshop on Algorithms and Data Structures, pp. 325–337 (2005)
10. Krozel, J., Lee, C., Mitchell, J.S.B.: Estimating time of arrival in heavy weather conditions. In: Proc. AIAA Guidance, Navigation, and Control, pp. 1481–1495 (1999)
11. Lanthier, M., Maheshwari, A., Sack, J.-R.: Approximating weighted shortest paths on polyhedral surfaces. In: Proc. 13th ACM Sympos. Computational Geometry, pp. 274–283 (1997)
12. Lanthier, M., Maheshwari, A., Sack, J.-R.: Approximating shortest paths on weighted polyhedral surfaces. Algorithmica 30(4), 527–562 (2001)
13. Mata, C., Mitchell, J.S.B.: A new algorithm for computing shortest paths in weighted planar subdivisions. In: Proc. 13th ACM Sympos. Computational Geometry, pp. 264–273 (1997)

14. Mitchell, J.S.B.: Geometric shortest paths and network optimization. In: Sack, J.-R., Urrutia, J. (eds.) Handbook of Computational Geometry, Elsevier Science, Amsterdam (2000)
15. Mitchell, J.S.B.: Shortest paths and networks. In: Goodman, J.E., O'Rourke, J. (eds.) Handbook of Discrete and Computational Geometry, ch. 27, 2nd edn., pp. 607–641. Chapman & Hall/CRC, Boca Raton (2004)
16. Mitchell, J.S.B., Papadimitriou, C.H.: The weighted region problem: Finding shortest paths through a weighted planar subdivision. Journal of the ACM 38(1), 18–73 (1991)
17. Mitchell, J.S.B., Piatko, C.D., Arkin, E.M.: Computing a shortest k-link path in a polygon. In: Proc. 33rd IEEE Sympos. Foundations Computer Science, pp. 573–582 (1992)
18. National Library of Medicine. The visible human project, `http://www.nlm.nih.gov/research/visible`
19. Piatko, C.D.: Geometric Bicriteria Optimal Path Problems. Ph.D. thesis, Computer Science, Cornell University (1993)
20. Shewchuk, J.R.: Triangle: Engineering a 2D Quality Mesh Generator and Delaunay Triangulator. In: Lin, M.C., Manocha, D. (eds.) FCRC-WS 1996 and WACG 1996. LNCS, vol. 1148, pp. 203–222. Springer, Heidelberg (1996)
21. Sun, Z., Reif, J.H.: Adaptive and compact discretization for weighted region optimal path finding. In: Lingas, A., Nilsson, B.J. (eds.) FCT 2003. LNCS, vol. 2751, pp. 258–270. Springer, Heidelberg (2003)

Low-Discrepancy Curves and Efficient Coverage of Space

Subramanian Ramamoorthy[1,3], Ram Rajagopal[2], Qing Ruan[3], and Lothar Wenzel[3]

[1] Department of Electrical and Computer Engineering, The University of Texas at Austin
s.ramamoorthy@mail.utexas.edu
[2] Department of Electrical Engineering and Computer Science,
The University of California at Berkeley
ramr@eecs.berkeley.edu
[3] National Instruments Corp.,
{qing.ruan,lothar.wenzel}@ni.com

Abstract. We introduce the notion of low-discrepancy curves and use it to solve the problem of optimally covering space. In doing so, we extend the notion of low-discrepancy sequences in such a way that sufficiently smooth curves with low discrepancy properties can be defined and generated. Based on a class of curves that cover the unit square in an efficient way, we define induced low discrepancy curves in Riemannian spaces. This allows us to efficiently cover an arbitrarily chosen abstract surface that admits a diffeomorphism to the unit square. We demonstrate the application of these ideas by presenting concrete examples of low-discrepancy curves on some surfaces that are of interest in robotics.

1 Introduction

Uniform coverage of space is an important requirement in several applications in robotics and allied areas involving motion planning. As noted in a recent survey article [1], coverage path planning is critical in applications such as robotic de-mining, spray painting, machine milling and non-destructive evaluation of complex industrial parts, to name just a few.

In a more general setting, uniform coverage is a natural requirement in problems involving search in abstract spaces. Common abstract spaces of interest include configuration spaces of mechanical systems, parameter spaces such as the space of coefficients of a rational transfer function or a probability distribution, etc. Often, one is interested in generating continuous curves that cover such spaces. This is a natural requirement for physical robots that move in a continuous world. However, such a need also arises in several other settings involving, e.g., continuous adaptation, learning and discovery in a complex dynamical system.

Motivated by such application needs, we are investigating the problem of generating low-discrepancy curves that provide an assurance of uniform coverage of space in an incremental setting, such that the length of the search path correlates to the quality of coverage.

Our approach to solving this problem involves two steps. First, we generate curves that uniformly and incrementally cover a model space, e.g., the unit square. We

S. Akella et al. (Eds.): Algorithmic Foundation of Robotics VII, STAR 47, pp. 203–218, 2008.
www.springerlink.com

generalize the well established theory of low-discrepancy sequences in such a way that sufficiently smooth curves with low discrepancy properties can be defined and generated. In addition to the types of curves that we present in this paper, one may also tap into a sizeable literature on ergodic theory [2] to construct alternate curves with different coverage properties. Based on such curves, induced low discrepancy curves in Riemannian spaces may be constructed. This is achieved through the definition and determination of an area and fairness-preserving diffeomorphism. Given a suitable parametrization of the space to be covered, this procedure yields a curve that can cover it uniformly, optimally in a low-discrepancy sense, and incrementally. This second step ensures that our algorithm is applicable in a wide variety of applications, requiring only that we have a description of the abstract space in the form of a suitable Riemannian metric.

Low discrepancy point sets and sequences [3] have a successful history within robotics. They have been successfully used in sampling based motion planning and area coverage applications. This work has been covered well in the past proceedings of the Workshop on the Algorithmic Foundations of Robotics, so we will not extensively survey it here. [4] contains an excellent discussion on the use of low-discrepancy sampling techniques in motion planning. In more recent work, [5], [6], techniques have been proposed for generating sequences in an incremental fashion, which is often a very important requirement. However, on the one hand, the generation of these sequences is based on a computationally expensive search for an optimal ordering [5] while on the other hand, even though some of these computational efficiency problems may be addressed by better algorithm design [6], we often seek the stronger result of a continuous curve in an abstract space.

The notion of using a diffeomorphism to induce a low-discrepancy curve in the abstract space bears a methodological resemblance to some prior work in robotics, e.g., [7], where sphere worlds are mapped to arbitrary convex spaces. However, we are not aware of prior attempts to define diffeomorphisms that address fairness of space coverage. This is crucial for our work.

The plan of the rest of the paper is as follows. In section 2, we begin with an overview of low-discrepancy sets and sequences. In section 3 we generalize the idea of low-discrepancy sequences to low-discrepancy curves. We show that such curves do exist. Section 4 deals with low-discrepancy curves on abstract surfaces and in Riemannian spaces. These curves can be derived from low-discrepancy curves in unit cubes. In Section 5, we apply the developed methods to the problem of covering various surfaces. In a certain sense to be defined, we will show that the proposed procedure yields optimal coverage. Finally, we conclude with some comments regarding future directions and open questions.

2 On the Notion of Low-Discrepancy Point Sets and Sequences

The definition of discrepancy of a finite set X was introduced to quantify the homogeneity of finite-dimensional point sets [8]:

$$D(X) = \sup_R |m(R) - p(R)| \tag{1}$$

In equation (1), R runs over all d-dimensional rectangles $[0, r]^d$ with $0 \leq r \leq 1$, $m(R)$ stands for the Lebesgue measure of R and $p(R)$ is the ratio of the number of points of X in R and the number of all points of X. The lower the discrepancy the better or more homogeneous is the distribution of the point set. The discrepancy of an infinite sequence $X = \{x_1, x_2, x_3, ..., x_n, ...\}$ is a new sequence of positive real numbers $D(X_n)$, where X_n stands for the first n elements of X.

There exists a point set of given length that realizes the lowest discrepancy. It is known (the Roth bound [9]) that the following inequality holds true for all finite sequences X_n of length n in the d-dimensional unit cube.

$$D(X_n) \geq B_d \frac{(\log n)^{\frac{d-1}{2}}}{n} \tag{2}$$

B_d depends only on d. Except for the trivial case $d = 1$, it is unknown whether the theoretical lower bound is attainable. Many schemes to build finite sequences X_n of length n do exist that deliver a slightly worse limit,

$$D(X_n) \geq B_d \frac{(\log n)^d}{n} \tag{3}$$

There are also infinite sequences X with the above lower bound, equation (3), for all subsequences consisting of the first n elements. The latter result leads to the definition of low-discrepancy infinite sequences X. The inequality (3) must be valid for all subsequences of the first n elements, where B_d is an appropriate constant.

Many low-discrepancy sequences in d-dimensional unit cubes can be constructed as combinations of 1-dimensional low-discrepancy sequences. Popular low-discrepancy sequences are based on schemes introduced by Corput [10], Halton [11], Sobol [12] and Niederreiter [8].

One of the primary motivations for investigations into these sequences arises from high-dimensional function approximation and Monte-Carlo integration. In this setting, there is a well-known [8] relationship between integrals I, approximations I_n, and an infinite sequence $X = \{x_1, x_2, ..., x_n, ...\}$ in d-dimensions, known as the Koksma-Hlawka inequality.

$$|I(f) - I_n(f)| \leq V(f)D(X_n) \tag{4}$$

$$I(f) = \int_0^1 f(x)dx \tag{5}$$

$$I_n(f) = \frac{1}{n}\sum_{i=1}^{n} f(x_i) \tag{6}$$

where $V(f)$ is the variation of the function in the sense of Hardy and Krause.

3 Low-Discrepancy Curves in the Unit Square

One of the earliest known quasi-random sequences is the Richtmyer sequence [13], [14], which illustrates a simple but general result in ergodic dynamics [15], [16]. Let

$x_n = \{n\alpha\}$ (i.e., $[n\alpha] \mod 1$) and $X = \{x_1, x_2, ..., x_n, ...\}$, where $\alpha = (\alpha_1, ..., \alpha_d)$ is irrational and $\alpha_1, ..., \alpha_d$ are linearly independent over the rational numbers. Then for almost all α in $\Re^d$ and for positive ϵ, with the exception of a set of points that has zero Lebesgue measure,

$$D(X_n) = O\Big(\frac{\log^{d+1+\epsilon} n}{n}\Big) \tag{7}$$

The Richtmeyer sequence is probably the only quasi-random sequence based on a linear congruential algorithm [17]. This is useful because it suggests a natural extension to the generation of curves. We will now provide such an extension.

Let C be a given piecewise smooth and finite curve in the unit square S. Furthermore, let R be an arbitrary aligned rectangle in S with lower left corner $(0,0)$. Let L be the length of the given curve in S and l be the length of the sub-curve of C that lies in R. In case of well-distributed curves, the ratio l/L should represent the area $A(R)$ of R reasonably well. This gives rise to the following definition of discrepancy of a given finite piecewise smooth curve in S:

$$D(C) = \sup_R \Big|\frac{l}{L} - \frac{A(R)}{A(S)}\Big| \tag{8}$$

It would be desirable to construct curves C with the property that the discrepancy is always small. More precisely, we will call an infinite and piecewise sufficiently smooth curve $C : \Re^+ \mapsto S$, in natural parametrization, a low-discrepancy curve if for all positive arc lengths L the curves $C_L = C/[0, L]$ satisfy the inequalities (the function F must be defined appropriately):

$$D(C_L) \leq F(L) \tag{9}$$

In fact, a piecewise smooth curve in natural parametrization generates sequences $\{x_1, x_2, ..., x_n, ...\}$ by setting $x_n = C_n(n\Delta)$ where Δ is a fixed positive number. The inequality 9 lets us hope for a similar formula for the derived sequence $\{x_1, x_2, ..., x_n, ...\}$. Because of equation 7, a realistic goal is:

$$F(L) = O\Big(\frac{\log^{3+\epsilon} L}{L}\Big), d = 2 \tag{10}$$

We have to show that this goal is attainable.

To this end, for $\alpha = (\alpha_1, \alpha_2)$, let $C_{\mathcal{A}}(\alpha)$ be the piecewise linear curve $(t\alpha_1 \mod 1, t\alpha_2 \mod 1) = (\{t\alpha_1\}, \{t\alpha_2\})$ where t is in $\Re^+$.

In fact, we can define three classes of curves in the unit square, as shown in figure 1. For the simplest type of curves, let us call this class $C_{\mathcal{A}}$, the right-left and top-bottom edges are identified so that the curve jumps from one edge to the other upon hitting it. In this scheme, all curves are parallel and continue indefinitely when the square tiles the plane. In the second type of curves, class $C_{\mathcal{B}}$, we introduce reflections at the top and bottom edges but preserve the same identification between right and left edges. The third type of curves, class $C_{\mathcal{C}}$, involves reflections on all edges. The latter curve is continuous.

Theorem 1. *For almost all numbers α in $\Re^2$, $C_{\mathcal{A}}(\alpha)$ is a low-discrepancy curve in the sense of equations 9 and 10.*

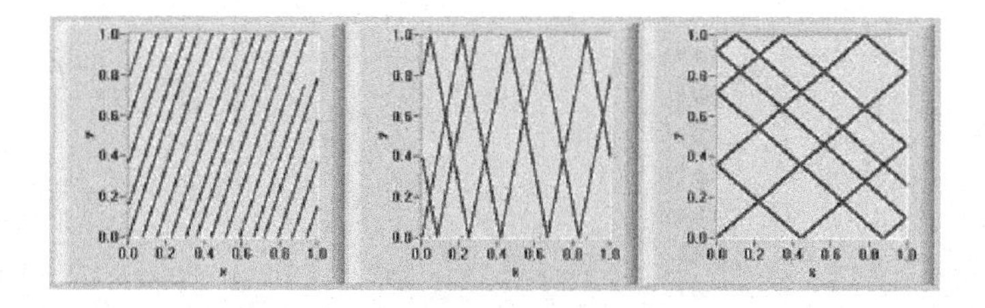

Fig. 1. Various low-discrepancy curves for the unit square: $C_{\mathcal{A}}, C_{\mathcal{B}}, C_{\mathcal{C}}$ from left to right

Proof

In order to prove this statement, we will establish that the ratio of the length of a curve segment to the total length is commensurate with the corresponding ratio of the area of an axis-aligned rectangle to the unit square that contains it, as suggested in equation 8.

Without loss of generality, we assume $\alpha_1, \alpha_2 > 0$. $C_{\mathcal{A}}(\alpha)$ intersects the axes at $(x = 0, y_n = \{\frac{n\alpha_2}{\alpha_1}\})$ and $(x_n = \{\frac{n\alpha_1}{\alpha_2}\}, y = 0)$, where n is an arbitrary natural number. For almost all α_1, α_2 all three of the quantities (α_1, α_2), $\frac{\alpha_1}{\alpha_2}$ and $\frac{\alpha_2}{\alpha_1}$ generate low-discrepancy sequences in the sense of equation 7, in $\Re^2$, $\Re$ and $\Re$ respectively. In other words, the aforementioned sequences x_n, y_n form low-discrepancy sequences in $[0, 1]$.

Now, for the class of curves $C_{\mathcal{A}}$, all curve segments between points of intersection with the edges of the square are parallel to each other. So, by reasoning about the distribution of these points of intersection, we may arrive at conclusions about the distribution of the curves themselves. With this in mind, we will define the average curve length, i.e., $A(R)$, in the form of integrals.

Let $\tan\phi = \frac{\alpha_2}{\alpha_1}$ and $[0, a] \times [0, b]$ be a rectangle with $0 < a, b < 1$. Depending on the relative values of $(\alpha_1, \alpha_2), a, b$ the integrals take on specific forms. We will explain the case when $\frac{b}{a} \le \tan\phi$, $b < 1 - \tan\phi$ (see Figure 2) in some detail, the other cases being similar.

We divide the unit square into three parts, as shown in figure 2. Then, I_1, I_2 and I_3 are real numbers that represent the average length that the (α_1, α_2) lines corresponding to these regions have in common with the rectangle $[0, a] \times [0, b]$. Asymptotically, with curve length $L \to \infty$, these quantities may be represented as follows (based on simple geometric considerations),

$$I_1 = \int_0^{a - \frac{b}{\tan\phi}} \frac{b}{\sin\phi} dx + \int_{a - \frac{b}{\tan\phi}}^{a} \frac{a - x}{\cos\phi} dx = \frac{ab}{\sin\phi} - \frac{b^2 \cos\phi}{2\sin^2\phi} \tag{11}$$

$$I_2 = \int_0^b \frac{b - y}{\sin\phi} dy = \frac{b^2}{2\sin\phi} \tag{12}$$

$$I_3 = 0 \tag{13}$$

$$I_1 \sin\phi + (I_2 + I_3)\cos\phi = ab \tag{14}$$

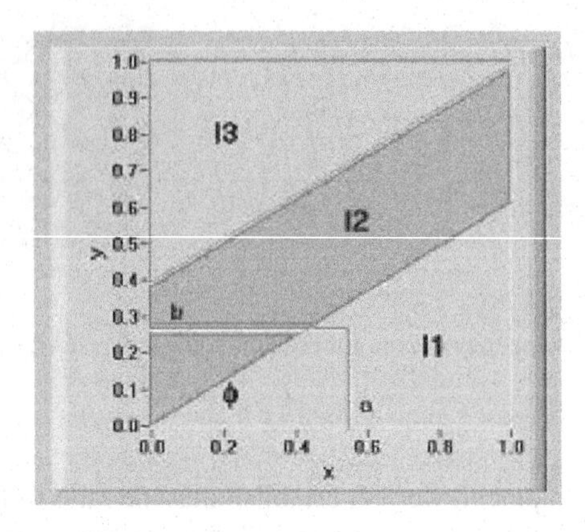

Fig. 2. Definition of the integrals I_1, I_2 and I_3

The final term stands for the average length that (α_1, α_2) lines in $[0, 1] \times [0, 1]$ with slope $\tan\phi$ have in common with the rectangle $[0, a] \times [0, b]$. As expected, it is exactly the area of the rectangle.

In practice, with a finite length curve, what is the discrepancy? We can estimate this using the Koksma-Hlawka inequality 4. For instance, I_1 is approximated using finite length curve segments of the form,

$$l_1(x_i) = \begin{cases} \dfrac{b}{\sin\phi} & : \quad 0 \le x_i \le a - \dfrac{b}{\tan\phi} \\ \dfrac{a - x_i}{\cos\phi} & : \quad a - \dfrac{b}{\tan\phi} \le x_i \le a \\ 0 & : \quad a \le x_i \le 1 \end{cases}$$

Using a finite number, n, of such segments, we have the discrepancy,

$$|I_1 - \sum_{i=1}^{n} l_1(x_i)| = O\Big(\frac{\log^{2+\epsilon} n}{n}\Big) \tag{15}$$

where $x_i = \{\frac{i\alpha_1}{\alpha_2}\}$

The sum stands for the length of that part of the given curve that lies in $[0, a] \times [0, b]$. The exact same argument may be made for the integrals I_2 and I_3. This means that a finitely generated curve segment of type $C_{\mathcal{A}}$ also has a very low discrepancy. Moreover, note that we are reasoning about a 1-dimensional sequence of points of intersection. So, the constant 3 (for $d = 2$) in equation 10 is now replaced with 2 ($d = 1$). ■

We can now develop two variations of Theorem 1.

Theorem 2. *For almost all numbers α in $\Re^2$, $C_{\mathcal{B}}(\alpha)$ and $C_{\mathcal{C}}(\alpha)$ are low-discrepancy curves in the sense of equations 9 and 10.*

Proof
We begin with a remark. One can show that Theorem 1 is still valid when in the definition of low-discrepancy curves a much broader class of rectangles R is considered, i.e., rectangles where we replace $[0, a] \times [0, b]$ with the more generic $[c, a] \times [d, b]$. We refer the reader to [3] for several related proofs and expanded discussion on such ideas.

Curves of type $\mathcal{B}$:
Such a curve can be translated into an equivalent version acting in $[0, 1] \times [0, 2]$, by simply mirroring the square. To this end, reflections at the upper edge (see Figure 1) are ignored. What results is an equivalent scheme as type $\mathcal{A}$ in $[0, 1] \times [0, 2]$. For almost all choices of α, the resulting curve in $[0, 1] \times [0, 2]$ is low-discrepancy. The relation between the original space and the new one is straightforward. The original curve goes through a rectangle $R = [0, a] \times [0, b]$ if and only if the derived curve in $[0, 1] \times [0, 2]$ goes through $[0, a] \times [0, b]$ or through $[0, a] \times [2 - b, 2]$ (see the remark at the beginning of this proof). The latter implies that $C_{\mathcal{B}}(\alpha)$ satisfies equations 9 and 10.

Curves of type $\mathcal{C}$:
We essentially repeat the arguments from type $\mathcal{B}$. For almost all α, curves of type $\mathcal{B}$ in $[0, 2] \times [0, 1]$ are low-discrepancy. Such curves can be generated when reflections at the right edge are ignored. The mirrored version of this curve goes through a rectangle $R = [0, a] \times [0, b]$ if and only if the original curve in $[0, 2] \times [0, 1]$ goes through $[0, a] \times [0, b]$ or $[2 - a, 2] \times [0, b]$ (see the remark at the beginning of this proof). The latter implies that $C_{\mathcal{C}}(\alpha)$ satisfies 9 and 10. ■

Curves $C_{\mathcal{C}}(\alpha)$ can be regarded as first examples of continuous trajectories in a unit square that offer low-discrepancy behavior. In real area coverage scenarios they are highly efficient compared to alternate techniques, as we will demonstrate in section 5.

Finally, these results can be generalized to higher dimensions, using the same style of argument. We state this theorem without proof.

Theorem 3. *For almost all numbers α in $\Re^d$, $C_{\mathcal{A}}(\alpha), C_{\mathcal{B}}(\alpha)$ and $C_{\mathcal{C}}(\alpha)$ are low-discrepancy curves in the generalized sense of equations 9 and 10 in d-dimensional unit cubes.*

In practice, the question arises as to how these curves can be realized on digital computers. For this purpose, it is quite reasonable to assume that we have the ability to generate rational numbers of user-specified arbitrary precision. In this case, the error due to the rational approximation of α in theorem 1, 2 and 3 is correspondingly small so that finite lengths of the resulting curves generate discrepancies that are very low.

Theorem 4. *If $\{n\alpha\}$ generates a low-discrepancy sequence, for some irrational number α, then $\exists N \gg 1$ such that $\{n\frac{p}{q}\}$, $n = 1, \ldots, N$, also generates a low-discrepancy sequence, where $p, q \in \aleph$.*

Proof
The Hurwitz theorem in number theory [18] states that there are infinitely many $\frac{p}{q}$ with the property $|\alpha - \frac{p}{q}| < \frac{1}{q^2}$.

Given any irrational number, it is possible to find a sufficiently large $q \in \aleph$ such that the error of the rational approximation is small. Let q_1 be such a number, with $q_1 > N \gg 1$. Then,

$$\left|n\left(\alpha - \frac{p_1}{q_1}\right)\right| < q_1.\frac{1}{q_1^2} = \frac{1}{q_1} < \frac{1}{N} \tag{16}$$

Now, assume that $\{(n\alpha) \mod 1\}$ generates a low-discrepancy point set, of discrepancy $D(\alpha)$ for $n = 1, \ldots, N$. When α is irrational, $\{(n\alpha) \mod 1\}$ does not intersect the vertices of the unit square. Then, there always exists a neighborhood where elements of $\{(nx) \mod 1\}$, $n = 1, \ldots, N$, are continuous functions of x. This implies that $D(x)$ is a continuous function of x in this neighborhood.

By selecting suitable p_1, q_1 that approximate α well, to within $\epsilon = \frac{1}{q_1}$, we arrive at the bound, $|D(x) - D(\alpha)| < \delta$, where δ is a suitably small constant. This implies that a curve constructed using these sequences, according to the procedure shown in theorem 1, and using rational approximations to α, is still low-discrepancy. ∎

4 Abstract Surfaces and Riemannian Spaces

In many common applications, we deal with spaces other than the unit square that may, nonetheless, be related to the unit square via a parametrization. This should allow us to extend our constructions to these new spaces of interest. However, the parametrization will only rarely, if ever, respect our stated requirements of fairness and low-discrepancy. In situations where the parametrization does not preserve low-discrepancy, we may suitably modify it through the use of some concepts from Riemannian geometry [19], [20]. In essence, Riemannian geometry allows us to deal with a space whose metric properties vary from point to point. This means that we can efficiently describe the warping of a model space into a desired space, i.e., a manifold, in a principled way through the notion of metrics, line, area and volume elements, etc. Our approach in this paper will be to define this warping such that the low-discrepancy properties are preserved.

Given an abstract surface S with a Riemannian metric defined for (u, v) in $[0, 1]^2$,

$$ds^2 = E(u, v)du^2 + F(u, v)dudv + G(u, v)dv^2 \tag{17}$$

where $E(u, v), F(u, v), G(u, v)$ are differentiable functions in u and v and $EG - F^2$ is positive. The area element dA is defined by,

$$dA = \sqrt{E(u, v)G(u, v) - F^2(u, v)}du \wedge dv \tag{18}$$

The function,

$$\Psi(u, v) = \sqrt{E(u, v)G(u, v) - F^2(u, v)} \tag{19}$$

is nonnegative in $[0, 1]^2$ and $\Psi^2(u, v)$ is differentiable.

Let $\alpha = (\alpha_1, \alpha_2)$ be a given irrational vector (direction) in $\Re^2$. According to the definitions 17 and 18, line and area elements of S for a specific direction $(du, dv) = (\alpha_1 du, \alpha_2 du)$ (i.e., with $dv = \frac{\alpha_2}{\alpha_1} du$) satisfy,

$$\frac{ds}{du} = \sqrt{E(u, v)\alpha_1^2 + F(u, v)\alpha_1\alpha_2 + G(u, v)\alpha_2^2} \tag{20}$$

$$\frac{dA}{du^2} = \sqrt{E(u, v)G(u, v) - F^2(u, v)}\alpha_1\alpha_2 \tag{21}$$

From this, we define the quantity Q which describes our notion of fairness,

$$Q = \frac{\frac{ds}{du}}{\frac{dA}{du^2}} = \frac{\sqrt{E(u,v)\alpha_1^2 + F(u,v)\alpha_1\alpha_2 + G(u,v)\alpha_2^2}}{\sqrt{E(u,v)G(u,v) - F^2(u,v)}\alpha_1\alpha_2} \tag{22}$$

Definition 1
A piecewise smooth curve $C : \Re^+ \mapsto S$ lying on an abstract surface S is called a low-discrepancy curve based on a vector $\alpha = (\alpha_1, \alpha_2)$, iff

1. C is S-filling, i.e. C comes arbitrarily close to any point of S.
2. There is a parametrization of S where Q in equation 22 is constant for all (u, v).
3. In any regular point of C, the tangent vector is parallel to $\alpha = (\alpha_1, \alpha_2)$.

The following algorithm is based on Definition 1.

Algorithm 1

1. Find a parametrization of S that satisfies conditions 2 and 3 in Definition 1. See also Remark 1 below.
2. Generate a curve in S based on the image of a low-discrepancy curve in the unit square according to Theorem 2.

Remark 1: The parametrization in step 2 is not unique. In all examples an originally given natural parametrization is modified using replacements $u \mapsto h(u)$ or $v \mapsto h(v)$ where h is a smooth diffeomorphism.

An abstract d-dimensional surface is defined by,

$$ds^2 = \sum_{i,j=1}^{d} g_{ij}(u_1, u_2, ..., u_d) du_i{}^2 du_j{}^2 \tag{23}$$

where the matrix consisting of $g_{ij} : [0, d]^d \mapsto \Re$ is always symmetric, differentiable, and positive semi-definite. An embedding of an abstract space as in equation 23 in an m-dimensional Euclidean space is a diffeomorphism f of the hypercube $[0, 1]^d$ with $f_1(u_1, u_2, ..., u_d), f_2(u_1, u_2, ..., u_d), ..., f_m(u_1, u_2, ..., u_d) : \Re^d \mapsto \Re^m$ where the Riemannian metric of this embedding is described by equation 23. Usually, this definition is too restrictive. Instead, local diffeomorphisms, i.e., coordinate patches, should be used where these patches cover the whole space under consideration.

Let $\alpha = (\alpha_1, \alpha_2, ..., \alpha_d)$ be a given vector (direction) in $\Re^d$. According to equation 23, the line and volume/content elements for a specific direction $(du_1, du_2, ..., du_d) = (\alpha_1 du, \alpha_2 du, ..., \alpha_d du)$ are:

$$\frac{ds}{du} = \sqrt{\sum_{i,j=1}^{d} g_{ij}(u_1, u_2, ..., u_d)\alpha_i\alpha_j} \tag{24}$$

$$\frac{dV}{du^n} = \sqrt{\det(g_{ij}(u_1, u_2, ..., u_d))\alpha_1\alpha_2...\alpha_d} \tag{25}$$

From this,

$$Q = \frac{\frac{ds}{du}}{\frac{dV}{du^n}} = \frac{\sqrt{\sum_{i,j=1}^{d} g_{ij}(u_1, u_2, ..., u_d)\alpha_i\alpha_j}}{\sqrt{\det(g_{ij}(u_1, u_2, ..., u_d))\alpha_1\alpha_2...\alpha_d}} \tag{26}$$

Definition 2

A piecewise smooth curve $C : \Re^+ \mapsto S$ in the given Riemannian space S is called low-discrepancy curve based on a vector direction $\alpha = (\alpha_1, \alpha_2, ..., \alpha_d)$ iff

1. C is S-filling, i.e. C comes arbitrarily close to any point of S.
2. There is a parametrization of S where Q in equation 26 is constant for all $(u_1, u_2, ..., u_d)$.
3. In any regular point of C, the tangent vector is parallel to $\alpha = (\alpha_1, \alpha_2, ..., \alpha_d)$.

5 Examples of the Coverage of Various Spaces

In this section we will demonstrate the use of the proposed technique to cover various surfaces. These examples are chosen to correspond to geometrical shapes and surfaces that are commonly encountered in robotics. We will present visualizations in the Euclidean space.

5.1 Covering the Surface of the Unit Cube

Consider the surface of a unit cube. This is the most natural generalization of the unit square in section 3. Figure 3 depicts an "opened out" version in the plane of the paper, a transformed version of the which can be used to tile the plane and a low-discrepancy curve drawn on this tiling.

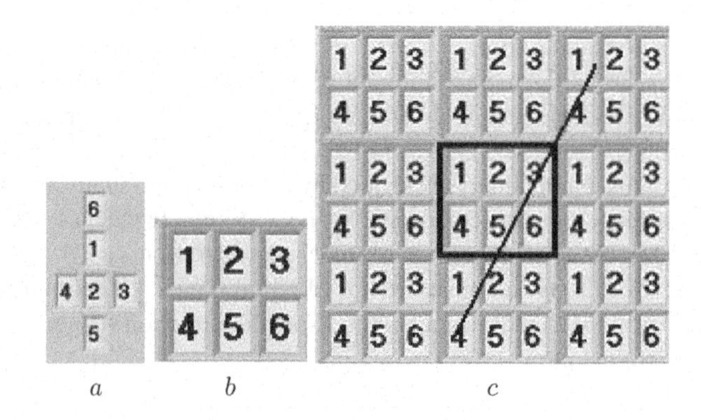

Fig. 3. Tiling of the plane and relation to the surface of the unit cube. Part a depicts the faces of an opened out cube (imagine a typical packing box). Part b depicts an equivalent version, rearranged in such a way that it is easier to tile the plane. Part c shows the low-discrepancy curve traversing the tiled and flattened cubes. The key point to note is that whenever two edges of a face in part c touch, they will also touch in the 3-dimensional cube. The depicted curve is a prescription of the sequence of faces, and points on that face, to visit.

5.2 Covering the Annulus

Consider an annulus, i.e., a ring, that has a standard parametrization given by $x(u, v) = (u \cos v, u \sin v)$, where $0 \leq u_0 \leq u \leq u_1$ and $0 \leq v \leq 2\pi$.

Let us consider a diffeomorphism such that $u \mapsto g(u)$ and $v \mapsto v$. Then the ring can be re-parameterized by $x(u, v) = (g(u) \cos v, g(u) \sin v)$. Now, we can apply the method developed earlier. Let g be sufficiently smooth, where g maps $[u_0, u_1]$ onto $[u_0, u_1]$. Then, according to our proposed method, $g(u)$ must satisfy an ordinary differential equation (for $\alpha_1, \alpha_2 > 0$),

$$g'(u) = \frac{\alpha_2 g(u)}{\sqrt{c^2 g^2(u)\alpha_1^2\alpha_2^2 - \alpha_1^2}} \tag{27}$$

where $g(u_0) = u_0$ and $g(u_1) = u_1$, and,

$$\frac{ds}{du} = \sqrt{g'^2(u)\alpha_1^2 + g^2(u)\alpha_2^2} \tag{28}$$

$$\frac{dA}{du^2} = g(u)g'(u)\alpha_1\alpha_2 \tag{29}$$

Equation 27 has the closed form solution,

$$\frac{\alpha_1}{\alpha_2}\left\{\sqrt{c^2 g^2(u)\alpha_2^2 - 1} + \arctan\left(\frac{1}{\sqrt{c^2 g^2(u)\alpha_2^2 - 1}}\right)\right\} = u + D \tag{30}$$

where D is the unknown constant of integration. Using boundary conditions that arise from the domain and image constraints of the mapping, $g(u_0) = u_0$ and $g(u_1) = u_1$, it follows that,

$$\frac{\alpha_1}{\alpha_2}\left\{\sqrt{c^2 u_0^2\alpha_2^2 - 1} + \arctan\left(\frac{1}{\sqrt{c^2 u_0^2\alpha_2^2 - 1}}\right)\right\} = u_0 + D \tag{31}$$

$$\frac{\alpha_1}{\alpha_2}\left\{\sqrt{c^2 u_1^2\alpha_2^2 - 1} + \arctan\left(\frac{1}{\sqrt{c^2 u_1^2\alpha_2^2 - 1}}\right)\right\} = u_1 + D \tag{32}$$

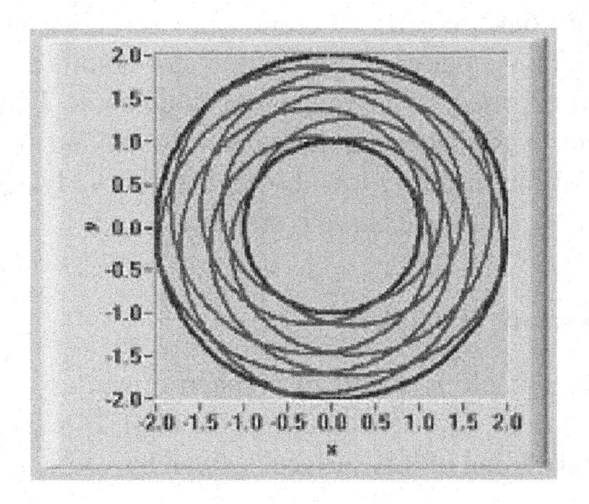

Fig. 4. A low-discrepancy curve in the annulus

We solve for c, D and then use equation 30 as an implicit definition of the required diffeomorphism. In general, obtaining such explicit solutions could become tedious and difficult. In such cases, one may solve the nonlinear differential equations numerically using shooting methods [21].

Figure 4 shows a resulting low-discrepancy curve filling the given ring for the specific case where $u_0 = 1, u_1 = 2$. The parameters α_1, α_2 and c were chosen appropriately by numerical experimentation.

5.3 Covering the Surface of a Torus

The torus is one of the most important non-trivial surfaces that appear in robotics, especially when working with configuration spaces of mechanical systems. We will use our algorithm to generate a low-discrepancy curve for this surface. We use the diffeomorphism $u \mapsto u$ and $v \mapsto g(v)$. Given the $\Re^3$ embedding of a torus ($b < a$),

$$x(u,v) = ((a + b\cos(2\pi g(v)))\cos(2\pi u),\\ (a + b\cos(2\pi g(v)))\sin(2\pi u), b\sin(2\pi g(v))) \quad (33)$$

$$ds^2 = 4\pi^2(a + b\cos(2\pi g(v)))^2 du^2 + 4\pi^2 b^2 g'^2(v) dv^2 \quad (34)$$

$$dA^2 = 16\pi^4 b^2 (a + b\cos(2\pi g(v)))^2 g'^2(v) du^2 dv^2 \quad (35)$$

The function g maps $[0,1]$ onto $[0,1]$ and is sufficiently smooth. Constant ratio of $\frac{ds}{du}$ and $\frac{dA}{du^2}$ in α-direction can be achieved if the following equation holds true (c is a constant, $\alpha_1 > 0$):

$$\frac{(a + b\cos(2\pi g(v)))^2\alpha_1^2 + b^2 g'^2(v)\alpha_2^2}{4\pi^2 b^2 g'^2(v)(a + b\cos(2\pi g(v)))^2\alpha_1^2\alpha_2^2} = c \Rightarrow$$

$$g'(v) = \frac{(a + b\cos(2\pi g(v)))\alpha_1}{\sqrt{4c\pi^2 b^2 (a + b\cos(2\pi g(v)))^2\alpha_1^2\alpha_2^2 - b^2\alpha_2^2}} \quad (36)$$

The boundary conditions are $g(0) = 0$ and $g(1) = 1$. A solution of equation 36 guarantees $g'(0) = g'(1)$. As before, this equation also admits a closed form solution. However, the resulting expressions are long and cumbersome. For the purposes of this example, the parameter c is chosen numerically with the aid of a shooting method. Figure 5 depicts part of the resulting low-discrepancy curve lying on the surface of a torus. Because of $g'(0) = g'(1)$, the curve is smooth.

5.4 Covering the Surface of a Sphere

A more sophisticated example is a part of a sphere given as an abstract surface by $ds^2 = 4\pi^2 \sin^2(\pi g(v)) du^2 + \pi^2 dv^2$ where (u,v) is in $[0,1] \times [v_0, v_1]$ with $v_0 < v_1$ in $(0,1)$. A Euclidean embedding of this surface is given by,

$$x(u,v) = (\sin(2\pi u)\sin(\pi g(v)), \cos(2\pi u)\sin(\pi g(v)), \cos(\pi g(v))) \quad (37)$$

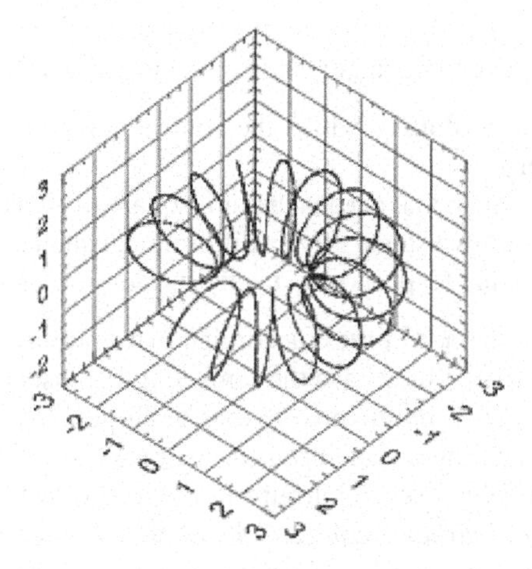

Fig. 5. Low-discrepancy curve filling the surface of a torus

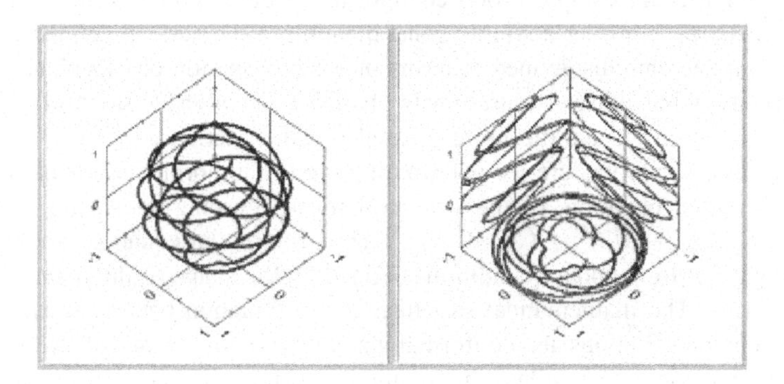

Fig. 6. A low-discrepancy curve on the surface of a sphere. The figure on the right shows the three 2-dimensional projections along the axis planes.

The function $g(v)$ is smooth and maps $[v_0, v_1]$ onto $[v_0, v_1]$. According to Definition 1 and equation 22, we have,

$$g'(v) = \frac{2\sin(\pi g(v))\alpha_1}{\sqrt{4c^2\pi^2\sin^2(\pi g(v))\alpha_1^2\alpha_2^2 - \alpha_2^2}} \tag{38}$$

The boundary conditions are $g(v_0) = v_0$ and $g(v_1) = v_1$. Figure 6 depicts the resulting low-discrepancy curve.

6 Discussion

Broadly speaking, there are three major steps in the procedure outlined above:

1. Define a criterion according to which the curve optimally covers space. This is a *fairness* requirement.
2. Define a diffeomorphism that carries the unit square to an abstract surface.
3. Based on the previous two steps, solve a differential equation whose solution yields the diffeomorphism that respects our stated criterion for fairness.

In this paper, we have defined fairness according to the requirement that any two arbitrary segments of the curve, if they have the same length, must cover the same amounts of area or volume. This is an intuitively obvious definition. There are several ways to extend the definition of fairness. For instance, we could ask that the diffeomorphism must be area preserving along several arbitrarily defined directions or for hyperplanes in n-dimensional space. Fairness requirements of this sort would be difficult to factor into the traditional techniques for generating low discrepancy sequences but do not substantially affect our proposed algorithm.

In this paper, for the purposes of exposition, we have considered somewhat regular and symmetric shapes. Our proposed algorithm would apply to other situations as well. For instance, consider a simple generalization to the surface defined in section 5.2, a generalized annulus defined in terms of the intersection of two planar surfaces with star-convex but otherwise arbitrarily shaped boundaries. Instead of $x(u, v) = (g(u)\cos v, g(u)\sin v)$, use the more general parametrization $x(u, v) = (M(u, v), N(u, v))$. Then our procedure for determining the diffeomorphism would generate a partial differential equation, of the structural form (ignoring constants), $(M_u N_v - M_v N_u)^2 = (M_u + M_v)^2 + (N_u + N_v)^2$. This differential equation would need to be solved subject to boundary conditions defined by the shape of the boundary of the planar surfaces. The mathematical structure of this equation bears a resemblance to Cauchy-Riemann equations and conformal mappings. However, our procedure applies to higher dimensions and handles other requirements including area preservation.

7 Conclusions

We have provided an extension of the theory of low discrepancy sequences to define curves that can uniformly cover space. In doing so, we take an approach that is more general than prior work that has depended on constructing and selecting points from lattices and grids. Our approach is incremental, well suited to search problems in abstract spaces and applicable to any problem where the abstract space can be described by an appropriate Riemannian metric. We have demonstrated the applicability of our proposed approach using examples that are of relevance to robotics.

In our current and future work, we are trying to extend these ideas in several directions. We already mentioned the connection between our technique and other techniques in the theory of conformal mappings and area-preserving mappings, which we are trying to clarify. From an applications standpoint, we view this approach as a powerful way to tackle search and coverage problems in abstract spaces including shape spaces (such as those describing reconfigurable robots or protein structures), parameter spaces and configuration spaces of complex dynamical systems, etc.

Acknowledgment

The authors wish to thank Prof. Harald Niederreiter for reading an early version of the manuscript and for providing valuable feedback.

References

1. Choset, H.: Coverage for robotics - A survey of recent results. Annals of Mathematics and Artificial Intelligence 31, 113–126 (2001)
2. Zaslavsky, G.M.: Hamiltonian Chaos and Fractional Dynamics. Oxford University Press, New York (2005)
3. Matoušek, J.: Geometric Discrepancy: An Illustrated Guide. Springer, Heidelberg (1999)
4. LaValle, S.M., Branicky, M.S., Lindemann, S.R.: On the relationship between classical grid search and probabilistic roadmaps. International Journal of Robotics Research 23(7-8), 673–692 (2004)
5. Lindemann, S.R., LaValle, S.M.: Incremental low-discrepancy lattice methods for motion planning. In: Proc. IEEE International Conference on Robotics and Automation (2003)
6. Lindemann, S.R., Yershova, A., LaValle, S.M.: Incremental grid sampling strategies in robotics. In: Proc. Workshop on Algorithmic Foundations of Robotics, pp. 297–312 (2004)
7. Rimon, E., Koditschek, D.E.: Exact robot navigation using artifical potential functions. IEEE Trans. Robotics and Automation 8(5), 501–518 (1992)
8. Niederreiter, H.: Random Number Generation and Quasi-Monte Carlo Methods. In: CBMS-NSF Regional Conference Series in Applied Math., vol. 63, SIAM, Philadelphia (1992)
9. Kocis, L., Whiten, W.J.: Computational investigations of low-discrepancy sequences. ACM Trans. Mathematical Software 23(2), 266–294 (1997)
10. van der Corput, J.G.: Vereilungsfunktionen I,II. Nederl. Akad. Wetensch. Proc. Ser. B 38, 813–821, 1058–1066 (1935)
11. Halton, J.H.: On the efficiency of certain quasi-random sequences of points in evaluating multi-dimensional integrals. Numer. Math. 2, 84–90 (1960)
12. Sobol', I.M.: The distribution of points in a cube and the approximate evaluation of integrals. Zh. Vychisl. Mat. i Mat. Fiz (USSR Computational Mathematics and Mathematical Physics 7, 784–802 (1967) (in Russian)
13. Richtmyer, R.D.: The evaluation of definite integrals, and quasi-Monte Carlo method based on the properties of algebraic numbers. Report LA-1342, Los Alamos Scientific Laboratory, Los Alamos, NM (1951)
14. Richtmyer, R.D.: A Non-random Sampling Method, Based on Congruences for Monte Carlo Problems, Report NYO-8674. New York: Institute of Mathematical Sciences, New York University (1958)
15. Hammersly, J.M., Handscomb, D.C.: Monte Carlo Methods. Methuen and Company, London (1964)
16. Morokoff, W., Caflisch, R.E.: Quasi-random sequences and their discrepancies. SIAM J. Sci. Stat. Computing 15, 1251–1279 (1994)
17. James, F., Hoogland, J., Kleiss, R.: Multidimensional sampling for simulation and integration: measures, discrepancies, and quasi-random numbers. Computer Physics Communications 99, 180–220 (1997)

18. Hardy, G.H., Wright, E.M.: An Introduction to the Theory of Numbers, 5th edn. Oxford University Press, Oxford (1980)
19. Gray, A.: Modern Differential Geometry Of Curves And Surfaces With Mathematica. CRC Press, Boca Raton (1998)
20. McCleary, J.: Geometry from a Differentiable Viewpoint. Cambridge University Press, Cambridge (1995)
21. Press, W.H., Teukolsky, S.A., Vetterling, W.T., Flannery, B.P.: Numerical Recipes in C: The Art of Scientific Computing, 2nd edn. Cambridge University Press, Cambridge (1992)

The Snowblower Problem*

Esther M. Arkin[1], Michael A. Bender[2], Joseph S.B. Mitchell[1], and Valentin Polishchuk[1]

[1] Department of Applied Mathematics and Statistics, Stony Brook University
{estie,jsbm,kotya}@ams.stonybrook.edu**

[2] Department of Computer Science, Stony Brook University
bender@cs.stonybrook.edu***

Abstract. We introduce the *snowblower problem* (*SBP*), a new optimization problem that is closely related to milling problems and to some material-handling problems. The objective in the SBP is to compute a short tour for the snowblower to follow to remove all the snow from a domain (driveway, sidewalk, etc.). When a snowblower passes over each region along the tour, it displaces snow into a nearby region. The constraint is that if the snow is piled too high, then the snowblower cannot clear the pile.

We give an algorithmic study of the SBP. We show that in general, the problem is NP-complete, and we present polynomial-time approximation algorithms for removing snow under various assumptions about the operation of the snowblower. Most commercially available snowblowers allow the user to control the direction in which the snow is thrown. We differentiate between the cases in which the snow can be thrown in any direction, in any direction except backwards, and only to the right. For all cases, we give constant-factor approximation algorithms; the constants increase as the throw direction becomes more restricted.

Our results are also applicable to robotic vacuuming (or lawnmowing) with bounded capacity dust bin and to some versions of material-handling problems, in which the goal is to rearrange cartons on the floor of a warehouse.

1 Introduction

During a recent major snowstorm in the northeastern USA, one of the authors used a snowblower to clear an expansive driveway. A snowblower is a "material shifting machine," which lifts snow and deposits it nearby. The goal is to dispose of all the snow, moving it outside the driveway. There is a skill in making sure that the deposited piles of snow do not grow higher than the maximum depth capacity of the snowblower. This experience crystallized into an algorithmic question, which we have called the *Snowblower Problem* (*SBP*): **How does one optimally use a snowblower to clear a given polygonal region?**

---

* Full version is available from http://arxiv.org/abs/cs/0603026.

** Partially supported by the U.S.-Israel Binational Science Foundation (2000160), NASA (NAG2-1620), NSF (CCF-0528209, ACI-0328930, CCF-0431030), and Metron Aviation.

*** Partially supported by NSF Grants EIA-0112849 and CCR-0208670.

S. Akella et al. (Eds.): Algorithmic Foundation of Robotics VII, STAR 47, pp. 219–234, 2008.
springerlink.com

The SBP shows up in other contexts: Consider a mobile robot that is equipped with a device that allows it to pick up a carton and then place the carton down again in a location just next to it, possibly on a stack of cartons. With each such operation, the robot shifts a unit of "material". The SBP models the problem in which the robot is to move a set of boxes to a specified destination in the most efficient manner, subject to the constraint that it cannot stack boxes higher than a capacity bound.

In a third motivating application, consider a robotic lawnmower or vacuum cleaner that has a catch basin for the clippings, leaves, dust, or other debris. The goal is to remove the debris from a region, with the constraint that the catch basin must be emptied (e.g., in the compost pile) whenever it gets full.

The SBP is related to other problems on milling, vehicle routing, and traveling salesman tours, but there are two important new features: (a) material must be moved (snow must be thrown), and (b) material may not pile up too high.

While the SBP arises naturally in these other application domains, for the rest of the paper, we use the terminology of snow removal.

The objective of the SBP is to find the shortest snowblower tour that clears a domain P, assumed to be initially covered with snow at uniform depth 1. An important parameter of the problem is the maximum snow depth $D > 1$ through which the snowblower can move. At all times no point of P should have snow of greater depth than D. The snow is to be moved to points outside of P. We assume that each point outside P is able to receive arbitrarily much snow (i.e., that the driveway is surrounded by a "cliff" over which we can toss as much snow as we want).[1]

Snowblowers offer the user the ability to control the direction in which the snow is thrown. Some throw directions are preferable over others; e.g., throwing the snow back into the user's face is undesirable. However, it can be cumbersome to change the throw direction too frequently during the course of clearing. Thus, we consider three *throw models*. In the *default* model throwing the snow backwards is allowed. In the *adjustable-throw* model the snow can be thrown only to the left, right, or forward. In the *fixed-throw* model the snow is always thrown to the right. Even though it seems silly to allow the throw direction to be back into one's face, the default model is the starting point for the analysis of other models and is equivalent to the vacuum cleaner problem (discussed later).

Results. In this paper we introduce the snowblower problem, model its variants, and give the first algorithmic results for its solution. We show that the SBP is NP-complete for multiply connected domains P. Our main results are constant-factor approximation algorithms for each of the three throw models, assuming $D \geq 2$; refer to Table 1. The approximation ratio of our algorithms increases as the throw direction becomes more restricted. We give extensions for clearing polygons with holes, both where the holes are obstacles and cliffs. Then we

[1] The "cliff" assumption accurately models the capacitated-vacuum-cleaner problem for which there is a (central) "dustpan vac" in the baseboard, where a robotic vacuum cleaner may empty its load [1] and applies also to urban snow removal using snow melters [2] or disposing off the snow into a river.

Table 1. Approximation factors of our algorithms

Default model, Thm. 2		
D	2 or 3	any $D \geq 4$
Apx.	6	8

Adjustable throw, Thm. 3	
D	any $D \geq 2$
Apx.	$4 + 3D/\lfloor D/2 \rfloor$

Fixed throw, Thm. 4	
D	any $D \geq 2$
Apx.	$34 + 24D/\lfloor D/2 \rfloor$

discuss how to adapt our algorithms for clearing nonrectilinear polygons and polygons with uneven initial distributions of snow. We conclude by giving a succinct representation of the snowblower tour, in which the tour specification is polynomial in the complexity of the input polygon.

Related Work. The SBP is closely related to milling and lawn-mowing problems, which have been studied extensively in the NC-machining and computational-geometry literatures; see e.g., [5, 4, 11]. The SBP is also closely related to material-handling problems, in which the goal is to rearrange a set of objects (e.g., cartons) within a storage facility; see [9, 8, 14]. The SBP may be considered as an intermediate point between the TSP/lawnmowing/milling problems and material-handling problems. Indeed, for $D = \infty$, the SBP is that of optimal milling. Unlike most material-handling problems, the SBP formulation allows the material (snow) to pile up on a single pixel of the domain, and it is this compressibility of the material that distinguishes the SBP from previously studied material-handling problems. With TSP and related problems, every pixel is visited only a constant number of times, whereas with material-handling problems, pixels may have to be visited a number of times exponential in the input size. For this reason, material-handling problems are not even known to be in NP [9, 8], in contrast with the SBP. Note that in material handling problems the objective is to minimize *workload* (distance traveled while loaded), while in the SBP (as in the milling/mowing problems) the objective is to minimize total travel distance (loaded or not).

The SBP is also related to the earth-mover's distance (EMD), which is the minimum amount of work needed to rearrange one distribution (of earth, snow, etc.) to another; see [7]. In the EMD literature, the question is explored mostly from an existential point of view, rather than planning the actual process of rearrangement. In the SBP, we are interested in optimizing the length of the tour, and we do not necessarily know in advance the final distribution of the snow after it has been removed from P.

The title of this paper coincides with that of [10] but the problems considered appear to be totally unrelated.

Notation. The input is a polygonal domain, P. Since we are mainly concerned with proving constant factor approximation algorithms, it suffices to consider distances measured according to the L_1 metric. We consider the snowblower to be an (axis-parallel) unit square that moves horizontally or vertically by unit steps. This justifies our assumption, in most of our discussion, that P is an integral-orthogonal simple polygon, which is comprised of a union of *pixels* – (closed) unit squares with disjoint interiors and integral coordinates. In Section 5 we remark how our methods extend to general (nonrectilinear) regions.

We say that two pixels are *adjacent* or *neighbors* if they share a side; the *degree* of a pixel is the number of its neighbors. For a region $R \subseteq P$ (subset of pixels), let G_R denote the *dual graph* of R, having a vertex in the center of each pixel of R and edges between adjacent pixels. A pixel of degree less than four is a *boundary pixel.* For a boundary pixel, a side that is also on the boundary of P is called a *boundary side.* The set of boundary sides, ∂P, forms the boundary of P. We assume that the elements of ∂P are ordered as they are encountered when the boundary of P is traversed counterclockwise.

An *articulation vertex* of a graph G is a vertex whose removal disconnects G. We assume that G_P has no articulation vertices. (Our algorithms can be adapted to regions having articulation vertices, at a possible increase in approximation ratio.)

Algorithms Overview. Our algorithms proceed by clearing the polygon Voronoi-cell-by-Voronoi-cell, starting from the Voronoi cell of the garage g — the pixel on the boundary of P at which the snowblower tour starts and ends. The order of the boundary sides in ∂P provides a natural order in which to clear the cells. We observe that the Voronoi cell of each boundary side is a tree of one of two special types, which we call *lines* and *combs.* We show how to clear the trees efficiently in each of the throw models. We prove that our algorithms give constant-factor approximations by charging the lengths of the tours produced by the algorithms to two lower bounds, described in the next section.

2 Preliminaries

Voronoi Decomposition. For a pixel $p \in P$ let $V(p)$ denote the element of ∂P closest to p. In case of ties, the tie-breaking rule (see below) is applied. Inspired by computational-geometry terminology, we call $V(p)$ the *Voronoi side* of p. We let $\delta(p)$ denote the length of the path from p to the pixel having $V(p)$ as a side. For a boundary side $e \in \partial P$ we let Voronoi(e) denote the (possibly, empty) set of pixels, having e is the Voronoi side: Voronoi(e) $= \{p \in P | V(p) = e\}$. We call Voronoi(e) the *Voronoi cell* of e. The Voronoi cells of the elements of ∂P form a partition of P, called the *Voronoi decomposition* of P.

A set of pixels $\mathcal{L}$ whose dual graph $G_{\mathcal{L}}$ is a straight path or a path with one bend, is called a *line.* Each line $\mathcal{L}$ has a *root* pixel p, which corresponds to one of the two leaves of $G_{\mathcal{L}}$, and a *base*, $e \in \partial P$, which is a side of p.

A *(horizontal) comb* $\mathcal{C}$ is a union of pixels consisting of a set of vertically adjacent (horizontal) rows of pixels, with all of the rightmost pixels (or all of the leftmost pixels) in a common column. (A vertical comb is defined similarly; however, by our tie breaking rules, we need consider only horizontal combs.) A comb is a special type of *histogram* polygon [6]. The common vertical column of rightmost/leftmost pixels is called the *handle* of comb $\mathcal{C}$, and each of the rows is called a *tooth*. A *leftward* comb has its teeth extending leftwards from the handle; a *rightward* comb is defined similarly. The pixel of a tooth that is furthest from the handle is the *tip* of the tooth. The topmost row is the *wisdom tooth* of the comb. The *root* pixel p of the comb is either the bottommost or topmost pixel of the handle, and its bottom or top side, $e \in \partial P$, is the *base* of the comb. See Fig. 1, left. The union of a leftward comb and a rightward comb having a common root pixel is called a *double-sided comb*.

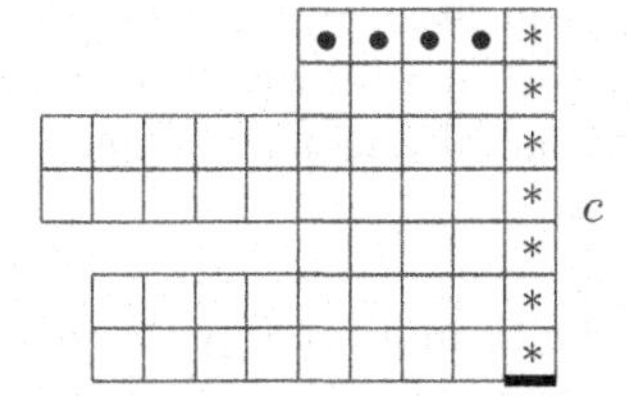

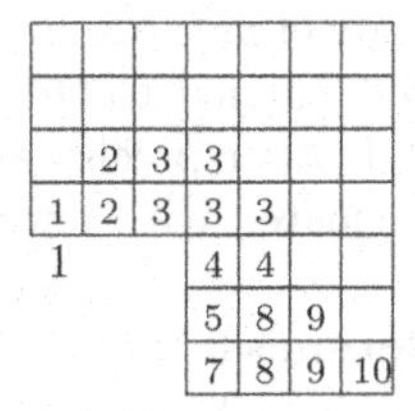

Fig. 1. Left: a comb. The base is bold. The pixels in the handle are marked with asterisks, the pixels in the wisdom tooth are marked with bullets. Right: Voronoi cells. The sides of ∂P are numbered $1 \dots 28$ counterclockwise. The pixels in the Voronoi cell of a side are marked with the corresponding number. Voronoi cell of side 3 is a comb; Voronoi cells of sides 6, 11, 17, 25, 28 are empty; cells of sides 1, 7, 10, 18, 24 are lines, comprised of just one pixel; cells of the other edges are lines with more than one pixel.

Tie Breaking. Our rules for finding $V(p)$ for a pixel p that is equidistant between two or more boundaries is based on the direction of the shortest path from p to $V(p)$; vertical edges are preferred to horizontal, going down has higher priority than going up, going to the right — than going left. In fact, any tie-breaking rule can be applied as long as it is applied consistently. The particular choice of the rule only affects the orientation of the combs.

Voronoi Cell Structure. An analysis of the structure of the Voronoi partition under our tie breaking rules gives:

Lemma 1. *For a side $e \in \partial P$, the Voronoi cell of e is either a line (whose dual graph is a straight path), or a comb, or a double-sided comb. By our tie-breaking rule, the combs may appear only as the Voronoi cells of horizontal edges. The double-sided combs may appear only as the Voronoi cells of (horizontal) edges of length 1.*

Let p be a boundary pixel of P, let $e \in \partial P$ be the side of p such that $p \in$ Voronoi(e). We denote Voronoi(e) by $\mathcal{T}(p)$ or $\mathcal{T}(e)$, indicating that it is a unique tree (a line or a comb) that has p as the root and e as the base.

Lower Bounds. We exhibit two lower bounds on the cost of an optimal tour, the *snow lower bound*, based on the number of pixels, and the *distance lower bound*, based on the Voronoi decomposition of the domain. At any time let

$s(R)$ be the set of pixels of R covered with snow and also, abusing notation, the number of these pixels. Let $d(R) = \frac{1}{D}\sum_{p\in s(R)} \delta(p)$.

Lemma 2. *Let R be a subset of P with the snowblower starting from a pixel outside R. Then $s(R)$ and $d(R)$ are lower bounds on the cost to clear R.*

Proof. For the snow lower bound, observe that region R cannot be cleared with fewer than $s(R)$ snowblower moves because each pixel of $s(R)$ needs to be visited.

For the distance lower bound, observe that, in order to clear the snow initially residing on a pixel p, the snowblower has to make at least $\delta(p)$ moves. When the snow from p is carried to the boundary of P and thrown away, the snow from at most $D-1$ other pixels can be thrown away simultaneously. Thus, a region R cannot be cleared with fewer than $d(R)$ moves. □

NP-Completeness. It is known [12,13] that the Hamiltonian path problem in cubic grid graphs is NP-complete. The problem can be straightforwardly reduced to SBP. If G is a cubic grid graph, construct an (integral orthohedral) domain P such that $G = G_P$. Since G_P is cubic, each pixel $p \in P$ is a boundary pixel, thus, the snowblower can throw the snow away from p upon entering it. Hence, SBP on P is equivalent to TSP on G, which has optimum less than $n+1$ iff G is Hamiltonian (where n is the number of nodes in G). The reduction works for any $D \geq 1$.

The algorithms proposed in this paper show that any domain can be cleared using a set of moves of cardinality polynomial in the number of pixels in the domain, assuming $D \geq 2$. Thus, we obtain

Theorem 1. *If $D \geq 2$, the SBP is NP-complete, both in the default model and in the adjustable throw model, for inputs that are domains with holes.*

3 Approximation Algorithm for the Default Model

In this section we give an 8-approximation algorithm for the case when the snow can be thrown in *all four* directions. We first show how to clear a line efficiently with the operation called *line-clearing*. We then introduce another operation, the *brush*, and show how to clear a comb efficiently with a sequence of line-clearings and brushes. Finally, we splice the subtours through each line and comb into a larger tour, clearing the entire domain. The algorithm for the default model, developed in this section, serves as a basis for the algorithms in the other models.

Clearing a Line. Let $\mathcal{L}$ be a line of pixels; let p and e be its root and the base. We are interested in clearing lines for which the base is a boundary side, i.e., $e \in \partial P$. Let $\ell = s(\mathcal{L})$; let the first J pixels of $\mathcal{L}$ counting from p be clear. We assume that p is already clear ($J > 0$); the snow from it was thrown away through the side e as the snowblower first entered pixel p. Let $\mathcal{L}|J$ denote $\mathcal{L}$ with the J pixels clear; let $\ell - J = kD + r$.[2] Denote by $(\mathcal{L}|J)_D$ the first kD pixels of $\mathcal{L}|J$ covered with snow; denote by $\mathcal{L}_r$ the last r pixels on $\mathcal{L}|J$. The idea of decomposing $\mathcal{L}|J$ into $(\mathcal{L}|J)_D$ and $\mathcal{L}_r$ is that the snow from $(\mathcal{L}|J)_D$ is thrown away with k "fully-loaded" throws, and the snow from $\mathcal{L}_r$ is thrown away with (at most one) additional "under-loaded" throw.

We clear line $\mathcal{L}$ starting at p by moving all the snow through the base e and returning back to p. The basic clearing operation is a back throw. In a back throw the snowblower, entering a pixel u from pixel v, throws u's snow backward onto v. Starting from p, the snowblower moves along $\mathcal{L}$ away from p until either the snowblower moves through D pixels covered with snow or the snowblower reaches the other end of $\mathcal{L}$; this is called the *forward pass.* Next, the snowblower makes a U-turn and moves back to p, pushing all the snow in front of it and over e; this is called the *backward pass.* A forward and backward pass that clears exactly D units of snow is called a *D-full pass.*

Lemma 3. *For arbitrary $D \geq 4$ the line-clearing cost is at most $2s(\mathcal{L} \setminus p) + 4d(\mathcal{L}|J)$. For $D = 2, 3$ the line-clearing cost is at most $2s(\mathcal{L} \setminus p) + 2d(\mathcal{L}|J)$. If every pass is D-full, the cost is $4d(\mathcal{L}|J)$ for $D \geq 4$ and $2d(\mathcal{L}|J)$ for $D = 2, 3$.*

Proof. The clearing cost is $c(\mathcal{L}|J) = c((\mathcal{L}|J)_D) + c(\mathcal{L}_r) = \sum_{i=1}^{k} 2(J - 1 + iD) + 2(\ell - 1) = 2kJ + Dk(k+1) - 2k + 2(\ell - 1)$. The *snow* lower bound of $\mathcal{L} \setminus p$ is $s(\mathcal{L} \setminus p) = \ell - 1$. The *distance* lower bound of $(\mathcal{L}|J)_D$ is $d((\mathcal{L}|J)_D) = \frac{1}{D}\sum_{i=1}^{kD}(J + i) = kJ + k(kD+1)/2$.

Thus,

$$c(\mathcal{L}|J) = 2s(\mathcal{L} \setminus p) + \left(2 + \frac{D-3}{J + (Dk+1)/2}\right) d((\mathcal{L}|J)_D)$$

If every pass is a D-full pass, then $c(\mathcal{L}_r) = 0$. Therefore, $c(\mathcal{L}|J) = c((\mathcal{L}|J)_D) = \left(2 + \frac{D-3}{J+(Dk+1)/2}\right) d((\mathcal{L}|J)_D)$. □

Clearing a Comb. Let $\mathcal{C}$ be a comb with the root p, base e, and handle $\mathcal{H}$ of length H. Let $\ell_1 \ldots \ell_H$ be the lengths of the teeth of the comb. Since we are interested in clearing combs for which the base e is a boundary side ($e \in \partial P$), we assume that pixel p is already clear — the snow from it was thrown away through e as the snowblower first entered p.

Our strategy for clearing $\mathcal{C}$ is as follows. While there exists a line $\mathcal{L} \subset \mathcal{C}$ rooted at p, such that $s(\mathcal{L}) \geq D$, we perform as many D-full passes on $\mathcal{L}$ as we can. When no such $\mathcal{L}$ remains, we call the comb *brush-ready* and we use another clearing operation, the *brush,* to finish the clearing.

A brush, essentially, is a "capacitated" depth-first-search. Among the teeth of a brush-ready comb that are not fully cleared, let t be the tooth, furthest from the base. In a brush, we move the snowblower from p through the handle, turn

[2] For ease of presentation, we adapt the following convention. For $d \in \{D, \lfloor D/2 \rfloor\}$ and an integer w we understand the equality $w = ad + b$ as follows: b and a are the remainder and the quotient, respectively, of w divided by d.

into t, reach its tip, U-turn, come back to the handle (pushing the pile of snow), turn onto the handle, move by the handle back towards p until we reach the next not fully cleared tooth, turn onto the tooth, and so on. We continue clearing the teeth one-by-one in this manner until D units of snow have been moved (or all the snow on the comb has been moved). Then we push the snow to p through the handle and across e. This tour is called a *brush* (Fig. 2).

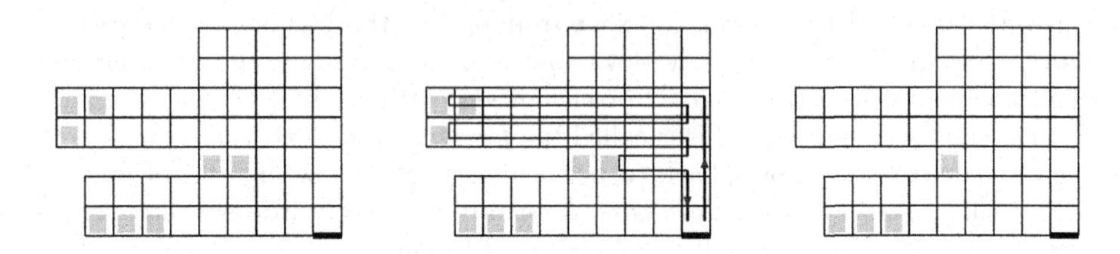

Fig. 2. Left: a brush-ready comb. The snow is shown in light gray. Center: a brush, $D = 4$; the part of the brush, traveling through the handle, is bold. Right: the comb after the brush.

Lemma 4. *For arbitrary $D \geq 4$ the comb $\mathcal{C}$ can be cleared at a cost of at most $4s(\mathcal{C} \setminus p) + 4d(\mathcal{C} \setminus p)$ (at most $4s(\mathcal{C} \setminus p) + 2d(\mathcal{C} \setminus p)$ for $D = 2, 3$).*

Proof. If $s(\mathcal{C} \setminus p) < D$, then the cost of clearing is just $2s(\mathcal{C} \setminus p)$, so suppose, $s(\mathcal{C} \setminus p) \geq D$. Let B be the number of brushes used; let $\mathcal{B}$ be the set of pixels cleared by the brushes. For $b = 1 \dots B$ let t_b and t'_b be the first and the last tooth visited during the bth brush. For $b \in \{1 \dots B-1\}$ the bth brush enters at least 2 teeth, so $t_b > t'_b \geq t_{b+1}$.

Each brush can be decomposed into two parts: the part traveling through the teeth and the part traveling through the handle (Fig. 2). Since each tooth is visited during at most 2 brushes, the length of the first part is at most 4 times the size of all teeth, that is, $4s(\mathcal{C} \setminus \mathcal{H})$. The total length of the second part of all brushes is $2\sum_{b=1}^{B}(t_b - 1)$. Thus, the cost of the "brushing" is

$$c(\mathcal{B}) \leq 2\sum_{b=1}^{B}(t_b - 1) + 4s(\mathcal{C} \setminus \mathcal{H}) \leq 2\sum_{b=2}^{B} t_b + 4s(\mathcal{C} \setminus p) - 2 \qquad (1)$$

since $t_1 \leq H$ and $H \geq 2$ (for otherwise $\mathcal{C}$ is a line).

There are exactly D pixels cleared during each brush $b \in \{0 \dots B-1\}$, and each of these pixels is at distance at least $t_{b'}$ from the base of the comb. Thus, the *distance* lower bound of the pixels, cleared during brush b, is at least $t_{b'}$. Consequently, the *distance* lower bound of $\mathcal{B}$

$$d(\mathcal{B}) \geq \sum_{b=1}^{B} t_{b'} \geq \sum_{b=1}^{B-1} t_{b+1} = \sum_{b=2}^{B} t_b \qquad (2)$$

From (1) and (2), $\mathcal{B}$ can be cleared at a cost of at most $2d(\mathcal{B}) + 4s(\mathcal{C} \setminus p)$.

Let $\mathcal{P} \subseteq \mathcal{C}$ be the pixels, cleared during the line-clearings. By our strategy, during each line-clearing, every pass is D-full; thus, by Lemma 3, $\mathcal{P}$ can be cleared at a cost of at most $4d(\mathcal{P})$ (or $2d(\mathcal{P})$ if $D = 2, 3$). Since $\mathcal{P}$ and $\mathcal{B}$ are snow-disjoint and $\mathcal{P} \cup \mathcal{B} = \mathcal{C} \setminus p$, the lemma follows. □

The above analysis is also valid in the case when the handle is initially clear. This is the case when the second side of a double-sided comb is being cleared. Thus, a double-sided comb can be cleared within the same bounds on the cost of clearing.

Clearing the Domain. Now that we have defined the operations which allow us to clear efficiently lines and combs, we are ready to present the algorithm for clearing the domain.

Theorem 2. *For arbitrary $D \geq 4$ (resp., $D = 2, 3$) an 8-approximate (resp., 6-approximate) tour can be found in polynomial time.*

Proof. Let $p_1, \ldots, p_M$ be the boundary pixels of P as they are encountered when going around the boundary of P counterclockwise starting from $g = p_1$; let $e_1, \ldots, e_M \in \partial P$ be the boundary sides of $p_1, \ldots, p_M$ such that $e_i = Ve(p_i)$, $i = 1 \ldots M$. The polygon P can be decomposed into disjoint trees $\mathcal{T}(p_1), \ldots, \mathcal{T}(p_M) = \mathcal{T}(e_1), \ldots, \mathcal{T}(e_M)$ with the bases $e_1 \ldots, e_M$, where each tree $\mathcal{T}(e_i)$ is either a line or a comb.

Our algorithm clears P tree-by-tree starting with $\mathcal{T}(e_1) = \mathcal{T}(g)$. By Lemmas 3 and 4, for $i = 1 \ldots M$, the tree $\mathcal{T}(p_i) \setminus p_i$ can be cleared at a cost of at most $4s(\mathcal{T}(p_i) \setminus p_i) + 4d(\mathcal{T}(p_i) \setminus p_i)$ starting from p_i and returning to p_i. Since $\bigcup_1^M \mathcal{T}(p_i) \setminus p_i = P \setminus \{p_1 \ldots p_M\}$, the interior of P can be cleared at a cost of at most $c(P \setminus \{p_1 \ldots p_M\}) = 4s(P \setminus \{p_1 \ldots p_M\}) + 4d(P \setminus \{p_1 \ldots p_M\}) \leq 4s(P \setminus g) + 4d(P \setminus g) - 4M + 4$. Finally, the tours clearing the interior of P can be spliced into a tour, clearing P at a cost of at most $2M$. Since the optimum is at least $s(P \setminus g)$ and is at least $d(P \setminus g)$, the theorem follows. □

4 Other Models

In this section we give approximation algorithms for the case when the throw direction is restricted. Specifically, we first consider the adjustable-throw-direction formulation. This is a convenient case for the snowblower operator who does not want the snow thrown in his face. We then consider the fixed-throw-direction formulation, which assumes that the snow is always thrown to the right.

We remark that the relatively low approximation factors of the algorithms for the default model, presented in the previous section, were due to a very conservative clearing: the snow from *every* pixel $p \in P$ was thrown through the Voronoi side $V(p)$. Unfortunately, it seems hard to preserve this appealing property if throwing back is forbidden. The reason is that the comb in the Voronoi cell Voronoi(e) of a boundary side $e \in \partial P$ often has a "staircase"-shaped boundary; clearing the first "stair" in the staircase cannot be done without throwing the snow onto a pixel of Voronoi(e'), where $e' \neq e$ is another boundary side. This is why the approximation factors of the algorithms in this section are higher than those in the previous one.

Adjustable Throw Direction

In the adjustable-throw model the snow cannot be thrown backward but can be thrown in the three other directions. To give a constant-factor approximation algorithm for this case, we show how to emulate line-clearings and brushes avoiding back throws (Fig. 3). The approximation ratios increase slightly in comparison with the default model.

Line-clearing. We can emulate a (half of a) pass by a sequence of moves, each with throwing the snow to the left, forward or to the right (Fig. 3, left and center). Thus, the line-clearing may be executed in the same way as it was done if the back throws were allowed. The only difference is that now the snow is moved to the base when the snow from only $\lfloor D/2 \rfloor$ pixels (as opposed to D pixels) of the line is gathered.

Lemma 5. *The line-clearing cost is at most* $3D/\lfloor D/2 \rfloor d(\mathcal{L}|J) + 2s(\mathcal{L} \setminus p)$. *If every pass is* $\lfloor D/2 \rfloor$*-full, the cost is* $3D/\lfloor D/2 \rfloor d(\mathcal{L}|J)$.

Proof. Let $\ell - J = k' \lfloor D/2 \rfloor + r'$. Let $(\mathcal{L}|J)_{\lfloor D/2 \rfloor}$ be the first $k' \lfloor D/2 \rfloor$ pixels of $\mathcal{L}|J$, let $\mathcal{L}_{r'}$ be its last r' pixels. Then the cost of the clearing of $\mathcal{L}|J$ is $c(\mathcal{L}|J) = c((\mathcal{L}|J)_{\lfloor D/2 \rfloor}) + c(\mathcal{L}_{r'}) = \sum_{i=1}^{k'} 2(J + i\lfloor D/2 \rfloor) + 2\ell = 2k'J + \lfloor D/2 \rfloor k'(k'+1) + 2\ell$. The lower bounds are given by $s(\mathcal{L} \setminus p) = \ell - 1$ and

$$d((\mathcal{L}|J)_{\lfloor \frac{D}{2} \rfloor}) = \frac{1}{D} \sum_{i=1}^{k' \lfloor \frac{D}{2} \rfloor} (J + i) = \frac{\lfloor \frac{D}{2} \rfloor}{D} \left[k'J + \frac{k'(k' \lfloor \frac{D}{2} \rfloor + 1)}{2} \right] \tag{3}$$

Thus,

$$c(\mathcal{L}|J) \leq \frac{D}{\lfloor \frac{D}{2} \rfloor} \left(2 + \frac{2 + \lfloor D/2 \rfloor k' - k'}{k'J + \frac{\lfloor D/2 \rfloor}{2} k'^2 + \frac{k'}{2}} \right) d(\mathcal{L} \setminus p) + 2s(\mathcal{L} \setminus p) \qquad \square$$

Brush. Brush also does not change too much from the default case. The difference is the same as with the line-clearing: now, instead of clearing D pixels with a brush, we prepare to clear only $\lfloor D/2 \rfloor$ pixels (Fig. 3, right). Consequently, the definition of a brush-ready comb is changed — now we require that there is less than $\lfloor D/2 \rfloor$ pixels covered with snow on each tooth of such a comb. Observe that together with each unit of snow, the snow from at most 1 other pixel is moved — thus (although the brush may go outside the comb, as, e.g., in Fig. 3), the brush is feasible.

Lemma 6. *A comb can be cleared at a cost of* $3D/\lfloor D/2 \rfloor d(\mathcal{C} \setminus p) + 4s(\mathcal{C} \setminus p)$.

Proof. In comparison with the default model (Lemma 4) several observations are in place. The number of brushes may go up; we still denote it by B. The

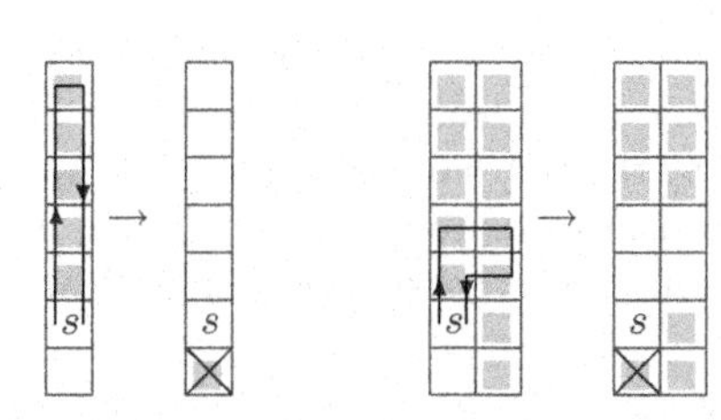

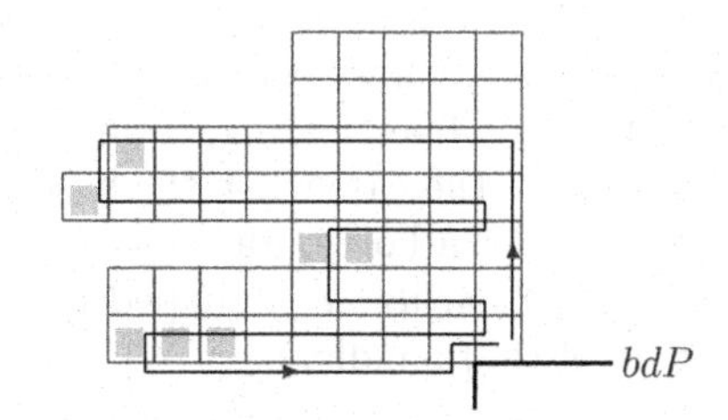

Fig. 3. Emulating line-clearing and brush. The (possible) snow locations are in light gray; s is the snowblower. Left: forward and backward passes in the default model; there are D units of snow on the checked pixel. Center: the passes emulation; there is (at most) $2\lfloor D/2 \rfloor$ units of snow on the checked pixel. Right: the snow to be cleared during a brush is in light gray; there are $\lfloor D/2 \rfloor$ light gray pixels.

cost of the brushes $1 \dots B-1$ does not change. If the Bth brush has to enter the first tooth, there may be 2 more moves needed to return to the root of the comb (see Fig. 3, right); hence, the total cost of the brushing (1) may go up by 2. The *distance* lower bound (2) goes down by $D/\lfloor D/2 \rfloor$. The rest of the proof is identical to the proof of Lemma 4 (with Lemma 5 used in place of Lemma 3). □

Observe that in fact the snow can be removed from *more than* $\lfloor D/2 \rfloor$ pixels during a brush; we just ignore it for now in our analysis. Note that a double-sided comb can also be cleared in the described way.

Clearing the Domain. As in the default case (Theorem 2),

Theorem 3. *A* $(4 + 3D/\lfloor D/2 \rfloor)$*-approximate tour can be found in polynomial time.*

Comment on the Parity of D. We remark that if D is even, the cost of the clearing is the same as it would be if the snowblower were able to move through snow of depth $D+1$ (the slight increase of $6/(D-1)$ in the approximation factor would be due to the decrease of the *distance* lower bound).

Fixed Throw Direction

In reality, changing the throw direction requires some effort. In particular, a snow *plow* does not change the direction of snow displacement at all. In this section we consider the fixed throw direction model, i.e., the case of the snowblower which can only throw the snow to the right. We exploit the same idea as in the previous subsection — reducing the problem in the fixed throw direction model to the problem in the default model. All we need is to show how to emulate line-clearing and brush.

In what follows we retain the notation from the previous section.

Lemma 7. *The line-clearing cost is at most* $24D/\lfloor D/2 \rfloor d(\mathcal{L}|J) + 25s(\mathcal{L} \setminus p)$*. If every pass is* $\lfloor D/2 \rfloor$*-full, the cost is* $24D/\lfloor D/2 \rfloor d(\mathcal{L}|J)$*.*

Proof. We first consider clearing a line whose dual graph is embedded as a single straight line segment and whose base is perpendicular to the segment; we describe the line-clearing, assuming that the line is vertical. Next, we extend the solution to the case when the base is parallel to the edges of the dual graph; this can only be a horizontal line — the first tooth in a (double-)comb. Finally, we consider clearing an L-shaped line; this can only by a tooth together with the (part of the) handle.

A Line $\mathcal{L}$ with $G_{\mathcal{L}} \perp e$. As in the adjustable-throw case (see Fig. 3, left and center), to clear $\mathcal{L}$ we will need to use the pixels to the right of $\mathcal{L}$ to throw the snow onto. Let p' be the boundary pixel, following p counterclockwise around the boundary of P. Before the line-clearing is begun, it will be convenient to have p' clear. Thus, the first thing we do upon entering $\mathcal{L}$ (through p) is clearing p'. Together with returning the snowblower to p it takes 2 or 4 moves (Fig. 4, left); we call these moves *the double-base setup.*

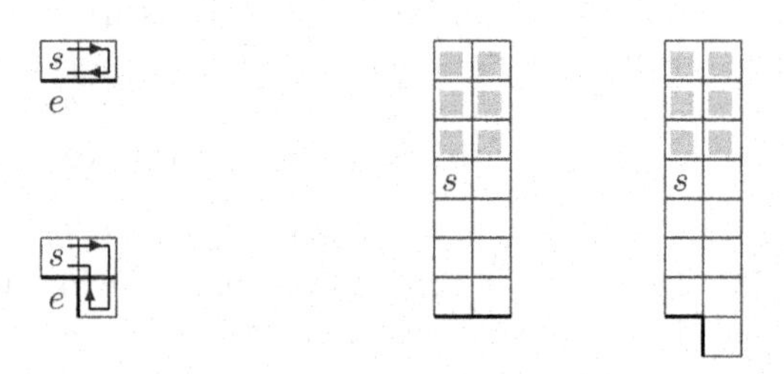

Fig. 4. Left: the double-base setup. Right: before the forward pass the snow below the snowblower is cleared on both lines.

Then, the following invariant is maintained during line-clearing. If the snowblower is at a pixel $q \in \mathcal{L}$ before starting the forward pass, all pixels on $\mathcal{L}$ from p to q are clear, along with the pixels to the right of them (Fig. 4, right). The invariant holds in the beginning of the line-clearing and our line-clearing strategy respects it.

Each back throw is emulated with 5 moves (Fig. 5, left). After moving up by $\lfloor D/2 \rfloor$ pixels (and thus, gathering $2\lfloor D/2 \rfloor$ units of snow on these $\lfloor D/2 \rfloor$ pixels), the snowblower U-turns and moves towards p "pushing" the snow in front of it; a push is emulated with 11 moves (Fig. 6).

The above observations already show that the cost of line-clearing increases only by a multiplicative constant in comparison with the adjustable-throw case. A more careful look at the Figs. 5, 6 reveals that: (1) in the push emulation the first two moves are the opposites of the last two, thus, all 4 moves may be omitted – consequently, a push may be emulated by a sequence of only 7 moves; (2) if the boundary side, following e, is vertical, the last push, throwing the snow away from P, may require 9 moves (Fig. 5, right); and, (3) when emulating the last back throw in a forward pass, the last 2 of the 5 moves (the move up and the move to the right in Fig. 5, left) can be omitted – indeed,

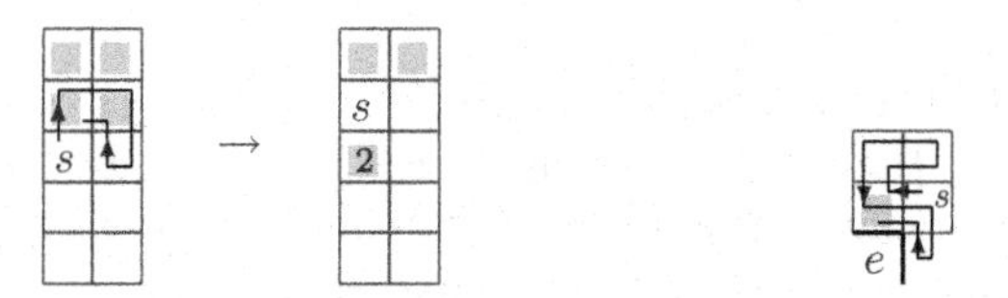

Fig. 5. Left: emulating back throw. Right: pushing the $2\lfloor D/2 \rfloor$ units of snow away from P and returning the snowblower to p may require 9 moves.

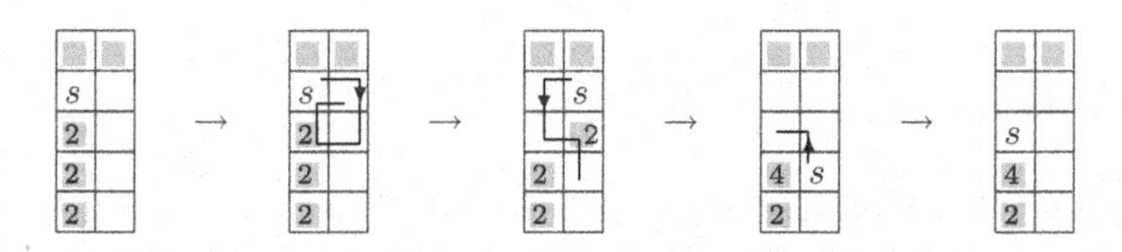

Fig. 6. Emulating pushing the snow in front of the snowblower

during the push emulation, the snowblower may as well start to the right of the snow (see Fig. 6). Thus, a line $\mathcal{L}|J$ can be cleared at a cost of $c(\mathcal{L}|J) \leq 4 + \sum_{i=1}^{k'} (J - 1 + (i-1)\lfloor D/2 \rfloor + 5\lfloor D/2 \rfloor + 7(J + i\lfloor D/2 \rfloor - 1)) + J + 5r' + 7(\ell - 1)$.

A Line $\mathcal{L}$ *with* $G_{\mathcal{L}}||e$. Consider a horizontal line, extending to the *left* of the base; such a line may represent the first tooth of a comb. The double-base can be cleared with 8 or 12 moves, the root can be cleared with 3 moves, left) instead of 9 moves; the rest of the clearing does not change. Consider now a horizontal line extending to the *right* of the base; such a line may appear as the first tooth in a double-sided comb. The double-base for such a line can be cleared with 3 moves; the rest of the clearing is the same as for the vertical line.

L-shaped Line. An L-shaped line $\mathcal{L}$ consists of a vertical and a horizontal segment. Each of the segments can be cleared as described above.

Thus, *any* line $\mathcal{L}|J$ can be cleared at a cost of at most $c(\mathcal{L}|J) \leq 12 + \sum_{i=1}^{k'} (J - 1 + (i-1)\lfloor D/2 \rfloor + 5\lfloor D/2 \rfloor + 7(J + i\lfloor D/2 \rfloor - 1)) + J + 5r' + 7(\ell - 1)$. Since the *snow* and *distance* (3) lower bounds do not change, the lemma follows. □

Lemma 8. *A comb can be cleared at a cost of* $34s(\mathcal{C} \setminus p) + 24D/\lfloor D/2 \rfloor d(\mathcal{C} \setminus p)$.

Proof. Brush in the fixed throw direction model can be described easily using analogy with: a) brush in default and adjustable-throw models and b) line-clearing in fixed-throw model. As in the adjustable-throw model, we prepare to clear $\lfloor D/2 \rfloor$ pixels during each brush. Same as with line-clearing, we setup the double-base for the comb with at most 12 moves; also, 9 moves per brush may be needed to push the snow away from P through the base. Back throw and push can be emulated with 5 and 7 moves (Fig. 5, left and Fig. 6). Thus, if the cost of a brush (1) in the default model was, say, c, the cost of the brush in the fixed-throw model is at most $7c+9$. Since any brush starts with the double-base setup, $c \geq 6$; this, in turn, implies $7c + 9 \leq (51/6)c$. Hence, the cost of clearing $\mathcal{B}$ increases by at most a factor of $51/6$.

By Lemma 7, the cost of clearing $\mathcal{P}$, $c(\mathcal{P}) \leq 24D/\lfloor D/2 \rfloor d(\mathcal{P})$. The *snow* and *distance* lower bounds do not change in comparison with the adjustable-throw case. The lemma now follows from simple arithmetic. □

As in the default and adjustable-throw models (Theorems 2, 3),

Theorem 4. *A* $(34 + \frac{24D}{\lfloor D/2 \rfloor})$*-approximate tour can be found in polynomial time.*

5 Extensions

Polygons with Holes. Our methods extend to the case in which P is a polygonal domain with holes. There are two natural ways that holes may arise in the model.

First, the holes may represent obstacles (e.g., walls of buildings that border the driveway). No snow can be thrown onto such holes; the holes' boundaries serve as walls for the motion of the snowblower and for the deposition of snow. Our algorithm for the default model extends immediately to this variation. The SBP in restricted-throw models, however, may become infeasible.

In the second variation, the holes' boundaries are assumed to be the same "cliffs" as the polygon's outer boundary. It is in fact this version of the problem that we proved to be NP-complete. With some modifications our algorithms work for this variation as well; see the full paper for details.

Nonrectilinear Polygonal Domains. If P is rectilinear, but not integral, we proceed as in [4]: first, the boundary of P is traversed once, and then our algorithms are applied to the remaining part, P', of the domain. Every time the snow is thrown away from P', a certain length (which depends on the throw model) may need to be added to the cost of the tour; thus, the approximation factors of our algorithms may increase by an additive constant.

We can also extend our methods to general nonrectilinear domains. Since the snowblower is not allowed to move outside the domain, care must be taken about specifying which portion of the domain is actually clearable. This portion can be found by traversing the boundary of the domain; then, the accessible portion can be cleared as described above.

Vacuum-Cleaner Problem. Consider the following problem. The floor — a polygonal domain, possibly with holes — is covered with dust and debris. The house is equipped with a central vacuum system, and certain places on the boundary of the floor (the baseboard) are connected to the "dustpan vac" — a dust dump location of infinite capacity [1]. The robotic vacuum cleaner has a dust/debris capacity D and must be emptied to a dump location whenever full. The described problem is equivalent to the SBP in the default throw model and provided the motivation to study the SBP with throwing backwards allowed.

Nonuniform Depth of Snow. Our algorithms generalize easily to the case in which some pixels of the domain initially contain more than one unit of snow. For a problem instance to be feasible it is required that there is less than D (less than $\lfloor D/2 \rfloor$ in restricted-throw direction models) units of snow on each pixel. The approximation ratios in this case depend (linearly) on D (or, in general, on the ratio of D to the minimum initial depth of snow on P).

Capacitated Disposal Region. If instead of "cliffs" at the boundary of P, there is a finite capacity (maximum depth) associated with each point in the complement of P, the SBP more accurately models some material handling problems, but also becomes considerably more difficult. The *snow* lower bound still applies, the *distance* lower bound transforms to a lower bound based on a minimum-cost matching between the pixels in P and the pixels in the complement of P. This problem represents a computational problem related to "earth-mover distance" [7] and is beyond the scope of this paper.

Possible Improvements. We opted for higher approximation factors in favor of more easily described algorithms. For instance, in the adjustable-throw case, the line-clearing cost could be reduced by going up for $D - 3$ pixels, making a small detour, and going back; in the fixed-throw model, instead of emulating each and every back throw with 5 moves, we could emulate a whole ($\lfloor D/2 \rfloor$-full) pass at once.

Open Problems. The complexity of the SBP in simple polygons and the complexity of the SBP in the fixed-throw model are open. We also do not have an algorithm for the case of holes as obstacles with restricted throw direction; the hardness of this version is also open.

One factor we did not address is the difficulty in *turning* a snowblower (see [3] for the discussion of the TSP-like problems with turn costs). Another factor is that a snowblower can throw much further than one cell away.

Acknowledgements. We would like to thank the anonymous referees for their helpful comments.

References

1. `http://www.centralvacuumstores.com/vacpan.htm`
2. `http://www.plowsunlimited.com/snow_melters.htm`
3. Arkin, E., Bender, M., Demaine, E., Fekete, S., Mitchell, J., Sethia, S.: Optimal covering tours with turn costs. SIAM J. on Computing 35(3), 531–566 (2005)
4. Arkin, E.M., Fekete, S.P., Mitchell, J.S.B.: Approximation algorithms for lawn mowing and milling. Comput. Geom. Theory Appl. 17, 25–50 (2000)
5. Arkin, E.M., Held, M., Smith, C.L.: Optimization problems related to zigzag pocket machining. Algorithmica 26(2), 197–236 (2000)
6. Chin, F., Snoeyink, J., Wang, C.A.: Finding the medial axis of a simple polygon in linear time. Discrete Comput. Geom. 21(3), 405–420 (1999)

7. Cohen, S., Guibas, L.: The earth mover's distance: Lower bounds and invariance under translation. Technical Report CS-TR-97-1597, Stanford Univ. Dept Of Computer Science (1997)
8. Culberson, J.: Sokoban is PSPACE-complete. In: Proc. Int. Conf. Fun with Algorithms, Elba, Italy, pp. 65–76 (June 1998)
9. Demaine, E.D., Demaine, M.L., Hoffmann, M., O'Rourke, J.: Pushing blocks is hard. Comput. Geom. Theory Appl. 26(1), 21–36 (2003); Special issue of selected papers from the 13th Canadian Conference on Computational Geometry (2001)
10. Eliazar, I.: The snowblower problem. Queueing Systems Theory and Applications 45(4), 357–380 (2003)
11. Held, M.: On the Computational Geometry of Pocket Machining. LNCS, vol. 500. Springer, Heidelberg (1991)
12. Itai, A., Papadimitriou, C.H., Szwarcfiter, J.L.: Hamilton paths in grid graphs. SIAM J. Comput. 11, 676–686 (1982)
13. Papadimitriou, C.H., Vazirani, U.V.: On two geometric problems related to the traveling salesman problem. J. Algorithms 5, 231–246 (1984)
14. Polishchuk, V.: The box mover problem. In: Proceedings of 16th Canadian Conference on Computational Geometry, pp. 36–39 (2004)

Stratified Deformation Space and Path Planning for a Planar Closed Chain with Revolute Joints*

L. Han, L. Rudolph, J. Blumenthal, and I. Valodzin

Department of Mathematics & Computer Science, Clark University
{lhan,lrudolph,jblumenthal,ivalodzin}@clarku.edu

Abstract. Given a linkage belonging to any of several broad classes (both planar and spatial), we have defined parameters adapted to a stratification of its deformation space (the quotient space of its configuration space by the group of rigid motions) making that space "practically piecewise convex". This leads to great simplifications in motion planning for the linkage, because in our new parameters the loop closure constraints are *exactly*, not approximately, a set of linear inequalities. We illustrate the general construction in the case of planar nR loops (closed chains with revolute joints), where the deformation space (link collisions allowed) has one connected component or two, stratified by copies of a single convex polyhedron via proper boundary identification. In essence, our approach makes path planning for a planar nR loop essentially no more difficult than for an open chain.

1 Overview

Motion planning is important to the study of robotics [13, 3, 15] and is also relevant to other fields as diverse as computer-aided design, computational biology, and computer animation. A unifying concept for motion planning is the set of all configurations of a system under study, called the configuration space of the system and here denoted *CSpace*. In terms of *CSpace*, motion planning amounts to finding a valid curve connecting two given points, where a system configuration is valid if it satisfies the underlying constraints of the system—*e.g.*, the collision free constraint for rigid objects, joint limit constraints for linkage systems, and loop closure constraints for closed chains. Thus all the complexity of motion planning is encoded in *CSpace* and its partition into subsets *CFree* and *CObstacle* of valid and invalid configurations.

In many practical systems, *CSpace* has high dimension and a complicated structure in its own right. For some constraints, the partition introduces much greater complication; for instance, the fastest complete planner taking into account the ubiquitous collision free constraint [2] has exponential running time complexity. The general impossibility of analytically computing *CSpace* and its partition has driven the development of sampling based methods like

* The authors acknowledge computing support from NSF award DBI-0320875, and thank the anonymous reviewers for their helpful comments.

S. Akella et al. (Eds.): Algorithmic Foundation of Robotics VII, STAR 47, pp. 235–250, 2008.
springerlink.com

Probabilistic Roadmap Methods (*PRM*) [11] and Rapidly-exploring Random Trees (*RRT*) [12], that try to capture the connectivity of *CSpace* or *CFree* using sampling and discrete data structures. These methods have been shown to perform very well for many difficult motion planning problems.

Knowledge of *CSpace*—hard as it is to compute—is invaluable for understanding the system in question and developing efficient motion planning algorithms. There has been renewed interest in studying and determining *CSpace* in the past few years. In particular, Trinkle, Milgram and Liu [22, 20, 18, 17] have made important discoveries for *CSpace* of a kinematic chain with fully rotatable joints, either n spherical joints in space (nS) or n revolute joints (nR) in the plane. Building on earlier work by geometers [16, 10], they obtained results on the geometry and topology of the set of closure configurations for a closed chain, initially without imposing the collision free constraint but recently allowing point obstacles. Using this information they develop complete path planners, such as an $O(n^3)$ accordion planner for a closed chain (ignoring collisions), and a planner for avoiding p point obstacles with conjectural lower and upper bounds $\Omega(p^{n-3})$ and $O(p^{2n-7})$. Their work, formulated with joint angle parameters, uses advanced topological tools.

The configuration of a multi-object system in the plane $\mathbb{R}^2$ or space $\mathbb{R}^3$ can be described by the configuration of the objects with respect to a local frame and a transformation from the local frame to a fixed reference frame. Unlike a rigid body which has fixed local coordinates for all points, a multi-body system has different local configurations. Thus a configuration of a multi-body system is described by a rigid body transformation together with a deformation of the system. Call the set of all deformations of the system its deformation space, *DSpace* for short, so that *DSpace* is *CSpace* modulo rigid motions of $\mathbb{R}^2$ or $\mathbb{R}^3$. For instance, for a kinematic chain the local coordinates of the joints are changed by deformations facilitated by the joint degrees of freedom, and restricted by constraints—*e.g.*, fixed link lengths and, for closed chains, the loop closure constraint—which are independent of rigid motions and so effectively are defined on *DSpace* of the chain. Here we focus on the *DSpace* of a loop and ignore the collision free constraint until section 3.3.

We have recently developed a new set of parameters—in this paper denoted by $\mathbf{r}$ and $\mathbf{s}$—to describe *DSpace* for many broad classes of planar and spatial linkages, including planar chains and loops with revolute joints, spatial chains and loops with spherical joints, chains with variable link lengths (which can model prismatic joints), and various kinematic structures more complicated than a chain or single loop. (Linkages can also be used to model a sequence of points under distance constraints.) Unlike the parameters used in earlier work, $\mathbf{r}$ and $\mathbf{s}$ are not joint parameters: $\mathbf{r}$ is a vector of inter-joint distances, and $\mathbf{s}$ is a vector of triangle orientation data to be described below. We use $\mathbf{r}$ and $\mathbf{s}$ to endow *DSpace* of a linkage with a stratification rendering it *practically piecewise convex* in a sense we will explain. For both planar and spatial linkages, $\mathbf{r}$ and $\mathbf{s}$ are uncoupled and $\mathbf{r}$ carries the complicated part of the "practical piecewise convexity" of *DSpace*; for planar linkages, $\mathbf{s}$ serves only to keep track of the

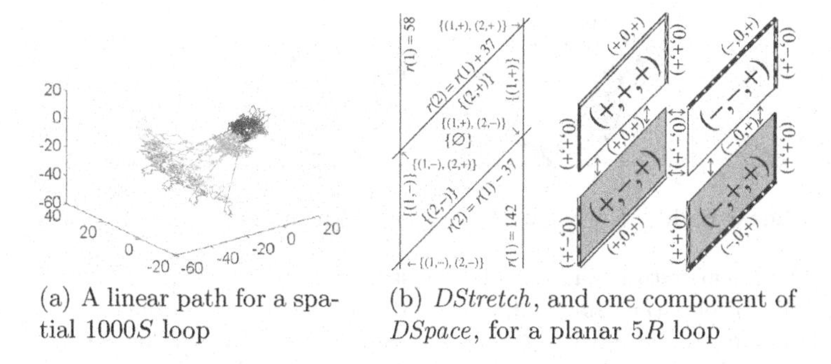

(a) A linear path for a spatial $1000S$ loop

(b) *DStretch*, and one component of *DSpace*, for a planar $5R$ loop

Fig. 1. The $5R$ loop in (b) has link lengths $(100, 42, 37, 95, 86)$

"pieces", whereas for spatial linkages there is just one "piece" and $\mathbf{s}$ serves only to contribute extra "practically convex" dimensions.

Complete treatments of our stratification and new parameters for various linkage types—including both spatial and planar chains—will appear in our future papers. Here we describe our approach for the special but representative case of a planar nR loop, which gives an excellent indication of one major computational and conceptual advantage of our new approach, namely, how constraints (*e.g.*, closure constraints on a loop) that are highly non-linear in terms of traditional joint angle parameters become *linear inequalities* in terms of $\mathbf{r}$ and $\mathbf{s}$. Our reformulation of the constraints and the resulting practical piecewise convexity of *DSpace* greatly simplify motion planning for both planar nR loops and spatial nS loops, as highlighted by the following examples.

Example 1. Consider the problems of generating and joining closure deformations for a loop. Fig. 1(a) illustrates a path between two deformations of a certain spatial $1000S$ loop with randomly chosen link lengths. The two ends of the path were generated by a method we call diagonal sweeping [8]. For each, we found a valid vector $\mathbf{r}$ of 997 positive inter-joint distances, and a vector $\mathbf{s}$ of 997 random angles in $[0, 2\pi]$ specifying triangle orientations (all angles are valid), in 19 milliseconds with Matlab on a desktop computer. Our space of valid vectors $\mathbf{r}$ and the cut-open 997-dimensional torus $[0, 2\pi]^{997}$ are both convex, so the path was then very easily calculated using *linear* interpolation.

Example 2. A planar nR loop has more complicated *DSpace* and path planning than a spatial nS loop. For a planar nR loop with generic link lengths, the set of feasible values of $\mathbf{r}$, which we call *DStretch*, is an $(n-3)$-dimensional convex polyhedron, and "almost all" of *DSpace* can be reconstructed from 2^{n-2} copies of *DStretch* glued together along parts of their boundaries into either one connected component or two (depending on the number of "long links", a technical term [22]; see section 2.4); the remainder of *DSpace* is comparatively low-dimensional and does not hinder motion planning. For one $5R$ loop, Fig. 1(b) shows the 2-dimensional *DStretch* and one component *DSpace*, comprising four copies of *DStretch* labeled with $\mathbf{s}$-values $(s(1), s(2), +)$ $(s(i) \in \{+, -\}$;

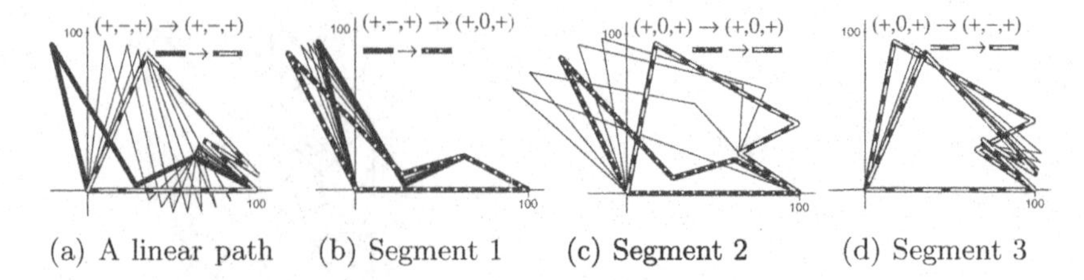

(a) A linear path (b) Segment 1 (c) Segment 2 (d) Segment 3

Fig. 2. The linear path in (a) stays in one stratum but has collisions. The 3-segment path in (b)–(d) joins the same endpoints and is collision free—segment 1 ends on the edge of one copy of *DStretch*, segment 2 crosses a second copy from one edge to its opposite (where **s** has the same value), and segment 3 returns to the first copy.

see section 2) and joined along the indicated pairs of edges; the copies in the other component are labeled $(s(1), s(2), -)$.

Example 3. If the collision free constraint is not imposed, motion planning in *DSpace* is straightforward. Within each copy of *DStretch* the convex structure provides a unique linear path joining any given start deformation to any given goal. Fig. 2(a) illustrates this for the copy of *DStretch* labeled $(+,-,+)$ in Fig. 1(b). Passing between copies is still straightforward, although uniqueness of paths is lost—a fact which can be advantageous for motion planning. In fact, any path joining deformations on different copies of *DStretch* necessarily passes through singular deformations. In section 3, we use a notion of singularity depth (defined in section 2.3) to give upper bounds on the number of singular deformations that must traversed by a path joining two given closure deformations of a planar nR loop. There is a trade-off between the number of singular deformations traversed and their depth. Essentially, the greater the singularity depth of a deformation, the more singular the deformation; a non-singular deformation has singularity depth 0. We show that any two closure deformations in the same component can be connected by a piecewise linear path traversing at most $n-2$ singular deformations, all of depth 1; they can also be connected via at most 2 singular deformations, one of which has singularity depth at least $n-3$ (it is a triangle deformation generalizing one devised by Lenhart and Whitesides [16]). Importantly, the singular deformations are reusable and easily computable. For $n = 1000$, we can find 998 singular $\mathbf{r}$ values of singularity depth 1 in about 20 seconds.

To find *collision free* paths like that in Fig. 2(b-d), we have developed a preliminary probabilistic planner that makes essential use of our efficient closure deformation generation and connection methods.

The efficient algorithms and nice geometry of our approach make many more systems available for use in robot design, where inverse kinematics (closely related to closure deformation generation), motion planning, and similar kinematic issues are very important (see [5, 21]).

2 *DSpace* for a Planar *nR* Loop

2.1 The Idea of a Stratification

We recall a few definitions from the mathematical theory of stratifications (see [6]). Suppose X is a subset of Euclidean space $\mathbb{R}^N$. A partition S of X into subsets $M_1, \ldots, M_K$ is a stratification in case: (1) $M_i \cap M_j = \emptyset$ for $i \neq j$; (2) each M_j is a connected smooth submanifold of $\mathbb{R}^N$; and (3) for each i the closure $\mathrm{cl}(M_i)$ of M_i is itself the union of some of the M_j. Each M_i is called an S-stratum. For $(i \neq j)$, M_i and M_j are incident if $M_i \subset \mathrm{cl}(M_j)$ or $M_j \subset \mathrm{cl}(M_i)$. The dimension $\dim(X)$ of X is $\max\{\dim(M_i) \mid i = 1, \ldots, K\}$; the codimension $\mathrm{codim}(M_i)$ of M_i is $\dim(X) - \dim(M_i)$. If $\mathrm{codim}(M) = 0$ then M is an open subset of X (in the topology induced on X by $\mathbb{R}^N$); if X is connected then X is the closure of the union of the codimension-0 S-strata.

Simple but paradigmatic examples of stratifications come from convexity theory. Let $P \subset \mathbb{R}^N$ be a convex polyhedron, *i.e.*, a closed bounded subset of $\mathbb{R}^N$ that is the intersection of finitely many closed half-spaces. The dimension $\dim(P)$ of P is the dimension of the unique smallest flat (*i.e.*, translated linear subspace of $\mathbb{R}^N$) containing P; the relative interior of P is its actual (topological) interior as a subspace of that flat—equivalently, the set of all points of P not contained in a face Q of P with $\dim(Q) < \dim(P)$. The partition of P into the relative interiors of all its faces is a stratification we call the face stratification *SFace* of P. Each *SFace*-stratum Q is convex, as is $\mathrm{cl}(Q)$; P has exactly one codimension-0 *SFace*-stratum. Below we make extensive use of *SFace* for a polyhedron we associate to a planar nR loop.

2.2 New Parameters

Consider a closed chain in the plane $\mathbb{R}^2$ consisting of n rigid links with consecutive link lengths $l_j > 0$ ($j = 0, \ldots, n-1$), connected by n revolute joints. Denote the consecutive joints of the chain by P_j, so link j is the vector P_jP_{j+1} (indices are modulo n). We call P_0 the anchor of the loop, and in general call an object "anchored" if it includes P_0. For $j = 1, \ldots, n-1$, we call the vector P_0P_j an anchored diagonal of the loop; the anchored diagonals P_0P_1 and P_0P_{n-1} are also links of the loop and thus have fixed non-zero lengths, but other anchored diagonal lengths can vary and may be 0. As illustrated in Fig. 3(a), for $j = 1, \ldots, n-2$ we denote by $\mathrm{Tri}(j)$ the anchored triangle with vertices at joints P_0, P_j, and P_{j+1}; one edge of $\mathrm{Tri}(j)$ is link j and the others are anchored diagonals. At a given point of *DSpace*, $\mathrm{Tri}(j)$ is degenerate (*i.e.*, reduces to an anchored line segment) if and only if its vertices are collinear, which can happen in two distinct ways: either $\mathrm{Tri}(j)$ has three distinct but collinear vertices, or $\mathrm{Tri}(j)$ has exactly two distinct vertices, in which case we call it doubly degenerate. (Since $l_j > 0$, $\mathrm{Tri}(j)$ cannot reduce to a point.) Note that $\mathrm{Tri}(j)$ is doubly degenerate in a given deformation if and only if one of P_j, P_{j+1} coincides with P_0, so that $\mathrm{Tri}(j-1)$ or $\mathrm{Tri}(j+1)$, respectively, is also doubly degenerate. We denote the subset of *DSpace* of deformations with no doubly degenerate anchored triangles by *NDD*, and its subset of deformations

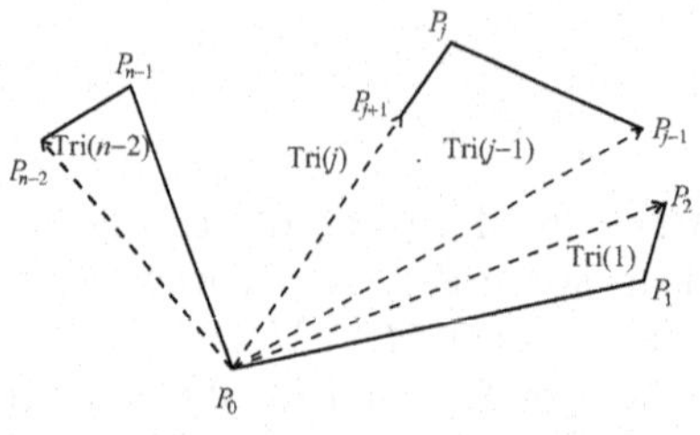

(a) Here $s(1) = s(j-1) = 1$, $s(j) = 0$, and $s(n-2) = -1$; Tri(j) is degenerate, with $r(j-1) = r(j) + l_j$.

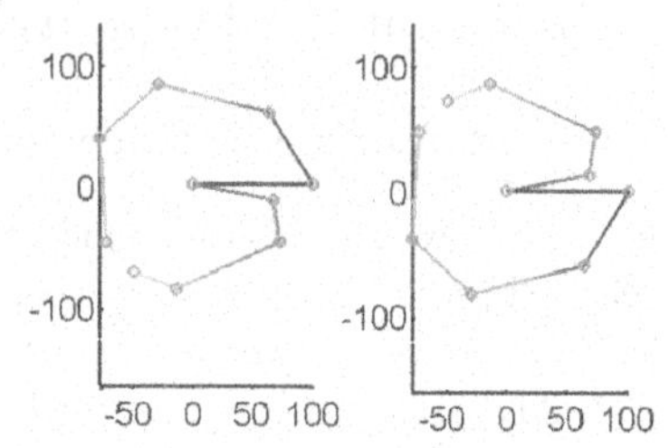

(b) Two 10-bar deformations with opposite orientations: all $s(j)$ are + at the left, − at the right.

Fig. 3. New parameters and deformation examples

with no degenerate anchored triangles by *ND*.(A loop deformation that is singular in the traditional sense but includes no degenerate *anchored* triangle poses no problems for our new parametrization, so for our purposes it is non-singular. We will discuss the role of anchor choice in future papers.)

In a sense, our new parameters for *DSpace* are the triangles Tri(j) themselves, as embedded in the plane modulo a single rigid motion. We extract more conventional parameters from them as follows (see Fig. 3).

Definitions. (1) For $j = 1, \ldots, n-3$, let $r(j) = \|P_0P_{j+1}\|$; the link lengths $l_0, \ldots, l_{n-1}$ and the vector $\mathbf{r} = (r(1), \ldots, r(n-3)) \in \mathbb{R}^{n-3}$ of lengths of anchored diagonals that are not links together encode Tri(1), …, Tri$(n-2)$ up to unoriented congruence. The first $n-3$ of our new parameters are $r(1), \ldots, r(n-3)$. A deformation belongs to *NDD* if and only if every $r(j)$ is strictly positive. (2) For $j = 1, \ldots, n-2$, let $s(j)$ be the sign of the determinant with first column P_0P_j and second column P_0P_{j+1}; so $s(j)$ is 0 if Tri(j) is degenerate, and otherwise it is + or − according as the vertices P_0, P_j, P_{j+1} are oriented counterclockwise or clockwise. The last $n-2$ of our new parameters are $s(1), \ldots, s(n-2)$. Let $\mathbf{s} = (s(1), \ldots, s(n-2))$.

On *NDD*, $\mathbf{r}$ and $\mathbf{s}$ (defined throughout *DSpace*) truly are parameters.

Theorem 1. *The restriction of* $(\mathbf{r}, \mathbf{s})$: *DSpace* $\to \mathbb{R}^{n-3} \times \{-, 0, +\}^{n-2}$ *to NDD is one-to-one onto its image.*

Proof. Given the value of $\mathbf{r}$ on a planar nR loop with fixed link lengths $l_0, \ldots, l_{n-1}$ and no doubly degenerate anchored triangles, we first reconstruct the triangles Tri(j) abstractly. Using the value of $\mathbf{s}$ on the deformation, and starting from an arbitrary placement of Tri(1) in $\mathbb{R}^2$, we then successively lay down Tri(2), …, Tri$(j-2)$ in positions that are well-defined because at each step the anchored edge along which the next triangle must match up has length $r(j-1) > 0$ and the orientation sign $\mathbf{s}(j)$ determines on which side of that edge (if either) that triangle must lie. The only indeterminacy in this construction is the initial placement of Tri(1); any two such placements differ by a rigid motion of $\mathbb{R}^2$ that is well-defined because $r(1) > 0$. □

Theorem 1 says that valid loop deformations in *NDD* correspond exactly to feasible values of $(\mathbf{r}, \mathbf{s})$. Moreover, given a feasible value of $(\mathbf{r}, \mathbf{s})$, the way we constructed the corresponding valid loop deformation makes it clear that we obtain other feasible values of $(\mathbf{r}, \mathbf{s})$ by changing an arbitrary set of non-zero entries of $\mathbf{s}$ from $+$ to $-$ or vice versa: reversing the sign of $s(j)$ corresponds to reversing the orientation of the non-degenerate triangle $\mathrm{Tri}(j)$ by flipping it across the anchored diagonal P_0P_j, and clearly any subset of the non-degenerate anchored triangles in a valid loop deformation can be flipped to create a new valid loop deformation—the entries of $\mathbf{s}$ are uncoupled from $\mathbf{r}$, and (for a given value of $\mathbf{r}$) from each other. We sum this up as follows.

Corollary 1. *If* $r(1), \ldots, r(n-3)$, $s(1), \ldots, s(n-2)$ *are the parameters of a valid deformation in NDD, then there is a valid deformation in NDD with parameters* $r(1), \ldots, r(n-3)$, $\varepsilon(1)s(1), \ldots, \varepsilon(n-2)s(n-2)$ *for every "triangle re-orientation" function* $\varepsilon\colon \{1, \ldots, n-2\} \to \{+, -\}$. □

Thus the problem of understanding the topology and geometry of *NDD* breaks into two subproblems: (A) What is the topology and geometry of the set $\mathbf{r}(NDD)$? (B) How can $\mathbf{s}$ be used to recover the topology and geometry of *NDD* from $\mathbf{r}(NDD)$? We answer (A) in section 2.3 and (B) in section 2.4.

2.3 The Set of Feasible Values of r

In [9], we denoted the set $\mathbf{r}(DSpace)$ of feasible values of $\mathbf{r}$ by *DStretch*; we keep that notation, and also write $DStretch^+$ for $\mathbf{r}(NDD)$. Our proof of Theorem 1 shows that, for a given nR loop, a value of $\mathbf{r}$ is feasible (corresponds to some valid loop deformation) if and only if that value and the given link lengths allow the successful construction of the $n-2$ anchored triangles: if some entries of $\mathbf{r}$ are too big or too small, one or more anchored triangles will be impossible to construct. More precisely, from basic geometry we know that $a, b, c \geq 0$ are the side lengths of a possibly degenerate triangle if and only if $a \leq b + c$, $b \leq c + a$, and $c \leq a + b$; furthermore, the triangle is non-degenerate if and only if all three inequalities are strict. In our case, taken together these inequalities for $\mathrm{Tri}(1), \ldots, \mathrm{Tri}(n-2)$ give an explicit description of *DStretch* in terms of the link lengths $l_0, \ldots, l_{n-1}$: it is the set of solutions $(r(1), \ldots, r(n-3))$ of the following system of linear inequalities $Ineq_j^{\sigma}$. (Here j indicates that $\mathrm{Tri}(j)$ contributed the inequality; $Ineq_j^+$ and $Ineq_j^-$ define anti-parallel half-spaces, to which that defined by $Ineq_j^{\perp}$ is perpendicular.)

$$
\begin{array}{lrl}
Ineq_1^+ : & r(1) \leq & l_0 + l_1 \\
Ineq_1^- : & -r(1) \leq & -|l_0 - l_1| \\
Ineq_j^+ : & r(j) - r(j-1) \leq & l_j \\
Ineq_j^- : & -r(j) + r(j-1) \leq & l_j \\
Ineq_j^{\perp} : & -r(j) - r(j-1) \leq & -\, l_j \\
Ineq_{n-2}^+ : & r(n-3) \leq & l_{n-2} + l_{n-1} \\
Ineq_{n-2}^- : & -r(n-3) \leq & -|l_{n-2} - l_{n-1}|
\end{array}
\quad \left.\right\} j = 2, \ldots, n-3 \qquad (1)
$$

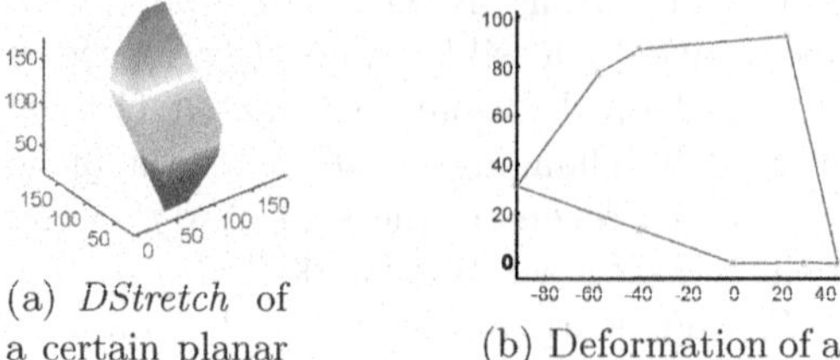

(a) *DStretch* of a certain planar $6R$ loop

(b) Deformation of a planar $8R$ loop of singularity depth 2

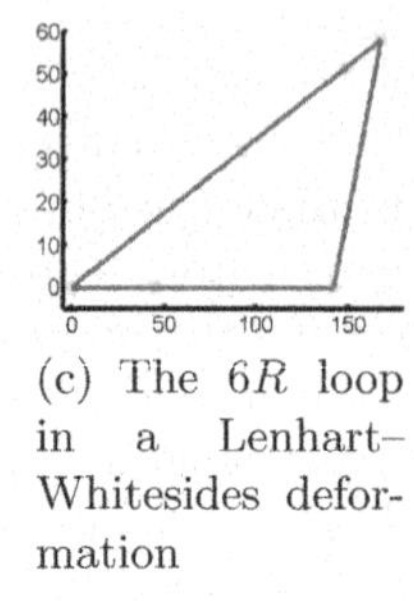

(c) The $6R$ loop in a Lenhart–Whitesides deformation

Fig. 4. The image in the 5-dimensional *DStretch* for the $8R$ loop of the deformation in (b) lies in a 3-dimensional face isometric to the polyhedron in (a)

Rewritten in matrix format, system (1) becomes $DStretch = \{\mathbf{r} \mid T\mathbf{r} \leq \mathbf{b}\}$, where $\mathbf{b} = (l_0 + l_1, \ldots, -|l_{n-2} - l_{n-1}|)$ and $\leq$ is applied termwise. Each row of T corresponds to an inequality $Ineq^{\tau}_j$ in (1), so restricts $\mathbf{r}$ to a closed half-space; thus *DStretch* is the intersection of at most $3n - 8$ closed half-spaces. The link lengths are fixed, so each $r(j)$ is bounded between zero and the sum of all link lengths. Thus *DStretch* is also bounded and is a *convex polyhedron.*

Example 4. Fig. 4(a) shows *DStretch* for a planar $6R$ loop with link lengths $(45, 97, 63, 20, 59, 98)$. It is a 3-dimensional polyhedron, with faces of codimension 0 (its interior), 1 (the interiors of its polygonal faces), 2 (the interiors of its edges), and 3 (its vertices). Since $r(1) \geq 97-45$, $r(3) \geq 98-59$, and $r(2) \geq r(3)-20 \geq 19$, (1) shows that for this loop $DStretch^+ = DStretch$. The same polyhedron also arises as the closure of a codimension-2 face of the 5-dimensional polyhedron *DStretch* of a planar $8R$ loop with link lengths $(31, 14, 97, 63, 20, 59, 56, 42)$, corresponding to deformations (like that in Fig. 4(b)) with Tri(1) and Tri(6) both degenerate.

Fig. 4(c) shows a special deformation of the $6R$ loop used in Example 4, generalized from the concept of "standard triangular form" used by Lenhart and Whitesides [16]. It is defined by finding joint index j satisfying $\sum_{i=0}^{j-1} l_i \leq L/2$ and $\sum_{i=0}^{j} l_i > L/2$, where L is the sum of all link lengths, and then using the subchain from joint 0 (the anchor) to joint j, link j, and the subchain from joint $j+1$ to 0 as three sides of a (possibly degenerate) triangle. It is easy to see that the r value of such a deformation is a vertex of *DStretch*, which we will call the *LW* vertex. We will call a deformation an *LW* deformation if the r image of the deformation is the *LW* vertex. Note that other *DStretch* vertices are also of great interest; our focus on the *LW* vertex in this paper is to facilitate the description of earlier work and illustrate the important roles of highly singular deformations in path planning.

Our earlier observations show the following.

Theorem 2. (a) *DStretch is a convex polyhedron.* (b) *$DStretch^+$ is the intersection of DStretch with $\{(x_1,\dots,x_{n-3}) \mid x_j > 0, j=1,\dots,n-3\}$, and is a union of open faces of DStretch.* □

Since *DStretch* is a convex polyhedron, by section 2.1 it has a natural face stratification. As a restriction of *DStretch*, $DStretch^+$ also has a natural stratification: its strata are exactly those open faces of *DStretch* not contained in (and therefore disjoint from) each of the $n-3$ coordinate hyperplanes $\{(x_1,\dots,x_{n-3}) \mid x_j = 0\} \subset \mathbb{R}^{n-3}$. For a planar nR loop that cannot have any doubly degenerate triangles, $DStretch^+ = DStretch$ (see Figs. 1(b) and 2).

Each stratum Q of $DStretch^+$ is characterized by the set of (j,σ), denoted by $E(Q)$, for which one of the two or three linear inequalities $Ineq^\sigma_j$ in (1) associated with triangle j is replaced with the corresponding equality Eq^σ_j. (Now we can explain the labels in Figs. 1(b): they are the values of $E(Q)$ on the *DStretch* strata of the $5R$ loop.) Let $e(Q)$ be the number of elements in $E(Q)$. For a stratum in $DStretch^+$, $e(Q)$ is also the number of degenerate anchored triangles that can be induced by the values of $\mathbf{r} \in Q$; we call $e(Q)$ the *singularity depth* of Q. For a loop with generic link lengths, the singularity depth of a stratum Q is equal to the co-dimension of the stratum; but the singularity depth of a stratum for a loop with non-generic link lengths may be different from its co-dimension. For example, the top-left subfigure in Fig. 5 shows *DStretch* for a planar $5R$ loop with link lengths $(2,3,4,2,3)$, with its $E(Q)$ labels; two of the codimension-2 strata (the vertices where $\mathbf{r} = (1,5)$ and $\mathbf{r} = (5,1)$) induce three degenerate triangles and have singularity depth 3. The following results show that $E(Q)$ can be used to label the *SFace*-strata for *DStretch* and derive their incidence relations.

Theorem 3. (a) $E(Q) \neq E(Q')$ *if* $Q \neq Q'$. (b) $E(Q) \subset E(Q')$ *if* $Q' \subset \mathrm{cl}(Q)$. (c) $\mathrm{codim}(Q) \leq e(Q)$. (d) $\mathrm{codim}(Q) = e(Q)$ *if* $e(Q) \leq 1$. □

2.4 The Stratification and Topology of *DSpace*

For a stratum Q of *DStretch*, $E(Q)$ identifies which triangles, if any, are degenerate for any given $\mathbf{r} \in Q$. In other words, if $(j,\sigma) \in E(Q)$, and $\mathbf{r} \in Q$, then $\mathrm{Tri}(j)$ is degenerate in any deformation with that value of $\mathbf{r}$, forcing $s(j) = 0$ for such a deformation. On the other hand, if $(j,\sigma) \notin Q$ for any $\sigma \in \{+,-,\perp\}$, then $\mathrm{Tri}(j)$ cannot be degenerate in any deformation with that value of $\mathbf{r}$, so $s(j)$ must be $+$ or $-$ for such a deformation. This observation leads to another way to label the open face Q, namely, by an $(n-2)$-vector reflecting the possible s values of loop deformations with $\mathbf{r} \in Q$: the jth component is the symbol $\pm$ if $Tri(j)$ cannot degenerate, 0 if it must. (The bottom left sub-figure of Fig. 5 is labeled in this way.) In fact, $DStretch^+$ with this type of labeling can also be viewed as a compact visualization of *NDD* itself: each label is to be understood as a template in which the $\pm$-signs take on all combinations of values $+$ and $-$, and the different ways to fill in each template represent different "convex tiles" of

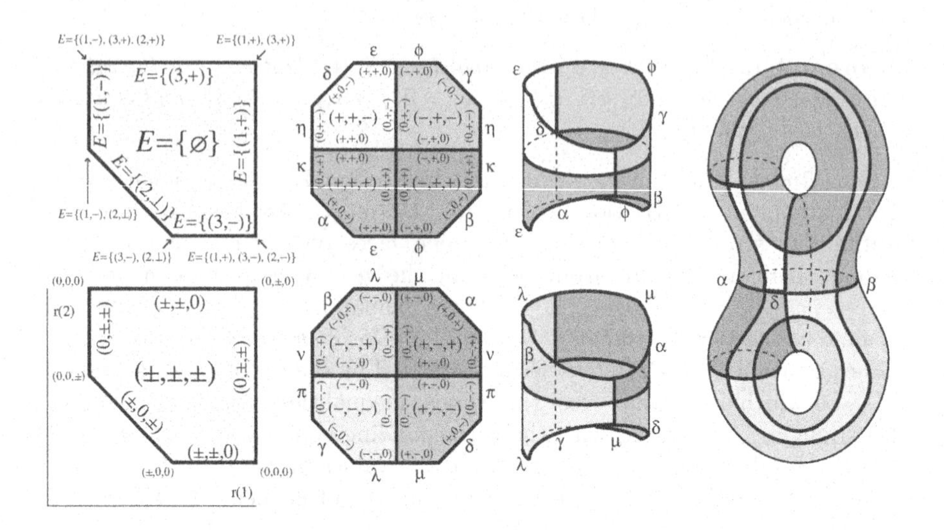

Fig. 5. *DSpace* for a planar $5R$ loop with link lengths $(2, 3, 4, 2, 3)$

NDD. From this viewpoint, each *DStretch*$^+$ stratum Q has $2^{n-2-e(Q)}$ embedded copies in *NDD* (which are the inverse images by $\mathbf{r}$), with 0 at the $e(Q)$ entries of $\mathbf{s}$ that correspond to the $e(Q)$ degenerate triangles, and $+$ or $-$ at the remaining $n - 2 - e(Q)$ entries that correspond to non-degenerate triangles. (Again, refer to Fig. 5 for an example.) We will show elsewhere that these embedded copies of *SFace*-strata of *DStretch*$^+$, which clearly form a partition of *NDD*, actually form a stratification (technical difficulties arise from doubly degenerate triangles, but can be overcome). We call it the "triangle orientation stratification" *STriO*.

Theorem 4. *If Q is an SFace-stratum of DStretch$^+$, then the STriO-strata mapped onto Q by $\mathbf{r}$ are in one-to-one correspondence via $\mathbf{s}$ with $2^{n-2-e(Q)}$, each stratum being distinguished by its unique pattern of $\{+, -\}$ orientation signs for the $n - 2 - e(Q)$ nondegenerate triangles.* □

The significance of Theorem 4 is that it renders *NDD*—that is, in every case "practically all" of *DSpace*, and in many cases literally all of it—*practically piecewise convex* in a strong sense: it shows how *NDD* can be decomposed *practically* into convex tiles labeled by values of $\mathbf{s}$, each of which is identified via $\mathbf{r}$ with an open face of the polyhedron *DStretch*.

As to how two *STriO*-strata can be directly joined, we have the following.

Theorem 5. *The closures of two STriO-strata of DSpace with triangle orientation signs that differ on some set of k triangles intersect each other if and only if those triangles can become singular simultaneously.* □

For instance, on the 3-dimensional, codimension-2 *STriO*-stratum Q for the planar $8R$ loop described in Example 4, $\mathbf{s}$ is $(0, +, +, +, +, 0)$; four codimension-0 *STriO*-strata (where $\mathbf{s}$ is $(s(1), +, +, +, +, s(6))$) are incident on Q, and a start deformation in any one of those strata can be joined to a goal in any other by a 2-segment stratum-wise linear path passing through Q as in Fig. 4(b) (or at any other point of Q).

Note that for an nR loop, if the $\mathbf{r}$ value of the LW vertex corresponds to a non-degenerate triangle, then the loop has two LW deformations with the same shape but opposite orientations. In this case, each LW deformation has singularity depth $n-3$ and connects half the convex tiles of *NDD*. If the LW triangle form degenerates into a line segment, then the loop has only one LW deformation, of singularity depth $n-2$, which connects all tiles of *NDD*.

Following [22], we say a planar nR loop ($n > 4$) has m long links provided m is the largest number for which there are link lengths $l_{j_1}, l_{j_2}, \ldots, l_{j_m}$ ($0 \leq j_1 < \cdots < j_m \leq n-1$) with the sum of any two of them being strictly greater than half the sum of all the loop's link lengths. Easily, if a loop has m long links, then $m \in \{0, 2, 3\}$. Prior results [16, 10] show that *DSpace* for a planar nR loop has two connected components or one, according as the loop has 3 long links or fewer. The $5R$ loops in Fig. 5 and Figs. 1(b) and 2 have 0 and 3 long links respectively; our figures show the correct reconstruction of the *DSpace* topology from the strata. Call an anchored triangle invertible if it is singular in some deformations. Such triangles are key to understanding the connectivity of *DSpace*, and lead to an alternative proof (short, but not short enough to include here!) of the *DSpace* connectivity results of [10, 16, 22].

Theorem 6. (a) *If a planar nR loop has* 0 *or* 2 *long links then every anchored triangle is invertible.* (b) *If l_{j_1}, l_{j_2} and l_{j_3} are three long links for a planar nR loop, then all but one anchored triangle is invertible, that being* $\mathrm{Tri}(j_2)$. □

3 Path Planning

3.1 Generation of Closure Deformations

We know of no prior closure deformation generation methods designed specifically for planar nR loops, though of course configuration generation methods for general closed chains [1, 4, 7, 14] apply in particular to planar nR loops. The more recent methods in [4] (random loop generator) and [1] (iterative constraint relaxation), designed with chains of many links in mind, considerably improve performance over earlier methods; but they still have difficulty for loops with many links (say, over 100), nor do they guarantee that every attempt will generate a closure deformation. Both these difficulties are overcome by using our new formulation of the loop closure constraint.

Our main task here is to compute a valid set $\mathbf{r}$ of diagonal lengths for a loop with given link lengths. Since all constraints on $\mathbf{r}$ are linear inequalities $Ineq_j^{\sigma}$ (or equalities Eq_j^{σ}, in case we wish to force $\mathbf{r}$ into particular strata of *DStretch*), we can easily find $\mathbf{r}$ with linear programming (LP). The problem size of an LP formulation for $\mathbf{r}$ is linear in n: the numbers of unknowns and constraints are both in $\Theta(n)$. Although there are as yet no theoretical bounds on the worst-case running time of LP, in practice [19] LP is considered a mature field with many efficient algorithms. We have also developed efficient methods other than *LP* (described in [8]) that take advantage of the kinematics of a planar nR loop and the particular simplicity of the constraints. Our fastest generation methods have linear time complexity $\Theta(n)$, which is optimal.

3.2 Connection of Closure Deformations

To the best of our knowledge, there are two complete planners for connecting deformations of a planar nR loop, ignoring collisions. The line tracking planner [16] of Lenhart and Whitesides generates a path in time $O(n)$ by using simple line tracking motions to move two given query deformations to their "standard triangle form", in two opposite orientations if needed, and then to move both triangle deformations to an appropriate singular deformation that allows the change of the triangle orientations. For some start and goal deformations, only one standard triangle form needs to be passed through. The accordion planner [22] of Trinkle and Milgram generates a smooth path between given deformation pairs and empirically exhibits cubic running time. It is also known [22, 20] that each component of *DSpace* for a planar nR loop with 3 long links is a $(n-3)$-dimensional torus parametrized by joint angles of the short links, so valid paths in a given component can be generated by linear interpolation of those angles (modulo 2π). Path planning for a planar nR loop using our new parameters is considerably simplified by the nice geometry of the *DSpace*, because (viewed through $\mathbf{r}$) all strata and their closures are convex, so two query deformations in the closure of a single stratum can be joined by a path on which $\mathbf{r}$ linearly interpolates their $\mathbf{r}$ values (or by a Manhattan path on which only one or a few entries of $\mathbf{r}$ change on each segment). If two deformations are in the same component of *DSpace* (easily checked), we can join them by a piecewise linear path once we determine critical singular deformations through which to pass successively between strata.

We sum this up in the nearly self-explanatory algorithm Fig. 6, where only Step 7 may require further comment.

Briefly, there are many ways to compute critical intermediate deformations. Assume the query deformations have different orientations for k triangles. Fig. 7 illustrates one extreme: we compute k codimension-1 strata, one for each triangle that needs to be inverted, then pass through them one at a time; feasibility is guaranteed by standard facts about stratified manifolds (cf. [6]). For a planar nR loop, we need at most $n-2$ singular deformations on codimension-1 strata, one for each triangle, so our path will be piecewise linear with at most $n-1$ segments, each on the closure of one codimension-0 stratum. With this approach, the running time of our algorithm is determined by the time needed to generate $O(n)$ singular deformations of depth 1, and has an upper bound of $O(n^2)$ when using deformation generation methods with linear running time [8].

Alternatively, we can look for a stratum of singularity depth at least k that corresponds to those triangles and directly joins the two strata. The extreme of this approach is to use the *LW* deformations mentioned earlier: we can (ignoring collisions) connect *any* two query deformations of an nR loop in the same

1. find the number and indices of the triangles, in which the two given deformations have opposite orientations
2. if the deformations do not have opposite orientation for any triangle
3. pathExistence=true; criticalIntCfgs=null;
4. elseif (the loop has 3 long links) and ...
   (the two cfgs have opposite orientations for the non-invertible triangle)
5. pathExistence=false;
6. else
7. pathExistence=true; find critical intermediate cfgs;
8. end;

Fig. 6. Algorithm for Connecting Two Closure Deformations of a Planar Chain

component of *DSpace* by using at most 2 critical deformations, one a *LW* deformation and the other a deformation singular in (at least) the unique non-singular anchored triangle (if one exists) of the *LW* deformation. Fig. 8(a) shows the two critical deformations (subgoals), the first subgoal being the *LW* deformation, as used to connect the same start and goal of a $5R$ loop as in Fig. 7. Clearly the running time for generating one such path, again determined by the generation time of the two special critical deformations, is $\Theta(n)$, which is optimal.

In Fig. 8(b), the same problem as in Figs 7 and 8(a) is solved with the *LW* deformation as the second subgoal: in fact, subgoal 1 (2) in Fig. 8(a) has the same $\mathbf{r}$ value as subgoal 2 (1) in Fig. 8(b). Recall that for a loop with given link lengths, a feasible $\mathbf{r}$ value completely specifies which triangles are singular or not. Denote by $dt(\mathbf{r})$ the (possibly empty) set of the indices of the anchored triangles that are singular under the given value of the diagonal lengths $\mathbf{r}$. So to connect two

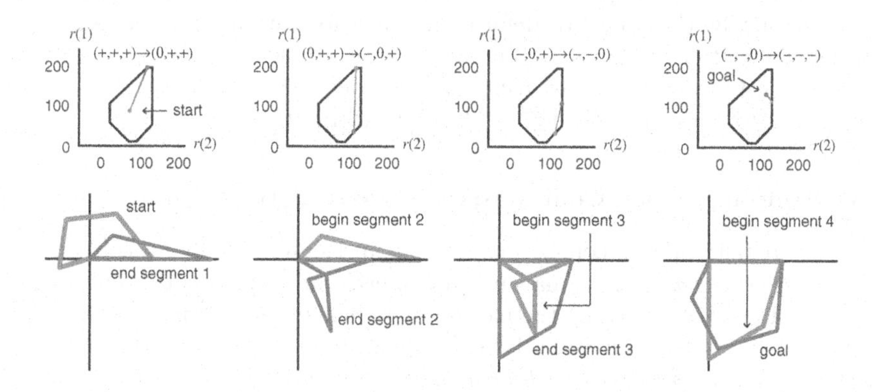

Fig. 7. Above, we show a piecewise-linear path in *DSpace* for a planar $5R$ loop link lengths $[100, 90, 80, 75, 50]$, connecting the closure deformations $(80, 70, +, +, +)$ and $(125, 110, -, -, -)$ by traversing four codimension-0 *STriO*-strata (each identified via $\mathbf{r}$ with a copy of *DStretch*, and labeled with its value of $\mathbf{s}$), crossing common boundary pieces of higher codimension. Below, we picture the loop itself, in its deformations at the beginning and end of each segment

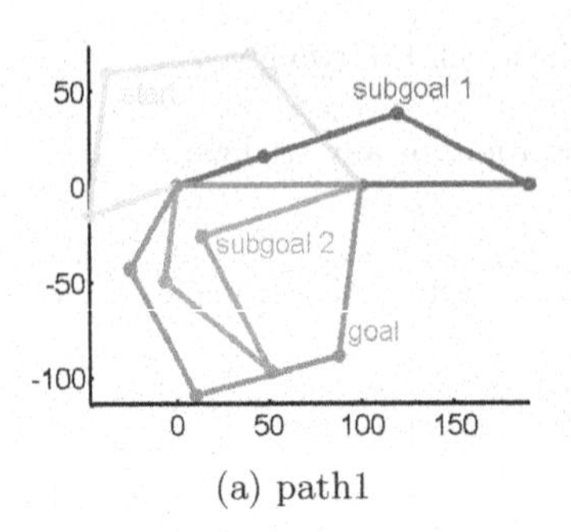

(a) path1

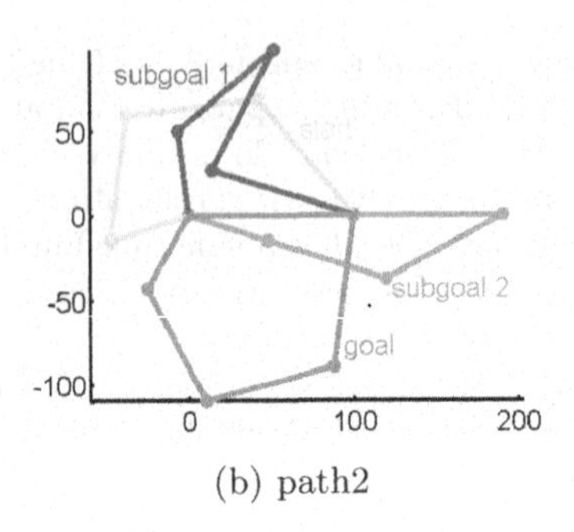

(b) path2

Fig. 8. Two paths for connecting the same start and goal deformations as in Fig. 7, using two singular **r** values—one the LW vertex (for the triangle deformations)—in two different orders, along with appropriate **s** values

deformations in the same connected component but with opposite orientations in k triangles of indices $\{j_1, j_2, \ldots, j_k\}$, we need to find **r** values $\{\mathbf{r}_1, \mathbf{r}_2, \ldots, \mathbf{r}_m\}$, such that $\{j_1, j_2, \ldots, j_k\} \subseteq \bigcup_{i=1}^{m} dt(\mathbf{r}_i)$. For such an **r** set having m members, these **r** values can be used in arbitrary orders, along with appropriate s values; we obtain $m!$ different but related paths. An example with $m = 2$ is shown in Fig. 8.

Our connection method is clearly complete and guarantees to find a path between any two closure deformations in one connected component. The complexity of this algorithm depends on the complexity of the generation of the critical singular deformations. But we note that the singular deformations can be reused. If it is known that a large number of path planning problems will be performed for a fixed loop, it will be worth preprocessing the loop to find critical deformations, be they the *LW* deformation or deformations of lower singularity depths. Thereafter we can solve any connection problem for any two deformations in constant time $\Theta(1)$ by using the *LW* deformation, or in time $O(n)$ by picking appropriate critical singular depth 1 deformations.

3.3 Sampling-Based Collision-Free Closure Path Planning

The closure deformation connection methods just described do not consider the *collision free constraint*, so may involve interference between the links, as in Fig. 7. Extensive research by the motion planning community has made it clear that the collision free constraint is very difficult to deal with, and that it is very hard to describe *CFree* and *CObstacle* analytically for general obstacles. Recent successes of randomized path planners suggest that sampling based planners like *PRM* and *RRT* may be an important framework in which to integrate efficient node generation and connection methods (including ours and previous ones) while also dealing with such difficult factors in planning as high dimensionality and complicated linkage constraints. Our preliminary strategy has been to capture the stratum connectivity with a roadmap or trees, from which we construct

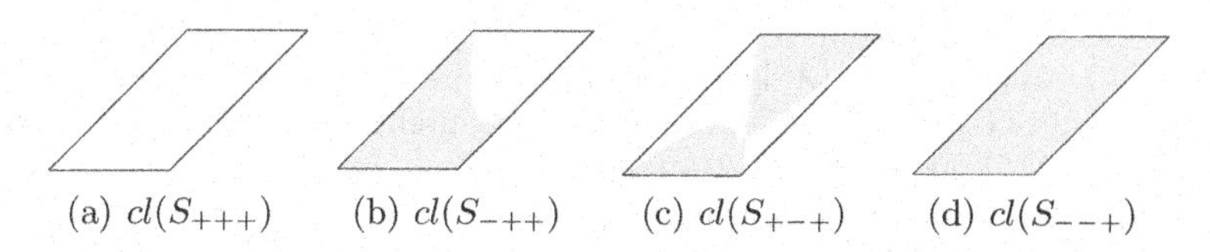

Fig. 9. *DSpace* for the 5-bar loop with 3 long links already familiar from Fig. 2

the global connectivity of *DFree* and solve for paths for given query deformation pairs. Fig. 9 suggests the daunting complexity of this problem. The shaded areas are the parts of *DObstacle* in the copies of *DStretch* that make up half of *DSpace* (the other half is its mirror image). The unobstructed part of the $(+,-,+)$ stratum has two connected components, which can however be connected via the $(+,+,+)$ stratum, as in Fig. 2(b-d).

4 Summary

In this paper, we used our new parameters—some inter-joint distances and triangle orientation data—to study the stratified deformation space and efficient path planning for a plainer closed chain with revolute joints. Instead of formulating the loop closure constraint as nonlinear equations in joint angles, we break a loop into an open chain of triangles then use the triangle inequality repeatedly to formulate the constraint as a set of linear inequalities. This new formulation endows the deformation space with a nice geometry; for a generic nR loop it is a stratified space of convex strata. This geometry (and its generalizations for more complex kinematic systems) greatly simplifies kinematics related issues including the generation and connection of closure deformations. In effect, our new parameters make path planning for a planar nR loop (or a spatial nS loop) no more difficult than path planning for an open chain.

References

1. Bayazit, O.B., Xie, D., Amato, N.M.: Iterative relaxation of constraints: A framework for improving automated motion planning. In: Proc. IEEE Int. Conf. Intel. Rob. Syst. IROS, Edmonton, Alberta, Canada, pp. 586–593 (August 2005)
2. Canny, J.: The Complexity of Robot Motion Planning. MIT Press, Cambridge (1988)
3. Choset, H., Burgard, W., Hutchinson, S., Kantor, G., Kavraki, L.E., Lynch, K., Thrun, S.: Principles of Robot Motion: Theory, Algorithms, and Implementation. MIT Press, Cambridge (2005)
4. Cortes, J., Simeon, T.: Sampling-based motion planning under kinematic loop closure constraints. In: Proc. of Workshop on Algorithmic Foundations of Robotics (2004)
5. Craig, J.J.: Introduction to Robotics: Mechanics and Control, 2nd edn. Addison-Wesley Publishing Company, Reading (1989)

6. Goresky, M., MacPherson, R.: Stratified Morse Theory. Springer, New York (1988)
7. Han, L., Amato, N.M.: A kinematics-based probabilistic roadmap method for closed chain systems. In: Algorithmic and Computational Robotics — New Directions WAFR 2000, pp. 233–246 (2000)
8. Han, L., Blumenthal, J., Valodzin, I., Rudolph, L.: Efficient methods for the nverse kinematics of a chain with maximal DOF rotatable joints (submitted, 2006)
9. Han, L., Rudolph, L.: Inverse kinematics for a serial chain with maximal DOF rotatable joints. In: Proc. of Robotics: Science and Systems, Philadelphia (2006)
10. Kapovich, M., Millson, J.: On the moduli spaces of polygons in the euclidean plane. Journal of Differential Geometry 42, 133–164 (1995)
11. Kavraki, L., Svestka, P., Latombe, J.C., Overmars, M.: Probabilistic roadmaps for path planning in high-dimensional configuration spaces. IEEE Trans. Robot. Automat. 12(4), 566–580 (1996)
12. Kuffner, J.J., LaValle, S.M.: RRT-Connect: An Efficient Approach to Single-Query Path Planning. In: Proc. IEEE Int. Conf. Robot. Autom (ICRA), pp. 995–1001 (2000)
13. Latombe, J.C.: Robot Motion Planning. Kluwer Academic Publishers, Boston (1991)
14. LaValle, S., Yakey, J., Kavraki, L.: A probabilistic roadmap approach for systems with closed kinematic chains. In: Proc. IEEE Int. Conf. Robot. Autom (ICRA) (1999)
15. La Valle, S.M.: Planning Algorithms. Cambridge University Press, Cambridge (2006), `http://msl.cs.uiuc.edu/planning/`
16. Lenhart, W., Whitesides, S.: Reconfiguring closed polygon chains in Euclidean d-space. Discrete and Computational Geometry 13, 123–140 (1995)
17. Liu, G., Trinkle, J.C.: Complete path planning for planar closed chains among point obstacles. In: Robotics: Science and Systems (2005)
18. Liu, G., Trinkle, J.C., Milgram, R.J.: Toward complete motion planning for planar $3R$-manipulators among point obstacles. In: Proc. Int. Workshop on Algorithmic Foundations of Robotics (WAFR) (2004)
19. Luenberger, D.: Linear and Nonlinear Programming. Kluwer Academic Publishers, Dordrecht (2004)
20. Milgram, R., Trinkle, J.: The geometry of configuration spaces for closed chains in two and three dimensions. Homology Homotopy Appl. (2002)
21. Murray, R.M., Li, Z., Sastry, S.S.: A Mathematical Introducation to Robotic Manipulation. CRC Press, Boca Raton (1994)
22. Trinkle, J., Milgram, R.: Complete path planning for closed kinematic chains with spherical joints. Int. J. Robot. Res. 21(9), 773–789 (2002)

Part V

Motion Planning

Competitive Disconnection Detection in On-Line Mobile Robot Navigation

Yoav Gabriely[1] and Elon Rimon[2]

[1] Technion, Israel Institute of Technology
yoavga@tx.technion.ac.il
[2] Technion, Israel Institute of Technology
elon@robby.technion.ac.il

Abstract. This paper concerns target unreachability detection during on-line mobile robot navigation in an unknown planar environment. Traditionally, *competitiveness* characterizes an on-line navigation algorithm in cases where the target is reachable from the robot's start position. This paper introduces a complementary notion of competitiveness which characterizes an on-line navigation algorithm in cases where the target is unreachable. The *disconnection competitiveness* of an on-line navigation algorithm measures the path length it generates in order to conclude target unreachability relative to the shortest off-line path that proves target unreachability from the same start position. It is shown that only competitive navigation algorithms can possess disconnection competitiveness. A competitive on-line navigation algorithm for a disc-shaped mobile robot, called *CBUG*, is described. This algorithm has a *quadratic* competitive performance, which is also the best achievable performance over all on-line navigation algorithms. The disconnection competitiveness of *CBUG* is analyzed and shown to be *quadratic* in the length of the shortest off-line disconnection path. Moreover, it is shown that quadratic disconnection competitiveness is the best achievable performance over all on-line navigation algorithms. Thus *CBUG* achieves optimal competitiveness both in terms of connection and disconnection paths. Examples illustrate the usefulness of connection-and-disconnection competitiveness in terms of path stability.

1 Introduction

This paper is concerned with target unreachability detection during mobile robot navigation in a planar environment populated by unknown obstacles. The robot has no apriori information about the environment, but may locally acquire this information using its on-board sensors. This class of on-line problems has a wide range of applications. Examples are navigation to various targets for mail and material delivery in offices and factories, and planetary exploration and sample acquisition. The most critical parameter in such tasks is physical travel time rather than on-board computation time. Under a uniform velocity assumption travel time corresponds to path length. Hence in this paper navigation algorithms are classified in terms of length of the path traveled by the robot during algorithm execution. Before discussing target unreachability detection, we summarize the relevant literature.

S. Akella et al. (Eds.): Algorithmic Foundation of Robotics VII, STAR 47, pp. 253–267, 2008.
springerlink.com

Mobile robot on-line algorithms are discussed in the robotics and computational geometry literature. Roboticists usually emphasize the type of sensors required to achieve a given task, and the problems considered here are referred to as sensor based motion planning [6, 7]. Notable early papers in this area describe the algorithms *BUG*1/*BUG*2 [17] and *ALG*1/*ALG*2 [20] for navigating a two degrees-of-freedom mobile robot in an unknown planar environment using position and tactile sensors. These works have been extended to navigation in planar environments using vision and laser sensors [15, 16, 18, 21]. However, the performance of these algorithms is typically characterized in terms of geometric parameters of the environment such as total obstacle perimeter, without any reference to the length of the optimal off-line solution, denoted l_{opt}. As a result, these algorithms may be fooled to generate lengthy paths in situations where the optimal off-line path is very short.

Computational geometry researchers introduced the notion of competitiveness. An algorithm for a task P is said to be *competitive* if its solution to every instance of P is bounded by a constant times the optimal off-line solution. An early influential paper investigates navigation of a point robot in an unknown planar environment consisting of m radial corridors [1]. (This problem has its origin with a simpler problem, where a cow seeks an entry to a pasture along an unknown fence which corresponds to two corridors [3].) However, a point robot cannot achieve any form of competitive navigation in general environments [1, 14, 19]. Subsequent papers discuss on-line navigation of point robots in specific classes of rectangular rooms [4, 5, 10], rooms with square-shaped obstacles [19], and generalized streets [8, 14]. All of these papers strive to achieve *linear* competitiveness (i.e. path length bounded by a constant times l_{opt}) in specific classes of environments.

In contrast, we depart from the point-robot paradigm and assume that the robot is a disc of physical size $D > 0$. While this assumption may seem obvious, only few papers make use of this assumption (e.g. [9]). We have recently reported on *CBUG*, an on-line navigation algorithm for a size D robot moving in a general planar environment [13]. This algorithm generates a path whose length is bounded from above by a *quadratic* function of l_{opt}. Moreover, we have shown that any on-line navigation algorithm generates in worst case a path whose length is bounded from below by a quadratic function of l_{opt}. Hence the quadratic bound of *CBUG* is tight.

This paper focuses on the performance of on-line navigation algorithms in cases where the target is unreachable from the robot's start position. This notion is known as *disconnection proofs* in off-line graph search algorithms. In the motion planning literature, disconnection proofs appear in the context of random path planning, where an off-line sampling technique detects target unreachability of a polyhedral robot [2]. The following on-line version of a disconnection proof is a contribution of this paper. Let a *disconnection path* be a path that starts at S and proves that a target T is unreachable (Figure 1). Let λ be the length of the disconnection path generated from S by an on-line algorithm. Let λ_{opt} be the length of the shortest off-line disconnection path which starts at the

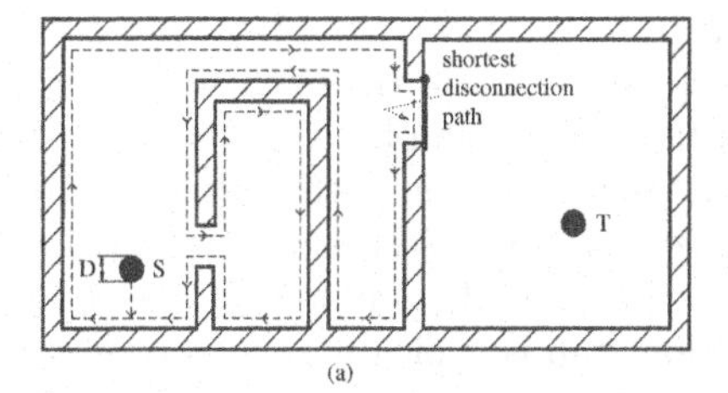

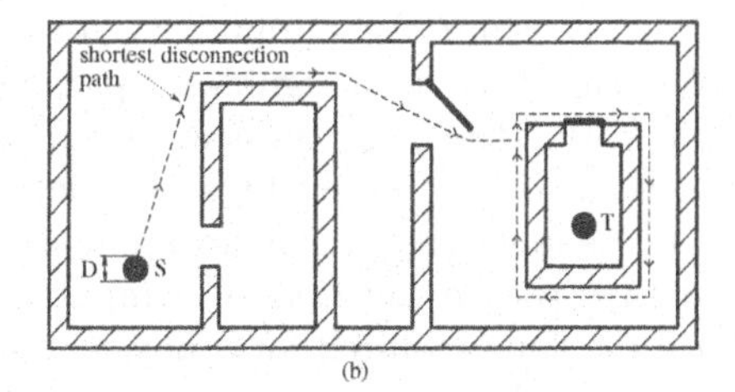

Fig. 1. All disconnection paths starting from S trace an obstacle boundary surrounding either S or T

same S. Then an algorithm is $h(\lambda_{opt})$ *disconnection competitive* of if $\lambda \leq h(\lambda_{opt})$ for all instances where T is unreachable from S. Based on this definition, we first show that a navigation algorithm must be competitive in order to be disconnection competitive. Then we establish that *CBUG* generates a disconnection path whose length is bounded from above by a *quadratic* function of λ_{opt}. Finally, we establish that any navigation algorithm in an unknown planar environment generates in worst case a disconnection path whose length is bounded from below by a quadratic function of λ_{opt}. The quadratic disconnection bound of *CBUG* is thus tight.

The structure of the paper is as follows. In the next section we define generalized competitiveness and introduce the notion of disconnection competitiveness. The *CBUG* algorithm is reviewed in Section 3. Its quadratic upper bound is summarized and shown to match the universal lower bound over all on-line navigation algorithms. The disconnection competitiveness of *CBUG* is analyzed in Section 4. It is shown that the length of the disconnection path traveled by the robot during execution of *CBUG* is at most quadratic in λ_{opt}. It is also shown that any on-line navigation algorithm generates in worst case a disconnection path whose length is at least quadratic in λ_{opt}, implying that up to constants *CBUG* has optimal disconnection competitiveness. Section 5 discusses the effect of connection-and-disconnection competitiveness on path stability, and compares the performance of *CBUG* relative to non-competitive algorithms. The concluding section mentions several open problems.

2 Definition of Disconnection Competitiveness

This section describes our basic setup, then proceeds with formal definitions of connection and disconnection competitiveness. We assume a planar unknown environment populated by stationary and compact obstacles. The mobile robot is a freely moving planar disc of size $D > 0$, where D is a given constant. The robot is equipped with two sensors which are assumed ideal. The first sensor measures the robot's position with respect to a fixed reference frame. The second is an obstacle detection tactile sensor which allows tracing of an obstacle boundary. In addition to sensors the robot has on-board memory in which information on the environment can be accumulated.

Next consider the parameters governing the performance of mobile robot tasks. The three most significant parameters are physical travel time, on-board computation time, and on-board memory. In order to simplify the ensuing analysis, we associate physical travel time with length l of the path traveled by the robot. As for on-board computation time, we limit our discussion to algorithms that take *polynomial time* to compute each physical motion step of the robot. Since the time required for a physical motion step is typically several orders of magnitudes longer than the execution time of an on-board computation step, we focus on l as the main performance parameter. Last, we limit the discussion to algorithms whose storage requirement is at most *linear* in the size of the environment.

Thus l denotes length of the path traveled by the robot, while l_{opt} denotes length of the optimal off-line path. The following definition generalizes the traditional notion of linear competitiveness to any functional relationship between l and l_{opt}.

Definition 1 (connection competitiveness). *An on-line navigation algorithm is* $\boldsymbol{f(l_{opt})}$ **competitive** *when its path length l is bounded from above by a scalable function $f(l_{opt})$ over all instances where the target is reachable. In particular, $l \leq c_1 l_{opt} + c_0$ is the traditional linear competitiveness, while $l \leq c_2 l_{opt}^2 + c_1 l_{opt} + c_0$ is quadratic competitiveness, where the c_i's are positive constants that depend on the robot size D.*

The meaning of *scalability* is as follows. When performance is measured in physical units such as meters m, one must ensure that both sides of the relationship $l \leq f(l_{opt})$ posses the same units, so that change of scale would not affect the bound. For instance, the coefficient c_2 in the relationship $l \leq c_2 l_{opt}^2 + c_1 l_{opt} + c_0$ must have units of m^{-1}, c_1 must be unitless, and c_0 must have units of m. Note that the definition of competitiveness focuses on a particular navigation algorithm. However, our objective is to characterize the least upper bound that can be achieved over all on-line navigation algorithms. This objective requires the following universal lower bound.

Definition 2. *A* **universal lower bound** *on the competitiveness of on-line navigation is a lower bound $g(l_{opt})$ such that $l \geq g(l_{opt})$ over all on-line navigation algorithms for this task.*

Note that the universal lower bound characterizes the on-line navigation task itself, not any specific algorithm for this task. When the competitive upper bound of a specific algorithm matches the universal lower bound up to constants, the bound itself becomes the *competitive complexity class* of the task [12]. Let us now define disconnection competitiveness. Recall that λ denotes length of the path traveled by the robot from a start S until it halts with a conclusion that the target T is unreachable. Recall, too, that λ_{opt} denotes length of the shortest off-line path which starts at the same S and proves that T is unreachable. The following definition is analogous to the definition of connection competitiveness.

Definition 3 (disconnection competitiveness). *An on-line navigation algorithm is* **$h(\lambda_{opt})$ disconnection competitive** *when its path length λ is bounded from above by a scalable function $h(\lambda_{opt})$ over all instances where the target is not reachable. In particular, $\lambda \leq c_2\lambda_{opt}^2 + c_1\lambda_{opt} + c_0$ is quadratic disconnection competitiveness, where the c_i's are positive constants that depend on the robot size D.*

The last definition concerns the least upper bound on disconnection competitiveness.

Definition 4. *A* **universal lower bound** *on the disconnection competitiveness of on-line navigation is a lower bound $e(\lambda_{opt})$ such that $\lambda \geq e(\lambda_{opt})$ over all on-line navigation algorithms for this task.*

Note that here, too, the universal lower bound is not associated with a specific on-line algorithm, but rather characterizes the on-line navigation task itself.

3 The CBUG Algorithm

For clarity of presentation, we describe the algorithm for a point robot equipped with position and tactile sensors, moving in a planar environment populated by unknown obstacles. The point robot represents the configuration of the disc robot, and the "obstacles" are c-space obstacles induced from the physical ones. The principle idea of *CBUG* is as follows. Given a start S and target T, the robot selects an initial ellipse with focal points S and T and area A_0, and searches for T in the portion of the ellipse accessible from S. The search is executed with the classical *BUG*1 algorithm reviewed below, which regards the bounding ellipse as a virtual obstacle[1]. If the target is detected the algorithm terminates. Otherwise the robot repeats the process in ellipses with areas $2^i A_0$ for $i=1,2,\ldots$ until the target is found or determined to be inaccessible from S (Figure 2). The basic algorithm treats the bounding ellipse as an obstacle whose boundary must be traced by the robot. A more advanced version of the algorithm described below does not require any tracing of the bounding ellipse. A description of the basic algorithm follows.

Basic *CBUG* Algorithm:
Sensors: Position and tactile sensors.
Input: A start S, a target T, an initial ellipse with focal points S and T and area A_0.
Initialization: Set $S_1 = S$. Set initial search area $A(1) = A_0$. Set $i = 1$.
Repeat
1. Starting at S_i, search for T using *BUG*1 in ellipse of area $A(i)$ with focal points S and T.

[1] Other sub-algorithms such as *ALG*1 [20] can be used, but the principle bounds reported here would remain the same. Simulations of *CBUG* with *BUG*1 and *ALG*1 as sub-algorithms are discussed below.

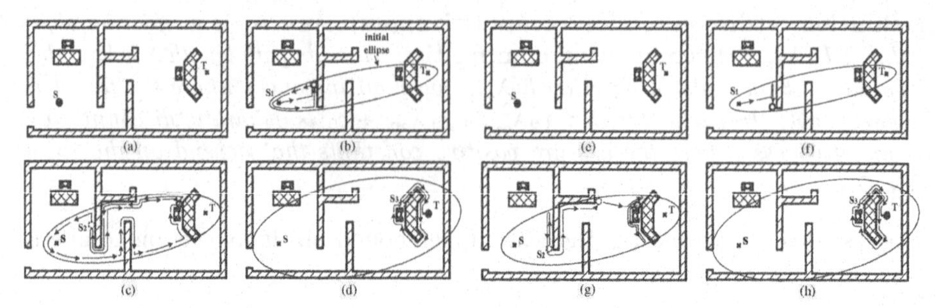

Fig. 2. (a)-(d) Execution example of the basic *CBUG*. (e)-(h) The modified *CBUG* does not require tracing of the bounding ellipses.

2. If $BUG1$ terminates at T: STOP, target is found.
3. If $BUG1$ determines that an obstacle boundary separates S from T (see text):
3.1 If obstacle boundary does not intersect i^{th} bounding ellipse: STOP, target is unreachable.
3.2 Set S_{i+1} at point where $BUG1$ terminated (see text).
4. Set $A(i+1) = 2A(i)$. Set $i = i + 1$.
(End of repeat loop)

First let us review the $BUG1$ sub-algorithm. Under $BUG1$ the robot moves from the ith start position towards the target until it hits an obstacle. Then it circumnavigates the obstacle in a clockwise direction while recording the closest point to the target along the current boundary as p_{min}. When the obstacle circumnavigation is complete, the robot returns to p_{min} along the shorter boundary segment. If the direction from p_{min} to T points into the current obstacle, the obstacle necessarily separates S from T and the target is unreachable [17]. Otherwise the robot resumes its motion to the target until the next obstacle is encountered or the target is found.

Based on the assumption of compact obstacles, the bounding ellipses of *CBUG* eventually contain a path to the target if one exists, or contain an entire obstacle boundary which separates the start from the target. At this stage the $BUG1$ sub-algorithm finds a path to the target if one exists or determines target unreachability. Note that *CBUG* determines target unreachability only when the separating obstacle boundary is wholly physical and does not contain portions of the bounding ellipse. Also note that *CBUG* requires *constant memory:* S and T, the current obstacle hit point, the current p_{min}, distances along the current obstacle boundary, and the current search area $A(i)$.

Example 1. Consider the execution of *CBUG* in the office-like environment shown in Figure 2(a). Starting at S, the disc robot determines that the initial ellipse blocks its path to T (Figure 2(b)). Hence it doubles the ellipse's area and resumes the search from the point S_2 which is closest to T (Figure 2(c)). The robot next determines that the new bounding ellipse still blocks its path to T.

Hence it doubles the ellipse's area for the second time and resumes the search at the point S_3 (Figure 2(d)). This last search ends successfully at T.

Next we describe a practical speedup of *CBUG* that eliminates the need to trace the bounding ellipses. When the robot hits an obstacle, either the entire obstacle boundary is contained inside the current bounding ellipse, or the boundary segment which contains the hit point has its two endpoints on the bounding ellipse. The modified algorithm requires that the robot trace an obstacle boundary until one of two events happens. Either the robot circumnavigates the entire obstacle boundary, or it reaches an endpoint of the boundary on the current bounding ellipse. In the latter case the robot *reverses* its boundary tracing direction and continues along the obstacle boundary until reaching the other endpoint of the boundary segment. The robot next moves to the closets point to the target along the boundary segment, and resumes execution of the sub-algorithm *BUG*1.

Example 2. An execution of the modified *CBUG* on the same office-like environment is shown in Figure 2(e)-(h). Each time the robot hits an obstacle, it initiates a clockwise circumnavigation of the obstacle boundary. In the first and second stages the robot encounters during boundary tracing the current bounding ellipse (Figure 2(f)-(g)). The robot consequently reverses its tracing direction until the other endpoint of the boundary segment is encountered. In both stages the path taken by the robot is significantly shorter than the path taken under the basic algorithm.

The following result asserts that the path generated by *CBUG* to an accessible target is bounded by a quadratic function of l_{opt}.

Proposition 3.1 ([13]). *If T is reachable from S, the basic CBUG algorithm finds the target using a path of length l satisfying the upper bound*

$$l \leq \frac{6\pi}{D} l_{opt}^2 + \|S - T\| + \frac{6A_0}{D}, \tag{1}$$

where l_{opt} is length of the shortest off-line path from S to T, D is the disc-robot size, and A_0 is area of the initial ellipse.

Note that the three summands in (1) have length units, so the upper bound is scalable. The next result asserts that the universal lower bound on connection competitiveness is also quadratic in l_{opt}.

Theorem 1 ([13]). *Any navigation algorithm in an unknown planar environment to a reachable target generates in worst case a path of length l satisfying the quadratic lower bound*

$$l \geq \frac{4\pi}{3(1+\pi)^2 D}(1-\epsilon) l_{opt}^2, \tag{2}$$

where l_{opt} is length of the shortest off-line path from S to T, D is the disc-robot size, and $\epsilon > 0$ is an arbitrary small constant.

The theorem implies that the connection competitiveness of *CBUG* is tight. On-line navigation of a disc robot in planar environments thus belongs to the *quadratic competitive complexity class.* Note that (1) and (2) approach infinity as D approaches zero. This is consistent with earlier observations that a point robot cannot achieve any form of competitive on-line navigation in general planar environments [1, 14, 19].

4 Disconnection Analysis of CBUG Algorithm

This section begins with a generic assertion that connection competitiveness is necessary for disconnection competitiveness. Then we derive an upper bound on the disconnection competitiveness of the basic *CBUG* algorithm. Finally, we derive a universal lower bound on disconnection competitiveness.

Proposition 4.1. *If an on-line navigation algorithm possesses an upper bound $h(\lambda_{opt})$ on its disconnection competitiveness, it possesses an upper bound of $h(l_{opt}+\epsilon)$ on its connection competitiveness (ϵ is an arbitrary small constant).*

Proof sketch: Consider a scenario where T can be reached from S. Let us assume that T can be surrounded by a small disc of radius δ which is free from obstacles. Let us further assume that the on-line algorithm guides the robot directly to the target within this small disc. We now render the target inaccessible by surrounding it with a disc-obstacle of radius δ. In this case the shortest disconnection path consists of the shortest path from S to the disc, and a loop around the disc. The length of the shortest disconnection path is $\lambda_{opt} = l_{opt} + \epsilon$, where l_{opt} is the original shortest off-line path from S to T and $\epsilon = 2\pi\delta - \delta$. By assumption any on-line disconnection path has length λ satisfying $\lambda \leq h(\lambda_{opt})$. Any disconnection path must circumnavigate the small disc surrounding T. Hence it can be converted to a path from S to T of length $l \leq h(\lambda_{opt}) - \epsilon$. Substituting $\lambda_{opt} = l_{opt} + \epsilon$ gives the upper bound $l \leq h(l_{opt}+\epsilon) - \epsilon \leq h(l_{opt}+\epsilon)$ on the connection paths. □

In the following analysis we treat the disc robot as a point equipped with position and tactile sensors, moving in a planar c-space amidst unknown c-obstacles. A c-space *disconnection path* starts at a configuration S and contains a c-obstacle boundary that separates S from T. The first lemma establishes that *CBUG* terminates once a disconnection path appears in its current bounding ellipse.

Lemma 4.2. *CBUG terminates with a conclusion that T is unreachable in the first bounding ellipse whose interior contains a disconnection path starting from S.*

The lemma is based on the following argument. The sub-algorithm *BUG*1 is known to be complete both in terms of its connection and disconnection paths [17]. Once a disconnection path which starts at S lies in the interior of the current bounding ellipse, *BUG*1 finds this path and terminates *CBUG* with a conclusion that T is unreachable from S.

The next lemma characterizes the shortest off-line disconnection path. Let $\mathcal{U}$ be the connected component of the free c-space containing the start S. In general, $\mathcal{U}$ is bounded from the outside by an outer c-obstacle, and is punctured from the inside by internal c-obstacles. Hence there are two possible cases of target unreachability. The first case occurs when T lies beyond the outer boundary. In this case any disconnection path must circumnavigate the outer boundary of $\mathcal{U}$. The second case occurs when T lies inside a puncture of $\mathcal{U}$. In this case any disconnection path must circumnavigate the internal c-obstacle boundary. The two cases are discussed in the following lemma.

Lemma 4.3. *Let α be the shortest off-line disconnection path starting from S, of length λ_{opt}. If T lies beyond the outer boundary of $\mathcal{U}$, α lies in a disc with center at S and radius $\lambda_{opt}/2$. If T lies inside a puncture of $\mathcal{U}$, α lies in an ellipse with focal points S and T and major axis of length λ_{opt}.*

Proof: First consider the case where T lies beyond the outer boundary of $\mathcal{U}$, denoted β. We may assume that β is a simple closed loop. Since T lies beyond β, any disconnection path from S must contain the entire loop β. Since the length of α is λ_{opt}, the length of β is at most λ_{opt}. Consider now the collection of all loops of length at most λ_{opt} surrounding S. Clearly, a disc with center at S and radius $\lambda_{opt}/2$ contains all such loops. In particular it contains the loop β, and consequently it contains the path α.

Next consider the case where T lies in a puncture of $\mathcal{U}$. Let γ denote the puncture's boundary, which we assume is a simple closed loop. The entire γ is part of any disconnection path starting at S. In particular, the shortest disconnection path α starts at S, contains the loop γ, and has total length λ_{opt}. Let p denote the point where α joins γ, and let L denote the length of γ. Any point x on γ satisfies $\|x - T\| \leq L/2$. Hence the length of the path from p to x along the shorter portion of γ, then from x straight toward T, has length bounded by L. Joining the latter path with the portion of α between S and p gives a continuous path with endpoints at S and T and total length bounded by λ_{opt}. All such paths are contained in an ellipse with focal points S and T and major axis of length λ_{opt}. Since x is an arbitrary point along the loop γ, the entire disconnection path α is contained in the ellipse. □

The next lemma gives an upper bound on the area of the first bounding ellipse of *CBUG* which contains a disconnection path starting from S.

Lemma 4.4. *CBUG terminates with a conclusion that T is unreachable in a bounding ellipse whose area $A(n)$ is bounded by $A(n) < \frac{\pi}{2}(\lambda_{opt} + \|S-T\|)(\lambda_{opt}^2 + 2\lambda_{opt}\|S-T\|)^{1/2}$, where λ_{opt} is length of the shortest off-line disconnection path which starts at S.*

Proof: Based on Lemma 4.2, *CBUG* terminates with a conclusion of target unreachability once its bounding ellipse contains a disconnection path starting at S. It can be verified that the disc and ellipse of Lemma 4.3 are contained in a larger ellipse with focal points S and T and major axis of length $2a = \lambda_{opt} + \|S-T\|$. The latter ellipse contains the shortest off-line disconnection path from S. It can be verified that the ellipse's minor axis has length $2b = (\lambda_{opt}^2 + 2\lambda_{opt}\|S-T\|)^{1/2}$. The ellipse's area is given by πab. Since $CBUG$ doubles the area of its bounding ellipse in each iteration, $A(n) \leq 2\pi ab$. Substituting for a and b in the latter inequality gives the bound on $A(n)$. □

Next we convert the bound on $A(n)$ to a bound on path length. The conversion is based on a key geometric fact for which we need the notion of traceable obstacles. Let $\mathcal{CB}_i$ be the c-space obstacle induced by an obstacle $\mathcal{B}_i$ for a disc robot of size D. The *traceable obstacle* induced by $\mathcal{B}_i$, denote $\boldsymbol{\mathcal{B}_i}$, is the physical obstacle obtained by filling any internal holes in $\mathcal{CB}_i$ and then shrinking $\mathcal{CB}_i$ inward by a distance of $D/2$. If two traceable obstacles overlap their union is considered a single traceable obstacle. An important property of $\boldsymbol{\mathcal{B}_i}$ is that the area swept by the disc robot during tracing of its boundary is precisely the area swept by the robot while tracing the boundary of the original obstacle $\mathcal{B}_i$.

Lemma 4.5 ([12]). *Let a planar environment contain individually traceable obstacles $\boldsymbol{\mathcal{B}_1}, \ldots, \boldsymbol{\mathcal{B}_k}$. Let a size-D disc robot trace the ith obstacle boundary, and let κ_i be the area swept by the robot during this tracing. Let $\mathcal{C}$ be any simple closed curve surrounding the k regions swept by the robot. Then $\sum_{i=1}^{k} \kappa_i \leq 4\mathcal{A}(\mathcal{C})$, where $\mathcal{A}(\mathcal{C})$ is the area of the obstacle-free points enclosed by $\mathcal{C}$.*

The following proposition establishes a quadratic upper bound on the disconnection paths generated by $CBUG$.

Proposition 4.6. *If T is not reachable from S, the basic CBUG algorithm concludes target unreachability along a path whose length λ satisfies the quadratic upper bound,*

$$\lambda \leq \frac{6\pi}{D}(\lambda_{opt} + \|S-T\|)^2 + \|S-T\| + \frac{6A_0}{D}, \tag{3}$$

where λ_{opt} is length of the shortest off-line disconnection path from S, D is the disc-robot size, and A_0 is area of the initial ellipse.

Note that the three summands have units of length, so the upper bound is scalable.

Proof: At the ith stage of $CBUG$ the robot executes the sub-algorithm $BUG1$ in an ellipse with focal points S and T and area $A(i)$. The regions swept by the robot during circumnavigation of obstacles in this ellipse (including the obstacle formed by the ellipse) are surrounded by the ellipse's boundary. Identifying the latter boundary with the curve $\mathcal{C}$ of Lemma 4.5, the total length of the robot's path during circumnavigation of the obstacles is at most $4A(i)/D$. Recall now that under $BUG1$ the robot circumnavigates the boundary of each obstacle at most 1.5 times. Hence the total length of the robot's path during boundary following is at most $6A(i)/D$. Under $BUG1$ motion between obstacles is always directly to the target. The total length of these motion segments equals to the net decrease of the robot's distance from T, which is $\|S_i - T\| - \|S_{i+1} - T\|$. Adding the two terms gives the ith stage path-length bound: $\lambda_i \leq 6A(i)/D + (\|S_i - T\| - \|S_{i+1} - T\|)$.

Suppose that $CBUG$ finds a disconnection path at the nth stage, such that $n > 1$. Since the area of the ellipses doubles in each step, $\sum_{i=1}^{n} A(i) = A_0 + 2A_0 + \cdots + 2^{n-1}A_0$, where A_0 is the area of the initial ellipse. Since $A(n) = 2^{n-1}A_0$, we see that $\sum_{i=1}^{n} A(i) = \sum_{i=1}^{n-1} A(i) + A(n) < 2A(n)$. According to Lemma 4.4, the area of the nth ellipse satisfies the inequality $A(n) < \frac{\pi}{2}(\lambda_{opt} + \|S-T\|)(\lambda_{opt}^2 + 2\lambda_{opt}\|S-T\|)^{1/2} \leq \frac{\pi}{2}(\lambda_{opt} + \|S-T\|)^2$. Hence the total length of the path traveled by the robot is bounded by $\lambda = \sum_{i=1}^{n} \lambda_i \leq \frac{6}{D}\sum_{i=1}^{n} A(i) + \sum_{i=1}^{n}(\|S_i - T\| - \|S_{i+1} - T\|) < \frac{6\pi}{D}(\lambda_{opt} + \|S-T\|)^2 + \|S - T\|$, where we substituted $S_1 = S$ and $S_{n+1} = T$. Finally, the term $6A_0/D$ bounds the path traveled by the robot in the case where the initial ellipse already contains a disconnection path starting from S. □

The final result is a universal lower bound on disconnection competitiveness.

Proposition 4.7. *Any navigation algorithm for unknown planar environments generates in worst case a disconnection path whose length satisfies the quadratic lower bound*

$$\lambda \geq \frac{4\pi}{3(1+2\pi)^2 D}(1-\epsilon)\lambda_{opt}^2, \tag{4}$$

where λ_{opt} *is length of the shortest off-line disconnection path starting from* S*,* D *is the disc-robot size, and* $\epsilon > 0$ *is an arbitrary small constant.*

Proof sketch: We use the environment which is used to prove the universal lower bound on connection competitiveness [13]. This environment consists of radial corridors emanating from S and having length r. The radial corridors are surrounded by a circular corridor such that only one radial corridor enters the circular corridor. The target is placed in the circular corridor. The shortest off-line path from S to T satisfies $l_{opt} \leq (1+\pi)r$, where πr is due to worst case motion in the circular corridor to a target located opposite the entry. Not knowing which radial corridor leads to the circular corridor, any on-line algorithm would guide the robot in worst case through all radial corridors before finding the entry to the circular corridor. It is shown in [13] that the worst case on-line path from S to T in this environment has length l satisfying $l \geq c(1-\epsilon)r^2$, where $c = 4\pi/3D$ and ϵ is an arbitrary small constant.

We now render the target inaccessible in two ways. First we place T outside the circular corridor so that it becomes inaccessible from S. In this case the shortest off-line disconnection path requires radial motion from S to the outer circular corridor, then a circumnavigation of the circular corridor. In the second case we place T in the circular corridor and surround it by walls located δ apart within the circular corridor. If the circular corridor is sufficiently wide, T lies in a puncture of the region accessible from S. In this case the shortest off-line disconnection path requires radial motion from S to the circular corridor, then motion in the circular corridor until the walls surrounding T are met. Combining the two cases, $\lambda_{opt} \leq (1+2\pi)r$ in this environment. Any on-line algorithm must explore in worst case all radial corridors before finding the entry to the circular corridor. Then it must circumnavigate in worst case the entire circular corridor. Hence $\lambda \geq c(1-\epsilon)r^2 + 2\pi r$. Since $r \geq \lambda_{opt}/(1+2\pi)$, we obtain the lower bound

$\lambda \geq c_1(1-\epsilon)\lambda_{opt}^2 + c_2\lambda_{opt}$, where $c_1 = 4\pi/3(1+2\pi)^2 D$ and $c_2 = 2\pi/(1+2\pi)$. The latter inequality implies that $\lambda \geq c_1(1-\epsilon)\lambda_{opt}^2$. □

The universal lower bounds (2) and (4) imply that *CBUG* has the lowest possible connection as well as disconnection competitiveness.

5 Significance of Double Competitiveness

This section discusses some useful properties of connection-and-disconnection competitiveness, termed *double* competitiveness. First and foremost, double competitiveness ensures that an on-line navigation algorithm would not stray away from the optimal path due to misleading sensory clues. Consider for instance the office floor environment shown in Figure 3. The paths generated by *ALG*1 (reviewed below) may stray along the outer walls arbitrarily far from S and T. When *CBUG* runs *ALG*1 as a sub-algorithm, the search is confined to the ellipses depicted in the figure.

Double competitiveness additionally provides some amount of path stability. A navigation algorithm possesses *path stability* when its path varies continuously with the position of S and T [11]. A weaker notion of path stability is as follows. A navigation algorithm possesses *path length stability* when its path length varies continuously with S and T. This means that small changes in S or T yield small changes in path length and hence travel time from S to T. Classical on-line navigation algorithms such as *BUG*1 and *ALG*1 respond to small changes in S or T with possibly unbounded path-length changes (Figure 3). However, the competitive quantities l_{opt} and λ_{opt} are approximately constant with respect to small changes of S and T. Hence competitive bounds in terms of l_{opt} and λ_{opt} automatically provide path-length bounds in response to small changes of S and T. Under *CBUG* the length of all connection paths from S to T vary in the bounded interval $[l_{opt}, c_1 l_{opt}^2 + c_2]$, where c_1 and c_2 are constants. Similarly, the length of all disconnection paths from S vary in the bounded interval $[\lambda_{opt}, c_1' \lambda_{opt}^2 + c_2']$, where c_1' and c_2' are constants.

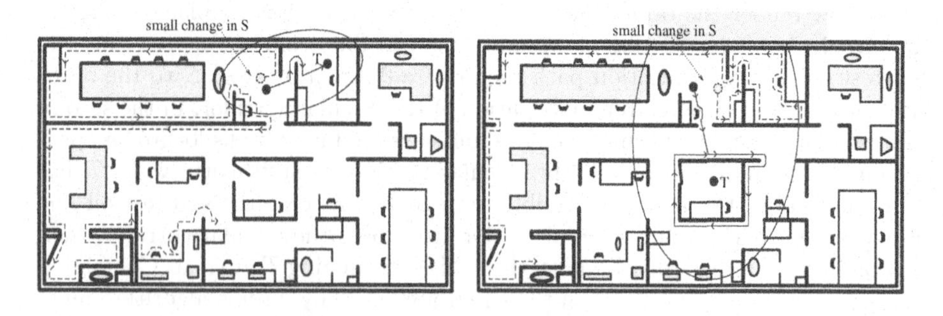

Fig. 3. The paths generated by *ALG*1 show significant path-length jumps in response to small changes in S. The ellipses show the total search area of *CBUG*.

Next we describe simulations comparing the non-competitive algorithms $BUG1$ and $ALG1$ with $CBUG$. The algorithm $ALG1$ relies on the straight line passing through S and T, and works as follows [20]. The robot moves from S along the S-T line towards T until it hits an obstacle. Then it circumnavigates the obstacle in a clockwise direction. Whenever the robot reaches an intersection point of the current obstacle boundary with the S-T line, denoted p, it leaves the obstacle if two conditions are met. The point p must be closer to T than all previous hit and leave points, and the direction from p to T must point away from the current obstacle. Once the robot leaves an obstacle it resumes motion along the S-T line until the next obstacle is encountered or the target is found. If p is not a valid leave point but is a previously defined hit or leave point, the robot *reverses* its boundary tracing direction at p. However, the robot is not allowed any further direction reversals along the current boundary segment. When the robot completes a loop around the current obstacle boundary without finding a suitable leave point, it halts with a conclusion of target unreachability. Several paths of $ALG1$ are depicted in Figure 3.

In general, the double competitiveness of $CBUG$ comes with an overhead incurred by its search ellipses. In order to study the effect of this overhead on average performance, we compared $BUG1$ and $ALG1$ to $CBUG$ implementing $BUG1$ and $ALG1$ as sub-algorithms. We tested the algorithms in the office floor environment depicted in Figure 3. We placed S and T in three distance ranges: $\|S-T\| \leq 10D$, $10D < \|S-T\| \leq 50D$, and $\|S-T\| > 50D$, where D is the robot size. Each distance range includes 30 runs with S and T varying within the prescribed range. Each 30 runs were subdivided into 10-run batches corresponding to three furniture occupancy levels of the office floor.

The results listed in Table 1 give the average ratio l/l_{opt}, where l is length of the path generated by the algorithm and l_{opt} is length of the shortest off-line path from S to T. In the highest distance range S and T are roughly at opposite corners of the office floor. In this case $CBUG$ is inferior to $BUG1$ and $ALG1$. In this case l_{opt} is not much shorter than the path along the outer walls persistently traced by $BUG1$. The superior performance of $ALG1$ is due to its boundary tracing rule, which typically limits its wall tracing to a single room. However, the advantage of $BUG1$ and $ALG1$ diminishes as T moves closer to S. In the lowest distance range S and T are typically in neighboring rooms. In this case $BUG1$ still follows the entire outer walls. Similarly, when T is located just on the other side of a wall, $ALG1$ guides the robot along the entire outer walls. In contrast, $CBUG$ always recognizes when both sub-algorithms stray away from S

Table 1. Summary of average l/l_{opt} results comparing $CBUG$ to $BUG1$ and $ALG1$

Distance Range	$BUG1$	$ALG1$	$CBUG$ with $BUG1$	$CBUG$ with $ALG1$
$\|S-T\| > 50D$	7.1	3.0	9.0	10.9
$10D < \|S-T\| \leq 50D$	11.5	4.6	7.2	12.0
$\|S-T\| \leq 10D$	28.8	14.1	3.5	7.3

and T. It cuts short their wall following with the bounding ellipses, thus ensuring paths whose average length is twice shorter than the paths of $ALG1$ and eight times shorter than the ones generated by $BUG1$. Simulations of cases where T is inaccessible from S are under preparation and will be discussed in an extended version of this paper.

6 Conclusion

The paper introduced a notion of disconnection competitiveness complimentary to the traditional notion of connection competitiveness. Connection competitiveness concerns cases where T is reachable from S, while disconnection competitiveness concerns cases where T cannot be reached from S. We described a tactile-sensor based navigation algorithm for a disc-shaped robot, called $CBUG$. The algorithm achieves connection as well as disconnection competitiveness by limiting its search to a series of expanding ellipses. The connection competitiveness of $CBUG$ is quadratic in l_{opt}, which matches up to constants the universal lower bound over all on-line navigation algorithms. The disconnection competitiveness of $CBUG$ is quadratic in λ_{opt}, where λ_{opt} is the shortest off-line disconnection path starting from S. The universal lower bound on disconnection competitiveness over all on-line navigation algorithms is also quadratic in λ_{opt}. Hence up to constants $CBUG$ has tight connection as well as disconnection competitiveness. However, competitiveness concerns worst case behavior, and does not necessarily indicate efficient average behavior. Simulations reveal that in practice $CBUG$ may incur significant overhead in certain situations.

Two over-simplifications of the navigation problem discussed in this paper are as follows. First, mobile robots are usually not disc-shaped but rather bodies having three degrees of freedom. Our preliminary work on three degrees-of-freedom mobile robots indicates that on-line navigation of simple shapes can be achieved with *cubic* competitiveness. Second, $CBUG$ assumes tactile senors. More sophisticated sensors such as vision and laser sensors do not have a significant advantage over tactile sensors in highly congested environments. However, practical environments tend to be reasonably sparse, and an adaptation of $CBUG$ to such sensors is an important open problem. Last, the constants in the quadratic upper bounds on $CBUG$ differ from the constants in the quadratic universal lower bounds by a factor of about 100. The closing of this gap is a major challenge that can yield algorithms with an improved average performance.

References

1. Baeza-Yates, R., Culderson, J., Rawline, G.: Searching in the plane. J. of Information and Computation 106, 234–252 (1993)
2. Basch, J., Guibas, L.J., Hsu, D., Nguyen, A.T.: Disconnection proofs for motion planning. In: IEEE Int. Conf. on Robotics and Automation, pp. 1765–1772 (2001)
3. Bellman, R.: Problem 63-9. SIAM Review 5(2) (1963)
4. Berman, P., Blum, A., Fiat, A., Karloff, H., Rosen, A., Saks, M.: Randomized robot navigation algorithms. In: SODA, pp. 75–84 (1996)

5. Blum, A., Raghavan, P., Schieber, B.: Navigating in unfamiliar terrain. STOC, 494–504 (1991)
6. Choset, H., Burdick, J.W.: Sensor based planning, part ii: Incremental construction of the generalized voronoi graph. In: IEEE Int. Conference on Robotics and Automation, pp. 1643–1649 (1995)
7. Choset, H., Lynch, K.M., Hutchinson, S., Kantor, G., Burgard, W., Kavraki, L.E., Thrun, S.: Principles of Robot Motion. MIT Press, Cambridge (2005)
8. Datta, A., Icking, C.: Competitive searching in a generalized street. In: 10th ACM Symp. on Computational Geometry, pp. 175–182 (1994)
9. Datta, A., Soundaralakshmi, S.: Motion planning in an unknown polygonal environment with bounded performance guarantee. In: IEEE Int. Conf. on Robotics and Automation, pp. 1032–1037 (1999)
10. Bar Eli, E., Berman, P., Fiat, A., Yan, P.: Online navigation in a room. J. of Algorithms 17(3), 319–341 (1996)
11. Farber, M.: Instabilities of robot motion. Topology and its Applications 140, 245–266 (2004)
12. Gabriely, Y., Rimon, E.: Competitive complexity of mobile robot on-line motion planning problems. In: 6'th Int. Workshop on Algorithmic Foundations of Robotics (WAFR), pp. 155–170 (2004)
13. Gabriely, Y., Rimon, E.: Cbug: A quadratically competitive mobile robot navigation algorithm. In: IEEE Int. Conf. on Robotics and Automation, pp. 954–960 (2005)
14. Icking, C., Klein, R., Langetepe, E.: An optimal competitive strategy for walking in streets. In: 16th Symp. on Theoretical Aspects of Computer Science, vol. 110, pp. 405–413 (1988)
15. Kamon, I., Rimon, E., Rivlin, E.: Tangentbug: A range-sensor based navigation algorithm. Int. Journal of Robotics Research 17(9), 934–953 (1998)
16. Laubach, S.L., Burdick, J.W., Matthies, L.: An autonomous path planner implemented on the rocky7 prototype microrover. In: IEEE Int. Conf. on Robotics and Automation, pp. 292–297 (1998)
17. Lumelsky, V.J., Stepanov, A.: Path planning strategies for point automaton moving amidst unknown obstacles of arbitrary shape. Algorithmica 2, 403–430 (1987)
18. Noborio, H., Yoshioka, T.: An on-line and deadlock-free path-planning algorithm based on world topology. In: Conf. on Intelligent Robots and Systems, IROS, pp. 1425–1430. IEEE/RSJ (1993)
19. Papadimitriou, C.H., Yanakakis, M.: Shortest paths without a map. Theoretical Computer Science 84, 127–150 (1991)
20. Sankaranarayanan, A., Vidyasagar, M.: Path planning for moving a point object amidst unknown obstacles in a plane: the universal lower bound on worst case path lengths and a classification of algorithms. In: IEEE Int. Conf. on Robotics and Automation, pp. 1734–1941 (1991)
21. Tovar, B., Lavalle, S.M., Murrieta, R.: Optimal navigation and object finding without geometric maps or localization. In: IEEE Int. Conf. on Robotics and Automation, pp. 464–470 (2003)

A Simple Path Non-existence Algorithm Using C-Obstacle Query

Liangjun Zhang[1], Young J. Kim[2], and Dinesh Manocha[1]

[1] Dept. of Computer Science, University of North Carolina at Chapel Hill, U.S.A.
{zlj,dm}@cs.unc.edu
[2] Dept. of Computer Science and Engineering, Ewha Womans University, Korea
kimy@ewha.ac.kr
http://gamma.cs.unc.edu/NOPATH

Abstract. We present a simple algorithm to check for path non-existence for a robot among static obstacles. Our algorithm is based on adaptive cell decomposition of configuration space or C-space. We use two basic queries: free cell query, which checks whether a cell in C-space lies entirely inside the free space, and C-obstacle cell query, which checks whether a cell lies entirely inside the C-obstacle region. Our approach reduces the path non-existence problem to checking whether there exists a path through cells that do not belong to the C-obstacle region. We describe simple and efficient algorithms to perform free cell and C-obstacle cell queries using *separation distance* and *generalized penetration depth* computations. Our algorithm is simple to implement and we demonstrate its performance on 3 DOF robots.

1 Introduction

Motion planning is a fundamental problem in robotics. The goal is to compute a collision-free path between two configurations of a given robot. This problem has been extensively studied in the field for more than three decades. At a broad level, prior algorithms for motion planning can be classified into roadmap methods, exact cell decomposition, approximate cell decomposition, potential field methods and randomized sampling-based methods [7, 14, 15]. In particular, planning algorithms such as the roadmap methods and exact cell decomposition are referred as complete motion planning algorithms. These approaches can compute a collision-free path if one exists; otherwise they report path non-existence between the two configurations. However, these methods are known to have a high theoretical complexity and are very difficult to implement. Their practical implementations are usually limited to planar robots, convex polytopes or special shapes such as spheres or ladders.

Practical algorithms for motion planning are based on approximate cell decomposition, potential field computation or sampling-based algorithms. The approximate cell decomposition based algorithms subdivide the configuration space into cells and can be made *resolution complete* based on a suitable choice of parameters. However, the number of subdivision in prior approaches grows quickly with the dimension of the configuration space. In practice, prior

S. Akella et al. (Eds.): Algorithmic Foundation of Robotics VII, STAR 47, pp. 269–284, 2008.
springerlink.com

cell-decomposition algorithms also suffer from the combinatorial complexity and robustness issues with respect to contact surface enumeration and computation. For a robot with high geometric complexity, generating and enumerating contact surfaces can be complicated and time consuming [8, 16].

On the contrary, randomized sampling methods such as probabilistic roadmap planners (PRMs) are relatively simple to implement and work quite well in practice [11]. The strength of PRMs lies in their simplicity and they can be easily applied to general robots with high DOF. However, PRMs have two major issues: path non-existence (i.e. no passage) and narrow passages. If there is no collision-free path, PRM algorithms may not terminate. Moreover, it is hard to distinguish whether such situations arise due to path non-existence or due to narrow passages and poor sampling.

Main Results

In this paper, we present a simple and efficient cell decomposition algorithm for path non-existence from the initial to the goal configuration. Our resulting motion planning algorithm is a complete algorithm for a rigid robot with translational and rotational DOF. Furthermore, our approach can also be extended to articulated robots.

We subdivide the configuration space into *empty*, $full$ and *mixed* cells. Unlike prior cell decomposition algorithms, we use efficient algorithms to label a given cell as $full$ or *empty* to check whether it lies entirely in the C-obstacle region or in the free space. Our algorithm uses two kind of queries: *free cell* query for identifying *empty* cells and *C-obstacle cell* query for identifying $full$ cells. We efficiently perform these queries in the workspace by computing the *separation distance* and *generalized penetration depth*. As a result, our cell query algorithms can be easily implemented for 2D or 3D rigid robots, or articulated robots.

In order to check for path non-existence, the algorithm searches for a sequence of adjacent *empty* or *mixed* cells to connect the cell containing the initial configuration to the cell containing the goal configuration. The non-existence of such a sequence is a sufficient condition for path non-existence between the initial and the goal configuration. We have implemented our algorithm and highlight its performance on 3 DOF robots.

Organization

The rest of the paper is organized as follows. In Section 2, we briefly survey related work on motion planning. We give an overview of our method in Section 3 and present our cell labelling algorithms in Section 4. We describe our implementation and discuss a few limitations of our approach in Section 5.

2 Previous Work

Motion planning has been extensively studied for more than three decades. Excellent surveys of this topic are available in [7, 14, 15]. In this section, we briefly review prior algorithms for exact and approximate motion planning.

2.1 Exact Motion Planning

The complete motion planning algorithms compute a collision-free path if one exists; otherwise they report path non-existence. These include criticality-based algorithms such as exact free-space computation for a class of robots [2, 9, 17, 12], roadmap methods [6], and exact cell decomposition methods [20]. The exact cell decomposition methods require an exact description of the configuration space consisting of the free space and the C-obstacle region. The boundary between the free space and C-obstacle region is described by a set of contact surfaces, each surface being the locus of configurations of a robot at which a specific boundary feature of the robot is in contact with a boundary feature of the obstacles. The exact cell decomposition approaches partition the free space into a collection of simpler geometric regions and compute a connectivity graph representing the adjacency between the regions.

In theory, these methods are quite general. However, in practice, it is quite challenging to implement them and no good implementations are known for general and high DOF robots. As a result, many variants have been proposed to deal with special cases of motion planning problems [14].

2.2 Approximate Cell Decomposition and Sampling-Based Approaches

A number of algorithms based on approximate cell decomposition have been proposed [4, 8, 27]. These methods partition the C-space into a collection of cells. They classify the cells into three types: *empty* cells that lie entirely in the free space, *full* cells that are entirely within the C-obstacle region, and *mixed* cells that correspond to the rest. Unlike exact cell decomposition, the cell used in approximate cell decomposition algorithms have a simpler, rectangloid shape, and the *empty* cells provide a conservative approximation of the free space. The planner searches through the empty cells to find a path. Moreover, approximate cell decomposition methods are *resolution complete*; i.e., they can find a path if one exists provided the resolution parameters are selected small enough [14]. In practice, approximate cell decomposition methods have been used for low DOF robots.

One of the main computational issues in approximate cell decomposition methods is cell labelling. In order to label a cell, most prior approaches rely on contact surface computations [27], which could be complicated and prone to degeneracies. Paden et al. [18] describe a method based on workspace distance computation. However, their method could be overly conservative in practice.

The probabilistic roadmap method (PRM) [11] is perhaps the most widely used path planning algorithm for different applications. It is relatively simple to implement and has been successfully applied to high DOF robots. Since PRM-based algorithms sample the free space randomly, they may fail to find paths, especially those passing through narrow passages. A number of extensions have been proposed to improve the sampling in terms of handling narrow passages [1, 10, 19] or use visibility-based techniques [22]. All these methods are *probabilistically complete.*

Recently, a deterministic sampling approach called star-shaped roadmaps has been proposed [24]. The free space is partitioned into star-shaped regions and connectors between star-shaped regions are computed for inter-region connectivity. This algorithm is complete as long as there are no tangential contacts in the boundary of the free space. Since star-shaped roadmaps compute the global connectivity of the free space, the number of regions can increase exponentially as a function of DOF. Moreover, this approach is based on contact surface enumeration.

2.3 Path Non-existence

Exact planning approaches such as exact cell decomposition and roadmap computation can check for path non-existence. However, these methods are not practical due to their theoretical complexity and implementation difficulty. Approximate cell decomposition approaches can also check for path non-existence, but they are also complicated because of contact surface computation. In general, a popular planning method such as PRM cannot deterministically guarantee the path non-existence as it is only probabilistically complete. An effort has been made to address the issue of path non-existence in PRM [3]. The authors have proposed a *disconnection prover*, probabilistically showing that the motion planning problem has no solution. However, this approach is restricted to the special problem of finding a path through a planar section. The deterministic sampling approach such as star-shaped roadmaps [24] is a complete approach but it may be overly conservative and can generate a large number of samples.

3 Overview

In this section, we give an overview of our algorithm. Following the basic framework of approximate cell decomposition, we use reliable algorithms to perform the cell labelling. Then we build the connectivity graph for the *empty* and *mixed* cells from the cell decomposition, and check for path non-existence between the initial and the goal configuration by performing a search in the connectivity graph.

3.1 Notation

We use a symbol $\mathcal{A}$ to denote a robot and $\mathcal{B}$ to represent a collection of all fixed obstacles. Let $\mathcal{C}$ denote the configuration space or C-space of the robot. $\mathcal{F}$ and $\mathcal{O} = \mathcal{C} \backslash \mathcal{F}$ represent the free space and the configuration space obstacle or C-obstacle region, respectively. A cell C in n-dimensional C-space is defined as a Cartesian product of real intervals:

$$C = [x'_1, x''_1] \times [x'_2, x''_2] \cdots \times [x'_n, x''_n].$$

We denote $\mathcal{A}(\mathbf{q})$ as a placement of the robot $\mathcal{A}$ at configuration $\mathbf{q}$. Let $\mathbf{q}_{init}$ and $\mathbf{q}_{goal}$ represent the initial and the goal configuration of the robot. A line

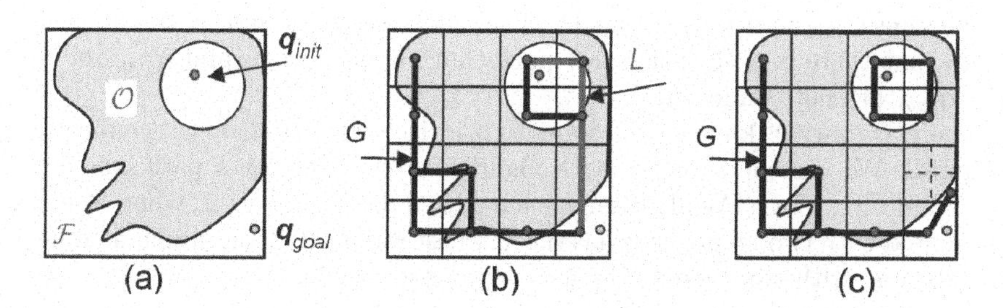

Fig. 1. Path non-existence between $\mathbf{q}_{init}$ and $\mathbf{q}_{goal}$. (b): A connectivity graph G is built. The path L, which connects the cells including $\mathbf{q}_{init}$ and $\mathbf{q}_{goal}$, is computed from G. Any mixed cell along L is further subdivided. (c): In the new connectivity graph, the cell containing $\mathbf{q}_{init}$ and the cell containing $\mathbf{q}_{goal}$ are not connected. This concludes that there is no collision-free path between $\mathbf{q}_{init}$ and $\mathbf{q}_{goal}$.

segment in C-space connecting configurations $\mathbf{q}_a$ and $\mathbf{q}_b$ is represented as $\pi_{\mathbf{q}_a,\mathbf{q}_b}$. Let $l(\mathbf{t}), t \in [0,1]$ be an arbitrary motion curve defined in the C-space. We denote $\mu(\mathbf{p}, l)$ as the trajectory length of a point $\mathbf{p}$ on $\mathcal{A}$ when $\mathcal{A}$ moves along the motion curve l.

3.2 Cell Decomposition and Labelling

Our approach to check for path non-existence is based on cell decomposition. The configuration space $\mathcal{C}$ is spatially subdivided into cells at successive levels of the subdivision. The cells are classified as *empty* or *full* depending on whether they lie entirely inside the free space $\mathcal{F}$, or entirely inside the C-obstacle $\mathcal{O}$. If they are neither *empty* nor *full*, they are labelled as *mixed*. Fig. 1 illustrates different type of cells. In Section 4, we present more details about our cell labelling algorithm.

3.3 Connectivity Graphs

For each level of subdivision, the connectivity graph G is built to represent the adjacency relationship between *empty* and *mixed* cells. Formally, the connectivity graph [14, 27, 28] associated with a decomposition $\mathcal{D}$ of $\mathcal{C}$ is an undirected graph, where:

- The vertices in G are the *empty* and *mixed* cells in $\mathcal{D}$.
- Two vertices in G are connected by an edge if and only if the corresponding cells are adjacent.

Intuitively, G captures the connectivity of both the identified free space, which is covered by the *empty* cells, and the 'uncertain' region, which is represented by the *mixed* cells.

In order to check for path non-existence, our algorithm first locates the cells C_{init} and C_{goal}, which contain $\mathbf{q}_{init}$ and $\mathbf{q}_{goal}$, respectively. Next, the algorithm searches G to find a path L, a sequence of adjacent *empty* and *mixed* cells connecting C_{init} and C_{goal} (Fig. 1). If no such path is found, it is sufficient to claim that there is no collision-free path that connects $\mathbf{q}_{init}$ and $\mathbf{q}_{goal}$, or $\mathbf{q}_{init}$ and $\mathbf{q}_{goal}$ are not connected.

There are various known techniques to prioritize the search on the connectivity graph G. We use the shortest path algorithm to search for a path connecting C_{init} and C_{goal} in G. We also assign each edge a different weight, where the edge associated with two *empty* cells has the smallest weight (0 in our implementation) and the one with two *mixed* cells has the largest weight.

Our algorithm terminates if we can prove path non-existence, or we can find a collision-free path. For this purpose, a subgraph G_e of G is also constructed. G_e represents the adjacency relationship among all the *empty* cells. Intuitively, G_e represents the connectivity of a part of the free space that has been identified till the current level of subdivision. If there is a path in G_e connecting C_{init} and C_{goal}, a collision-free path can be easily extracted and optimized [27].

3.4 Guided Subdivision

When a path L is reported after searching the connectivity graph G, it is not clear whether $\mathbf{q}_{init}$ and $\mathbf{q}_{goal}$ are not connected. If so, we need to further explore the 'uncertain' regions - the union of *mixed* cells, to acquire more information about their connectivity. Considering the fact that not all 'uncertain' regions contribute to separating $\mathbf{q}_{init}$ from $\mathbf{q}_{goal}$, we employ the first-cut algorithm [14, 27] to first subdivide some of the 'uncertain' regions. More specifically, all the *mixed* cells on the path L are assigned higher priorities for the next level of the subdivision than other *mixed* cells. Our algorithm is recursively applied until it finds a collision-free path or concludes path non-existence.

4 Cell Labelling

Compared to prior cell decomposition approaches, one of our distinct features is that during cell decomposition, we use reliable algorithms for cell labelling. As a result, our algorithm does not need to compute the contact surfaces. In this section, we present our *free cell query* and *C-obstacle cell query* algorithms for labelling the cells in $\mathcal{C}$. Formally speaking, the free cell query checks whether a given cell C is *empty* or the following predicate P_f is true:

$$P_f(\mathcal{A}, \mathcal{B}, C): \ \forall \mathbf{q} \in C, \ interior(\mathcal{A}(\mathbf{q})) \cap interior(\mathcal{B}) = \emptyset,$$

where $\mathcal{A}$ is a robot, $\mathcal{B}$ represents the obstacles and the operator *interior* is the interior of a set. Similarly, the C-obstacle cell query or C-obstacle query checks whether a given cell C is *full* or the following predicate P_o is true:

$$P_o(\mathcal{A}, \mathcal{B}, C): \ \forall \mathbf{q} \in C, \ interior(\mathcal{A}(\mathbf{q})) \cap interior(\mathcal{B}) \neq \emptyset.$$

The collision detection algorithms can check whether a single configuration lies in $\mathcal{F}$ or $\mathcal{O}$. However, these predicates need to check whether a spatial cell lies in $\mathcal{F}$ or $\mathcal{O}$, which corresponds to the collision detection for a set of continuous configurations. Therefore, it is relatively harder to perform these queries as compared to checking a single configuration.

In our algorithms, we place the robot at $\mathbf{q}_c$ - the center of the cell and compute the 'extent' of the motion that the robot can undergo as it moves away from $\mathbf{q}_c$ while still being confined within the cell C. To answer the predicate P_f, we compute the *separation distance* between the robot $\mathcal{A}(\mathbf{q}_c)$ and the obstacle $\mathcal{B}$. This distance describes the 'clearance' between the robot and the obstacle. If this 'clearance' is greater than the amount of the maximal motion that the robot can make, the robot will not collide with the obstacle, and the cell C will be declared as a free cell.

Similarly, in order to perform the C-obstacle cell query, we measure the amount of inter-penetration between the robot and the obstacle, and compare it with the extent of the robot's bounding motion. In subsection 4.2, we shall present our method for formulating and computing the inter-penetration between the robot and the obstacle.

4.1 Motion Bound Calculation

Bounding Motion for a Line Segment

In order to formulate the bounding motion for a C-space cell, we first introduce a case when a robot moves along a line segment in C-space. Schwarzer *et al.* [21] define the bounding motion λ when a robot moves along a line segment $\pi_{\mathbf{q}_a,\mathbf{q}_b}$ as the maximal trajectory length over all points on the moving robot:

$$\lambda(\mathcal{A}, \pi_{\mathbf{q}_a,\mathbf{q}_b}) = \text{Upper Bound}(\mu(\mathbf{p}, \pi_{\mathbf{q}_a,\mathbf{q}_b}) \mid \mathbf{p} \in \mathcal{A}).$$

For 2D planar robots with translational and rotational DOF, the bounding motion λ can be computed as a weighted sum of the difference between $\mathbf{q}_a$ and $\mathbf{q}_b$ for translational components x, y and the rotational component ϕ:

$$\lambda(\mathcal{A}, \pi_{\mathbf{q}_a,\mathbf{q}_b}) = |\mathbf{q}_{b.x} - \mathbf{q}_{a.x}| + |\mathbf{q}_{b.y} - \mathbf{q}_{a.y}| + R_\phi \times |\mathbf{q}_{b.\phi} - \mathbf{q}_{a.\phi}|,$$

where the weight R_ϕ is defined as the maximum Euclidean distance between every point on $\mathcal{A}$ and its rotation center. In this case, we can achieve a tighter bound:

$$\lambda(\mathcal{A}, \pi_{\mathbf{q}_a,\mathbf{q}_b}) = \sqrt{|\mathbf{q}_{b.x} - \mathbf{q}_{a.x}|^2 + |\mathbf{q}_{b.y} - \mathbf{q}_{a.y}|^2} + R_\phi \times |\mathbf{q}_{b.\phi} - \mathbf{q}_{a.\phi}|. \quad (1)$$

We can also extend this bound to 3D rigid objects.

Bounding Motion for a Cell

Now, we define the bounding motion λ of a robot when it is restricted within a cell C, instead of a line segment, as:

$$\lambda(\mathcal{A}, C) = \max\{\lambda(\mathcal{A}, \pi_{\mathbf{q}_a,\mathbf{q}_b}) \mid \mathbf{q}_b \in \partial C\}, \quad (2)$$

where $\mathbf{q}_a$ is the center of C, and $\mathbf{q}_b$ is any point on ∂C or the boundary of C.

Among all line segments $\pi_{\mathbf{q}_a,\mathbf{q}_b}$, the diagonal line segments have the maximum difference on each component between these two configurations. According to Eq. (1), we can infer that the maximum of the bounding motion $\lambda(\mathcal{A}, \pi_{\mathbf{q}_a,\mathbf{q}_b})$ is achieved by any diagonal line segment of the cell. Therefore, the bounding motion for the cell C is equivalent to the bounding motion over any diagonal line segment $\pi_{\mathbf{q}_a,\mathbf{q}_c}$:

$$\lambda(\mathcal{A}, C) = \lambda(\mathcal{A}, \pi_{\mathbf{q}_a,\mathbf{q}_c}), \tag{3}$$

where $\mathbf{q}_a$ is the center of the cell and $\mathbf{q}_c$ is any corner vertex of the cell.

4.2 C-Obstacle Cell Query

In order to perform the C-obstacle cell query, we can measure the extent of interpenetration between the robot and the obstacle, and compare it with a bound on the robot's motion. If the robot only has translational DOF, we can use translational PD, PD^t, which is defined as the minimum translational distance to separate the robot from the obstacle:

$$PD^t(\mathcal{A}, \mathcal{B}) = \min(\{\| \mathbf{d} \| \,|interior(\mathcal{A} + \mathbf{d}) \cap \mathcal{B} = \emptyset\}).$$

However, PD^t is only useful to perform C-obstacle cell query, when the robot has only translational DOF. This is because PD^t only considers the translational motion to separate the robot from the obstacle. When the robot is allowed to both translate and rotate, Fig. 2 shows that the robot can be separated from the obstacle with 'less' amount of motion by making use of rotational motion.

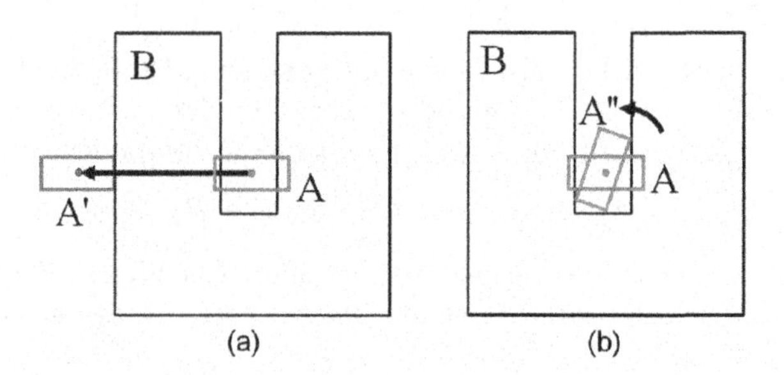

Fig. 2. An example shows that, to separate A from B, the amount of the 'motion' when both translational and rotational transformation are allowed (b) is much smaller than the amount of the 'motion' when only translation is allowed (a)

Generalized Penetration Depth

In order to deal with a robot with translational and rotational DOF, we adopt the notion of generalized penetration depth, PD^g, proposed by [25]. PD^g takes both translational and rotational motion into account and can be defined using the notion of the separating path. A separating path l is such a motion curve in C-space when a robot moves along l, the robot can be completely separated from the obstacle.

Given a set L of all possible candidates of separating paths, PD^g between a robot $\mathcal{A}$ and an obstacle $\mathcal{B}$ is defined as:

$$PD^g(\mathcal{A},\mathcal{B}) = \min\{\max\{\mu(\mathbf{p},l)|\mathbf{p} \in \mathcal{A}\}|l \in L\}. \quad (4)$$

A useful property related to PD^g is as follows:

Lemma 1. *For two convex polytopes $\mathcal{A}$ and $\mathcal{B}$, we have*

$$PD^g(\mathcal{A},\mathcal{B}) = PD^t(\mathcal{A},\mathcal{B}).$$

The proof of this lemma can be found in [26].

The exact computation of PD^g between non-convex objects is a difficult problem [26]. In our C-obstacle cell query algorithm, we compute a lower bound on PD^g, which guarantees the correctness of the query. Using Lemma 1, we efficiently compute a lower bound on PD^g by (1) decomposing non-convex models into convex pieces and (2) for each convex pair, compute the PD^t as its PD^g, (3) take the maximum value of PD^g's between all pairwise combinations of convex pieces. Many efficient algorithms are known to compute the PD^t between two convex polytopes [5, 23, 13]. The resulting PD^g computation algorithm is described in Algorithm 1.

Algorithm 1. Lower bound on PD^g computation
Input: The robot $\mathcal{A}$, the obstacle $\mathcal{B}$ and the configuration $\mathbf{q}$
Output: The lower bound on PD^g between $\mathcal{A}(\mathbf{q})$ and $\mathcal{B}$.

1: {During preprocessing}
2: Decompose $\mathcal{A}$ and $\mathcal{B}$ into m and n convex pieces; i.e., $\mathcal{A} = \cup\mathcal{A}_i$ and $\mathcal{B} = \cup\mathcal{B}_j$.
3: {During run-time query}
4: **for** each pair of $(\mathcal{A}_i(\mathbf{q}),\mathcal{B}_j)$ **do**
5: $\quad k = (i-1)n + j$
6: $\quad$ **if** $\mathcal{A}_i(\mathbf{q})$ collides with $\mathcal{B}_j$ **then**
7: $\quad\quad PD^g_k = PD^t((\mathcal{A}_i(\mathbf{q}),\mathcal{B}_j)$
8: $\quad$ **else**
9: $\quad\quad PD^g_k = 0$
10: $\quad$ **end if**
11: **end for**
12: **return** $\max(PD^g_k)$ for all k.

4.3 C-Obstacle Cell Query Criterion

We now state a sufficient condition for C-obstacle cell query; i.e., checking whether $\mathcal{A}$ and $\mathcal{B}$ overlap at every configuration $\mathbf{q}$ within a cell C.

Theorem 1: *For a cell C with a center at $\mathbf{q}_a$, the predicate $P_o(\mathcal{A},\mathcal{B},C)$ is true if:*

$$PD^g(\mathcal{A}(\mathbf{q}_a),\mathcal{B}) > \lambda(\mathcal{A},C). \quad (5)$$

Proof. Our goal is to show that Eq. (5) implies that there is no free configuration along any line segment $\pi_{\mathbf{q}_a,\mathbf{q}_b}$, where $\mathbf{q}_b$ is any configuration on the boundary of the cell C. According to the definition of PD^g, the maximum trajectory length for every point on a robot $\mathcal{A}$ moving along a possible separating path should be greater than or equal to $PD^g(\mathcal{A}(\mathbf{q}_a), \mathcal{B})$. Moreover, according to Eq. (2), the trajectory length of the robot when it moves along $\pi_{\mathbf{q}_a,\mathbf{q}_b}$ is less than or equal to $\lambda(\mathcal{A}, C)$. Since $PD^g(\mathcal{A}(\mathbf{q}_a), \mathcal{B}) > \lambda(\mathcal{A}, C)$, the minimum motion required to separate the robot $\mathcal{A}$ from obstacle $\mathcal{B}$ is larger than the maximum motion the robot $\mathcal{A}$ can undergo. Therefore, there are no free configurations along any line segment $\pi_{\mathbf{q}_a,\mathbf{q}_b}$.

Since there is no free configuration along every line segment between $\mathbf{q}_a$ to $\mathbf{q}_b$, we conclude that every configuration in the cell C lies inside the C-obstacle region, and therefore, the predicate $P_o(\mathcal{A}, \mathcal{B}, C)$ holds. □

We use *Theorem* 1 to conservatively decide whether a given cell C lies inside the C-obstacle region. The C-obstacle cell query algorithm consists of two parts:

1. Compute a lower bound on PD^g for the robot $\mathcal{A}(\mathbf{q}_a)$ and the obstacle $\mathcal{B}$ by the Algorithm 1.
2. Compute an upper bound on motion, $\lambda(\mathcal{A}, C)$ by Eq.'s (3) and (1).

Our C-obstacle cell query algorithm is general for both 2D and 3D rigid objects. We have implemented the query for both types of objects. The main computational component is to compute PD^t between convex objects.

4.4 Free Cell Query Criterion

Similar to C-obstacle cell query, we compare the separation distance between the robot $\mathcal{A}(\mathbf{q}_c)$ and the obstacle $\mathcal{B}$ with the bounding motion of the cell $\lambda(\mathcal{A}, C)$. If the distance is greater than the bounding motion, then the cell is classified as a free cell.

4.5 Extension to Articulated Robots

Our free cell and C-obstacle cell queries based method for checking path non-existence can be extended to articulated robots. The main modifications for articulated robots are in the components: generalized penetration depth, separation distance, and bounding motion computations.

The definition of generalized penetration depth PD^g in Eq. (4) is also applicable to articulated robots. In this case, the separating path in C-space is defined as a curve such that when the articulated robot $\mathcal{A}$ moves along it, $\mathcal{A}$ will be completely separated from the obstacle. In order to compute a lower bound on PD^g between $\mathcal{A}$ and the obstacles, we regard each link of $\mathcal{A}$ as a rigid robot with translational and rotational DOF. The maximum of lower bounds PD^g between each link of $\mathcal{A}$ and the obstacles yields a lower bound on PD^g between $\mathcal{A}$ and the obstacles. In order to compute the separation distance and bounding motion for the articulated robots, we use the algorithms introduced by Schwarzer et al. [21].

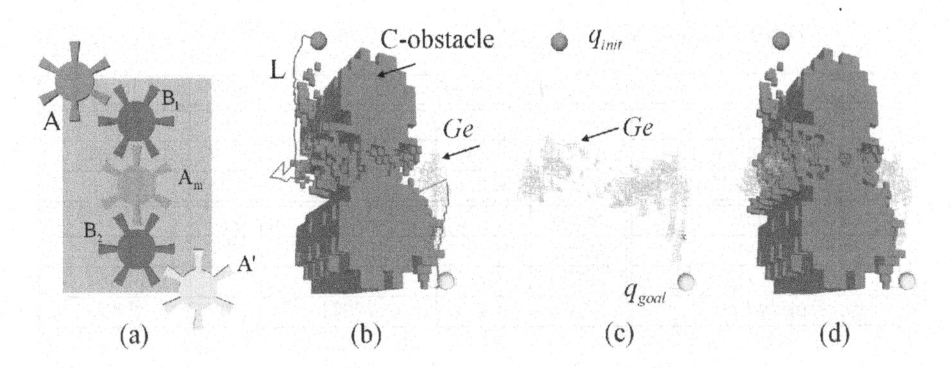

Fig. 3. Application of our algorithm to the gear benchmark: (a) The goal of this example is to move a gear-shaped robot from A to A' through the two gear-shaped obstacles B_1 and B_2. It is uncertain whether there is a path for these configurations, even though the robot at A_m is collision-free. (b, c) shows the graph G_e built from empty cells, and the region of full cells (shaded volumes). Since no path is found when searching the G_e, we search the graph G for a guiding path L, which indicates the next level of subdivision. (d) After the subdivision is recursively applied, the algorithm finally concludes that no path exists. This is because the initial and the goal configuration are separated by full cells (shaded volumes in (d)).

Table 1. Performance: This table highlights the performance of our algorithm on different benchmarks

	two-gear	five-gear	five-gear,narrow	puzzle	narrow puzzle
Total timing(s)	3.356	6.317	85.163	7.898	15.751
Free cell query(s)	0.858	1.376	6.532	2.174	2.993
C-obstacle cell query(s)	0.827	1.162	4.675	2.021	2.612
G searching(s)	0.389	1.409	30.687	1.991	5.685
G_e searching(s)	0.077	0.332	7.169	0.309	1.035
Subdivision,Overhead(s)	1.205	2.038	36.100	1.403	3.426

5 Experimental Results

In this section, we describe the implementation of our algorithm and highlight its performance on several motion planning scenarios. All timings are measured on a 2.8 GHz Pentium IV PC with 2G RAM. Our current implementation is not optimized.

We illustrate the running process of our algorithm for the 'two-gear' example in Fig. 3. In order to find whether the gear-shaped robot can pass through the passage among star-shaped obstacles, the algorithm performs cell decomposition, and builds the connectivity graph G for *empty* and *mixed* cells as well as its subgraph G_e for *empty* cells. The cell decomposition, which is performed in the region indicated by the guiding path from the search on the connectivity

Table 2. Application of our algorithm to different benchmarks

	two-gear	five-gear	narrow five-gear	puzzle	narrow puzzle
# of iterations	41	67	237	66	107
# of free cell queries	32329	44649	192009	59121	77297
# of C-obstacle cell queries	30069	41177	176685	55683	70438
# of cells	28288	39068	168008	51731	67635
# of empty cells	2260	3472	15324	3438	6859
# of full cells	12255	16172	74713	26295	30351
# of mixed cells	13773	19424	77971	21998	30425

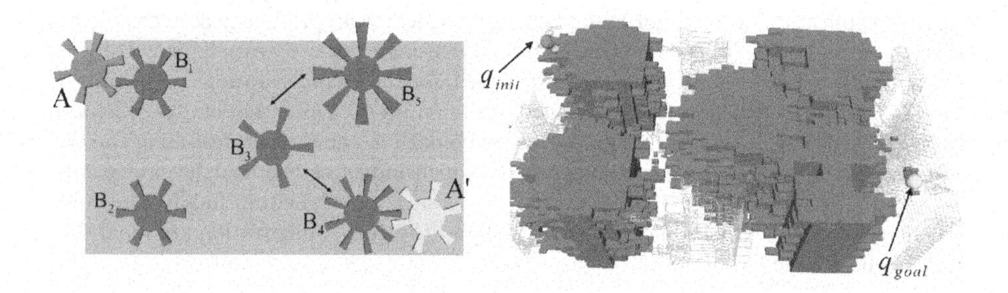

Fig. 4. 'Five-gear' example. (Left) The goal of this example is to move a gear-shaped robot from A to A' through the five gears B_1, ... and B_5. (Right) There does not exist a collision-free path for this example. This is because the initial and the goal configuration are separated by full cells, which correspond to shaded volumes. The right figure also highlights that to find path non-existence for this example, it is unnecessary to classify the entire configuration space.

graph G, is iterated by 40 times until the initial and the goal configuration are found to be separated by *full* cells. The entire computation takes 3.356s.

We have applied our algorithm to more complex examples of: 'five-gear', 'five-gear with narrow passage', '2D puzzle' and '2D puzzle with narrow passage'. Table 1 highlights the performance of our algorithm on these examples. According to Table 1, our approach can report path non-existence for these examples within 10s. In particular, for the 'five-gear' example, the total timing is 6.317s with 1.162s and 1.376s for the C-obstacle cell query and free cell query, respectively.

Table 2 gives details about application of our algorithm to different benchmarks. For the 'five-gear' example, the cell decomposition, which is restricted in the region indicated by the guiding path, is iterated 67 times. The final cell-decomposition includes 39068 cells, with 3473 *empty* cells, 16172 *full* cells and 19424 *mixed* cells.

Since our algorithm uses cell decomposition, the algorithm is applicable to finding a collision-free paths even when a narrow passage exists. Finding a collision-free path through a narrow passage has been considered as a difficult

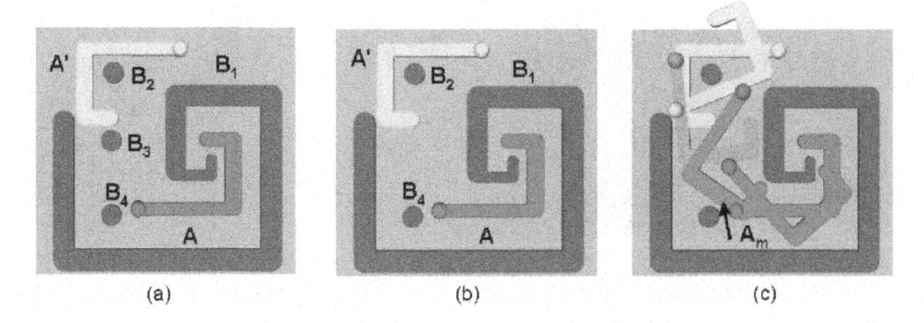

Fig. 5. '2D puzzle' example. (a) Our algorithm can report the path non-existence for the problem to move A to A' in 7.898s. (b) is a modified version of (a) without the obstacle B_3. Our algorithm can find a collision-free path through a narrow passage among the obstacles. (c) shows intermediate configurations A_m of the robot along the collision free path.

task for probabilistic methods, such as PRM. Fig. 6 shows such an example. According to the Table 1, for this example, our un-optimized method achieves about 1.3 times speedup over a deterministic sampling approach, the star-shaped roadmap [26].

5.1 Comparison

We compare our algorithm for path non-existence with star-shaped roadmap algorithm, especially because our approach shares similarities with the star-shaped roadmap algorithm. Star-shaped roadmap method partitions the free space into star-shaped regions and for each star-shaped region computes a single point called a guard which can see every point in the region. In our approach, the *empty* cells are a special case of star-shaped regions where any configuration in the cell can be always considered as a guard. Moreover, our method can label *empty* cells in a simpler way than the star-shaped roadmap, as the star-shaped roadmap is based on expensive contact surfaces enumeration.

Finally, star-shaped roadmap method needs to explicitly capture the intra-connectivity between two adjacent regions, which can be computed using a similar way to the guard computation, but in one dimension less. In our method, the intra-connectivity between two adjacent *empty* cells is implicit and an edge connecting the guards of two adjacent cells can represent such a connectivity.

5.2 Analysis

The computational complexity of our C-obstacle cell query is bounded by generalized penetration depth PD^g computation. We only compute a lower bound to PD^g and its complexity is governed by the number of convex pieces that are obtained from the convex decomposition, and the geometric complexity of these convex pieces. Let m, n denote the number of convex pieces of the robot $\mathcal{A}$ and

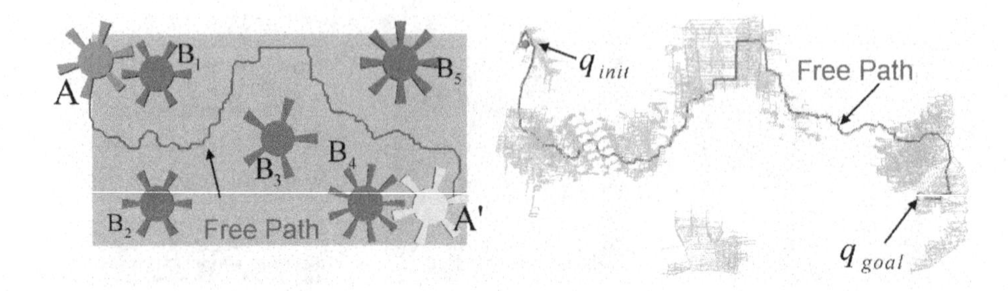

Fig. 6. Finding a passage through narrow passages for the modified 'five-gear' example. (Left) this planning problem is almost the same as Fig. 4 except that the obstacle B_5 is slightly modified as well as translated. (Right) our method can find a path under the existence of narrow passages, which are challenging for probabilistic methods, such as PRM. The collision-free path, passing through the narrow passage in the free space, is derived from empty cells.

the obstacle $\mathcal{B}$, respectively. Let the geometric complexity of all convex pieces of $\mathcal{A}$ and $\mathcal{B}$ be a and b, respectively. Then, the average numbers of features in each piece of $\mathcal{A}$ and $\mathcal{B}$ are $\frac{a}{m}$ and $\frac{b}{n}$, respectively. Using computational complexity of translational PD, we can derive that the computational complexity of PD^g for 2D rigid objects $\mathcal{A}$ and $\mathcal{B}$ is $O(an + bm)$, and for 3D rigid objects is $O(ab)$.

Our algorithm for checking path non-existence is based on adaptive decomposition of the configuration space. At each step, the number of decomposition depends on the number of the *mixed* cells indicated by the guiding path L.

5.3 Limitations

Our approach has a few limitations. Our free cell and C-obstacle cell queries are conservative, which stems from the conservativeness of PD^g and bounding motion computations. Secondly, our algorithm assumes that are no tangential contacts in the boundary of the free space, otherwise, our path non-existence algorithm may not terminate. As a result, our algorithm can not deal with compliant motion planning, where a robot cannot pass through obstacles when the robot is not allowed to touch them. The complexity of our adaptive subdivision algorithm varies as a function of the dimension of the configuration space. Our current implementation is limited to 3 DOF robots.

6 Conclusion and Future Work

In this paper, we present a simple approach to check for path non-existence for low DOF robots. Our approach uses two basic queries to efficiently check whether a cell in C-space lies entirely inside free space (free cell query) or inside the C-obstacle region (C-obstacle cell query). We describe simple and efficient algorithms to perform these queries using *separation distance* and *generalized penetration depth* computations. Our query algorithms are general for 2D or 3D rigid robots, or articulated robots. Using these queries, our approach for path non-existence computation is simpler and more efficient than prior cell decomposition methods.

There are several directions to pursue for future work. We are interested in extending our approach for higher DOF motion planning problems, such as 6 DOF rigid robots. Our algorithms to compute various queries are directly applicable and the main challenge is to perform spatial cell decomposition in higher dimensions. Moreover, we are interested in combining our algorithm with probabilistic sampling algorithms to design a hybrid planner, which is not only able to find a collision-free path, but can also check for path non-existence and handle narrow passages.

Acknowledgment

This project was supported in part by ARO Contracts DAAD19-02-1-0390 and W911NF-04-1-0088, NSF awards 0400134 and 0118743, ONR Contract N00014-01-1-0496, DARPA/RDECOM Contract N61339-04-C-0043 and Intel. Young J. Kim was supported in part by the grant 2004-205-D00168 of KRF, the STAR program of MOST and the ITRC program.

References

1. Amato, N., Bayazit, O., Dale, L., Jones, C., Vallejo, D.: Obprm: An obstacle-based prm for 3d workspaces. In: Proceedings of WAFR 1998, pp. 197–204 (1998)
2. Avnaim, F., Boissonnat, J.-D.: Practical exact motion planning of a class of robots with three degrees of freedom. In: Proc. of Canadian Conference on Computational Geometry, p. 19 (1989)
3. Basch, J., Guibas, L.J., Hsu, D., Nguyen, A.T.: Disconnection proofs for motion planning. In: Proc. IEEE International Conference on Robotics and Automation (2001)
4. Brooks, R.A., Lozano-Pérez, T.: A subdivision algorithm in configuration space for findpath with rotation. IEEE Trans. Syst. SMC-15, 224–233 (1985)
5. Cameron, S.: Enhancing GJK: Computing minimum and penetration distance between convex polyhedra. In: IEEE International Conference on Robotics and Automation, pp. 3112–3117 (1997)
6. Canny, J.: The Complexity of Robot Motion Planning. ACM Doctoral Dissertation Award. MIT Press, Cambridge (1988)
7. Choset, H., Lynch, K., Hutchinson, S., Kantor, G., Burgard, W., Kavraki, L., Thrun, S.: Principles of Robot Motion: Theory, Algorithms, and Implementations. The MIT Press, Cambridge
8. Donald, B.R.: Motion planning with six degrees of freedom. Master's thesis, MIT Artificial Intelligence Lab, AI-TR-791 (1984)
9. Halperin, D.: Robust geometric computing in motion. International Journal of Robotics Research 21(3), 219–232 (2002)
10. Hsu, D., Kavraki, L., Latombe, J., Motwani, R., Sorkin, S.: On finding narrow passages with probabilistic roadmap planners. In: Proc. of 3rd Workshop on Algorithmic Foundations of Robotics, pp. 25–32 (1998)

11. Kavraki, L., Svestka, P., Latombe, J.C., Overmars, M.: Probabilistic roadmaps for path planning in high-dimensional configuration spaces. IEEE Trans. Robot. Automat. 12(4), 566–580 (1996)
12. Kedem, K., Sharir, M.: An automatic motion planning system for a convex polygonal mobile robot in 2-d polygonal space. In: ACM Symposium on Computational Geometry, pp. 329–340 (1988)
13. Kim, Y., Lin, M., Manocha, D.: Deep: Dual-space expansion for estimating penetration depth between convex polytopes. In: Proc. IEEE International Conference on Robotics and Automation (May 2002)
14. Latombe, J.: Robot Motion Planning. Kluwer Academic Publishers, Dordrecht (1991)
15. LaValle, S.M.: Planning Algorithms. Cambridge University Press, Cambridge (2006), http://msl.cs.uiuc.edu/planning/
16. Lozano-Pérez, T.: Spatial planning: A configuration space approach. IEEE Trans. Comput. C-32, 108–120 (1983)
17. Lozano-Pérez, T., Wesley, M.: An algorithm for planning collision-free paths among polyhedral obstacles. Comm. ACM 22(10), 560–570 (1979)
18. Paden, B., Mess, A., Fisher, M.: Path planning using a jacobian-based freespace generation algorithm. In: Proceedings of International Conference on Robotics and Automation (1989)
19. Pisula, C., Hoff, K., Lin, M., Manocha, D.: Randomized path planning for a rigid body based on hardware accelerated voronoi sampling. In: Proc. of 4th International Workshop on Algorithmic Foundations of Robotics (2000)
20. Schwartz, J.T., Sharir, M.: On the piano movers probelem ii, general techniques for computing topological properties of real algebraic manifolds. Advances of Applied Maths 4, 298–351 (1983)
21. Schwarzer, F., Saha, M., Latombe, J.: Adaptive dynamic collision checking for single and multiple articulated robots in complex environments. IEEE Tr. on Robotics 21(3), 338–353 (2005)
22. Simeon, T., Laumond, J.P., Nissoux, C.: Visibility based probabilistic roadmaps for motion planning. Advanced Robotics Journal 14(6) (2000)
23. van den Bergen, G.: Proximity queries and penetration depth computation on 3d game objects. In: Game Developers Conference (2001)
24. Varadhan, G., Manocha, D.: Star-shaped roadmaps - a deterministic sampling approach for complete motion planning. In: Proceedings of Robotics: Science and Systems, Cambridge, USA (June 2005)
25. Zhang, L., Kim, Y., Varadhan, G., Manocha, D.: Generalized penetration depth computation. In: ACM Solid and Physical Modeling Symposium (SPM 2006), pp. 173–184 (2006)
26. Zhang, L., Kim, Y., Varadhan, G., Manocha, D.: Fast c-obstacle query computation for motion planning. In: IEEE International Conference on Robotics and Automation (ICRA 2006), pp. 3035–3040 (2006)
27. Zhu, D., Latombe, J.: Constraint reformulation in a hierarchical path planner. In: Proceedings of International Conference on Robotics and Automation, pp. 1918–1923 (1990)
28. Zhu, D., Latombe, J.: New heuristic algorithms for efficient hierarchical path planning. IEEE Trans. on Robotics and Automation 7(1), 9–20 (1991)

RESAMPL: A Region-Sensitive Adaptive Motion Planner

Samuel Rodriguez, Shawna Thomas, Roger Pearce, and Nancy M. Amato

Parasol Lab, Department of Computer Science, Texas A&M University, College Station, TX USA
{sor8786,sthomas,rap2317,amato}@cs.tamu.edu

Abstract. Automatic motion planning has applications ranging from traditional robotics to computer-aided design to computational biology and chemistry. While randomized planners, such as probabilistic roadmap methods (PRMs) or rapidly-exploring random trees (RRT), have been highly successful in solving many high degree of freedom problems, there are still many scenarios in which we need better methods, e.g., problems involving narrow passages or which contain multiple regions that are best suited to different planners.

In this work, we present RESAMPL, a motion planning strategy that uses local region information to make intelligent decisions about how and where to sample, which samples to connect together, and to find paths through the environment. Briefly, RESAMPL classifies regions based on the *entropy* of the samples in it, and then uses these classifications to further refine the sampling. Regions are placed in a region graph that encodes relationships between regions, e.g., edges correspond to overlapping regions. The strategy for connecting samples is guided by the region graph, and can be exploited in both multi-query and single-query scenarios. Our experimental results comparing RESAMPL to previous multi-query and single-query methods show that RESAMPL is generally significantly faster and also usually requires fewer samples to solve the problem.

1 Introduction

The general *motion planning* problem consists of finding a valid path for an object from a start configuration to a goal configuration. Traditionally, a valid path is any path that is collision-free, e.g., avoiding collision with obstacles in the environment and avoiding self-collision. Motion planning has applications in robotics, games/virtual reality, computer-aided design (CAD), virtual prototyping, and bioinformatics.

While an exact motion planning algorithm exists, its complexity grows exponentially in the complexity of the robot [17]. Instead, research has turned towards randomized algorithms. One widely used and quite successful randomized algorithm is the Probabilistic Roadmap Method (PRM) [11]. PRMs operate in *configuration space* (C-space), where each point in C-space corresponds to a specific robot configuration/placement. While not guaranteed to find a solution,

S. Akella et al. (Eds.): Algorithmic Foundation of Robotics VII, STAR 47, pp. 285–300, 2008.
springerlink.com

PRMs are probabilistically complete, i.e., the probability of finding a solution given one exists approaches 1 as the number of samples in the roadmap approaches ∞.

Issues: The motion planning problem is significantly more challenging when there are difficult or narrow areas in C-space that must be explored. While there have been many attempts to generate samples in difficult or interesting areas of C-space [1, 4, 21, 7, 5], they are typically applied over the entire C-space and do not allow for the identification and refinement of particular areas of C-space.

Motion planning problems typically come in one of two types: multi-query path planning and single-query path planning. The goal of a multi-query planner is to efficiently model the entire free C-space so as to answer any query in that space. A single-query planner, however, is only concerned about the portion of free C-space needed for the query, so it is generally faster than a multi-query planner. Most randomized motion planners are well-suited to one of these problem types, but not to both.

Our Contribution: In this work, we propose RESAMPL, a motion planning strategy that uses local region information to make intelligent decisions about how and where to sample, which samples to connect together, and to find paths through the environment. Based on an initial set of samples, we classify regions of C-space according to the *entropy* of their samples. We then use these classifications to further refine the sampling. For example, we increase sampling in "narrow" regions and decrease sampling in "free" regions. Regions are placed in a region graph that encodes relationships between regions, e.g., edges correspond to overlapping regions. We use the region graph to determine appropriate connection strategies for multi-query planning and to extract a sequence of regions on which to focus sampling and connection for single-query planning.

Our experimental results comparing RESAMPL to previous multi-query and single-query methods show that it is generally significantly faster and also usually requires fewer samples to solve the problem. Hence, RESAMPL's region-based approach to motion planning addresses both issues mentioned above.

- Regions: Considering local information when deciding where and how to refine sampling and connection enables us to focus on difficult areas instead of continuously searching in the entire space as is done by most previous methods.
- Region Graph: The relationships between regions can be exploited during connection in both multi-query and single-query situations.

2 Related Work

There has been extensive work on randomized motion planners for both multi-query and single-query problems. In this section we give an overview of some of the methods that have been proposed.

Multi-Query Planning. One widely used and quite successful multi-query randomized planner is the Probabilistic Roadmap Method (PRM) [11]. PRMs consist of two phases, a preprocessing/roadmap construction phase and a query phase. During roadmap construction, robot configurations are first randomly sampled from C-space. Samples are kept if they are in the feasible region of C-space (C-free). Connections are then attempted using a simple local planner between neighboring configurations. Valid connections are stored as edges in the roadmap.

Although PRMs have been successful in solving previously unsolvable problems, they have difficulty when the solution path must pass through a narrow passage in the C-space. Attempts have been made to generate configurations in interesting areas of C-space that are difficult to discover using uniform random sampling. For example, [1,4,9,16] attempt to generate samples near the surface of C-space obstacles. In [21], samples are generated and then pushed toward the approximate medial axis of C-free, and in [7], the roadmap is generated from a discrete approximation of the workspace medial axis.

In addition, machine learning techniques have been used to improve planner performance. In [14], regions of C-space are classified as either free, cluttered, narrow, or non-homogeneous using features obtained from a coarse sampling and a decision tree. Regions classified as non-homogeneous are further subdivided until properly classified or a maximum number of subdivisions has occurred. Specific node generation methods that were manually selected to work well in a given type of region are then applied in each region.

In [5, 6], entropy is used to build a model of C-space. To generate a new sample, the expected information gain is computed over a set of random samples. The sample with the greatest information gain is added to the model and also added to the roadmap if it is valid. An important difference from our work is that sampling and evaluation is done on a global basis, rather than focusing on particular regions.

Adaptive sampling [10] is proposed to select node generation methods in order to generate nodes that have been classified as more useful. Again, node generation is done on a global level reducing the likelihood of generating nodes in the narrow passages of C-space.

Finally, [20] is a complete, deterministic planner that partitions the free space into star-shaped regions such that a single sample can see every point in the region. Then, these samples are connected together to form a roadmap for planning. This method performs well for low dof robots, but because the complexity grows exponentially with the robot's dof, it may be impractical for high dof robots.

Single-Query Planning. Various single-query techniques have been developed that attempt to limit planning to the portions of the environment needed to solve the query. RRT (Rapidly-Exploring Random Tree) [13] is a tree-based method that attempts to explore C-space beginning from a start configuration until it reaches the goal configuration. The tree grows by biasing sampling towards unexplored regions. In [12], a variation to RRT was developed that biases the growth of two trees initiated from the start and the goal configurations toward each other for faster solution of a particular query.

Lazy Evaluation Methods. Several PRM variants have been proposed that delay some or all node/edge validation until they are needed in the query phase. These methods can be used as multi-query or single-query methods. Lazy PRM [3] initially assumes all nodes and edges to be valid during roadmap construction. To process a query, nodes and edges are checked. Invalid portions are removed from the roadmap and a new path is extracted. This repeats until a valid path is found or a path no longer exists in the roadmap. Fuzzy PRM [15] validates nodes during roadmap construction but postpones edge validation until the query phase. It uses a priority-based evaluation scheme to validate edges along the path. Finally, Customizable PRM [19] performs a coarse validation of nodes and edges during roadmap construction and completely validates nodes and edges as necessary to solve the query.

3 Model Overview

Our general strategy is to learn about local regions of C-space and to exploit that information during planning. For example, regions are classified to determine how they should be treated in other planning phases such as node generation or connection. The local regions may also be put in a *region graph* that approximately describes the connectivity of the C-space. This region graph can be used in both node connection and in single-shot planning.

To improve an existing model, our objective is to identify regions where additional sampling will lead to significant gains in C-space knowledge. For example, transition areas between C-free and C-obstacles may represent areas on the surface of C-obstacles or narrow passages in C-space. We want to identify these transition areas and bias our sampling to increase our knowledge of these areas. Similarly, we can limit sampling in regions that are completely in C-free or in C-obstacles as more samples in these areas will be unlikely to yield benefit. In this way, we focus on areas of C-space that are interesting in both node generation and connection. In this section we describe how the model is constructed, regions are classified, and sampling is both biased and filtered. An example of how our method works can be seen in Figure 1.

3.1 Region Construction

The model is initialized with a set of samples from the C-space, as in Figure 1(b), including both free and collision configurations. These samples may be generated by any method. Regions are then defined by a representative sample (e.g., the center of the region) and neighboring samples (used to compute region statistics such as the entropy and radius), as in Figure 1(c).

There are many ways to construct a set of regions. Algorithm 3.1 describes a simple region construction technique. Each new region center is randomly selected from the set of initial samples that are not already in another region.

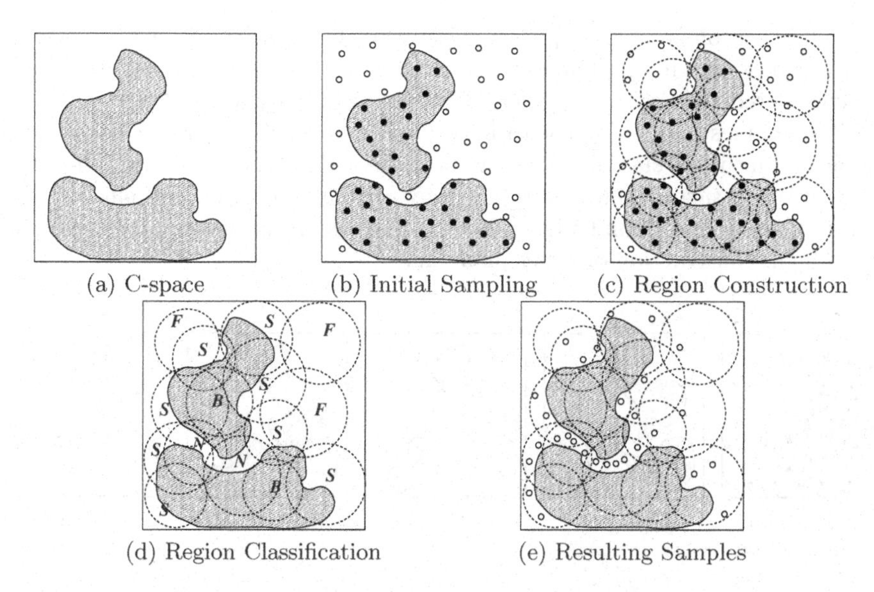

(a) C-space (b) Initial Sampling (c) Region Construction

(d) Region Classification (e) Resulting Samples

Fig. 1. Overview of model creation and usage. (a) Given an initial C-space, and (b) an initial sampling, (c) local regions can be constructed and (d) classified as free (F), blocked (B), surface (S), or narrow (N). Region classification results in (e) further sampling or filtering.

Neighboring samples are selected from all initial samples. Samples may be selected as part of multiple regions. In this way, the region radii are relatively similar. Region construction is complete when each sample is either a region center, part of a region, or both. As is discussed below, the quality of this type of region construction is somewhat dependent on the initial sample coverage.

Algorithm 3.1. Region Construction

Require: Model $\mathcal{M}$, initial samples S, and k.
1: **while** there exists an unmarked sample in S **do**
2: Let c be a randomly selected unmarked sample $\in S$.
3: Set $N = \{k$ nearest neighbors to $c\}$.
4: Set $R =$ a new region with center c and neighbors N.
5: Add R to $\mathcal{M}$.
6: Flag c and N as marked.
7: **end while**
8: **return** $\mathcal{M}$

3.2 Entropy Biased Region Classification

In order to identify the transition regions of C-space, we need a model that calculates how "interesting" the region is. We use the region's *entropy* to determine if it is a transition area. Entropy is a measure of the disorder of the region's samples. Regions containing samples that are completely free or completely blocked are considered to have *low entropy*. Regions containing a mixture of free and blocked samples are considered to have *high entropy*. As described below, four simple and intuitive classifications can be obtained based on entropy values: *free* (low entropy), *surface* (high entropy), *narrow* (high entropy), and *blocked* (low entropy), see Figure 2 and Figure 1(d). These classifications can later be used in roadmap construction or other planning.

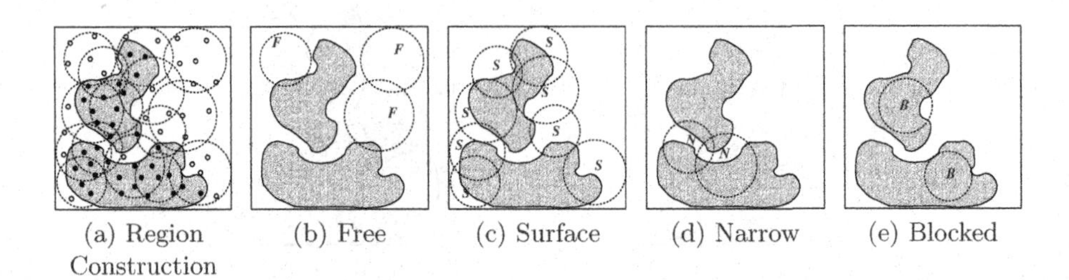

(a) Region Construction (b) Free (c) Surface (d) Narrow (e) Blocked

Fig. 2. Classifications based off of region construction

Algorithm 3.2 describes one way to classify regions based on entropy. For each region, we iteratively evaluate the region's entropy, attempt to classify, and add additional samples if a classification cannot be made.

Free regions can be identified by computing the percentage of blocked samples in the region. When this percentage (or entropy) is low enough, the region is classified as free. Experience indicates that it is unlikely to misclassify a region as free with this method. If the initial, coarse sampling in a region contains mostly free samples, then it is likely that a finer sampling will also contain mostly free samples. Thus, in every iteration, we first attempt to classify the region as free.

Blocked regions can be identified in a similar manner, i.e., if the percentage of free samples in the region (or entropy) is low enough, the region is classified as blocked. Note that unlike a *low entropy* free region, a *low entropy* blocked region should not automatically be considered blocked and then disregarded. This is because a blocked region could potentially become a *high entropy* region with additional sampling, e.g., when the region contains some volume of C-free which has not yet been sampled. For example, see Figure 1(c) and 1(d) in which a region constructed does not initially contain any free nodes, but it is classified as narrow since free nodes are discovered during the classification process. Thus, we do not classify a region as blocked until several attempts have been made to classify and add additional samples.

A region is classified as surface if sub-regions within the given region have *low entropy*. One way to define the sub-regions is as follows. Let c_F be the centroid of all the free samples and c_B be the centroid of all the blocked samples in the parent region. We then define two regions with centers c_F and c_B and we assign each sample in the parent region to the sub-region whose center it is closest to.

Algorithm 3.2. Region Classification

Require: A region R, threshold e_{low}, threshold e_{high}, number of attempts to classify t, and number of samples to add in each classification attempt k.
1: **for** t attempts to classify R **do**
2: Let e_R be the entropy of R (% of blocked samples in R).
3: **if** $e_R < e_{low}$ **then**
4: **return** *free*
5: **end if**
6: Add k additional samples to R and recompute e_R.
7: Partition R into two subregions, R_{free} and $R_{blocked}$.
8: Let e_{free} be the entropy of R_{free} (% of blocked samples in R_{free}).
9: Let $e_{blocked}$ be the entropy of $R_{blocked}$ (% of free samples in $R_{blocked}$).
10: **if** $e_{free} < e_{low}$ and $e_{blocked} < e_{low}$ **then**
11: **return** *surface*
12: **end if**
13: **end for**
14: **if** $e_R == 1$ **then**
15: **return** *blocked*
16: **end if**
17: **if** $e_R > e_{high}$ **then**
18: **return** *narrow*
19: **end if**
20: **return** *surface*

Then, if both sub-regions have *low entropy*, we classify the parent region as a surface region.

Regions are classified as narrow if they are *high entropy* regions that cannot be partitioned into two *low entropy* regions. Like blocked regions, narrow regions are more difficult to classify because of the risk of misclassification. Thus, we do not attempt to classify a region as narrow until several attempts have been made to classify and add additional samples.

Finally, when a transition region cannot be classified as described above, then it is considered as a surface region. Empirical testing showed this was the best assignment for such regions.

3.3 Region Graph

To complete the model construction, we build a region graph that approximately describes the connectivity of the local C-space regions. In our current implementation, vertices correspond to regions and an edge is placed between two regions if they overlap. We assign an edge weight based on the types of regions connected. With this region graph, we can extract region paths to aid single-query planning or refine it to aid multiple-query planning.

The region graph may be refined by merging adjacent regions of the same type or splitting regions that were not clearly classified. Regions may be combined if the resulting parent region is also of the same type. In addition to resulting in fewer regions, region merging is useful in obtaining larger portions of C-space of the same type. This is important when adding or removing samples based on the region type.

4 Multi-query Planning

In multiple query planning, a single roadmap must support many varied queries so one desires a roadmap that efficiently characterizes the connectivity of as much of the free C-space as possible. For this type of planning, we construct the regions and region graph as outlined in Section 3.

To reduce roadmap construction costs, we only keep "important" samples from the regions in the roadmap. This greatly reduces construction time by focusing connection on difficult/narrow areas of C-space and less on large, open areas of C-space. We keep a sample in a free region with a low probability p_F, a sample in a surface region with a higher probability p_S, and a sample in a narrow region with a high probability p_N. We do not keep any samples from blocked regions since they do not contain any valid samples. We then perform a user-selected connection strategy only on these samples.

In addition, we can use the region classification to further improve the roadmap. For example, we can use RRT to explicitly explore narrow passages because we have already identified them with the region classification. Thus, for each connected component in a narrow region, we allow RRT to expand the component by a user-defined number of iterations. This exploits RRT's ability to rapidly search confined regions of C-space by starting it in the difficult to find narrow passages.

5 Single-Query Planning

Single-shot motion planning involves finding a path for a given query from a start to goal configuration. Ideally, it involves exploring only the portions of the space needed to solve the query. An effective single-shot planner should be able to focus on portions of the path that will be used to solve the query.

We are able to use the model that we have constructed to first find an approximate region path connecting the start and goal configurations. The region path is extracted from the region graph and approximates a path through regions that the robot should travel through to move from region to region. In the following we will describe how the paths are obtained and connected to result in a path for a given robot from a start to a goal configuration.

5.1 Path Extraction and Improvement

The first step in region path extraction is to find the regions that the start and goal configurations can connect to. The nearest unblocked (free, surface, or narrow) region that the start and the goal configurations can connect to are set as the start and end regions, respectively, of the region path. A path is then found through the region graph that connects the start and end regions. The region graph is weighted such that a path is extracted through unblocked (free, surface or narrow) regions if possible and uses blocked regions only if needed, see Figure 3(a). Blocked regions found in the path can be reclassified in order to have a continuous sequence of unblocked path regions.

The region path extracted as described above is simply a minimal path of neighboring regions. While it is generally simple to extract an actual path from the region path that connects two adjacent free regions, it can sometimes be difficult to extract an actual path when the region path passes through more difficult (surface, narrow or blocked) regions. To improve our ability to extract paths in the latter case, we apply a simple region path improvement step that expands the volume of the region path by including neighboring unblocked regions in difficult areas. In particular, given neighboring path regions R_i and R_{i+1}, region path improvement is achieved by including unblocked regions in the path that neighbor both R_i and R_{i+1}. If both R_i and R_{i+1} are classified as *free* regions, then the region path improvement step can be omitted. An example of this process can be seen in Figure 3 in which the resulting region path covers a larger volume in the difficult and narrow regions. Though this is a simple process, it was shown to be quite effective during the connection phase in our experiments.

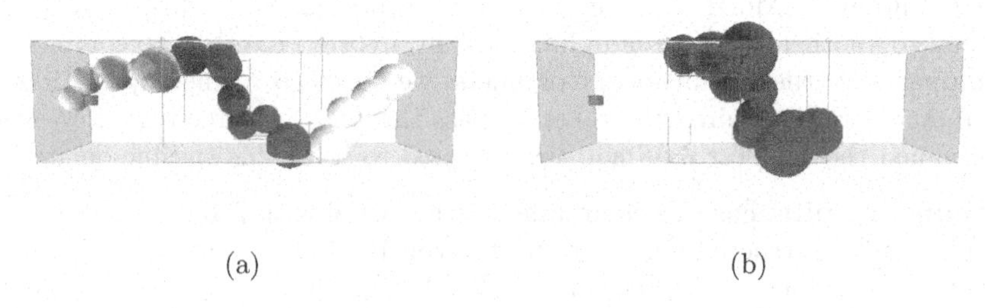

(a) (b)

Fig. 3. Region paths extracted from s-tunnel environment (a) a minimal path extracted and (b) an improved region path resulting in better connection

5.2 Path Connection

Although the connection strategy proposed here is very simple, it has proven sufficient for our purposes. For a given region path, nodes can be sampled as described in Section 4. The samples obtained can then be connected using a simple k-closest connection strategy. If necessary a simple component connection method can be applied that connects l-pairs from neighboring unconnected components.

As a final step, the path obtained should connect the start and goal configurations of the query. If a path cannot be found, then more connection attempts between neighboring unconnected regions can be attempted.

6 Results and Discussion

In this section we report on the performance of our region based motion planner as both a multi-query and a single-query planner. All planners were implemented using the Parasol Lab motion planning library developed at Texas A&M University. RAPID [8] is used to provide collision detection. Two types of local planners, straight-line and rotate-at-0.5 [2], are used to connect sampled configurations. Unless otherwise stated, connections were attempted only between $k = 20$ "nearby" nodes according to some selected distance metric. All experiments were run on a 700MHz Intel PIII Xeon processor and results are averaged over 10 runs.

6.1 Multi-query Planning

For multi-query planning, we tested two rigid body environments with narrow passages, L-Tunnel (Figure 4(a)), where traversing the passage requires mainly translational motion, and Hook (Figure 4(b)) where traversing the passage requires mainly orientational motion. We also tested an articulated linkage with 12 dof in an environment similar to the Hook environment (Figure 4(c)). We compare our method to some common PRM methods: uniform random sampling [11], obstacle-based sampling (OBPRM) [1], gauss-based sampling [4], medial axis-based sampling (MAPRM) [21], and bridge test sampling [9]. We also compare our method to another adaptive sampling method, hybrid PRM [10]. To compare the performance of these multi-query planners, we specified a single query in each environment that required the robot to pass through each free region. We then determined the smallest roadmap size required to solve this specific query.

L-Tunnel results. For this environment, we started with 2500 uniform random samples and constructed regions using the the 15 closest samples to the region center, as described in Algorithm 3.1. To classify the regions, we defined low entropy as 0.1 (i.e., at most 10% of the region samples are of one type and the remaining are of the other type) and attempted to classify each region at most 10 times by adding 45 random samples to the region. We then filtered the nodes by only keeping samples from narrow regions. We used the region graph to aid connection. Within each region, we perform a quick but sparse connection by attempting the $k = 2$ nearest unconnected neighbors. Then we connect components in overlapping regions (i.e., adjacent regions in the region graph) by attempting to connect the 5 closest pairs of samples between the regions. We then enhanced the roadmap by using RRT to further explore narrow regions and attempted 10 additional connections between these components. We determined the appropriate roadmap size to solve the query by varying the amount RRT could explore.

Figure 5(a-d) shows how our method classifies local regions as free, surface, narrow, and blocked. While not perfect, it is able to successfully identify the two key narrow passages in the central obstacle. Figure 5(e-f) shows the effect of exploiting region classification to increase and filter samples in local areas.

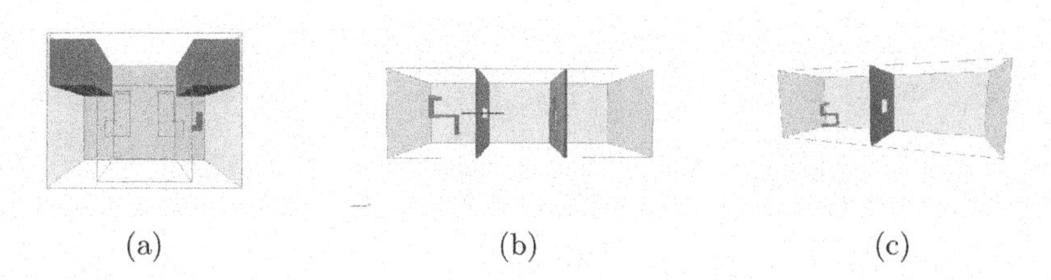

(a) (b) (c)

Fig. 4. (a) L-Tunnel environment. The robot (on the right) must pass through corridors in the central obstacle. (b) Hook environment. The robot (on the left) must twist through both plates to reach the other end of the environment. (c) Articulated linkage environment. The robot (on the left) must pass through the plate to reach the other end of the environment.

The initial distribution of uniform random samples cannot find any free samples in the two narrow passages and has oversampled the three large free regions. However, we then classified local regions based on these initial samples, reduced samples in free and surface regions, and increased samples in narrow regions. The resulting distribution is much more biased to highly constrained regions of the environment, namely the two narrow passages in the central obstacle and near the surfaces of all three obstacles.

Results for the L-Tunnel environment can be seen in Table 1. For this environment, uniform random sampling did not solve the query with 32,000 samples in any of the 10 runs. Region-based sampling was able to solve the query with the fewest nodes, time, and collision detection calls. While bridge test sampling requires approximately the same number of nodes, it takes nearly three times as long and roughly ten times more collision detection calls on average to find a solution. The local adaptable behavior of region-based sampling enables it to out-perform global adaptive strategies like hybrid PRM.

Hook results. Figure 6(a-d) shows how our method classifies local regions as free, surface, narrow, and blocked. We scaled the region diameters down for visualization clarity since regions are mostly defined by orientational differences than positional differences. Again, the method was able to successfully identity the different region types, even with a coarser model than the one used in the L-Tunnel environment. (Here, we reduced the number of initial model samples down to 400 and only added 25 samples to a region during a classification iteration; all other method parameters were kept the same.) Figure 6(e-f) shows how our method altered the initial sampling distribution to find more samples in the narrow passages.

Results for the hook environment can be seen in Table 1. For this environment, uniform random sampling was only able to solve the query 90% of the time requiring an extremely large number of samples. In terms of time, region-based sampling out-performed all methods except OBPRM, while making the fewest collision detection calls of all methods. This latter fact could prove significant in environments were collision detection is more expensive.

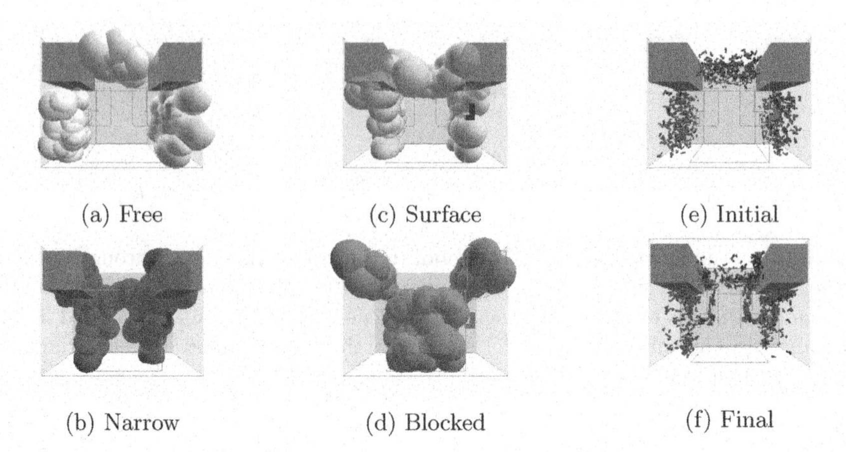

Fig. 5. (a-d) Region classification for the L-Tunnel environment. (e-f) Sampling distributions for L-Tunnel environment. The robot is scaled to 30% of its original size for visualization clarity.

Table 1. Multi-query planner performance

Multi-Query Planning					
Environment	**Method**	**Nodes**	**Time (s)**	**CD Calls**	**% Solved**
L-Tunnel	Region-Based	2,086	136	205,636	100
	OBPRM	8,400	1,036	354,706	100
	Gauss	6,150	672	356,177	100
	MAPRM	7,750	1,294	2,015,507	100
	Bridge Test	2,500	473	1,963,948	100
	Hybrid PRM	3,710	1,401	1,233,573	100
	Uniform Sampling	32,000	13,186	554,160	0
Hook	Region-Based	1,352	62	64,500	100
	OBPRM	925	36	175,711	100
	Gauss	1,625	68	66,953	100
	MAPRM	2,450	202	429,814	100
	Bridge Test	1,175	416	698,786	100
	Hybrid PRM	1,892	454	413,690	100
	Uniform Sampling	28,440	12,840	306,131	90
Articulated linkage	Region-Based	3,134	845	151,716	100
	OBPRM	30,600	29,826	1,406,023	10
	Gauss	32,000	31,994	1,440,563	10
	MAPRM	15,400	14,647	4,187,952	90
	Bridge Test	29,700	29,163	4,538,367	20
	Uniform Sampling	30,400	29,015	1,327,550	10

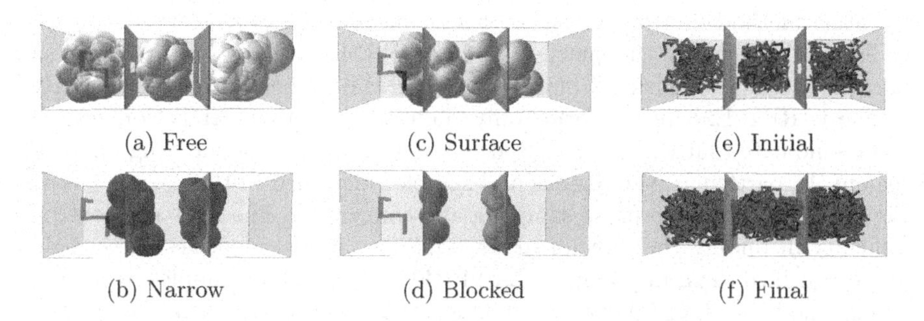

Fig. 6. (a-d) Region classification for the Hook environment. Region diameters are scaled down for visualization clarity since regions are mostly defined by orientational differences than positional differences. (e-f) Sampling distributions for Hook environment. The robot is scaled to 50% of its original size for visualization clarity.

Articulated linkage results. In this environment, we only used a straight-line local planner because rotate-at-0.5 [2] did not perform well. We used the same parameters as in the Hook environment to build a coarse model of C-space. Table 1 gives the results for this environment. Region-based sampling and MAPRM were the only methods to consistently solve the query. However, region-based sampling required significantly fewer nodes, collision detection calls, and consequently time. We believe that the reason why region-based sampling and MAPRM were the only methods to consistently solve the query is that these are the only two methods that target both free and narrow regions of C-space. In addition, because only a simple local planner was used, more pressure is placed on the sampling method to perform well. Note we did not compare to hybrid PRM because all of the component samplers performed poorly.

6.2 Single-Query Planning

In single-query planning, the planner tries to explore only the portions of C-space relevant for a given query. The approximate representation of C-free in our region-based approach allows us to focus our search in these relevant portions, offsetting the cost of building the initial model.

The planners tested for single-query planning are RRT Connect [12], RRT Expand [13], LazyPRM [3], and our region-based single-query planner. For each environment, each method is run until a given query can be solved. The environments tested for these single query methods are the S-Tunnel and Maze environment. Both of these environments have narrow passages that the robot must travel through when moving from the start to goal configuration.

S-Tunnel results. The S-Tunnel environment can be seen in Figure 3. The start and goal configurations are on opposite sides of the environment, such that the robot has to travel through the narrow passage connecting the query configurations. As seen in Table 2, the region-based single query planner outperforms the other methods in the number of nodes, time and collision detection calls (CD Calls) needed to solve the query. While LazyPRM and RRT Connect have similar performance in this environment, LazyPRM does perform better than RRT Connect. RRT Expand performs only slightly worse than RRT Connect. Our results indicate that the region-based strategy succeeds in identifying important regions. In particular, the regions and samples identified in the narrow passage, Figure 3, are used to improve sampling and connection. The RRT methods have difficulty in finding the difficult areas, while our model identifies these regions and can utilize them in planning. LazyPRM is able to find samples in these difficult regions but has difficulty in making valid connections when in the difficult region.

Table 2. Single-query planner performance

Single-Query Planning				
Environment	**Method**	**Nodes**	**Time (sec)**	**CD Calls**
S-Tunnel	Region-Based	573	36	82,298
	RRT Connect	7,939	958	360,513
	RRT Expand	7,774	1,170	473,735
	LazyPRM	1,173	609	361,889
Maze	Region-Based	334	32	27,493
	RRT Connect	2,131	79	48,775
	RRT Expand	2,947	187	69,639
	LazyPRM	561	364	127,747

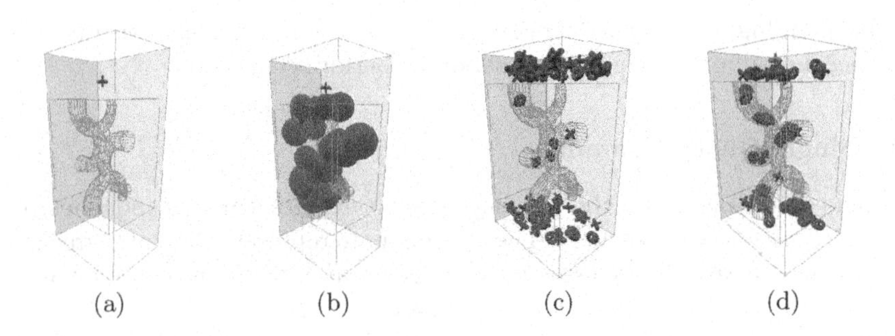

Fig. 7. (a) Maze environment. The robot must pass through a series of passages from the start to the goal configuration. (b) Narrow local regions of C-space found. (c) Initial collision-free sampling distribution and (d) final sampling distribution.

Maze results. The Maze environment (Figure 7(a)), consists of a series of passages that the robot must travel through from the start to the goal configuration. These configurations are on opposite ends of the maze. Though this environment is less difficult than the S-Tunnel environment, it is difficult for the RRT and LazyPRM methods. As seen in Table 2, utilizing information about the local regions of C-space enables our region-based method perform significantly better than the other methods. These important regions, extracted from the region map are shown in Figure 7(b). Here again, RRT Connect performs better than RRT Expand and the RRT approaches again have difficulty of finding the narrow passages. A difference in this case is that LazyPRM spends much more time trying to solve the query. Although LazyPRM only uses a small number of samples, it spends a large amount of time verifying edges and finding configurations in the narrow passage.

7 Conclusion

In this work we have shown how a region-based approach can be applied to both multi-query and single-query motion planning problems. It has also been shown to perform better, in many cases, than existing techniques. By focusing on regions that have been appropriately classified, we are able to better explore and sample the space. Additional figures are available in [18].

Acknowledgments. This research supported in part by NSF Grants EIA-0103742, ACR-0081510, ACR-0113971, CCR-0113974, ACI-0326350, and by the DOE. Rodriguez supported in part by a National Physical Sciences Consortium Fellowship. Thomas supported in part by an NSF Graduate Research Fellowship and a PEO Scholarship.

References

1. Amato, N.M., Bayazit, O.B., Dale, L.K., Jones, C.V., Vallejo, D.: OBPRM: An obstacle-based PRM for 3D workspaces. In: Robotics: The Algorithmic Perspective. Proc. Third Workshop on Algorithmic Foundations of Robotics (WAFR), Houston, TX, Natick, MA, pp. 155–168. A.K. Peters (1998)
2. Amato, N.M., Bayazit, O.B., Dale, L.K., Jones, C.V., Vallejo, D.: Choosing good distance metrics and local planners for probabilistic roadmap methods. IEEE Trans. Robot. Automat. 16(4), 442–447 (2000)
3. Bohlin, R., Kavraki, L.E.: Path planning using Lazy PRM. In: Proc. IEEE Int. Conf. Robot. Autom. (ICRA), pp. 521–528 (2000)
4. Boor, V., Overmars, M.H., van der Stappen, A.F.: The Gaussian sampling strategy for probabilistic roadmap planners. In: Proc. IEEE Int. Conf. Robot. Autom. (ICRA), vol. 2, pp. 1018–1023 (1999)
5. Burns, B., Brock, O.: Sampling-based motion planning using predictive models. In: Proc. IEEE Int. Conf. Robot. Autom. (ICRA) (2005)
6. Burns, B., Brock, O.: Toward optimal configuration space sampling. In: Proc. Robotics: Sci. Sys. (RSS) (2005)
7. Foskey, M., Garber, M., Lin, M., Manocha, D.: A voronoi-based hybrid motion planner. In: Proc. IEEE/RSJ International Conf. on Intelligent Robots and Systems (IROS 2001) (2001)
8. Gottschalk, S., Lin, M.C., Manocha, D.: OBB-tree: A hierarchical structure for rapid interference detection. Comput. Graph. 30, 171–180 (1996); Proc. SIGGRAPH 1996

9. Hsu, D., Jiang, T., Reif, J., Sun, Z.: Bridge test for sampling narrow passages with proabilistic roadmap planners. In: Proc. IEEE Int. Conf. Robot. Autom. (ICRA), pp. 4420–4426 (2003)
10. Hsu, D., Sánchez-Ante, G., Sun, Z.: Hybrid PRM sampling with a cost-sensitive adaptive strategy. In: Proc. IEEE Int. Conf. Robot. Autom. (ICRA), pp. 3885–3891 (2005)
11. Kavraki, L.E., Svestka, P., Latombe, J.C., Overmars, M.H.: Probabilistic roadmaps for path planning in high-dimensional configuration spaces. IEEE Trans. Robot. Automat. 12(4), 566–580 (1996)
12. Kuffner, J.J., LaValle, S.M.: RRT-Connect: An Efficient Approach to Single-Query Path Planning. In: Proc. IEEE Int. Conf. Robot. Autom. (ICRA), pp. 995–1001 (2000)
13. La Valle, S.M., Kuffner, J.J.: Rapidly-Exploring Random Trees: Progress and Prospects. In: Proc. Int. Workshop on Algorithmic Foundations of Robotics (WAFR), pp. SA45–SA59 (2000)
14. Morales, M., Tapia, L., Pearce, R., Rodriguez, S., Amato, N.M.: A machine learning approach for feature-sensitive motion planning. In: Proc. Int. Workshop on Algorithmic Foundations of Robotics (WAFR), Utrecht/Zeist, The Netherlands, pp. 316–376 (July 2004)
15. Nielsen, C.L., Kavraki, L.E.: A two level fuzzy PRM for manipulation planning. Technical Report TR2000-365, Computer Science, Rice University, Houston, TX (2000)
16. Redon, S., Lin, M.C.: Practical local planning in the contact space. In: Proc. IEEE Int. Conf. Robot. Autom. (ICRA) (April 2005)
17. Reif, J.H.: Complexity of the mover's problem and generalizations. In: Proc. IEEE Symp. Foundations of Computer Science (FOCS), San Juan, Puerto Rico, pp. 421–427 (October 1979)
18. Rodriguez, S., Thomas, S., Pearce, R., Amato, N.M.: Resampl: A region-sensitive adaptive motion planner. Technical Report TR06-004, Parasol Lab, Dept. of Computer Science, Texas A&M University (March 2006)
19. Song, G., Miller, S.L., Amato, N.M.: Customizing PRM roadmaps at query time. In: Proc. IEEE Int. Conf. Robot. Autom. (ICRA), pp. 1500–1505 (2001)
20. Varadhan, G., Manocha, D.: Star-shaped roadmaps: A deterministic sampling approach for complete motion planning. In: Proc. Robotics: Sci. Sys. (RSS) (2005)
21. Wilmarth, S.A., Amato, N.M., Stiller, P.F.: MAPRM: A probabilistic roadmap planner with sampling on the medial axis of the free space. In: Proc. IEEE Int. Conf. Robot. Autom. (ICRA), vol. 2, pp. 1024–1031 (1999)

Motion Planning for a Six-Legged Lunar Robot

Kris Hauser[1], Timothy Bretl[1], Jean-Claude Latombe[1], and Brian Wilcox[2]

[1] Computer Science Department, Stanford University
{khauser,tbretl}@stanford.edu,
latombe@cs.stanford.edu

[2] Jet Propulsion Laboratory, California Institute of Technology
Brian.H.Wilcox@jpl.nasa.gov

Abstract. This paper studies the motion of a large and highly mobile six-legged lunar vehicle called ATHLETE, developed by the Jet Propulsion Laboratory. This vehicle rolls on wheels when possible, but can use the wheels as feet to walk when necessary. While gaited walking may suffice for most situations, rough and steep terrain requires novel sequences of footsteps and postural adjustments that are specifically adapted to local geometric and physical properties. This paper presents a planner to compute these motions that combines graph searching techniques to generate a sequence of candidate footfalls with probabilistic sample-based planning to generate continuous motions to reach them. The viability of this approach is demonstrated in simulation on several example terrains, even one that requires rappelling.

1 Introduction

In this paper we describe the design and implementation of a motion planner for a six-legged lunar vehicle called ATHLETE (All-Terrain Hex-Limbed Extra-Terrestrial Explorer), shown in Fig 1. This large and highly mobile vehicle was developed by the Jet Propulsion Laboratory (JPL).[1] It can roll rapidly on rotating wheels over flat smooth terrain and walk carefully on fixed wheels over irregular and steep terrain. In particular, ATHLETE is designed to scramble across terrain so rough that a fixed gait (for example, an alternating tripod gait) may prove insufficient. Such terrain is abundant on the Moon, most of which is rough, mountainous, and heavily cratered – particularly in the polar regions, a likely target for future surface operations. These craters can be of enormous size, filled with scattered rocks and boulders of a few centimeters to several meters in diameter (Fig. 2). Crater walls are sloped at angles of between 10-45°, and sometimes have sharp rims [19].

On this type of terrain, ATHLETE's walking motion is governed largely by two interdependent constraints: *contact* (keep wheels, or *feet*, at a carefully chosen set of footfalls) and *equilibrium* (apply forces at these footfalls that exactly compensate for gravity without causing slip). The range of forces that may be applied at the footfalls without causing slip depends on their geometry (for example,

[1] The views presented in this paper do not reflect those of NASA or JPL.

S. Akella et al. (Eds.): Algorithmic Foundation of Robotics VII, STAR 47, pp. 301–316, 2008.
springerlink.com

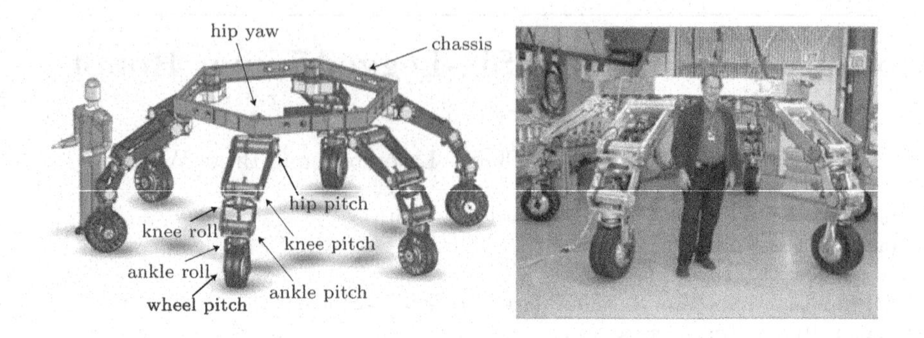

Fig. 1. The ATHLETE lunar vehicle (developed by JPL)

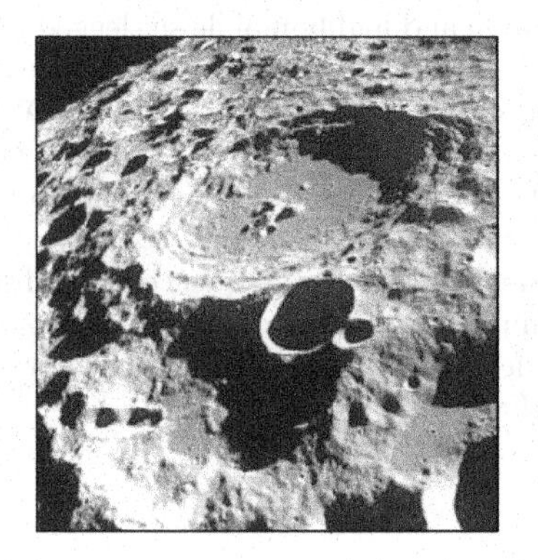

Fig. 2. Pictures of lunar terrain from Apollo missions [19]

average slope) and their physical properties (for example, coefficient of friction), both of which vary across the terrain. So every time ATHLETE takes a step, it faces a dilemma: it can't know the constraints on its subsequent motion until it chooses a footfall, a choice it can't make until it knows where it will step next. Direct teleoperation does not help to resolve this dilemma – on the contrary, teleoperation can be difficult and painfully slow for robots like ATHLETE [6].

To handle this dilemma in our planner, we make a key design choice (Section 3) – *to choose footfalls before computing motions.* We begin by identifying a number of potentially useful footfalls across the terrain. Each mapping of ATHLETE's feet to a set of footfalls is a *stance*, associated with a (possibly empty) set of feasible configurations that satisfy all motion constraints (including contact and equilibrium). ATHLETE can take a step from one stance to another if they differ by a single footfall and if they share some feasible configuration, which we call a *transition.* Our planner proceeds in two stages: first, we generate a candidate sequence of footfalls by finding transitions between stances; then, we refine this sequence into a feasible, continuous trajectory by finding paths between subsequent transitions. We do this because ATHLETE's motion on irregular and steep terrain is most constrained just as it places a foot at or removes a foot from a footfall. At this instant, ATHLETE must be able to reach the footfall (contact) but can not use it to avoid falling (equilibrium). So footfalls are the "bottleneck" of any motion – if we can find two subsequent transitions, it is likely we can find a path between them. This statement has been verified in our experiments.

We implement our planner using an approach similar to [6] and [18] that combines graph searching techniques to generate a sequence of candidate footfalls with probabilistic sample-based planning to generate continuous motions to reach them. But several key tools embedded in this framework (Section 4) are tailored specifically to ATHLETE. We need a method of sampling feasible configurations (from scratch as well as via perturbation) and of connecting pairs of configurations with local paths, hard since ATHLETE has many degrees of freedom and many closed-loop chains. We also need a heuristic to generate footfalls and to guide our search through the collection of stances, hard since lunar terrain is difficult (so careful selection of footfalls is important) but not extreme (so the number of candidate stances is enormous). Finally, we need to smooth ATHLETE's motion both to look natural when interacting with a human operator – hard since the robot is not anthropomorphic – and to help avoid disturbing the ground (for example, by toppling rock).

Simulation results (Section 5) demonstrate the viability of our approach. We also show the flexibility of our implementation by adapting it to rappelling as well as walking motions of ATHLETE.

2 Related Work

2.1 Application

Some humanoids are capable of walking over somewhat uneven terrain [49, 28]. Other legged robots are capable of walking over rougher terrain, including quadrupeds [20], hexapods [43, 26], parallel walkers [48], and spherically symmetric robots [35]. Wheeled robots with active or rocker-bogie suspension can also traverse rough terrain by changing wheel angles and center of mass position [14, 23, 29]. Careful descent is possible by rappelling as well, using either legs [3, 21, 46] or wheels [32]. The terrain we consider for ATHLETE is even more irregular and steep than in most previous applications, although not as steep as for free-climbing robots [6].

Careful walking also resembles dexterous manipulation. ATHLETE grasps the terrain like a hand grasps an object, placing and removing footfalls rather than finger contacts. ATHLETE has to remain in equilibrium as it moves (only the object must remain in equilibrium during manipulation), and uses fewer contact modes while walking (no sliding or rolling), but still faces similar challenges [4, 34]. Manipulation planning, involving the rearrangement of many objects with a simple manipulator, is another related application. A manipulator takes a sequence of motions with and without a grasped object (different states of contact) just like ATHLETE takes a sequence of steps [2].

2.2 Planning

In order to walk, ATHLETE must plan both a sequence of footfalls and continuous motions to reach them. Previous approaches differ primarily in which part of the problem they consider first:

(a) *Motion before footfalls.* When it does not matter much where a robot contacts its environment, it makes sense to compute the robot's (or object's) overall motion first. For example, a manipulation planner might generate a trajectory for the grasped object ignoring manipulators, then compute manipulator trajectories that achieve necessary re-grasps [25]. Similarly, a humanoid planner might generate a 2-D collision-free path of a bounding cylinder, then follow this path with a fixed gait [27, 36]. A related strategy is to plan a path for the center of mass, then to compute footfalls and limb motions that keep the center of mass stable [13]. These techniques are fast, but do not extend well to irregular and steep terrain.

(b) *Footfalls before motion.* When the choice of contact location is critical, it makes sense to compute a sequence of footfalls first. Most work is based on the approach to manipulation planning proposed by [2], which expresses connectivity between different states of contact as a graph. For "spider-robots" walking on horizontal terrain, the exact structure of this graph can be computed quickly using analytical techniques [5]. For more general systems, the graph can sometimes be simplified by assuming partial gaits, for example restricting the order in which limbs are moved [40] or restricting footsteps to a discrete set [28]. But when motion is distinctly non-gaited (as in manipulation planning [33, 37], free-climbing [6], or for ATHLETE), each step requires the exploration of configuration space. This motivates the two-stage search strategy we adopt in Section 3.

2.3 Key Tools

Each of the tools embedded in our planner improves and extends previous techniques to satisfy the specific needs of ATHLETE:

(a) *Sampling and local connection.* We use a variant of the Probabilistic-Roadmap (PRM) approach (see Chap. 7 of [11]) to generate transitions between stances (configurations that are feasible at both one stance and another) as well as paths between transitions. A PRM planner samples configurations at random, retaining feasible ones as milestones and connecting close milestones if possible with feasible local paths. Its performance depends on fast methods of sampling and local connection, either from scratch across all of configuration space [24] or via perturbation by growing trees from existing milestones [1, 22, 30]. Closed kinematic chains (ATHLETE has many) make both of these operations harder because there is zero probability that an arbitrary configuration will satisfy the closure constraints. One approach breaks chains into "active" and "passive" joints, sampling a configuration of the active joints and using analytical inverse kinematics to solve for the rest [12, 17]. Another approach uses numerical optimization to move a configuration onto the constraint manifold [18, 45, 47]. We use a combination of these two methods.

(b) *Heuristics for footfall selection.* A variety of heuristics have been proposed for estimating the usefulness of a footfall. Most are geometric criteria that determine how flat a footfall is [9, 10, 31]. On irregular and steep terrain, however, the usefulness of a footfall also depends on its location with respect to other footfalls – in particular, on how these footfalls are combined in each stance. We use these heuristics to guide the search for a candidate sequence of stances to reach a goal position, similar to [41, 10].

(c) *Path smoothing.* Paths generated by a PRM planner are feasible, but not necessarily optimal. A number of methods have been suggested to improve the result, including "short-cut" heuristics [24, 42] and gradient descent algorithms [44, 15]. We use a similar approach. But in addition to being safe and efficient, ATHLETE's motions must also "look good" to human operators.

3 Design of the Motion Planner

3.1 Motion Constraints

A *configuration* of ATHLETE, denoted q, is a parameterization of the robot's placement in 3-D space. In the following, q consists of 6 parameters defining the position and orientation of the robot's hexagonal chassis and a list of 36 joint angles (each leg has six actuated, revolute joints). The set of all such q is the *configuration space*, denoted $\mathcal{Q}$, of dimensionality 42.

When ATHLETE is walking, a brake is applied to each wheel so it can not roll. In this case, we call each wheel a *foot*. Whenever a foot is placed in contact with the terrain, we call this placement (the fixed position and orientation of a wheel in 3-D space) a *footfall*. Since all feet are identical, potentially any foot could be placed at any footfall. We call a specific mapping of feet to footfalls a *stance*. Consider a stance σ with $3 \leq N \leq 6$ footfalls (in general, at least three are required to achieve statically stable equilibrium). The feasible space $\mathcal{F}_\sigma$ is the set of all feasible configurations of the robot at stance σ. To be in $\mathcal{F}_\sigma$, a configuration q must satisfy several constraints:

(a) *Contact.* The N legs whose feet are in contact with the ground form a linkage with multiple closed-loop chains. So, q must satisfy inverse kinematic equations. Let $\mathcal{Q}_\sigma \subset \mathcal{Q}$ be the set of all configurations q that satisfy these equations. This set $\mathcal{Q}_\sigma$ is a sub-manifold of $\mathcal{Q}$ of dimensionality $42 - 6N$, which we call the *stance manifold.* This manifold is empty if it is impossible for the robot to achieve the contacts specified by σ, for example if two contact points are farther apart than the maximum span of two legs.

(b) *Static equilibrium.* To remain balanced, ATHLETE must be able to apply forces with its feet on the terrain that compensate for gravity without slipping. A necessary condition is that ATHLETE's center of mass (CM) lie above a *support polygon.* But on irregular and steep terrain, the support polygon does not always correspond to the base of ATHLETE's feet. For example, ATHLETE will slip off a flat and featureless slope that is too steep, regardless of its CM position. To compute the support polygon, we model the contact interface at each footfall as a frictional point. Let $r_1, \ldots, r_N \in \mathbb{R}^3$ be the position, $\nu_i \in \mathbb{R}^3$ be the normal vector, μ_i be the static coefficient of friction, and $f_i \in \mathbb{R}^3$ be the reaction force acting on the robot at each point. We decompose each force f_i into a component $\nu_i^T f_i \nu_i$ normal to the terrain surface (in the direction ν_i) and a component $(I - \nu_i \nu_i^T) f_i$ tangential to the surface. Let $c \in \mathbb{R}^3$ be the position of ATHLETE's CM (which varies with its configuration). Assume ATHLETE has mass m, and the acceleration due to gravity is $g \in \mathbb{R}^3$. All vectors are defined with respect to a global coordinate system with axes e_1, e_2, e_3, where $g = -\|g\| e_3$. Then ATHLETE is in static equilibrium if

$$\sum_{i=1}^{N} f_i + mg = 0 \qquad \text{(force balance)} \qquad (1)$$

$$\sum_{i=1}^{N} r_i \times f_i + c \times mg = 0 \qquad \text{(torque balance)} \qquad (2)$$

$$\|(I - \nu_i \nu_i^T) f_i\|_2 \leq \mu_i \nu_i^T f_i \text{ for all } i = 1, \ldots, N. \qquad \text{(friction cones)} \qquad (3)$$

These constraints are jointly convex in $f_1, \ldots, f_N$ and c. In particular, (1)-(2) are linear and (3) is a second-order cone constraint. In practice we approximate (3) by a polyhedral cone, so the set of jointly feasible contact forces and CM positions is a high-dimensional polyhedron [8, 6, 7]. Finally, since

$$c \times mg = m\|g\| \begin{bmatrix} -c \cdot e_2 \\ c \cdot e_1 \\ 0 \end{bmatrix}$$

then (1)-(2) do not depend on $c \cdot e_3$ (the CM coordinate parallel to gravity), so the support polygon is the projection of this polyhedron onto the coordinates e_1, e_2. There are many ways to compute this projection and to test the membership of c. An approach that works well for our application is [7].

(c) *Joint torque limits.* The above equilibrium test assumes ATHLETE is a rigid body, "frozen" at configuration q. In reality, to maintain q each joint must exert a torque, which in turn must not exceed a given bound. Let τ be the vector of all joint torques exerted by the robot, and let $\|\cdot\|$ be a weighted L_∞ norm where $\|\tau\| < 1$ implies that each joint torque is within bounds. Then we check joint torque limits by computing τ that achieves equilibrium with minimum $\|\tau\|$ (a linear program), and verify $\|\tau\| < 1$.

(d) *Collision.* In addition to satisfying joint angle limits, the robot must avoid collision with the environment (except at contact points) and with itself. We use techniques based on bounding volume hierarchies to perform collision checking, as in [16, 39].

3.2 Two-Stage Search

To walk from once place to another, ATHLETE has to take a sequence of steps. Formally, we define a *step* as any continuous motion at a fixed stance that

```
EXPLORE-STANCEGRAPH(q_initial, σ_initial, σ_final)
 1  Q ← {σ_initial}
 2  while Q is nonempty do
 3      unstack a node σ from Q
 4      if σ = σ_final then
 5          construct a path [σ_1, ..., σ_n] from σ_initial to σ_final
 6          i ← EXPLORE-TRANSITIONGRAPH(σ_1, ..., σ_n, q_initial)
 7          if i = n then
 8              return the multi-step motion
 9          else
10              delete the edge (σ_i, σ_{i+1}) from the stance graph
11      else
12          for each unexplored stance σ' adjacent to σ do
13              if FIND-TRANSITION(σ, σ') then
14                  add a node σ' and an edge (σ, σ')
15                  stack σ' in Q
16  return "failure"

EXPLORE-TRANSITIONGRAPH(σ_i, ..., σ_n, q)
1  i_max ← i
2  for q' ← FIND-TRANSITION(σ_i, σ_{i+1}) in each component of F_{σ_i} ∩ F_{σ_{i+1}} do
3      if FIND-PATH(σ_i, q, q') then
4          i_cur ← EXPLORE-TRANSITIONGRAPH(σ_{i+1}, ..., σ_n, q')
5          if i_cur = n then
6              return n
7          elseif i_cur > i_max then
8              i_max = i_cur
9  return i_max
```

Fig. 3. Algorithms to explore the stance graph and the transition graph

terminates by either placing or removing a foot. In particular, let σ and σ' be the stances before and after a step, respectively. Then this step is a continuous path from the robot's current configuration $q_{\text{initial}} \in \mathcal{F}_\sigma$ to some configuration $q_{\text{final}} \in \mathcal{F}_\sigma \cap \mathcal{F}_{\sigma'}$ that we call a *transition*. During this step, ATHLETE may move all legs simultaneously, but we assume that no two feet are placed or removed simultaneously. Therefore, σ and σ' differ only by a single footfall, which is present in only one of the two stances.

We encode the connectivity among stances as a *stance graph*. Each node of this graph is a stance. Two nodes σ and σ' are connected by an edge if there is a transition between $\mathcal{F}_\sigma$ and $\mathcal{F}_{\sigma'}$. So the existence of an edge in the stance graph is a necessary condition for ATHLETE to take a step from one stance to another. Both necessary and sufficient conditions are provided by a *transition graph*. Each node of this graph is a transition. Two nodes $q \in \mathcal{F}_\sigma \cap \mathcal{F}_{\sigma'}$ and $q' \in \mathcal{F}_\sigma \cap \mathcal{F}_{\sigma''}$ are connected by an edge if there is a continuous path between them in $\mathcal{F}_\sigma$. The stance and transition graphs represent the connectivity of ATHLETE's configuration space at coarse and fine resolutions, respectively.

Our planner interweaves exploration of the stance graph and the transition graph, based on the method of [6]. The algorithm EXPLORE-STANCEGRAPH searches the stance graph (Fig. 3). It maintains a priority queue Q of nodes to explore. When it unstacks σ_{final}, it computes a candidate sequence of nodes and edges from σ_{initial}. The algorithm EXPLORE-TRANSITIONGRAPH verifies that this candidate sequence corresponds to a feasible motion by searching a subset of the transition graph (Fig. 3). It explores a transition $q \in \mathcal{F}_\sigma \cap \mathcal{F}_{\sigma'}$ only if (σ, σ') is an edge along the candidate sequence, and a path between $q, q' \in \mathcal{F}_\sigma$ only if σ is a node along this sequence. We say that EXPLORE-TRANSITIONGRAPH has *reached* a stance σ_i if some transition $q \in \mathcal{F}_{\sigma_{i-1}} \cap \mathcal{F}_{\sigma_i}$ is connected to q_{initial} in the transition graph. The algorithm returns the index i of the farthest stance reached along the candidate sequence. If this is not σ_{final}, then the edge (σ_i, σ_{i+1}) is removed from the stance graph, and EXPLORE-STANCEGRAPH resumes exploration.

The effect of this two-stage search strategy is to postpone the generation of one-step paths (a costly computation) until after generating transitions. It works well because, as we mentioned in Section 1, ATHLETE's motion on irregular and steep terrain is most constrained just as it places or removes a foot. In our experiments we have observed that if we can find $q \in \mathcal{F}_\sigma \cap \mathcal{F}_{\sigma'}$ and $q' \in \mathcal{F}_\sigma \cap \mathcal{F}_{\sigma''}$, then a path between q and q' likely exists in $\mathcal{F}_\sigma$.

A number of tools are embedded in this framework (the subroutines FIND-TRANSITION and FIND-PATH, a heuristic for ordering Q, and a method of smoothing the resulting motion) that we discuss in the following section.

4 Tools to Support the Motion Planner

4.1 Generating Transitions

Both EXPLORE-STANCEGRAPH and EXPLORE-TRANSITIONGRAPH require the subroutine FIND-TRANSITION to generate transitions $q \in \mathcal{F}_\sigma \cap \mathcal{F}_{\sigma'}$ between pairs of stances σ and σ'. To implement FIND-TRANSITION, we use a sample-based approach. The basic idea is to sample configurations randomly in $q \in \mathcal{Q}$ and reject them if they are not in $\mathcal{F}_\sigma \cap \mathcal{F}_{\sigma'}$. But since $\mathcal{Q}_\sigma$ has zero measure in $\mathcal{Q}$, this approach will never generate a feasible transition. So like [12, 45, 47], we spend more time trying to generate configurations that satisfy the contact constraint at σ (hence, at σ' if $\sigma' \subset \sigma$) before rejecting those that do not satisfy other constraints. Like [18], we do this in two steps:

(a) *Create a candidate configuration that is close to $\mathcal{Q}_\sigma$.* First, we create a nominal position and orientation of the chassis: (1) given a stance σ, we fit a plane to the footfalls in a least-squares sense; (2) we place the chassis in this plane, minimizing the distance from each hip to its corresponding footfall; (3) we move the chassis a nominal distance parallel to the plane-fit and away from the terrain. Then, we sample a position and orientation of the chassis in a Gaussian distribution about this nominal placement. Finally, we compute the set of joint angles that either reach or come closest to reaching each footfall. Note that a footfall fixes the intersection of the ankle pitch and ankle roll joints relative to the chassis (Fig. 1). The hip yaw, hip pitch, and knee pitch joints determine this position. There are up to four inverse kinematic solutions for these joints – or, if no solutions exist, there are two configurations that are closest (straight-knee and completely bent-knee). The knee roll, ankle roll, and ankle pitch determine the orientation of the foot, for which there are two inverse kinematic solutions. We select a configuration that satisfies joint-limit constraints; if none exists, we reject the sample and repeat.

(b) *Repair the candidate configuration using numerical inverse kinematics.* We move the candidate configuration to a point in $\mathcal{Q}_\sigma$ using an iterative Newton-Raphson method. We represent the error in position and orientation of each foot i as a differentiable function $f_i(q)$ of the configuration q. Let

$$g(q) = \begin{bmatrix} f_1(q) \\ \vdots \\ f_N(q) \end{bmatrix}$$

so we can write the contact constraint as the equality $g(q) = 0$. Assume we are given a candidate configuration q_1. Then at each iteration k, we transform this configuration by taking the step

$$q_{k+1} = q_k - \alpha_k \nabla g(q_k)^{-\dagger} g(q_k),$$

where $\nabla g(q_k)^{-\dagger}$ is the pseudo-inverse of the gradient of the error function, and α_k is the step size (computed using backtracking line search). The algorithm terminates with success if at some iteration $\|g(q_k)\| < \varepsilon$ for some tolerance ε, or with failure if a maximum number of iterations is exceeded.

The first step rarely generates configurations in $\mathcal{Q}_\sigma$, but it quickly generates configurations that are close to $\mathcal{Q}_\sigma$. Conversely, the primary cost of the second step is in computing $\nabla g(q_k)^{-\dagger}$ at every iteration, but if candidate configurations are sufficiently close to $\mathcal{Q}_\sigma$ then few iterations are necessary. So, it is the combination of these two methods that makes our sampler fast.

Note that EXPLORE-TRANSITIONGRAPH additionally requires that we sample a single transition in each connected component of $\mathcal{F}_\sigma \cap \mathcal{F}_{\sigma'}$. Our approach is not guaranteed to do this, but the probability that it samples at least one in each component increases with the number of samples.

4.2 Generating Paths between Transitions

EXPLORE-TRANSITIONGRAPH requires the subroutine FIND-PATH to generate paths in $\mathcal{F}_\sigma$ between pairs of transitions $q \in \mathcal{F}_\sigma \cap \mathcal{F}_{\sigma'}$ and $q' \in \mathcal{F}_\sigma \cap \mathcal{F}_{\sigma''}$. We use a variant of the probabilistic roadmap approach called SBL that is bi-directional (growing trees from both q and q') and lazy (delaying the creation of local paths until a candidate sequence of milestones is found) [38].

```
FREE-PATH(q, q')
 1  if the distance from q to q' is less than ε then
 2      return TRUE
 3  q_mid ← (q + q')/2
 4  if Newton-Raphson from q_mid results in q_mid ∈ Q_σ then
 5      if q_mid ∈ F_σ then
 6          return (FREE-PATH(q, q_mid) & FREE-PATH(q_mid, q'))
 7      else
 8          return FALSE
 9  else
10      return FALSE
```

Fig. 4. Algorithm to connect close configurations with a local path

To sample configurations in $\mathcal{F}_\sigma$, we face the same challenge discussed in the previous section (that a random configuration has zero probability of being in $\mathcal{Q}_\sigma$), and so we use a similar approach. However, in this case we can focus our search on a small part of feasible space, near existing milestones in each tree of the roadmap. Rather than sample a candidate configuration $q \in \mathcal{Q}$ at random, we sample it in a neighborhood of an existing configuration q_0. Close to q_0, the shape of $\mathcal{Q}_\sigma$ is approximated well by the hyperplane

$$\{ p \in \mathcal{Q} \mid \nabla g(q_0)^T p = \nabla g(q_0)^T q_0 \}.$$

So before applying the iterative method to repair the sampled configuration, we first project it onto this hyperplane (as in [47]).

To connect milestones with local paths, we face a similar challenge, since the straight-line path between any two configurations q and q' will not (in general) lie in $\mathcal{Q}_\sigma$. So, we deform this straight-line path into $\mathcal{Q}_\sigma$ using the bisection method FREE-PATH (Fig. 4). At each iteration, FREE-PATH first applies Newton-Raphson (see Section 4.1) to the midpoint of q and q' to generate $q_{\text{mid}} \in \mathcal{Q}_\sigma$, then it checks that $q_{\text{mid}} \in \mathcal{F}_\sigma$. If both steps succeed, the algorithm continues to recurse until a desired resolution has been reached; otherwise, the algorithm returns failure. The advantage of this approach is that it does not require a direct local parameterization of $\mathcal{Q}_\sigma$, as it may be difficult to compute such a parameterization that covers both q and q'.

4.3 Ordering the Graph Search

Our two-stage search strategy can be improved by ordering the stances in Q according to a heuristic cost function $g(\sigma) + h(\sigma)$ in EXPLORE-STANCEGRAPH, where stances with lower cost are given higher priority. We define $g(\sigma)$ as the minimum number of steps required to reach σ from σ_{initial} in the stance graph. We define $h(\sigma)$ as a weighted sum of several criteria:

- *Planning time.* We increase the cost of σ proportional to the amount of time spent trying to sample a transition $q \in \mathcal{F}_{\sigma'} \cap \mathcal{F}_{\sigma}$ to reach it [33].
- *Distance to goal.* We increase the cost of σ proportional to the distance between the centroid of its footfalls and those of the goal stance σ_{final}.
- *Footfall distribution.* We increase the cost of σ proportional to the difference (in a least-squares sense) between its footfalls and those of a nominal stance on flat ground (with footfalls directly under each hip).
- *Equilibrium criteria.* We increase the cost of σ inversely proportional to the area of its support polygon.

This heuristic reduces planning time and improves the resulting motion. It also allows us to relax an implicit assumption – that FIND-TRANSITION and FIND-PATH always return "failure" correctly. Because we implement these subroutines using a probabilistic, sample-based approach, we are unable to distinguish between impossible and difficult queries. So on failure of FIND-TRANSITION in EXPLORE-STANCEGRAPH, we still add σ to the stance graph but give σ a high cost. Likewise, rather than delete (σ, σ') on failure of FIND-PATH, we increase the cost of σ and σ'.

4.4 Path Smoothing

Because we use probabilistic sample-based methods to sample transitions and plan paths between them, the motions we generate are feasible (given an accurate terrain model) but not necessarily high-quality. To improve the result, we apply a method of smoothing similar to [44, 15], which uses gradient descent to achieve criteria like minimum path length and maximum clearance (or safety margin). However, we modify this approach in two ways. First, ATHLETE's motion consists of a sequence of short paths (steps) through separate feasible spaces rather than a single path through one feasible space. We consider this entire sequence of paths at once (deforming transitions as well as paths) rather than each one individually. So during the optimization, different parts of ATHLETE's motion are subject to different constraints. Second, because ATHLETE is expected to interact with humans, we try to make its motion "look good" to human operators. We do this by allowing the operator to select, ahead of time, a small set of nominal configurations (for example, standing on six legs, standing on three legs, or crouching). Then, in addition to minimizing path length and maximizing clearance, we also minimize deviation from any point q along the path to the closest nominal configuration q'. Even a small number of iterations (taking about 10 minutes on a 2GHz PC) makes a noticeable difference in motion quality.

5 Implementation and Results

We tested our planner in simulation on several example terrains. Each terrain is a height-map of the form $z = f(x, y)$, created using a fractal generation method and represented by a triangular mesh consisting of 32768 triangles, each about

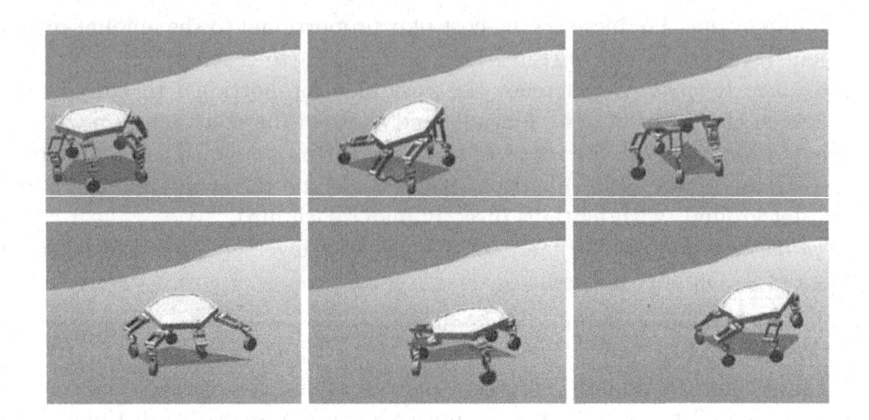

Fig. 5. Walking on smooth, undulating terrain with no fixed gait

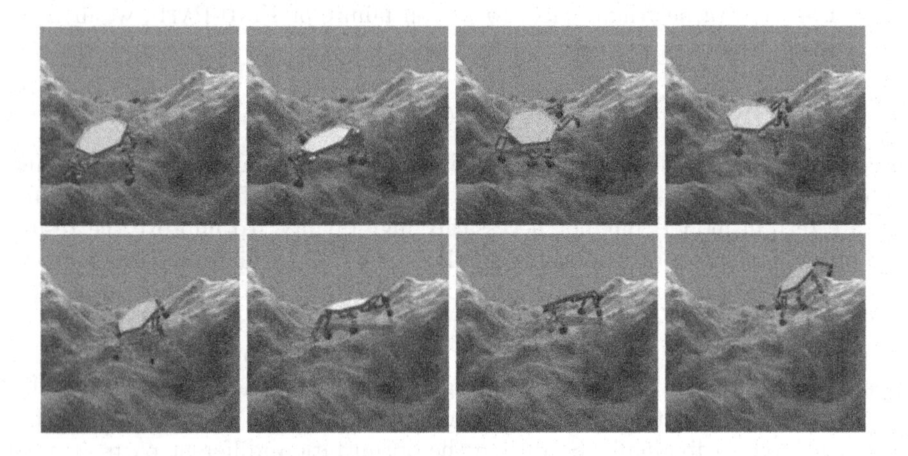

Fig. 6. Walking on steep, uneven terrain with no fixed gait

the size of one of ATHLETE's wheels. Currently, we randomly sample 200 footfalls in each terrain to use in our planner, relying on our graph search heuristic (Section 4.3) to identify which of these footfalls are useful. We are working on ways to better refine our selection of footfalls (for example, during incremental sensing), but right now the benefit is marginal.

First, we show that our planner enables ATHLETE to walk across varied terrain. Fig. 5 shows motion on smooth, undulating ground (where all contacts are modeled with the same coefficient of friction). The initial and final stances are at a distance of about twice the radius of ATHLETE's chassis. The resulting motion consisted of 66 steps. Total computation time was 14 minutes. Fig. 6 shows motion on irregular and steep ground. The resulting motion consisted of 84 steps. Total computation time was 26 minutes. For comparison, Fig. 7 shows

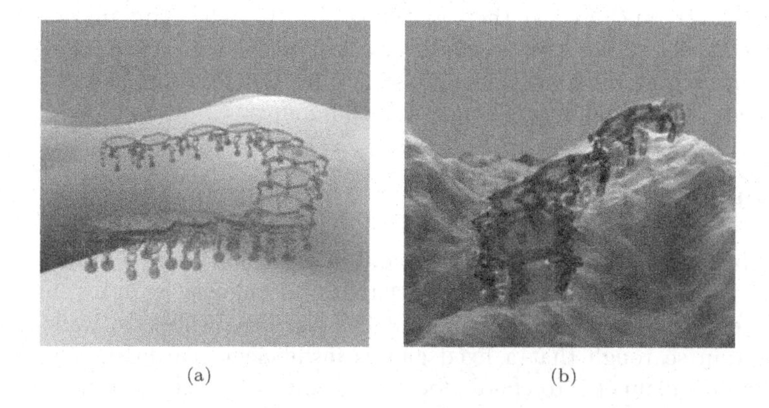

Fig. 7. Walking with an alternating tripod gait is (a) feasible on smooth terrain but (b) infeasible on uneven terrain. Infeasible configurations are highlighted red

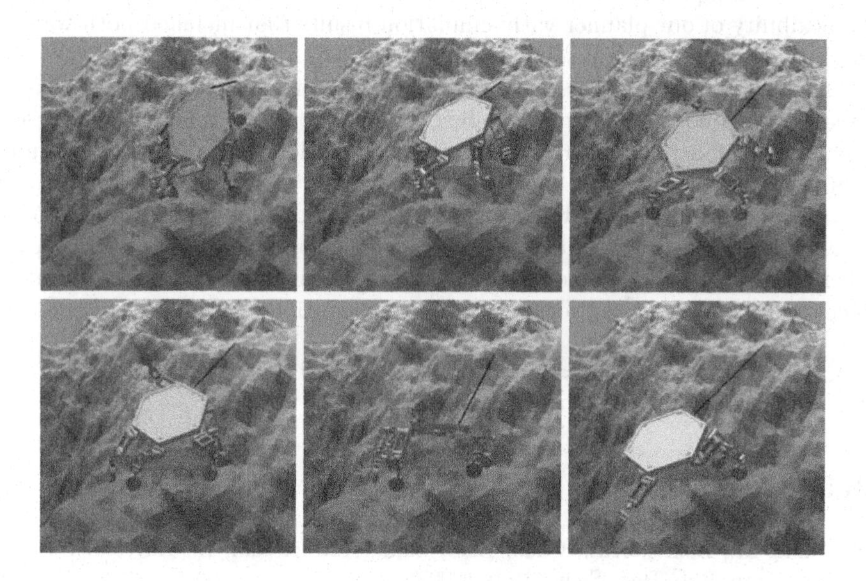

Fig. 8. Rappelling down an irregular 60° slope with no fixed gait

the result of applying a common fixed gait (an alternating-tripod) to both of these terrains. On smooth ground, the gait works well – it is simpler to plan, and results in more efficient motion (Fig. 7(a)). On irregular and steep ground, however, the gait does not work at all – it causes ATHLETE to lose balance or exceed torque limits at several locations (Fig. 7(b)).

Our results also demonstrate that the planner is flexible enough to handle different robot morphologies. Fig. 8 shows motion to descend irregular and steep terrain at an average angle of about 60°. In this example, ATHLETE is rappelling, using a tether (anchored at the top of the cliff) to help maintain equilibrium. We included the tether with no modification to our planner, treating it as an additional leg with a different kinematic structure. The resulting motion consisted of 32 steps. Total computation time was 16 minutes.

6 Conclusion

In this paper we described the design and implementation of a motion planner for a six-legged lunar vehicle called ATHLETE, developed by JPL. This vehicle has wheels on the end of each leg, but can fix these wheels to walk carefully over terrain so rough that a fixed gait is insufficient. We made a key design choice in our planner – to choose footfalls before computing motions – because on this type of terrain, ATHLETE's motion is most constrained just as it places or removes a foot. We presented several tools embedded in our planner (for sampling, local connection, search heuristics, and path smoothing) that extend previous techniques to satisfy the specific needs of ATHLETE. We demonstrated the flexibility of our planner with simulation results that included both walking and rappelling motions on several example terrains.

There are many opportunities for future work. For example, our planner takes a reasonable amount of time for off-line computation (less than one hour), so it may help human pilots at JPL design difficult motions more quickly. A similar approach was used to plan motions for the recent Mars rovers. However, our planner is still too slow to be used on-the-fly (which may require computation times of less than five minutes). We are working to derive motion strategies or other methods of model reduction to address this problem. Other important issues include incremental sensing and a consideration of dynamics.

Acknowledgments. This work was supported by the RTLSM grant from NASA-JPL, specifically for the ATHLETE project.

References

1. Akinc, M., Bekris, K.E., Chen, B.Y., Ladd, A.M., Plaku, E., Kavraki, L.E.: In: Int. Symp. Rob. Res., Siena, Italy (2003)
2. Alami, R., Laumond, J.-P., Siméon, T.: Two manipulation planning algorithms. In: Goldberg, K., Halperin, D., Latombe, J.-C., Wilson, R. (eds.) Alg. Found. Rob., pp. 109–125. A.K. Peters, Wellesley (1995)
3. Bares, J.E., Wettergreen, D.S.: Dante II: Technical description, results and lessons learned. Int. J. Rob. Res. 18(7), 621–649 (1999)
4. Bicchi, A., Kumar, V.: Robotic grasping and contact: A review. In: IEEE Int. Conf. Rob. Aut., San Francisco, pp. 348–353 (2000)
5. Boissonnat, J.-D., Devillers, O., Lazard, S.: Motion planning of legged robots. SIAM J. Computing 30(1), 218–246 (2000)
6. Bretl, T.: Motion planning of multi-limbed robots subject to equilibrium constraints: The free-climbing robot problem. Int. J. Rob. Res. 25(4), 317–342 (2006)

7. Bretl, T., Lall, S.: A fast and adaptive test of static equilibrium for legged robots. In: IEEE Int. Conf. Rob. Aut., Orlando, FL (2006)
8. Bretl, T., Latombe, J.-C., Rock, S.: Toward autonomous free-climbing robots. In: Int. Symp. Rob. Res., Siena, Italy (2003)
9. Caillas, C., Hebert, M., Krotkov, E., Kweon, I., Kanade, T.: Methods for identifying footfall positions for a legged robot. In: Int. Work. Int. Rob. Sys., pp. 244–250 (1989)
10. Chestnutt, J., Kuffner, J., Nishiwaki, K., Kagami, S.: Planning biped navigation strategies in complex environments. In: IEEE Int. Conf. Hum. Rob., Munich, Germany (2003)
11. Choset, H., Lynch, K., Hutchinson, S., Kanto, G., Burgard, W., Kavraki, L., Thrun, S.: Principles of Robot Motion: Theory, Algorithms, and Implementations. MIT Press, Cambridge (2005)
12. Cortés, J., Siméon, T., Laumond, J.-P.: A random loop generator for planning the motions of closed kinematic chains using prm methods. In: IEEE Int. Conf. Rob. Aut., Washington, D.C. (2002)
13. Eldershaw, C., Yim, M.: Motion planning of legged vehicles in an unstructured environment. In: IEEE Int. Conf. Rob. Aut., Seoul, South Korea (2001)
14. Estier, T., Crausaz, Y., Merminod, B., Lauria, M., Pguet, R., Siegwart, R.: An innovative space rover with extended climbing abilities. In: Space and Robotics, Albuquerque, NM (2000)
15. Geraerts, R., Overmars, M.: Clearance based path optimization for motion planning. In: IEEE Int. Conf. Rob. Aut., New Orleans, LA (2004)
16. Gottschalk, S., Lin, M., Manocha, D.: OBB-tree: A hierarchical structure for rapid interference detection. In: ACM SIGGRAPH, pp. 171–180 (1996)
17. Han, L., Amato, N.M.: A kinematics-based probabilistic roadmap method for closed chain systems. In: WAFR (2000)
18. Hauser, K., Bretl, T., Latombe, J.-C.: Non-gaited humanoid locomotion planning. In: Humanoids, Tsukuba, Japan (2005)
19. Heiken, G.H., Vaniman, D.T., French, B.M.: Lunar Sourcebook: A User's Guide to the Moon. Cambridge University Press, Cambridge (1991)
20. Hirose, S., Kunieda, O.: Generalized standard foot trajectory for a quadruped walking vehicle. Int. J. Rob. Res. 10(1), 3–12 (1991)
21. Hirose, S., Yoneda, K., Tsukagoshi, H.: Titan VII: Quadruped walking and manipulating robot on a steep slope. In: IEEE Int. Conf. Rob. Aut., Albuquerque, NM, pp. 494–500 (1997)
22. Hsu, D., Latombe, J.-C., Motwani, R.: Path planning in expansive configuration spaces. In: IEEE Int. Conf. Rob. Aut., pp. 2219–2226 (1997)
23. Iagnemma, K., Genot, F., Dubowsky, S.: Rapid physics-based rough-terrain rover planning with sensor and control uncertainty. In: IEEE Int. Conf. Rob. Aut., Detroit, MI (1999)
24. Kavraki, L.E., Svetska, P., Latombe, J.-C., Overmars, M.: Probabilistic roadmaps for path planning in high-dimensional configuration spaces. IEEE Trans. Robot. Automat. 12(4), 566–580 (1996)
25. Koga, Y., Latombe, J.-C.: On multi-arm manipulation planning. In: IEEE Int. Conf. Rob. Aut., San Diego, CA, pp. 945–952 (1994)
26. Krotkov, E., Simmons, R.: Perception, planning, and control for autonomous walking with the ambler planetary rover. Int. J. Rob. Res. 15, 155–180 (1996)
27. Kuffner, Jr., J.J.: Autonomous Agents for Real-Time Animation. PhD thesis, Stanford University (1999)

28. Kuffner Jr., J.J., Nishiwaki, K., Kagami, S., Inaba, M., Inoue, H.: Motion planning for humanoid robots. In: Int. Symp. Rob. Res., Siena, Italy (2003)
29. Lauria, M., Piguet, Y., Siegwart, R.: Octopus: an autonomous wheeled climbing robot. In: CLAWAR (2002)
30. LaValle, S.M., Kuffner Jr., J.J.: Randomized kinodynamic planning. Int. J. Rob. Res. 20(5), 379–400 (2001)
31. Low, K., Bai, S.: Terrain-evaluation-based motion planning for legged locomotion on irregular terrain. Adv. Rob. 17(8), 761–778 (2003)
32. Mumm, E., Farritor, S., Pirjanian, P., Leger, C., Schenker, P.: Planetary cliff descent using cooperative robots. Autonomous Robots 16, 259–272 (2004)
33. Nielsen, C.L., Kavraki, L.E.: A two level fuzzy prm for manipulation planning. In: IEEE/RSJ Int. Conf. Int. Rob. Sys., Takamatsu, Japan, pp. 1716–1721 (2000)
34. Okamura, A., Smaby, N., Cutkosky, M.: An overview of dexterous manipulation. In: IEEE Int. Conf. Rob. Aut., pp. 255–262 (2000)
35. Pai, D.K., Barman, R.A., Ralph, S.K.: Platonic beasts: Spherically symmetric multilimbed robots. Autonomous Robots 2(4), 191–201 (1995)
36. Pettré, J., Laumond, J.-P., Siméon, T.: A 2-stages locomotion planner for digital actors. In: Eurographics/SIGGRAPH Symp. Comp. Anim. (2003)
37. Sahbani, A., Cortés, J., Siméon, T.: A probabilistic algorithm for manipulation planning under continuous grasps and placements. In: IEEE/RSJ Int. Conf. Int. Rob. Sys., Lausanne, Switzerland, pp. 1560–1565 (2002)
38. Sánchez, G., Latombe, J.-C.: On delaying collision checking in PRM planning: Application to multi-robot coordination. Int. J. of Rob. Res. 21(1), 5–26 (2002)
39. Schwarzer, F., Saha, M., Latombe, J.-C.: Exact collision checking of robot paths. In: WAFR, Nice, France (December 2002)
40. Shapiro, A., Rimon, E.: PCG: A foothold selection algorithm for spider robot locomotion in 2d tunnels. In: IEEE Int. Conf. Rob. Aut., Taipei, Taiwan, pp. 2966–2972 (2003)
41. Singh, S., Simmons, R., Smith, T., Stentz, A.T., Verma, V., Yahja, A., Schwehr, K.: Recent progress in local and global traversability for planetary rovers. In: IEEE Int. Conf. Rob. Aut. (2000)
42. Song, G., Miller, S., Amato, N.M.: Customizing PRM roadmaps at query time. In: IEEE Int. Conf. Rob. Aut., Seoul, Korea, pp. 1500–1505 (2001)
43. Song, S.-M., Waldron, K.J.: Machines that walk: The adaptive suspension vehicle. The MIT Press, Cambridge (1989)
44. Vougioukas, S.G.: Optimization of robot paths computed by randomized planners. In: IEEE Int. Conf. Rob. Aut., Barcelona, Spain (2005)
45. Wang, L., Chen, C.: A combined optimization method for solving the inverse kinematics problem of mechanical manipulators. IEEE Trans. Robot. Automat. 7(4), 489–499 (1991)
46. Wettergreen, D., Thorpe, C., Whittaker, W.: Exploring mount erebus by walking robot. Robotics and Autonomous Systems 11, 171–185 (1993)
47. Yakey, J.H., LaValle, S.M., Kavraki, L.E.: Randomized path planning for linkages with closed kinematic chains. IEEE Trans. Robot. Automat. 17(6), 951–958 (2001)
48. Yoneda, K., Ito, F., Ota, Y., Hirose, S.: Steep slope locomotion and manipulation mechanism with minimum degrees of freedom. In: IEEE/RSJ Int. Conf. Int. Rob. Sys., pp. 1897–1901 (1999)
49. Zheng, Y.F., Shen, J.: Gait synthesis for the SD-2 biped robot to climb sloping surface. IEEE Trans. Robot. Automat. 6(1), 86–96 (1990)

Part VI

Applications in Medicine and Biology

Constant-Curvature Motion Planning Under Uncertainty with Applications in Image-Guided Medical Needle Steering

Ron Alterovitz[1], Michael Branicky[2], and Ken Goldberg[3]

[1] IEOR Department, University of California, Berkeley
ron@ieor.berkeley.edu
[2] EECS Department, Case Western Reserve University
mb@case.edu
[3] IEOR and EECS Departments, University of California, Berkeley
goldberg@berkeley.edu

Abstract. We consider a variant of nonholonomic motion planning for a Dubins car with no reversals, binary left/right steering, and uncertainty in motion direction. We develop a new motion planner and apply it to *steerable needles*, a new class of flexible bevel-tip medical needles that clinicians can steer through soft tissue to reach targets inaccessible to traditional stiff needles. Our method explicitly considers uncertainty in needle motion due to patient differences and the difficulty in predicting needle/tissue interaction: the planner computes optimal turning points to maximize the probability that the needle will reach the desired target. Given a medical image with segmented obstacles and target, our method formulates the planning problem as a Markov Decision Process (MDP) based on an efficient discretization of the state space, models motion uncertainty using probability distributions, and computes turning points to maximize the probability of successfully reaching the target using infinite horizon Dynamic Programming (DP). This approach has three features particularly beneficial for medical planning problems. First, the planning formulation only requires parameters that can be directly extracted from images. Second, we can compute the optimal needle insertion point by examining the DP look-up table of optimal controls for every needle state. Third, intra-operative medical imaging can be combined with the pre-computed DP look-up table to permit optimal control of the needle in the operating room without requiring time-consuming intra-operative re-planning. We apply the method to generate motion plans for steerable needles to reach targets inaccessible to stiff needles and illustrate the importance of considering uncertainty during motion plan optimization.

1 Introduction

Advances in medical imaging such as x-ray fluoroscopy, ultrasound, and MRI are now providing physicians with real-time patient-specific information as they perform medical procedures such as extracting tissue samples for biopsies, injecting drugs for anesthesia, or implanting radioactive seeds for brachytherapy cancer treatment. These diagnostic and therapeutic medical procedures require insertion of a needle to a specific location in soft tissue. We are developing motion

S. Akella et al. (Eds.): Algorithmic Foundation of Robotics VII, STAR 47, pp. 319–334, 2008.
springerlink.com

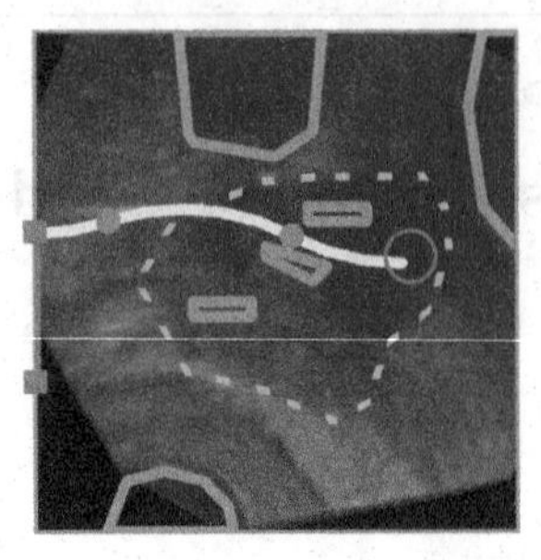

(a) Minimize path length
$P_s = 36.7\%$

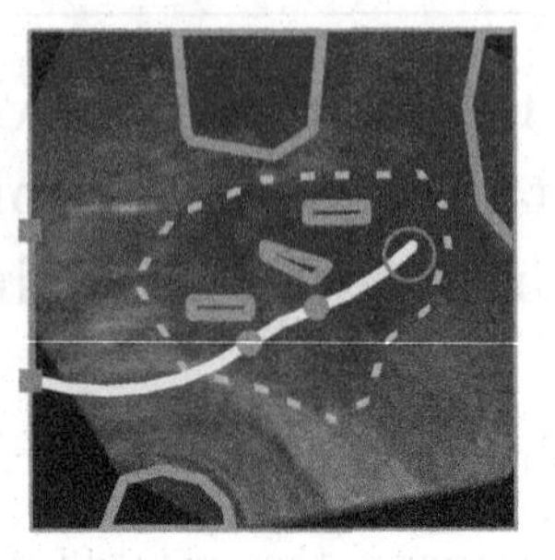

(b) Maximize probability of success
$P_s = 73.7\%$

Fig. 1. Our motion planner computes controls (insertions and direction changes, indicated by dots) to steer the needle from an insertion entry region (vertical line on left between the solid squares) to the target (open circle) inside soft tissue, without touching critical areas indicated by polygonal obstacles in the imaging plane. The motion of the needle is not known with certainty; the needle tip may be deflected during insertion due to tissue inhomogeneities or other unpredictable soft tissue interactions. We explicitly consider this uncertainty to generate motion plans to maximize the probability of success, P_s, the probability that the needle will reach the target without colliding with an obstacle or exiting the workspace boundary. Relative to minimizing path length, our planner can generate longer paths with greater clearance from obstacles to maximize P_s.

planning algorithms for medical needle insertion procedures that can utilize the information obtained by real-time imaging to accurately reach desired locations.

We consider a new class of medical needles, composed of a flexible material and with a bevel-tip, that can be steered to targets in soft tissue that are inaccessible to traditional stiff needles [30, 31, 3, 4]. Steerable needles are controlled by 2 degrees of freedom actuated at the needle base: insertion distance and bevel direction. Webster et al. experimentally demonstrated that, under ideal conditions, a flexible bevel-tip needle cuts a path of constant curvature in the direction of the bevel and the needle shaft bends to follow the path cut by the bevel tip [30]. In a plane, this nonholonomic constraint based on bevel direction is equivalent to a Dubins car that cannot go straight; it can only steer its wheels far left or far right.

The steerable needle motion planning problem is to determine a sequence of controls (insertions and direction changes) so the needle tip reaches the specified target while avoiding obstacles and staying inside the workspace. Given a segmented medical image of the target, obstacles, and starting location, the feasible workspace for motion planning is defined by the soft tissues through which the needle can be steered. Obstacles represent tissues that cannot be cut by the needle, such as bone, or sensitive tissues that should not be damaged, such as nerves or arteries. In this paper we consider motion plans in an imaging plane since the speed/resolution trade-off of 3D imaging modalities is generally poor for 3D real-time interventional applications. In future work, we will explore the natural extension of our planning approach to 3D as imaging modalities continue to improve.

Clinicians performing medical needle insertion procedures must consider uncertainty in the needle's motion through tissue due to patient differences and the difficulty in predicting needle/tissue interaction. These sources of uncertainty may result in deflections of the needle's orientation, which is a type of slip in the motion of a Dubins car. Real-time imaging in the operating room can measure the needle's current position and orientation, but this measurement by itself provides no information about the effect of future deflections during insertion. We develop a new motion planning approach for steering flexible needles through soft tissue that explicitly considers uncertainty: our method formulates the planning problem as a Markov Decision Process (MDP) based on an efficient discretization of the state space, models motion uncertainty using probability distributions, and computes optimal controls (within error due to discretization) using infinite horizon Dynamic Programming (DP).

To define optimality for a needle steering plan, we introduce a new objective for image-guided motion planning: maximizing probability of success. In the case of needle steering, the needle is controlled until it reaches the target (success) or until failure occurs, where failure is defined as hitting an obstacle, exiting the feasible workspace, or reaching a state in which it is impossible to prevent the former two outcomes. Since the motion response of the needle is not deterministic, success of the procedure can rarely be guaranteed. Our objective function value for a particular plan has physical meaning: it is the probability that the needle insertion will succeed assuming optimal control of the needle. In addition to this intuitive interpretation of the objective, our formulation has a secondary benefit: all data required for planning can be measured directly from imaging data without requiring tweaking of user-specified parameters. Rather than assigning costs to insertion distance, needle rotation, etc., which are difficult to estimate or quantify, our method only requires the probability distributions of the needle response to each feasible control, which can be estimated from previously obtained images.

Solving the MDP using DP has key benefits particularly relevant for medical planning problems where feedback is provided at regular time intervals using medical imaging or other sensor modalities. Like a well-constructed navigation field, the DP solver provides an optimal control for any state in the workspace. We use the DP look-up table to automatically optimize the needle insertion point. Integrated with intra-operative medical imaging, this DP look-up table can be used to optimally control the needle in the operating room without requiring costly intra-operative re-planning. Hence, the planning solution can serve as a means of control under real-time medical imaging.

In Fig. 1, we apply our motion planner in simulation to prostate brachytherapy, a medical procedure in which physicians implant radioactive seeds at precise locations inside the prostate under ultrasound image guidance to treat prostate cancer. In this ultrasound image of the prostate (segmented by a dotted line), obstacles correspond to bones, the rectum, the bladder, the urethra, and previously implanted seeds. Brachytherapy is currently performed using rigid needles; here we consider steerable needles capable of obstacle avoidance. We compare the output of our new method to previous work on shortest path planning for steerable needles [4]. Our method improves the expected probability of success by over 30% compared to shortest path planning, illustrating the importance of explicitly considering uncertainty in needle motion.

2 Related Work

Nonholonomic motion planning has a long history in robotics and related fields [10, 20, 21, 23]. Past work has addressed deterministic curvature-constrained path planning where a mobile robot's path is, like a car, constrained by a minimum turning radius. Dubins showed that the optimal curvature-constrained trajectory in open space from a start pose to a target pose can be described from a discrete set of canonical trajectories composed of straight line segments and arcs of minimum radius of curvature [15]. Jacobs and Canny considered polygonal obstacles and constructed a configuration space for a set of canonical trajectories [18], and Agarwal et al. developed a fast algorithm for a shortest path inside a convex polygon [1]. For Reeds-Shepp cars with reverse, Laumond et el. developed a nonholonomic planner using recursive subdivision of collision-free paths generated by a lower-level geometric planner [22] and Bicchi et al. proposed a technique that provides the shortest path for circular unicycles [8]. Sellen developed a discrete state-space approach; his discrete representation of orientation using a unit circle inspired our discretization approach [27].

Our planning problem considers steerable needles currently under development that are subject to a *constant* magnitude turning radius rather than a *minimum* turning radius. Webster et al. showed experimentally that, under ideal conditions, steerable bevel-tip needles follow paths of constant curvature in the direction of the bevel tip [30], and that radius of curvature of the needle path is not significantly affected by insertion velocity [31].

Park et al. formulated the planning problem for steerable bevel-tip needles in stiff tissue as a nonholonomic kinematics problem based on a 3D extension of a unicycle model and used a diffusion-based motion planning algorithm to numerically compute a path [25]. Park's method searches for a feasible path in full 3D space using continuous control, but it does not consider obstacle avoidance or the uncertainty of the response of the needle to insertion or direction changes, both of which are emphasized in our method.

Past work has investigated needle insertion planning in situations where soft tissue deformations are significant and can be modeled. Our past work addressed planning optimal insertion location and insertion distance for rigid symmetric-tip needles to compensate for 2D tissue deformations [5, 6]. Past work has also addressed steering slightly flexible symmetric-tip needles by translating and orienting the needle base to explicitly cause tissue deformations that will guide the needle around point obstacles with oval-shaped potential fields [14]. Glozman and Shoham also address symmetric-tip needles and approximate the tissue using springs [17]. We previously developed a different 2D planner for bevel-tip needles to explicitly compensate for the effects of tissue deformation by combining finite element simulation with numeric optimization [3]. This previous approach assumed that bevel direction can only be set once prior to insertion and employed local optimization that can fail to find a globally optimal solution in the presence of obstacles.

In preliminary work, we proposed an MDP formulation for needle steering [4] to find a stochastic shortest path from a start position to a target, subject to user-specified "cost" parameters for direction changes, insertion distance, and obstacle collisions. However, the formulation was not targeted at image-guided procedures, did not include insertion point optimization, and optimized an objective function that has no physical meaning. In this paper, we develop a 2D motion planning approach for image-guided needle steering that explicitly considers motion uncertainty to maximize the probability of success based on parameters that can be extracted from medical imaging without requiring user-specified "cost" parameters that may be difficult to determine.

MDP's are ideally suited for medical planning problems because of the variance in characteristics between patients and the necessity for clinicians to make decisions at discrete time intervals based on limited known information. In the context of medical procedure planning, MDP's have been developed to assist in timing decisions for liver transplants [2], discharge of severe sepsis cases [19], and start dates for HIV drug cocktail treatment [28].

Integrating motion planning with intra-operative medical imaging requires real-time localization of the needle in the images. Methods are available for this purpose for a variety of imaging modalities [11, 12]. X-ray fluoroscopy, a relatively low-cost imaging modality capable of obtaining images at regular discrete time intervals, is ideally suited for our application because it generates 2D projection images from which the needle can be cleanly segmented [11].

Medical needle insertion procedures may also benefit from the more precise control of needle position and velocity made possible through robotic surgical assistants [29]. Dedicated hardware for needle insertion is being developed for

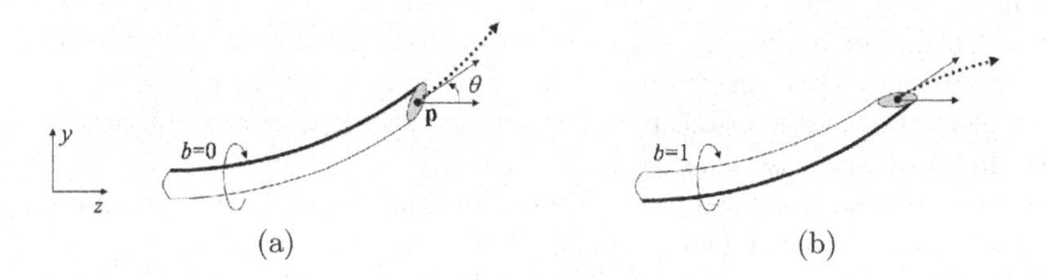

Fig. 2. The state of a steerable needle during insertion is characterized by tip position $\mathbf{p}$, tip orientation angle θ, and bevel direction b (a). Rotating the needle about its base changes the bevel direction but does not affect needle position (b). The needle will cut soft tissue along an arc (dotted vector) based on bevel direction.

stereotactic neurosurgery [24], MR compatible surgical assistance [9, 13], and prostate biopsy and therapeutic interventions [16, 26].

3 Motion Planning Method

3.1 Problem Definition

Steerable bevel-tip needles are controlled by 2 degrees of freedom actuated at the needle base: insertion distance and rotation angle about the needle axis. Insertion pushes the needle deeper into the tissue, while rotation re-orients the bevel at the needle tip. For a sufficiently flexible needle, Webster et al. experimentally demonstrated that rotating the needle base will change the bevel direction without changing the needle's position in the tissue [30]. In the plane, the needle base can be rotated 180° about the insertion axis at the base so the bevel points in either the bevel-left or bevel-right direction. When inserted, the asymmetric force applied by the bevel causes the needle to bend and follow a curved path through the tissue [30]. Under ideal conditions, the curve will have a constant radius of curvature r, which is a property of the needle and tissue. We assume the tissue is stiff relative to the needle and that the needle is thin, sharp, and low-friction so the tissue does not significantly deform. While the needle can be partially retracted and re-inserted, the needle is likely to follow the path in the tissue cut by the needle prior to retraction. Hence, we only consider insertion, not retraction, of the needle in this paper.

We define the workspace as a 2D rectangle of depth z_{max} and height y_{max}. We do not consider motion by the needle out of the imaging plane. Obstacles in the workspace are defined by (possibly nonconvex) polygons. The obstacles can be expanded using a Minkowski sum to specify a minimum clearance [23]. The target region is defined by a circle with center point $\mathbf{t}$ and radius r_t.

As shown in Fig. 2, the state w of the needle during insertion is fully characterized by the needle tip's position $\mathbf{p} = (p_y, p_z)$, orientation angle θ, and bevel direction b, where b is either bevel-left (b=0) or bevel-right (b=1).

We assume imaging occurs at discrete time intervals and the motion planner obtains needle tip position and orientation information only at these times. Between images, we assume the needle moves at constant velocity and is inserted a distance δ. In our model, direction changes can only occur at discrete *decision points* separated by the insertion distance δ. One of two controls, or actions, u can be selected at any decision point: insert the needle a distance δ ($u = 0$), or change direction and insert a distance δ ($u = 1$).

During insertion, the needle tip orientation may be deflected by inhomogeneous tissue, small anatomical structures not visible in medical images, or local tissue displacements. Additional deflection may occur during direction changes due to stiffness along the needle shaft. These deflections are due to an unknown aspect of the tissue structure or needle/tissue interaction, not errors in measurement of the needle's orientation, and can be considered a type of noise parameter in the plane. We model uncertainty in needle motion due to such deflections using probability distributions. The orientation angle θ may be deflected by some angle β, which we model as normally distributed with mean 0 and standard deviations σ_i for insertion ($u = 0$) and σ_r for direction changes followed by insertion ($u = 1$). Since σ_i and σ_r are properties of the needle and tissue, we plan in future work to automatically estimate these parameters by retrospectively analyzing images of needle insertion.

The goal of motion planning is to compute an optimal control u for every state w in the workspace to maximize the probability of success P_s. We define $P_s(w)$ to be the probability of success given that the needle is currently in state w. If the position of state w is inside the target, $P_s(w) = 1$. If the position of state w is inside an obstacle, $P_s(w) = 0$. Given a control u for some other state w, the probability of success will depend on the response of the needle to the control (the next state) and the probability of success for that next state. The expected probability of success is $P_s(w) = E[P_s(v)|w, u]$, where the expectation is over v, a random variable for the next state. The goal of motion planning is to compute an optimal control u for every state w:

$$P_s(w) = \max_u \{E[P_s(v)|w, u]\}. \tag{1}$$

3.2 Problem Formulation

To evaluate Eq. 1, we approximate needle state $w = \{\mathbf{p}, \theta, b\}$ using a discrete representation. To make this approach tractable, we must round $\mathbf{p}$ and θ without generating an unwieldy number of states while simultaneously bounding error due to discretization. We describe our approximation approach, which results in N discrete states, in Sec. 3.3.

For N discrete states, the motion planning problem is to determine the optimal control u_i for each state $i = 1, \ldots, N$. We re-write Eq. 1 using the discrete approximation and expand the expected value to a summation:

$$P_s(x_i) = \max_{u_i} \left\{ \sum_{j=1}^{N} P_{ij}(u_i) P_s(x_j) \right\}, \tag{2}$$

where $P_{ij}(u_i)$ is the probability of entering state x_j after executing control u_i at current state x_i.

We observe that the needle steering motion planning problem is a type of MDP. In particular, Eq. 2 has the form of the Bellman equation for a stochastic shortest path problem [7]:

$$J^*(x_i) = \max_{u_i} \sum_{j=1}^{N} P_{ij}(u_i) \left(g(x_i, u_i, x_j) + J^*(x_j) \right). \tag{3}$$

where $g(x_i, u_i, x_j)$ is a "reward" for transitioning from state x_i to x_j after control u_i. In our case, $g(x_i, u_i, x_j) = 0$ for all x_i, u_i, and x_j, and $J^*(x_i) = P_s(x_i)$. Stochastic shortest path problems of this form can be optimally solved using infinite horizon DP, as we describe in Sec. 3.4.

3.3 State Space Discretization

Our discretization of the planar workspace is based on a grid of points with a spacing Δ horizontally and vertically. We approximate a point $\mathbf{p} = (p_y, p_z)$ by rounding to the nearest point $\mathbf{q} = (q_y, q_z)$ on the grid. For a rectangular workspace bounded by depth z_{max} and height y_{max}, this results in $N_s = \lfloor z_{max} y_{max} / \Delta^2 \rfloor$ position states aligned at the origin.

Rather than directly approximating θ by rounding, which would incur a cumulative error with every transition, we take advantage of discrete insertion distances δ. We define a *control circle* of radius r, the radius of curvature of the needle. Each point $\mathbf{c}$ on the control circle represents an orientation θ of the needle, where θ is the angle of the tangent of the circle at $\mathbf{c}$ with respect to the z-axis. The needle will trace an arc of length δ along the control circle in a counter-clockwise direction for $b = 0$ and in the clockwise direction for $b = 1$. Direction changes correspond to rotating the point $\mathbf{c}$ by 180° about the control circle origin and tracing subsequent insertions in the opposite direction, as shown in Fig. 3(a). Since the needle traces arcs of length δ, we divide the control circle into N_c arcs of length $\delta = 2\pi r / N_c$. The endpoints of the arcs generate a set of N_c control circle points, each representing a discrete orientation state, as shown in Fig. 3(b). We require that N_c is a multiple of 4 to facilitate the orientation state change after a direction change.

At each position on the Δ grid, the needle may be in any of the N_c orientation states. To define transitions for each orientation state, we overlay the control circle on a regular grid of spacing Δ and round the positions of the control circle points to the nearest grid point, as shown in Fig. 3(c). The displacements between rounded control circle points encode the transitions of the needle tip. This discretization results in 0 discretization error in orientation when the needle is controlled at δ intervals.

Using this discretization, a needle state $w = \{\mathbf{p}, \theta, b\}$ can be approximated as a discrete state $s = \{\mathbf{q}, \Theta, b\}$, where $\mathbf{q} = (q_y, q_z)$ is the discrete point closest to $\mathbf{p}$ on the Δ-density grid and Θ is the integer index of the discrete control circle point with tangent angle closest to θ. The total number of discrete states is $N = 2N_s N_c$.

Deterministic paths designated using this discrete representation of state will incur error due to discretization, but the error is bounded. At any decision point, the position error due to rounding to the Δ workspace grid is $E_0 = \Delta\sqrt{2}/2$. When the bevel direction is changed, a position error is also incurred because the distance in centers of the original control circle and the center of the control circle after the direction change will be in the range $2r \pm \Delta\sqrt{2}$. Hence, for a needle path with h direction changes, the final orientation is precise but the error in position is bounded above by $E_h = h\Delta\sqrt{2} + \Delta\sqrt{2}/2$.

Due to motion uncertainty, actual needle paths will not always exactly trace the control circle. The deflection angle β defined in Sec. 3.1 must also be approximated as discrete. We define discrete transitions from a state x_i, each separated by an angle of deflection of $\alpha = 360°/N_c$, and store the transition probability in $P_{ij}(u)$. In this paper, we model β using a normal distribution with mean 0 and

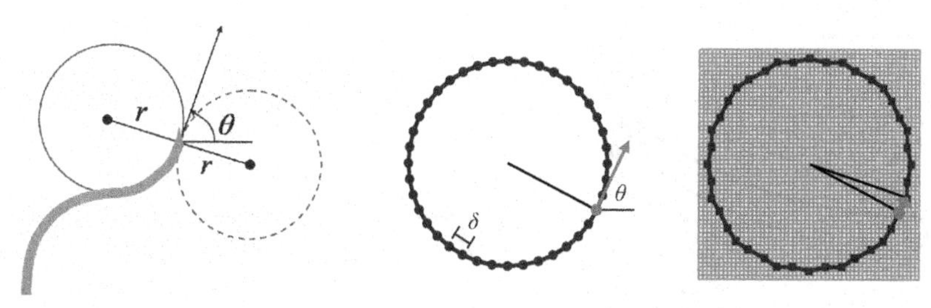

(a) Needle tracing control circle (b) Control circle (c) Rounded control circle

Fig. 3. A needle in the bevel-left direction with orientation θ is tracing the solid control circle with radius r (a). A direction change would result in tracing the dotted circle. The control circle is divided into $N_c = 40$ discrete arcs of length δ (b). The control circle points are rounded to the nearest point on the Δ-density grid, and transitions for insertion of distance δ are defined by the vectors between rounded control circle points (c).

standard deviation σ_i or σ_r, and compute the probability for each discrete transition by integrating the corresponding area under the normal curve, as shown in Fig. 4. We set the number of discrete transitions N_{p_i} such that the areas on the left and right tails of the normal distribution sum to less than 1%. The left and right tail probabilites are added to the left-most and right-most transitions, respectively.

Certain states and transitions must be handled as special cases. States inside the target region and states inside obstacles are absorbing states. If the transition arc from feasible state x_i exits the workspace or intersects an edge of a polygonal obstacle, a transition to an obstacle state is used.

3.4 Optimization Using Infinite Horizon Dynamic Programming

Infinite horizon dynamic programming is a type of dynamic programming in which there is no finite time horizon [7]. For stationary problems, this implies that the optimal control at each state is purely a function of the state without explicit dependence on time. In the case of needle steering, once a state transition is made, the next control is computed based on the current position, orientation, and bevel direction without explicit dependence on past controls.

To solve the infinite horizon DP problem defined by the Bellman Eq. 3, we use the value iteration algorithm [7], which iteratively updates $P_s(x_i)$ for each state i by evaluating Eq. 2. This generates a DP look-up table containing the optimal control u_i and the probability of success $P_s(x_i)$ for $i = 1, \ldots, N$.

Termination of the algorithm is guaranteed in N iterations if the transition probability graph corresponding to some optimal stationary policy is acyclic [7]. Violation of this requirement will be rare in motion planning since it implies that an optimal control sequence results in a path that, with probability greater than 0, loops and passes through the same point at the same orientation more

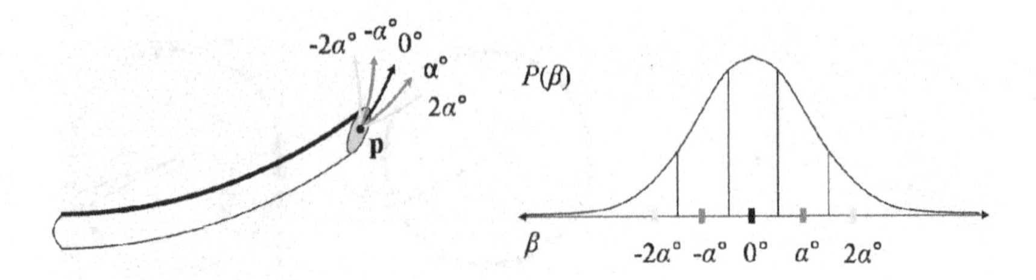

Fig. 4. When the needle is inserted, the insertion angle θ may be deflected by some angle β. We model the probability distribution of β using a normal distribution with mean 0 and standard deviation σ_i for insertion or σ_r for direction change. For a discrete sample of deflections ($\beta = \{-2\alpha, -\alpha, 0, \alpha, 2\alpha\}$), we obtain the probability of each deflection by integrating the corresponding area under the normal curve.

than once. Each iteration requires matrix-vector multiplication. To improve performance, we take advantage of the sparsity of the matrices $P_{ij}(u)$ for $u = 0$ and $u = 1$. Although $P_{ij}(u)$ has N^2 entries, each row of $P_{ij}(u)$ has only k nonzero entries, where $k << N$ since the needle will only transition to a state j in the spatial vicinity of state i. Hence, $P_{ij}(u)$ has at most kN nonzero entries. By only accessing nonzero entries of $P_{ij}(u)$ during computation, each iteration of the value iteration algorithm requires only $O(kN)$ rather than $O(N^2)$ time and memory. Thus, the total algorithm's complexity is $O(kN^2)$. To further improve performance, we terminate value iteration when the maximum change ϵ over all states is less than 10^{-3}, which in our test cases occurred in far fewer than N iterations, as described in Sec. 4.

4 Computational Results

We implemented the motion planner in C++ and tested it on a 2.21GHz Athlon 64 PC. In Fig. 1, we set the needle radius of curvature $r = 5.0$, defined the workspace by $z_{max} = y_{max} = 10$, and used discretization parameters $N_c = 40$, $\Delta = 0.1$, and $\delta = 0.785$. The resulting DP problem contained $N = 800{,}000$ states. In all further examples, we set $r = 2.5$, $z_{max} = y_{max} = 10$, $N_c = 40$, $\Delta = 0.1$, and $\delta = 0.393$, resulting in $N = 800{,}000$ states.

Optimal plans and probability of success P_s depend on the level of uncertainty in needle motion. As shown in Figs. 1 and 5, explicitly considering the variance of needle motion significantly affects the optimal plan relative to shortest path plan generated under the assumption of deterministic motion. We also vary the variance during direction changes independently from the variance during insertions without direction changes. Optimal plans and probability of success P_s are highly sensitive to the level of uncertainty in needle motion due to direction changes. As shown in Fig. 6, the number of direction changes decreases as the variance during direction changes increases.

By examining the DP look-up table, we can optimize the initial insertion location, orientation, and bevel direction, as shown in Figs. 1, 5, and 6. In these

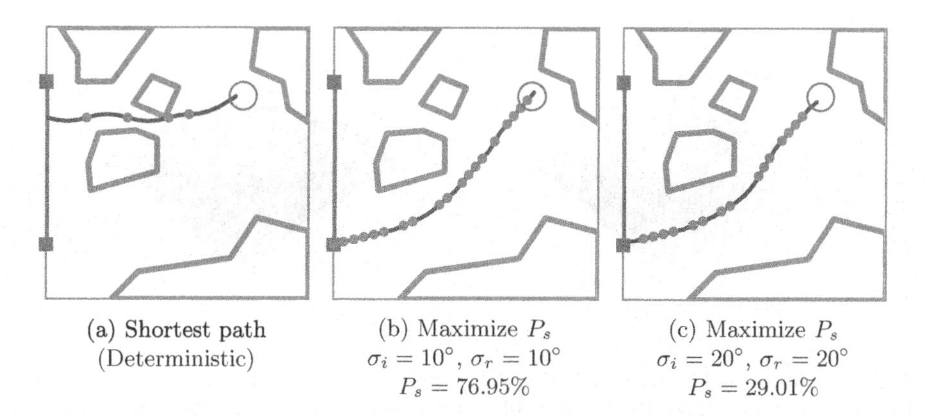

(a) **Shortest path** (Deterministic)

(b) Maximize P_s, $\sigma_i = 10°$, $\sigma_r = 10°$, $P_s = 76.95\%$

(c) Maximize P_s, $\sigma_i = 20°$, $\sigma_r = 20°$, $P_s = 29.01\%$

Fig. 5. As in Fig. 1, optimal plans maximizing the probability of success P_s illustrate the importance of considering uncertainty in needle motion. The shortest path plan passes through a narrow gap between obstacles (a). Since maximizing P_s explicitly considers uncertainty, the optimal expected path has greater clearance from obstacles, decreasing the probability that large deflections will cause failure to reach the target. Here we consider medium (b) and large (c) variance in tip deflections for a needle with smaller radius of curvature than in Fig. 1.

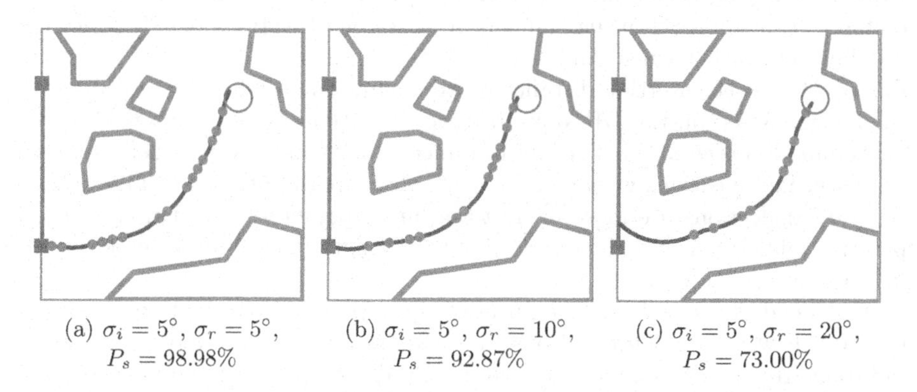

(a) $\sigma_i = 5°$, $\sigma_r = 5°$, $P_s = 98.98\%$

(b) $\sigma_i = 5°$, $\sigma_r = 10°$, $P_s = 92.87\%$

(c) $\sigma_i = 5°$, $\sigma_r = 20°$, $P_s = 73.00\%$

Fig. 6. Optimal plans demonstrate the importance of considering uncertainty in needle motion, where σ_i and σ_r are the standard deviations of needle tip deflections that can occur during insertion and direction changes, respectively. For higher σ_r relative to σ_i, the optimal plan includes fewer direction changes. Needle motion uncertainty at locations of direction changes may be substantially higher than uncertainty during insertion due to transverse stiffness of the needle.

examples, the set of feasible start states was defined as a subset of all states on the left edge of the workspace. By linearly scanning the DP look-up table, the method identifies the bevel direction b, insertion point (height y on the left edge of the workspace), and starting orientation angle θ (which varies from $-90°$ to $90°$) that maximizes probability of success, as shown in Fig. 7.

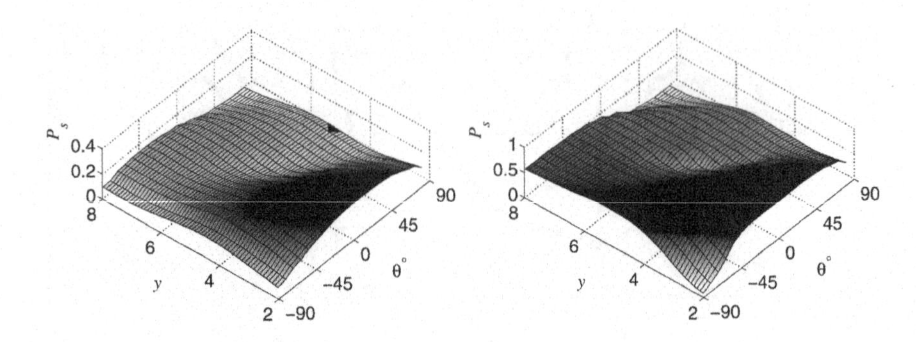

Fig. 7. The optimal needle insertion location y, angle θ, and bevel direction b are found by scanning the DP look-up table for the feasible start state with maximal P_s. Here we plot optimization surfaces for $b = 0$. The low regions correspond to states from which the needle has high probability of colliding with an obstacle or exiting the workspace, and the high regions correspond to better start states.

Integrating intra-operative medical imaging with the pre-computed DP look-up table could permit optimal control of the needle in the operating room without requiring costly intra-operative re-planning. We demonstrate the potential of this approach using simulation of needle deflections based on normal distributions with mean 0 and standard deviations $\sigma_i = 5°$ and $\sigma_r = 20°$ in Fig. 8. After each insertion distance δ, we assume the needle tip is localized in the image. Based on the DP look-up table, the needle is either inserted or the bevel direction is changed. The effect of uncertainty can be seen as deflections in the path, i.e., locations where the tangent of the path abruptly changes. Since $\sigma_r > \sigma_i$, deflections are more likely to occur at points of direction change. In practice, clinicians could monitor P_s, insertion length, and self-intersection while performing needle insertion.

As defined in Sec. 3.4, the computational complexity of the motion planner is $O(kN^2)$. Fewer than 300 iterations were required for each example, with fewer iterations required for smaller σ_i and σ_r. In all examples, the number of transitions per state $k \leq 25$. Computation time to solve the MDP for the examples ranged from 67 sec to 110 sec on a 2.21GHz AMD Athlon 64 PC, with higher computation times required for problems with greater variance, due to the increased number of transitions from each state. As computation only needs to be performed at the pre-procedure stage, we believe this computation time is reasonable for the intended applications. Intra-operative computation time is effectively instantaneous since only a memory access to the DP look-up table is required to retrieve the optimal control after the needle has been localized in imaging.

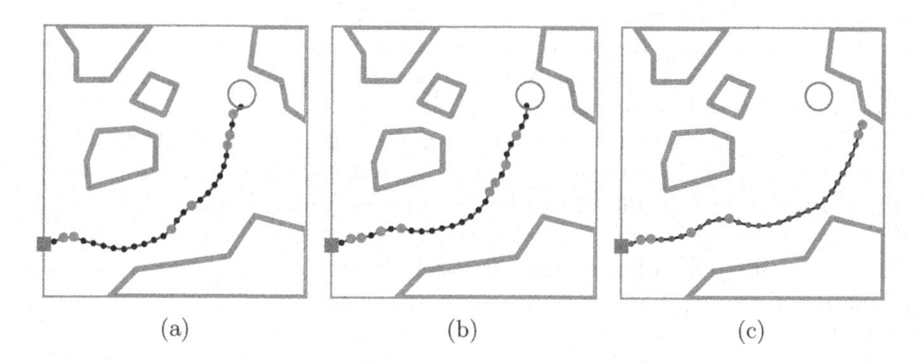

Fig. 8. Three simulated image-guided needle insertion procedures from a fixed starting point with needle motion uncertainty standard deviations of $\sigma_i = 5^\circ$ during insertion and $\sigma_r = 20^\circ$ during direction changes. After each insertion distance δ, we assume the needle tip is localized in the image and identified using a dot. Based on the DP look-up table, the needle is either inserted (small dots) or a direction change is made (larger dots). The effect of uncertainty can be seen as deflections in the path, i.e., locations where the tangent of the path abruptly changes. Since $\sigma_r > \sigma_i$, deflections are more likely to occur at points of direction change. In all cases, $P_s = 72.35\%$ at the initial state. In (c), multiple deflections and the nonholonomic constraint on needle motion prevent the needle from reaching the target.

5 Conclusion and Future Work

We developed a new motion planning approach for steering flexible needles through soft tissue that explicitly considers uncertainty: the planner computes optimal controls to maximize the probability that the needle will reach the desired target. Motion planning for steerable needles, which can be controlled by 2 degrees of freedom at the needle base (bevel direction and insertion distance), is a variant of nonholonomic planning for a Dubins car with no reversals, binary left/right steering, and uncertainty in motion direction.

Given a medical image with segmented obstacles and target, our method formulates the planning problem as a Markov Decision Process (MDP) based on an efficient discretization of the state space, models motion uncertainty using probability distributions, and computes controls to maximize the probability of success using infinite horizon DP. We implemented the motion planner and ran test problems of 800,000 states on a 2.21GHz Athlon 64 PC. The method generated motion plans for steerable needles to reach targets inaccessible to stiff needles and illustrated the importance of considering uncertainty in needle motion, as shown in Figs. 1, 5, and 6.

Our approach has key features particularly beneficial for medical planning problems. First, the planning formulation only requires parameters that can be directly extracted from images (the variance of needle orientation after insertion with or without direction change). Second, we can locate the optimal needle insertion point by examining the DP look-up table of optimal controls for every needle state, as demonstrated in Fig. 7. Third, intra-operative medical imaging can be combined with the pre-computed DP look-up table to permit optimal control of the needle in the operating room without requiring time-consuming intra-operative re-planning, as shown in Fig. 8.

In future work, we plan to extend the motion planner to 3D. Although the mathematical formulation can be naturally extended, substantial effort will be required to specify 3D state transitions and improve solving methods to handle the larger state space. We also plan to develop automated methods to estimate curvature and variance properties from images and explore the inclusion of multiple tissue types in the workspace with different needle/tissue interaction properties.

Our motion planner has implications outside the needle steering domain. We can directly extend the method to motion planning problems with a bounded number of discrete turning radii where current position and orientation can be measured but future motion response to controls is uncertain. For example, mobile robots subject to motion uncertainty with similar properties can receive periodic "imaging" updates from GPS or satellite images. Optimization of "insertion location" could apply to automated guided vehicles in a factory setting, where one machine is fixed but a second machine can be placed to maximize the probability that the vehicle will not collide with other objects on the factory floor. By identifying a relationship between needle steering and infinite horizon DP, we developed a motion planner capable of rigorously computing plans that are optimal in the presence of uncertainty.

Acknowledgment

This work was supported in part by the NIH under grant R21 EB003452 and a NSF Graduate Research Fellowship to Ron Alterovitz. We thank Andrew Lim and A. Frank van der Stappen for their suggestions, and clinicians Leonard Shlain of CPMC and I-Chow Hsu and Jean Pouliot of UCSF for their input on medical aspects of this work. We particularly thank Allison Okamura, Greg Chirikjian, Noah Cowan, and Robert Webster, our collaborators at JHU studying steerable needles, and the organizers of WAFR 2006, Srinivas Akella, Nancy Amato, Wes Huang, and Bud Mishra.

References

1. Agarwal, P.K., Biedl, T., Lazard, S., Robbins, S., Suri, S., Whitesides, S.: Curvature-constrained shortest paths in a convex polygon. SIAM J. Comput. 31(6), 1814–1851 (2002)
2. Alagoz, O., Maillart, L.M., Schaefer, A.J., Roberts, M.: The optimal timing of living-donor liver transplantation. Management Science 50(10), 1420–1430 (2005)
3. Alterovitz, R., Goldberg, K., Okamura, A.: Planning for steerable bevel-tip needle insertion through 2D soft tissue with obstacles. In: Proc. IEEE Int. Conf. on Robotics and Automation, pp. 1652–1657 (April 2005)

4. Alterovitz, R., Lim, A., Goldberg, K., Chirikjian, G.S., Okamura, A.M.: Steering flexible needles under Markov motion uncertainty. In: Proc. IEEE/RSJ Int. Conf. on Intelligent Robots and Systems, pp. 120–125 (August 2005)
5. Alterovitz, R., Pouliot, J., Taschereau, R., Hsu, I.-C., Goldberg, K.: Needle insertion and radioactive seed implantation in human tissues: Simulation and sensitivity analysis. In: Proc. IEEE Int. Conf. on Robotics and Automation, vol. 2, pp. 1793–1799 (September 2003)
6. Alterovitz, R., Pouliot, J., Taschereau, R., Hsu, I.-C., Goldberg, K.: Sensorless planning for medical needle insertion procedures. In: Proc. IEEE/RSJ Int. Conf. on Intelligent Robots and Systems, vol. 3, pp. 3337–3343 (October 2003)
7. Bertsekas, D.P.: Dynamic Programming and Optimal Control, 2nd edn. Athena Scientific, Belmont (2000)
8. Bicchi, A., Casalino, G., Santilli, C.: Planning shortest bounded-curvature paths for a class of nonholonomic vehicles among obstacles. In: Proc. IEEE Int. Conf. on Robotics and Automation, pp. 1349–1354 (1995)
9. Chinzei, K., Hata, N., Jolesz, F.A., Kikinis, R.: MR compatible surgical assist robot: System integration and preliminary feasibility study. In: Medical Image Computing and Computer Assisted Intervention, pp. 921–930 (October 2000)
10. Choset, H., Lynch, K.M., Hutchinson, S., Kantor, G., Burgard, W., Kavraki, L.E., Thrun, S.: Principles of Robot Motion: Theory, Algorithms, and Implementations. MIT Press, Cambridge (2005)
11. Cleary, K., Ibanez, L., Navab, N., Stoianovici, D., Patriciu, A., Corral, G.: Segmentation of surgical needles for fluoroscopy servoing using the insight software toolkit (itk). In: Proc. Int. Conf. of the IEEE Engineering In Medicine and Biology Society, pp. 698–701 (2003)
12. DiMaio, S.P., Kacher, D.F., Ellis, R.E., Fichtinger, G., Hata, N., Zientara, G.P., Panych, L.P., Kikinis, R., Jolesz, F.A.: Needle artifact localization in 3T MR images. In: Westwood, J.D., et al. (eds.) Medicine Meets Virtual Reality 14, pp. 120–125. IOS Press, Amsterdam (2006)
13. DiMaio, S.P., Pieper, S., Chinzei, K., Hata, N., Balogh, E., Fichtinger, G., Tempany, C.M., Kikinis, R.: Robot-assisted needle placement in open-MRI: System architecture, integration and validation. In: Westwood, J.D., et al. (eds.) Medicine Meets Virtual Reality 14, Long Beach, CA, pp. 126–131. IOS Press, Amsterdam (2006)
14. DiMaio, S.P., Salcudean, S.E.: Needle steering and model-based trajectory planning. In: Medical Image Computing and Computer Assisted Intervention, pp. 33–40 (2003)
15. Dubins, L.: On curves of minimal length with a constraint on average curvature and with prescribed initial and terminal positions and tangents. American J. of Mathematics 79, 497–516 (1957)
16. Fichtinger, G., DeWeese, T.L., Patriciu, A., Tanacs, A., Mazilu, D., Anderson, J.H., Masamune, K., Taylor, R.H., Stoianovici, D.: System for robotically assisted prostate biopsy and therapy with intraoperative CT guidance. Academic Radiology 9(1), 60–74 (2002)
17. Glozman, D., Shoham, M.: Flexible needle steering and optimal trajectory planning for percutaneous therapies. In: Medical Image Computing and Computer Assisted Intervention (September 2004)
18. Jacobs, P., Canny, J.: Planning smooth paths for mobile robots. In: Proc. IEEE Int. Conf. on Robotics and Automation, pp. 2–7 (May 1989)

19. Kreke, J., Schaefer, A.J., Roberts, M., Bailey, M.: Optimizing testing and discharge decisions in the management of severe sepsis. In: Annual Meeting of INFORMS (November 2005)
20. Latombe, J.-C.: Robot Motion Planning. Kluwer Academic Pub., Dordrecht (1991)
21. Latombe, J.-C.: Motion planning: A journey of robots, molecules, digital actors, and other artifacts. Int. J. of Robotics Research 18(11), 1119–1128 (1999)
22. Laumond, J.-P., Jacobs, P.E., Taïx, M., Murray, R.M.: A motion planner for non-holonomic mobile robots. IEEE Trans. on Robotics and Automation 10(5), 577–593 (1994)
23. LaValle, S.M.: Planning Algorithms. Cambridge University Press, Cambridge (2006)
24. Masamune, K., Ji, L., Suzuki, M., Dohi, T., Iseki, H., Takakura, K.: A newly developed stereotactic robot with detachable drive for neurosurgery. In: Medical Image Computing and Computer Assisted Intervention (1998)
25. Park, W., Kim, J.S., Zhou, Y., Cowan, N.J., Okamura, A.M., Chirikjian, G.S.: Diffusion-based motion planning for a nonholonomic flexible needle model. In: Proc. IEEE Int. Conf. on Robotics and Automation, pp. 4611–4616 (April 2005)
26. Schneider, C., Okamura, A.M., Fichtinger, G.: A robotic system for transrectal needle insertion into the prostate with integrated ultrasound. In: Proc. IEEE Int. Conf. on Robotics and Automation, pp. 2085–2091 (May 2004)
27. Sellen, J.: Approximation and decision algorithms for curvature-constrained path planning: A state-space approach. In: Agarwal, P.K., Kavraki, L.E., Mason, M.T. (eds.) Workshop on the Algorithmic Foundations of Robotics, pp. 59–67. AK Peters, Ltd., Houston (1998)
28. Shechter, S., Schaefer, A.J., Braithwaite, S., Roberts, M., Bailey, M.: The optimal time to initiate HIV therapy. In: Annual Meeting of INFORMS (November 2005)
29. Taylor, R.H., Stoianovici, D.: Medical robotics in computer-integrated surgery. IEEE Trans. on Robotics and Automation 19(5), 765–781 (2003)
30. Webster III, R.J., Cowan, N.J., Chirikjian, G., Okamura, A.M.: Nonholonomic modeling of needle steering. In: Proc. 9th Int. Symp. on Experimental Robotics (June 2004)
31. Webster III, R.J., Memisevic, J., Okamura, A.M.: Design considerations for robotic needle steering. In: Proc. IEEE Int. Conf. on Robotics and Automation, pp. 3599–3605 (April 2005)

Extended Abstract: Structure Determination of Symmetric Protein Complexes by a Complete Search of Symmetry Configuration Space Using NMR Distance Restraints

Shobha Potluri[1], Anthony K. Yan[1], James J. Chou[2], Bruce R. Donald[1,3,4,5], and Chris Bailey-Kellogg[1,5]

[1] Department of Computer Science, Dartmouth College, Hanover, NH 03755, USA
[2] Department of Biological Chemistry and Molecular Pharmacology, Harvard Medical School, Boston, MA 02115, USA
[3] Department of Chemistry, Dartmouth College, Hanover, NH 03755, USA
[4] Department of Biological Sciences, Dartmouth College, Hanover, NH 03755, USA
[5] Corresponding authors. 6211 Sudikoff Laboratory, Hanover, NH 03755
{brd,cbk}@cs.dartmouth.edu

Symmetric homo-oligomers are protein complexes with similar subunits arranged symmetrically [10]. Figure 1 illustrates the structure of a symmetric homo-oligomer called phospholamban. Phospholamban is a membrane protein that helps regulate the calcium level inside the cell and hence aids in muscle contraction and relaxation [7]; ion conductance studies [5] also suggest that phospholamban might have a separate role as an ion channel. A detailed molecular-level understanding of homo-oligomeric structures provides insights into their functions and, in some cases, how to design appropriate drugs. Nuclear Magnetic Resonance (NMR) spectroscopy underlies many structural studies of homo-oligomers, but poses significant computational challenges in inferring three-dimensional structures from indirect (and often sparse) measurements of geometry.

We use two types of information in homo-oligomeric structure determination: distance restraints from nuclear Overhauser effect (NOE) data, and biophysical modeling terms evaluating packing quality. An inter-subunit distance restraint is of the form $\|\mathbf{p} - \mathbf{q}'\| \leq d$, where $\mathbf{p}$ and $\mathbf{q}'$ are atoms in different subunits of the complex, and d is the given distance for the restraint. We say that a structure is consistent with a distance restraint if $\mathbf{p}$ and $\mathbf{q}'$ are within d Å of each other. The experimental data are complemented by biophysical models of the (non-covalent) interactions that stabilize complexes. Figure 1(b,c) illustrates that the atoms of adjacent subunits of phospholamban are well-packed, interacting at just the right distance to hold the complex together. Packing interactions are typically evaluated with functions that model the van der Waals (vdW) energies between the atoms forming the complex [1, 4]. Our approach separately accounts for experimental data and biophysical modeling terms, and ultimately finds structures of symmetric homo-oligomers that are consistent with the inter-subunit distance restraints and that display high-quality inter-subunit packing interactions.

Traditional protocols [6] for structure determination of protein complexes from NMR data use simulated annealing and molecular dynamics to optimize a pseudo-potential

S. Akella et al. (Eds.): Algorithmic Foundation of Robotics VII, STAR 47, pp. 335–340, 2008.
www.springerlink.com

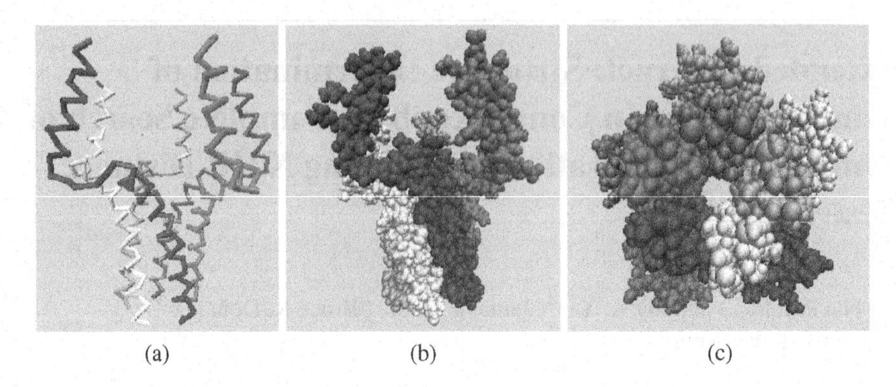

(a) (b) (c)

Fig. 1. Structure of Phospholamban. The five subunits are shown in different colors. (a) Wireframe (backbone trace) representation. (b) Van der Waals sphere representation of all the atoms. (c) Van der Waals sphere representation, viewed down the symmetry axis and illustrating the 5-fold symmetry.

combining both biophysical terms (including packing interactions) and terms evaluating consistency with experimental data. The goal is to find low-energy conformations, but these techniques may become trapped in local minima and miss structures consistent with the data. The precision in the determined structure is also strongly affected by the annealing temperature. Further, since these approaches combine data and packing, they cannot identify the contribution to the structure from the experimental data alone versus both data and packing. Alternative docking-based approaches [2, 3, 8] for structure determination typically involve a two-stage approach: generate a set of possible docked structures, and then score them. The possible structures are generated by a heuristic and/or grid-based sampling of the space of rotations and translations of one subunit with respect to another. The generated structures are scored by geometric/energetic functions, and can be filtered based on symmetry. However, the sampling in the generation step does not account for consistency with the data and thus may miss consistent structures. Wang et al. [11] developed a branch-and-bound algorithm to compute rigid body transformations satisfying potentially ambiguous inter-subunit distance restraints. In contrast to this approach, our algorithm exploits the kinematics of the 'closed-ring' constraint due to symmetry, and thereby derives an analytical bound for pruning, which is tighter and more accurate than the previous randomized numerical techniques.

Our approach, described in detail in [9], is *complete* in that it tests *all* possible structures, and it is *data-driven* in that our algorithm has two separate stages where the first stage only tests structures for consistency with the data, and the second stage evaluates the consistent structures for vdW packing. Completeness ensures that our algorithm does not miss any solutions because it returns a superset of all structures which are consistent with the data. This avoids bias in the search, as well as any potential for becoming trapped in local minima. The data-driven nature of our method allows us to independently quantify the amount of structural constraint provided by data alone, versus both data and packing. This avoids over-reliance on subjective choices of parameters for energy minimization [1], and consequent false precision in determined structures.

Given a set of inter-subunit NOE restraints, the subunit structure and oligomeric number (number of subunits forming the complex) as input, our approach determines the 3D structure of a symmetric homo-oligomer. (We note that it is possible to experimentally determine the subunit structure prior to computing the complex [7].) Given a single (fixed) sub-unit structure, the entire structure of the homo-oligomer is determined by the position and orientation of the symmetry axis. We take a configuration space-based approach and represent each possible structure of the symmetric homo-oligomer by a point in the four-dimensional space of symmetry-axis parameters, which we call the *symmetry configuration space* (SCS), $S^2 \times \mathbb{R}^2$. Geometrically, a point in $\mathbb{R}^2$ represents the position of the symmetry axis, and a unit vector in S^2 gives the orientation of the symmetry axis. We must identify all points in SCS representing symmetry axes that lead to structures consistent with the given set of inter-subunit distance restraints. Let $R_{\mathbf{a}}(\theta) \in SO(3)$ be a rotation around the unit vector $\mathbf{a}$ by $\theta = 2\pi/n$ radians, where n is the oligomeric number. Let $\mathbf{t} \in \mathbb{R}^2$ be the point where the axis of rotation pierces the xy-plane, specifying the location of the symmetry axis. For an atom $\mathbf{q}$ in the fixed subunit, the corresponding atom in the adjacent subunit, $\mathbf{q}'$, when the symmetry axis is at $(\mathbf{a}, \mathbf{t})$, is obtained as $\mathbf{q}' = T_{\mathbf{at}}(\mathbf{q}) = R_{\mathbf{a}}(\theta)(\mathbf{q} - \mathbf{t}) + \mathbf{t}$. We wish to find the set

$$M = \{(\mathbf{a}, \mathbf{t}) \mid \mathbf{a} \in S^2, \mathbf{t} \in \mathbb{R}^2, \|\mathbf{p} - T_{\mathbf{at}}(\mathbf{q})\| \leq d \; \forall \text{ ordered triples } (\mathbf{p}, \mathbf{q}, d) \in D\}, \quad (1)$$

where D is the set of inter-subunit distance restraints, each specifying atoms $\mathbf{p}$ and $\mathbf{q}$ in the fixed subunit and distance d. A restraint constrains the maximum distance between $\mathbf{p}$ and $T_{\mathbf{at}}(\mathbf{q})$, the atom corresponding to $\mathbf{q}$ in the adjacent subunit when the symmetry axis is at $(\mathbf{a}, \mathbf{t})$. The set M corresponds to all points in SCS that satisfy all the restraints.

In order to compute the set M, we perform a search over the SCS. The SCS is too large to search naïvely or exhaustively. Therefore, we have developed a novel branch-and-bound algorithm to search the SCS that is efficient and provably conservative in that it examines and conservatively eliminates regions in SCS inconsistent with the data. Without this algorithm, a complete, data-driven search would not be computationally feasible. The branch-and-bound search performs a search of the SCS by hierarchically subdividing it. Each node in the tree is a SCS *cell*—a 4-dimensional hypercuboid defined by values representing extrema along each of the four dimensions. At each node of the hierarchical subdivision, we test whether any point in the cell represents a consistent structure. If such a point possibly exists, we *branch* and partition the cell into smaller sub-cells. We continue branching until we can either *eliminate* or *accept* each cell. We *eliminate* a cell when all the structures represented by the cell violate at least one restraint (see below) or contain several atoms that significantly clash with each other. We conservatively *accept* a cell as part of the consistent regions when all the structures it represents either provably satisfy all the restraints or are within an RMSD (root mean square deviation) of τ_0 Å (a user-defined similarity level) of each other and each restraint is satisfied by at least one structure represented by the cell. At the end of the branch-and-bound search, we return regions in SCS, the *consistent regions*, which provably contain all structures that are consistent with the data.

To test whether we can eliminate a cell G due to restraint violation, we independently consider each restraint, $\|\mathbf{p} - \mathbf{q}'\| \leq d$. We would like to compute $G\mathbf{q}$ (recall that $\mathbf{q}$ corresponds to $\mathbf{q}'$ in the fixed subunit), the set of all possible positions of $\mathbf{q}'$ under

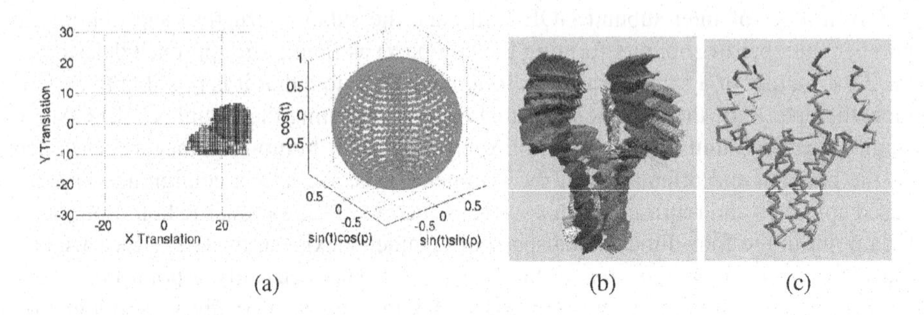

Fig. 2. Phospholamban Results: (a) Region of 4D space output by our branch and bound approach using the nine experimental restraints and knowledge of C_5 symmetry. The solution space of translation parameters and rotation parameters (theta angle denoted by t and phi angle denoted by p) on a sphere is shown. (b) The set of WPS structures after alignment to the structure with lowest packing score. Different subunits are in different colors. (c) Variance of the atoms illustrated by a color scale with blue indicating maximum variance and red minimum variance.

the symmetries defined by G. Since the region $G\mathbf{q}$ is characterized by high-degree polynomials and it is computationally expensive to test for intersections with $G\mathbf{q}$, we approximate $G\mathbf{q}$ by a *conservative bounding region* that completely contains $G\mathbf{q}$. If there is an empty intersection between the conservative bounding region and the ball of radius d centered at $\mathbf{p}$, then all the structures represented by G violate the restraint and we eliminate G.

Figure 2(a) shows the consistent regions in SCS for phospholamban based on the nine experimentally-determined distance restraints. For the sake of illustration, we show the consistent regions as separate 2-d projections into S^2 and $\mathbb{R}^2$. The volume of the consistent regions in the SCS is 1.24 Å^2-radian^2. This volume indicates the constraint on structure provided by data alone. The larger the volume, the lesser the constraint.

Once the consistent regions have been identified, we choose *representative structures* from them such that every structure in the consistent regions is within an RMSD of τ_0 Å to at least one representative structure. Note that this sampling is different from the sampling in docking-based approaches in that the native structures are always within τ_0 Å to at least one of the representative structures. Due to the conservative bounds used in our search, the representative structures might contain structures that are inconsistent with the data. The set of *satisfying structures* includes only those representative structures with restraint satisfaction scores below a chosen threshold. We then evaluate each of the satisfying structures by energy-minimizing and scoring them based on van der Waals packing. The set of *well-packed satisfying (WPS) structures* includes those energy-minimized satisfying structures with van der Waals packing scores below a chosen threshold. Thus, we ultimately return a set of structures consistent with data and packing representing any consistent, well-packed structure to within an RMSD of τ_0 Å.

The structural uncertainty in a set of structures can be quantified by the average variance in the positions of the atoms. The satisfying structures of phospholamban have a variance of 12.32 Å; the incorporation of vdW packing reduces this to 6.80 Å for the well-packed satisfying structures. Figure 2(b) illustrates the set of WPS structures for phospholamban. Figure 2(c) illustrates the variance of the atoms in the set of WPS structures. There is less uncertainty in the lower half of each subunit than in the upper half, since there are more experimental restraints in the lower half. Our complete approach hence allows us to identify the atoms of the complex that have high structural uncertainty. Further, it allows us to separately quantify the amount of structural constraint provided by data alone (satisfying structures), versus data and packing (WPS structures).

Our approach also provides for an independent verification of the oligomeric number, which is typically determined using experiments such as chemical cross-linking followed by SDS-PAGE, or by equilibrium sedimentation. We determine the oligomeric number by extending our search space to include a search over possible oligomeric numbers. We refer to this extended space as the *extended symmetry configuration space* (ESCS), $\mathbb{Z}_9 \times S^2 \times \mathbb{R}^2$, where $\mathbb{Z}_9$ is the set of possible oligomeric numbers of 2 to 9. We first obtain the set of WPS structures for each oligomeric number. We immediately prune out those oligomeric numbers that have no WPS structures. This allows us to determine the oligomeric number with high certainty when only a single oligomeric number has WPS structures. When several oligomeric numbers have WPS structures, we determine the oligomeric number as follows. Let $E_l(m)$ and $E_l(n)$ represent the lowest packing scores of the WPS structures from oligomeric numbers of m and n respectively. If $E_l(m) < E_l(n)$, the difference $E_l(n) - E_l(m)$ indicates the confidence we have in preferring m versus n as the oligomeric number. On applying this approach to determine the oligomeric number of phospholamban, the pentamer has the lowest packing score causing us to correctly conclude that the pentamer is the most feasible oligomeric number.

In summary, we have developed a novel approach that performs a complete, data-driven search to identify all structures of a homo-oligomeric complex that are consistent with NOE restraints and display high-quality vdW packing. Our tests on phospholamban and four other proteins demonstrate the power of our method in determining and evaluating homo-oligomeric complex structures. Our approach is particularly important in sparse-data cases, where relying on an incomplete, biased search may result in missing well-packed, satisfying conformations. Examination of the entire solution space further enables objective evaluation of the amount of structural uncertainty. Finally, we show that it is possible to determine the oligomeric number directly from NMR data. The details of our methods and results are available in our paper [9].

Acknowledgments

We would like to acknowledge members of the CBK lab and Ivelin Georgiev from the BRD lab for helpful discussions. This work was supported in part by the following grants, to BRD: National Institutes of Health (R01 GM 65982) and National Science Foundation (EIA-9802068 and EIA-0305444); CBK: National Science Foundation (IIS-0444544 and IIS-0502801); JJC: Smith Family Award for Young Investigators. JJC is a Pew scholar.

References

1. Brunger, A.T., Adams, P.D., Clore, G.M., DeLano, W.L., Gros, P., Grosse-Kunstleve, R.W., Jiang, J.S., Kuszewski, J., Nilges, M., Pannu, N.S., Read, R.J., Rice, L.M., Simonson, T., Warren, G.L.: Crystallography and NMR system: A new software suite for macromolecular structure determination. Acta. Cryst. D54, 905–921 (1998)
2. Comeau, S.R., Camacho, C.J.: Predicting oligomeric assemblies: N-mers a primer. J. Struct. Biol. 150, 233–244 (2005)
3. Duhovny, D., Nussinov, R., Wolfson, H.J.: Efficient unbound docking of rigid molecules. In: Guigó, R., Gusfield, D. (eds.) WABI 2002. LNCS, vol. 2452, pp. 185–200. Springer, Heidelberg (2002)
4. Dunfield, L.G., Burgess, A.W., Scheraga, H.A.: Energy parameters in polypeptides. 8. Empirical potential energy algorithm for the conformational analysis of large molecules. J. Phys. Chem. 82, 2609–2616 (1978)
5. Kovacs, R.J., Nelson, M.T., Simmerman, H.K., Jones, L.R.: Phospholamban forms Ca2+-selective channels in lipid bilayers. J. Biol. Chem. 263, 18364–18368 (1988)
6. Nilges, M.: A calculation strategy for the structure determination of symmetric dimers by 1H NMR. Proteins 17, 297–309 (1993)
7. Oxenoid, K., Chou, J.J.: The structure of phospholamban pentamer reveals a channel-like architecture in membranes. PNAS 102, 10870–10875 (2005)
8. Pierce, B., Weng, Z.: M-ZDOCK: A grid-based approach for C_n symmetric multimer docking. Bioinformatics 21, 1472–1476 (2005)
9. Potluri, S., Yan, A.K., Chou, J.J., Donald, B.R., Bailey-Kellogg, C.: Structure determination of symmetric protein complexes by a complete search of symmetry configuration space using NMR distance restraints. Proteins 65, 203–219 (2006)
10. Goodsell, D.S., Olson, A.J.: Structural symmetry and protein function. Annu. Rev. Biophys. Biomol. Struct. 29, 105–153 (2000)
11. Wang, C.E., Pérez, T.L., Tidor, B.: AMBIPACK: A systematic algorithm for packing of macromolecular structures with ambiguous distance constraints. Proteins 32, 26–42 (1998)

Part VII

Control and Planning for Mechanical Systems

The Minimum-Time Trajectories for an Omni-Directional Vehicle

Devin J. Balkcom[1], Paritosh A. Kavathekar[1], and Matthew T. Mason[2]

[1] Dartmouth Computer Science Department
{devin,paritosh}@cs.dartmouth.edu
[2] Carnegie Mellon Robotics Institute
matt.mason@cs.cmu.edu

Abstract. One common mobile robot design consists of three 'omniwheels' arranged at the vertices of an equilateral triangle, with wheel axles aligned with the rays from the center of the triangle to each wheel. Omniwheels, like standard wheels, are driven by the motors in a direction perpendicular to the wheel axle, but unlike standard wheels, can slip in a direction parallel to the axle. Unlike a steered car, a vehicle with this design can move in any direction without needing to rotate first, and can spin as it does so. We show that if there are independent bounds on the speeds of the wheels, the fastest trajectories for this vehicle contain only spins in place, circular arcs, and straight lines parallel to the wheel axles. We classify optimal trajectories by the order and type of the segments; there are four such classes, and there are no more than 18 control switches in any optimal trajectory.

1 Introduction

This paper presents the time-optimal trajectories for a simple model of the common mobile-robot design shown in figure 1(b). [1] The three wheels are "omni-wheels"; the wheels not only rotate forwards and backwards when driven by the motors, but can also slip sideways freely. Such a robot can drive in any direction instantaneously.

The only other ground vehicles for which the fastest trajectories are known explicitly are steered cars and differential-drives. Although our results are specific to the particular vehicle studied, we hope that expanding the set of vehicles for which the optimal trajectories are known will eventually lead to a more unified understanding of the relationship between robot mechanism design and the use of resources.

We show that the time-optimal trajectories consist of spins in place, circular arcs, and straight lines parallel to the wheel axles. We label each segment type by a letter: P, C, S, respectively. There are specific sequences of segments that may be optimal; we call the four possible classes of trajectories *spin*, *roll*, *shuffle*, and *tangent*. Figures 3(a), 4(a), 4(b), and 4(c) show an example of each type.

1. *Spin* trajectories consist of a spin in place through an angle no greater than π, and are described by the single-letter control sequence P.
2. *Roll* trajectories consist of a sequence of up to five circular arcs of equal radius separated by spins in place, and are described by the control sequence CPCPC.

[1] A version of this paper with complete proofs of the theorems appears in [1].

S. Akella et al. (Eds.): Algorithmic Foundation of Robotics VII, STAR 47, pp. 343–358, 2008.
www.springerlink.com

The centers of the arcs all fall on a straight line. With the possible exception of the first and last segments, the arcs all encompass the same angle, as do the spins, and the sum of the angular displacement of a complete arc and a complete spin is 120°.

3. *Shuffle* trajectories are composed of sequences of three circular arcs followed by a spin, CCCP, and contain no more than seven control switches. A complete period of a *shuffle* moves the vehicle 'sideways' in a direction parallel to a line connecting two wheels.
4. *Tangent* trajectories consist of a sequence of arcs of circles and spins in place separated by arbitrarily long translations in a direction parallel to the line containing the center of the robot and one of its wheels. All straight segments are colinear. The control sequence is CSCSCP, and trajectories contain no more than 18 switches. Intuitively, the robot 'lines up' in its fastest direction of translation, translates, and then follows arcs of circles to arrive at its final position and orientation.

Why study optimal trajectories? Knowledge of the shortest or fastest paths between any two configurations of a particular robot is fundamental. Robots expend resources to achieve tasks. Possibly the simplest resource is time; the amount of time that must be expended to move the robot between configurations is a basic property of the mechanism, and a fundamental metric on the configuration space.

Knowledge of the time optimal trajectories is also useful. Mechanisms should be designed so that common tasks can be achieved efficiently. Furthermore, the time-optimal metric is independent of software-design decisions, and therefore provides a benchmark to compare planners or controllers. Finally, the metric derived from the optimal trajectories may be used as a heuristic to guide sampling in complete planning systems that permit obstacles or a more complex dynamic model of the mechanism.

We do not argue that controllers should be designed to drive robots to follow the 'optimal' trajectories we derive.In fact, resources other than time may also be important, including energy consumption, safety, simplicity of programming, sensing opportunities, and accuracy. Tradeoffs must be made, but understanding the relative payoffs of each design requires an understanding of the fundamental behavior of the mechanism. The knowledge that great circles are geodesics on the sphere does not require that airplanes must strictly follow great circles, but may nonetheless influence the choice of flight paths.

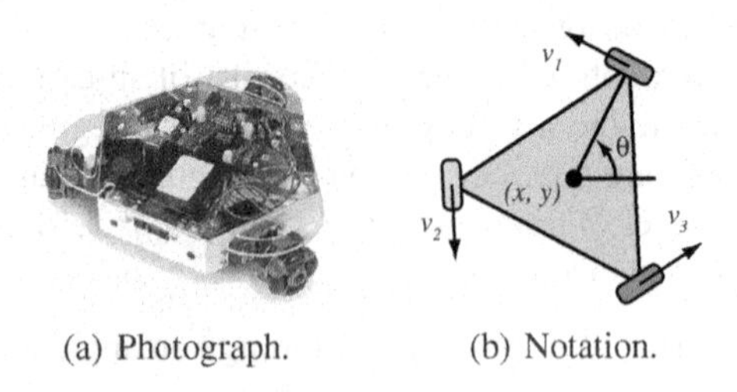

(a) Photograph. (b) Notation.

Fig. 1. The Palm-Pilot Robot Kit, an example of an omni-directional vehicle. Photograph used by permission of Acroname, Inc., `www.acroname.com`.

1.1 Related Work

Most of the work on time-optimal control for vehicles has focused on bounded-velocity models of steered cars. Dubins [7] determined the shortest paths between two configurations of a car that can only move forwards at constant speed, with bounded steering angle. Reeds and Shepp [10] found the shortest paths for a steered car that can move backwards as well as forwards. Sussmann and Tang [15] further refined these results, reducing the number of families of trajectories thought to be optimal by two, and Souères and Boissonnat [13], and Souères and Laumond [14] discovered the mapping from pairs of configurations to optimal trajectories for the Reeds and Shepp car. Desaulniers [6] showed that in the presence of obstacles shortest paths may not exist between certain configurations of steered cars. Furthermore, in addition to the straight lines and circular arcs of minimum radius discovered by Dubins, the shortest paths may also contain segments that follow the boundaries of obstacles. Vendittelli *et al.* [16] used geometric techniques to develop an algorithm to obtain the shortest non-holonomic distance from a robot to any point on an obstacle.

Recently, the optimal trajectories have been found for vehicles that are not steered cars, and metrics other than time. The time-optimal controls for bounded-velocity differential-drives were discovered by Balkcom and Mason [2]. Chitsaz *et al.* [3] determined the trajectories for a differential-drive that minimize the sum of the rotation of the two wheels. The optimal paths have also been explored for some examples of vehicles without wheels. Coombs and Lewis [5] consider a simplified model of a hovercraft, and Chyba and Haberkorn [4] consider underwater vehicles. We know of no previous closed-form solutions for the optimal trajectories for any wheeled omni-directional vehicle.

Bounded-velocity models of the type we study capture the kinematics of a vehicle, but not the dynamics. The results of this paper strongly depend on the analytical solution of differential equations describing the optimal trajectories. Analysis of dynamic models, for which analytical solutions are not typically available, is a very difficult problem. Results include numerical techniques and geometric characterization rather than complete closed-form solutions; see papers by Reister and Pin [11], Renaud and Fourquet [12], and Kalmár-Nagy *et al.* [8].

2 Model, Assumptions, Notation

Let the state of the robot be $q = (x, y, \theta)^T$; the location of the center of the robot, and the angle that the line from the center to the first wheel makes with the horizontal, as shown in figure 1(b). Without loss of generality, we assume that the distance from the center of the robot to the wheels is one. We further assume that each of the three wheel-speed controls v_1, v_2, and v_3 is in the interval $[-1, 1]$. We define the control region

$$U = [-1, 1] \times [-1, 1] \times [-1, 1], \tag{1}$$

and consider the class of *admissible controls* to be the measurable functions $u(t)$ mapping the time interval $[0, T]$ to U: $u(t) = (v_1(t), v_2(t), v_3(t))^T$.

To simplify notation, we define $c_i = \cos\theta_i$, and $s_i = \sin\theta_i$, where $\theta_i = \theta + (i - 1)120°$, the angle of the ith wheel measured from the horizontal. Define the matrix S to be the Jacobian that transforms between configuration-space velocities of the vehicle, and velocities of the wheels in the controlled direction:

$$S = \begin{bmatrix} -s_1 & c_1 & 1 \\ -s_2 & c_2 & 1 \\ -s_3 & c_3 & 1 \end{bmatrix}, \qquad S^{-1} = \frac{2}{3} \begin{bmatrix} -s_1 & -s_2 & -s_3 \\ c_1 & c_2 & c_3 \\ 1/2 & 1/2 & 1/2 \end{bmatrix}. \tag{2}$$

We define the state trajectory $q(t) = (x(t), y(t), \theta(t))$ for any initial state q_0 and admissible control $u(t)$ using Lebesgue integration, with the standard measure:

$$q(t) = q_0 + \int S^{-1} u\,. \tag{3}$$

It may be easily verified that the kinematic equations and bounds on the controls satisfy the conditions of theorem 6 of Sussmann and Tang [15]; an optimal trajectory exists between every pair of start and goal configurations.

3 Pontryagin's Maximum Principle

This section uses Pontryagin's Maximum Principle [9] to derive necessary conditions for time-optimal trajectories. The Maximum Principle states that if the trajectory $q(t)$ with corresponding control $u(t)$ is time-optimal then the following conditions must hold:

1. There exists a non-trivial (not identically zero) *adjoint function*: an absolutely continuous $\mathbf{R}^3$-valued function of time, $\lambda(t)$, defined by a differential equation, the *adjoint equation*, in the configuration and in time-derivatives of the configuration:

$$\dot{\lambda} = -\frac{\partial}{\partial q}\langle \lambda, \dot{q}(q,u)\rangle \quad \text{a.e.} \tag{4}$$

   We call the inner product appearing in equation 4 the *Hamiltonian*:

$$H(\lambda, q, u) = \langle \lambda, \dot{q}(q,u)\rangle. \tag{5}$$

2. The control $u(t)$ minimizes the Hamiltonian:

$$H(\lambda(t), q(t), u(t)) = \min_{z \in U} H(\lambda(t), q(t), z) \quad \text{a.e.} \tag{6}$$

   Equation 6 is called the *minimization equation*.
3. The Hamiltonian is constant and non-positive over the trajectory. We define λ_0 as the negative of the value of the Hamiltonian; λ_0 is constant and non-negative for any optimal trajectory.

3.1 Application of the Maximum Principle

We solve for the adjoint vector by direct integration: $\lambda_1 = 3k_1$, $\lambda_2 = 3k_2$, and $\lambda_3 = 3(k_1 y - k_2 x + k_3)$, where $3k_1$, $3k_2$, and $3k_3$ are constants of integration. (The constant factor of 3 will simplify the form of equation 9 below.)

We now substitute the adjoint function into the minimization equation to determine necessary conditions for time-optimal trajectories. To simplify notation, we define three functions,

$$\varphi_i(t) = \langle \lambda(t), f_i(q(t)) \rangle, \tag{7}$$

where f_i is the ith column of S^{-1}. Explicitly, if we define the function

$$\eta(x, y) = k_1 y - k_2 x + k_3, \tag{8}$$

then the functions are:

$$\varphi_i = 2(-k_1 s_i + k_2 c_i) + \eta(x, y) \tag{9}$$

We may now write the equation for the Hamiltonian in terms of these functions and the controls v_1, v_2, and v_3:

$$H = \varphi_1 v_1 + \varphi_2 v_2 + \varphi_3 v_3. \tag{10}$$

The minimization condition of the Maximum Principle (condition 2, above) applied to equation 10 implies that if the function φ_i is negative, then v_i should be chosen to take its maximum possible value, 1, in order to minimize H. If the function φ_i is positive, then v_i should be chosen to be -1. Since the controls switch whenever one of the functions φ_i changes sign, we refer to the functions φ_i as *switching functions*.

Theorem 1. *For any time-optimal trajectory of the omni-directional vehicle, there exist constants k_1, k_2, and k_3, with $k_1^2 + k_2^2 + k_3^2 \neq 0$, such that at almost every time t, the value of the control v_i is determined by the sign of the switching function φ_i:*

$$v_i = \begin{cases} 1 & \text{if } \varphi_i < 0 \\ -1 & \text{if } \varphi_i > 0, \end{cases} \tag{11}$$

where the switching functions φ_1, φ_2, and φ_3 are given by equations 8, 9. Furthermore, the quantity λ_0 defined by

$$\lambda_0 = -H(\varphi_1, \varphi_2, \varphi_3) = |\varphi_1| + |\varphi_2| + |\varphi_3| \tag{12}$$

is constant along the trajectory.

Proof: Application of the Maximum Principle. ■

The Maximum Principle does not directly give information about the optimal controls in the case that one or more of the switching functions φ_i is zero. Theorems 7 and 8 in section 4 specifically address this case. The Maximum Principle also does not give information about the constants of integration, as these depend on the initial and final configurations of the robot. In this paper, we give the structure of trajectories as a function of these constants, but do not describe how to determine the constants except in a few cases.

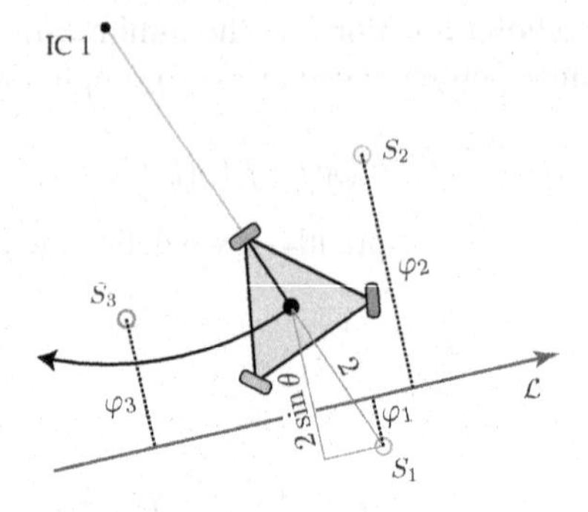

Fig. 2. Geometric interpretation of the switching functions. For the case shown, $\varphi_1 < 0$, $\varphi_2 > 0$, and $\varphi_3 > 0$, so the controls are $v_1 = 1$, $v_2 = -1$, and $v_3 = -1$.

3.2 Geometric Interpretation of the Switching Functions

The switching functions are not independent, and have a geometric interpretation. Consider the function $\eta(x, y)$:

$$\eta(x, y) = k_1 y - k_2 x + k_3. \tag{13}$$

$\eta(x, y)$ gives the signed distance of the point (x, y) from a line in the plane whose location is determined by the constants k_1, k_2, and k_3, scaled by the factor $k_1^2 + k_2^2$. (If $k_1^2 + k_2^2 = 0$, we may consider the line to be 'at infinity'; the robot spins in place indefinitely. Since this control is identical to the *spin* trajectories described in section 5, we do not consider this case separately.) We will call this line the *switching line*. We also associate a direction with the switching line such that any point (x, y) is to the left of the switching line if $\eta(x, y) > 0$, and to the right of the switching line if $\eta(x, y) < 0$.

Theorem 2. *Define the points S_1, S_2, and S_3 rigidly attached to the vehicle, with distance 2 from the center of the vehicle, and making angles of $180°$, $300°$, and $60°$ with the ray from the center of the vehicle to wheel 1, respectively (refer to figure 2). For any time-optimal trajectory, there exist constants k_1, k_2, and k_3, and a line (the switching line)*

$$\mathcal{L} = \{(a, b) \in \mathbf{R}^2 : k_1 b - k_2 a + k_3 = 0\},$$

such that the controls of the vehicle v_1, v_2, and v_3 depend on the location of the points S_1, S_2, and S_3 relative to the line. Specifically, for $i \in \{1, 2, 3\}$,

$$v_i = \begin{cases} 1 & \text{if } S_i \text{ is to the right of the switching line,} \\ -1 & \text{if } S_i \text{ is to the left of the switching line.} \end{cases}$$

Proof: Let (x_{S_i}, y_{S_i}) be the coordinates of S_i. We compute the signed, scaled distance of the point S_i from the line $\mathcal{L}$, and observe from the definition of the switching functions that $\varphi_i(x, y, \theta) = \eta(x_{S_i}, y_{S_i})$. ■

We will call S_1, S_2, and S_3 the *switching points*. For any optimal trajectory, the location of the switching line is fixed by the choice of constants, and the controls at any point depend on the signs, but not on the magnitudes, of the switching functions. Figure 2 shows an example. Two of the switching points (S_2 and S_3) are to the left of the switching line, so the corresponding switching functions are positive, and wheels 2 and 3 spin at full speed in the negative direction. The remaining switching point (S_1) is to the right of the switching line, so wheel 1 spins at full speed in the positive direction. As a result of these controls, the robot will follow a clockwise circular arc. The center of the arc is a distance of four from the robot, and along the line containing the center of the robot and wheel 1.

In general, if all three switching functions have the same sign, the controls all take either their maximum or minimum value, and the robot spins in place. The center of rotation is the center of the robot; we call this point IC 0. If the switching functions are non-zero but do not all have the same sign, the vehicle rotates in a circular arc. The rotation center is a distance of four from the center of the robot, on the ray connecting the center of the robot and the wheel corresponding to the 'minority' switching function. We call these rotation centers IC 1, IC 2, and IC 3.

The switching functions are invariant to translation of the vehicle parallel to the switching line (see figure 2), and scaling the switching functions by a positive constant does not affect the controls. Therefore, for any optimal trajectory, we may without loss of generality choose a coordinate frame with x-axis on the switching line, and an appropriate scaling, such that y gives the distance from the switching line, and θ gives the angle of the vehicle relative to the switching line. With this choice of coordinates, the switching functions become

$$\varphi_i = y - 2s_i \tag{14}$$

We will use these coordinates for the remainder of the paper.

4 Properties of Extremals

We will say that any trajectory that satisfies the conditions of theorem 1 (or equivalently, theorem 2) is *extremal*. In this section, we will enumerate several properties of extremal trajectories. The primary result is that every extremal trajectory contains only a finite number of control switches with an upper bound determined by λ_0.

We say that an extremal trajectory is *generic* on some interval if none of the switching functions φ_i is zero at any point contained in the interval. We say that a trajectory is *singular* on some interval if exactly one of the switching functions is identically zero on that interval, and no other switching function is zero at any point on the interval. We say that an extremal trajectory is *doubly singular* on an interval if exactly two of the switching functions are zero on that interval, and the third switching function is never zero on the interval. We will call a trajectory *singular* if it contains any singular interval of non-zero width.

Detailed proofs of the following properties are omitted due to space limitations, but may be found in [1] . Most of the proofs are based on differential analysis of the switching functions.

Theorem 3. *At no point along an extremal trajectory does* $\varphi_1 = \varphi_2 = \varphi_3 = 0$.

Theorem 4. *If an extremal trajectory contains any doubly-singular point, then every point of the trajectory is doubly-singular.*

Theorem 5. *Every pair of singular points of an extremal trajectory is contained in a single singular interval, or is separated by a generic point.*

Theorem 6. *The number of control switches in an extremal trajectory is finite, and upper-bounded by a constant that depends only on* λ_0.

In section 7, we will show that for *optimal* trajectories, a much stronger property holds: the number of control switches is never greater than 18.

Theorem 7. *Consider a singular interval of non-zero duration, with* $\varphi_i = 0$. *At every point of the interval,* $y = \sin\theta_i = 0$, *and the controls are constant:* $v_i = 0$, *and* $v_j = -v_k = \pm 1$.

Theorem 8. *Consider a doubly-singular interval of non-zero duration, with* $\varphi_i = \varphi_j = 0$. *Along the interval, (i)* $y = \pm 1$, $\cos\theta_k = 0$, *and (ii) the controls are constant, with* $v_k = \pm 1$, *and* $v_i = v_j = \mp .5$.

5 Extremal Controls

Theorems 1, 6, 7, and 8 imply that optimal trajectories are composed of a finite number of segments, along each of which the controls are constant. Considering all possible combinations of signs and zeros of the switching functions allows the twenty extremal controls to be enumerated; table 1 shows the results. The vehicle may spin in place, follow a circular arc, translate in a direction perpendicular to the line joining two wheels, or translate in a direction parallel to the line joining two wheels. We denote each control by a symbol: $\mathrm{P}_{\pm}$, $\mathrm{C}_{i\pm}$, $\mathrm{S}_{i,j}$, or $\mathrm{D}_{k\pm}$, respectively. The subscripts depend on the specific signs of the switching functions.

Theorem 2 gives a more geometric interpretation of the extremal controls. The controls depend on the location of the switching points relative to the switching line. There are four cases:

- **Spin in place.** If the vehicle is far from the switching line, all of the switching points are on the same side of the line, and all of the wheels spin in the same direction. Figure 3(a) shows an example. If the robot is to the left of the switching line, the robot spins clockwise (P_-); if the robot is to the right of the switching line, the robot spins counterclockwise (P_+).
- **Circular arc.** Figure 3(b) shows an example of a counterclockwise arc around IC2 (C_{2+}). If two switching points are on one side of the line, and one switching point is on the other, two wheels spin in one direction at full speed, and one wheel spins in the opposite direction at full speed. These controls cause the vehicle to follow a circular arc of radius four; the center of the arc is the IC corresponding to the switching point that is not on the same side of the switching line as the others, and the direction of rotation depends on whether this switching point is to the left or right of the line.

Table 1. The twenty extremal controls

Symbol	φ	u	λ_0
P_-	+++	-1, -1, -1	$3y$
P_+	---	1, 1, 1	$-3y$
C_{1-}	-++	1, -1, -1	$y + 4\sin\theta_1$
C_{2-}	+-+	-1, 1, -1	$y + 4\sin\theta_2$
C_{3-}	++-	-1, -1, 1	$y + 4\sin\theta_3$
C_{1+}	+--	-1, 1, 1	$-y - 4\sin\theta_1$
C_{2+}	-+-	1, -1, 1	$-y - 4\sin\theta_2$
C_{3+}	--+	1, 1, -1	$-y - 4\sin\theta_3$

Symbol	φ	u	λ_0
$S_{1,3}$	-0+	1, 0, -1	$2\sqrt{3}$
$S_{1,2}$	-+0	1, -1, 0	$2\sqrt{3}$
$S_{3,2}$	0+-	0, -1, 1	$2\sqrt{3}$
$S_{3,1}$	+0-	-1, 0, 1	$2\sqrt{3}$
$S_{2,1}$	+-0	-1, 1, 0	$2\sqrt{3}$
$S_{2,3}$	0-+	0, 1, -1	$2\sqrt{3}$
D_{3+}	00+	.5, .5, -1	3
D_{1-}	-00	1, -.5, -.5	3
D_{2+}	0+0	.5, -1, .5	3
D_{3-}	00-	-.5, -.5, 1	3
D_{1+}	+00	-1, .5, .5	3
D_{2-}	0-0	-.5, 1, -.5	3

- **Singular translation.** Figure 3(c) shows an example, $S_{1,3}$, where the second switching point slides along the switching line. If two switching points are an equal distance from the switching line but on opposite sides of the line, two of the wheels spin at full speed, but in opposite direction. If the last switching point falls exactly on the switching line, theorem 2 does not provide any information about the speed of the last wheel. If the wheel does not spin, then the vehicle translates along the switching line, as described by theorem 7. Otherwise, the singular translation is only instantaneous.
- **Doubly-singular translation.** Figure 3(b) shows an example, D_{3+}, where the first and second switching points slide along the switching line. If two switching points fall on the switching line, the speeds of the corresponding wheels cannot be determined from theorem 2. If these wheels spin at half speed, in a direction opposite to that of the third wheel, both switching points slide along the switching line, and the vehicle translates. It turns out that that doubly-singular controls, although extremal, are *never* optimal; see section 7.

6 Classification of Extremal Trajectories

Every extremal trajectory is generated by a sequence of constant controls from table 1. However, not every sequence is extremal. This section geometrically enumerates the five structures of extremal trajectories.

First consider an example, shown in figure 4(a). Initially, switching points 1 and 3 fall to the left of the switching line, and switching point 2 falls to the right of the switching line. The vehicle rotates in the clockwise direction about IC 2. After some amount of rotation, switching point 2 crosses the switching line. Now all three switching points are to the left of the switching line, the velocity of wheel 2 changes sign, and the vehicle spins in place. When switching point 3 crosses the switching line, the vehicle begins to rotate about IC 3. When switching point 3 crosses back to the left side, the vehicle

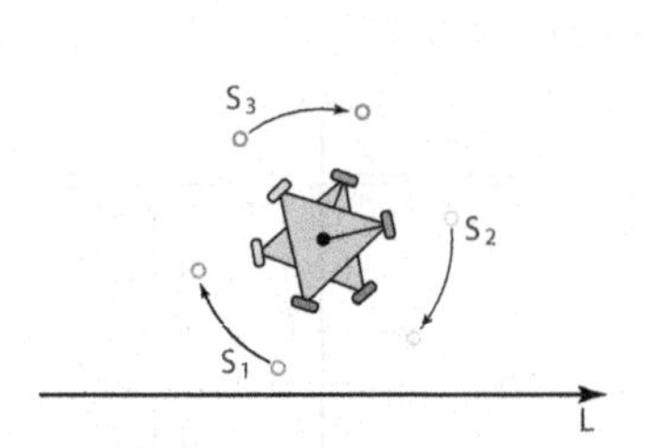

(a) An example clockwise *spin* control, P_-.

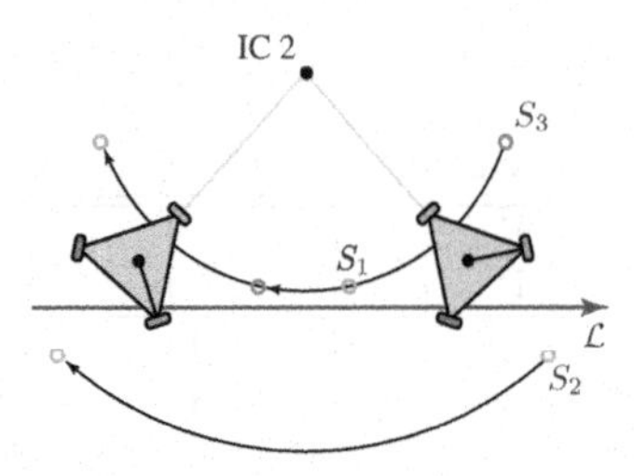

(b) An example clockwise *circular arc* control, C_{2-}.

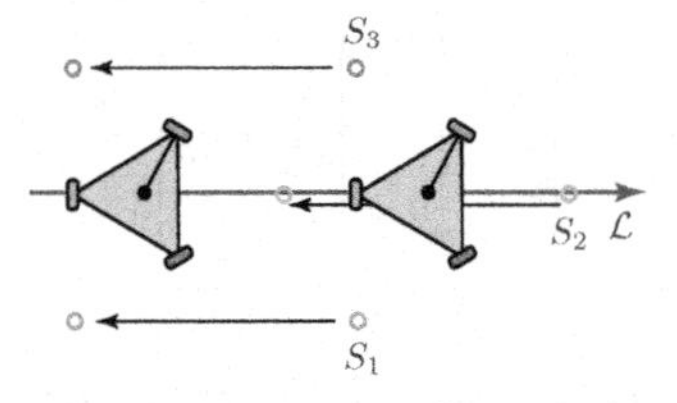

(c) An example *singular translation* control, $S_{1,3}$.

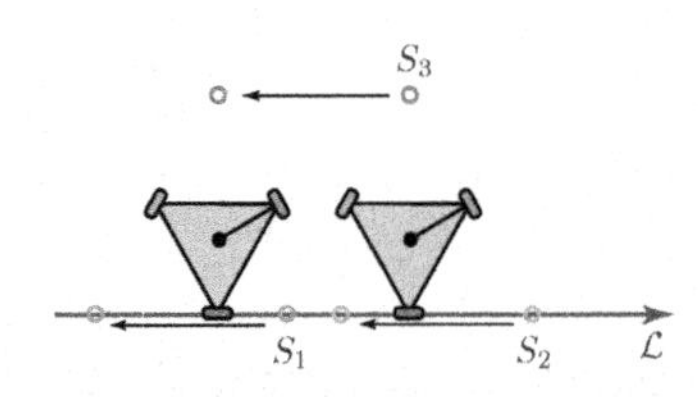

(d) An example *doubly-singular translation* control, D_{3+}.

Fig. 3. Extremal controls for an omni-directional robot

spins in place again until switching point 1 crosses the line. The pattern continues in this form; we describe the trajectory by the sequence of symbols $C_{3+}C_{2-}C_{1+}P_+\ldots$.

In general, if no switching points fall on the switching line (the generic case), then the controls are completely determined by theorem 2, and the vehicle either spins in place or rotates around a fixed point. When one of the switching points crosses the switching line, the controls change. For some configurations for which one or two of the switching points fall exactly on the switching line (the singular and doubly-singular cases), there exist controls that allow the switching points to slide along the switching line.

We will define these classes more rigorously in sections 6.1 and 6.2. However, we can see geometrically that there are five cases:

- **SpinCW** and **SpinCCW.** If the vehicle is far from the switching line, the switching points are on the same side of the switching line and never cross it; the vehicle spins in place indefinitely. The structure off the trajectory is either P_- (if the vehicle is to the left of the switching line) or P_+ (if the vehicle is to the right of the switching line). An example is shown in figure 3(a).
- **RollCW** and **RollCCW.** If the switching points either straddle the switching line, or the vehicle is close enough to the switching line that spinning in place will eventually cause the switching points to straddle the line, the trajectory is a sequence of circular arcs and spins in place. If the vehicle is far enough from the switching line that every switching point crosses the switching line and returns to the same side before the next switching point crosses the line, the structure of the trajectory

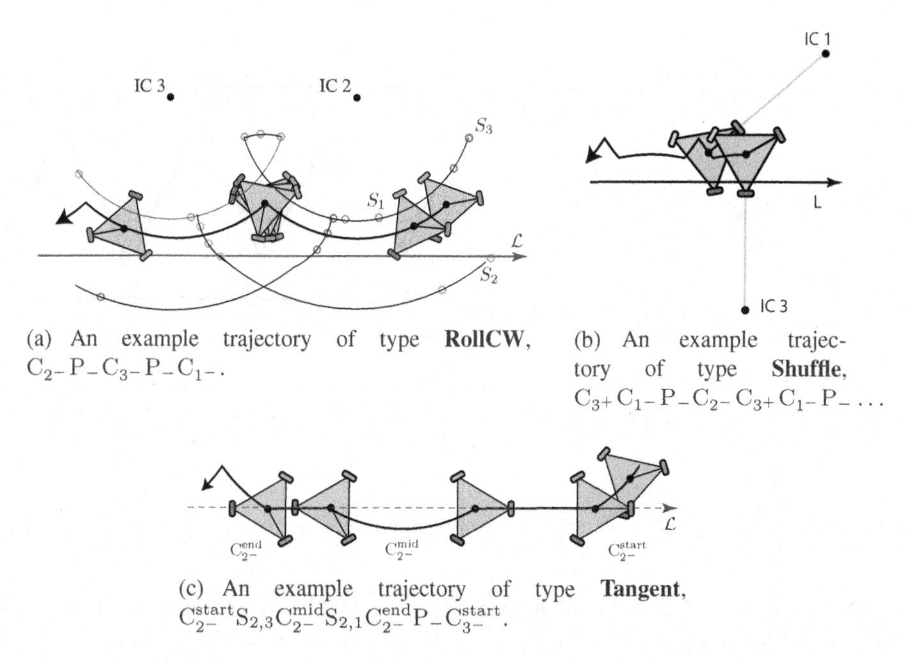

(a) An example trajectory of type **RollCW**, $C_{2-}P_{-}C_{3-}P_{-}C_{1-}$.

(b) An example trajectory of type **Shuffle**, $C_{3+}C_{1-}P_{-}C_{2-}C_{3+}C_{1-}P_{-}\ldots$

(c) An example trajectory of type **Tangent**, $C_{2-}^{\mathrm{start}}S_{2,3}C_{2-}^{\mathrm{mid}}S_{2,1}C_{2-}^{\mathrm{end}}P_{-}C_{3-}^{\mathrm{start}}$.

Fig. 4. Extremal trajectories for an omni-directional robot

is as described in the example above and in figure 4(a). θ is monotonic during the trajectory (except where θ wraps around 0).

- **Shuffle.** If the vehicle is close enough to the switching line that two switching points cross the switching line before the first returns to its initial side, the sign of $\dot{\theta}$ changes during the trajectory. An example is shown in figure 4(b).
- **Tangent.** As the vehicle spins in place or follows a circular arc, the switching points follow circular arcs. If one of these arcs is tangent to the switching line, a singular control becomes possible at the point of tangency, and the vehicle may translate along the switching line for an arbitrary duration before returning to following a circular arc. An example is shown in figure 4(c). A single circular arc is divided into three segments in a *tangent* trajectory. These segments are separated by the singular S curves, possibly of zero duration. We call these segments C^{start}, C^{mid}, and C^{end}, as shown in figure 4(c). The robot rotates through 60° during a complete C^{mid} segment.
- **Slide.** If two switching points fall on the switching line, the trajectory is doubly singular. The vehicle slides along the switching line in a pure translation; an example of this trajectory type is shown by figure 3(d). Although *slide* trajectories are extremal, we will show in section 7 that they are never optimal.

6.1 Configuration Space

In order to show that the above list of trajectory classes is exhaustive, it is useful to consider the structure of trajectories in configuration space. The configuration of the

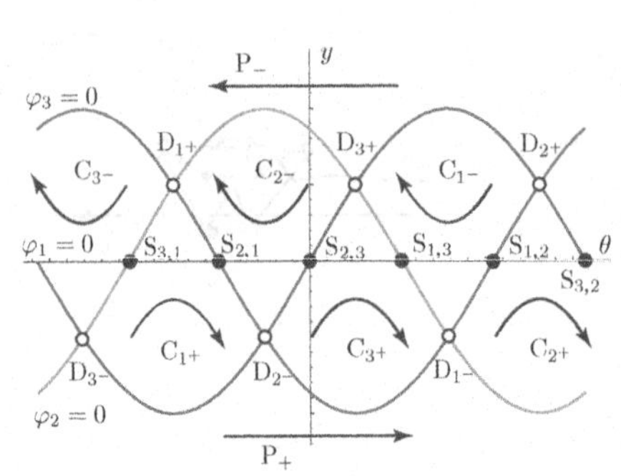

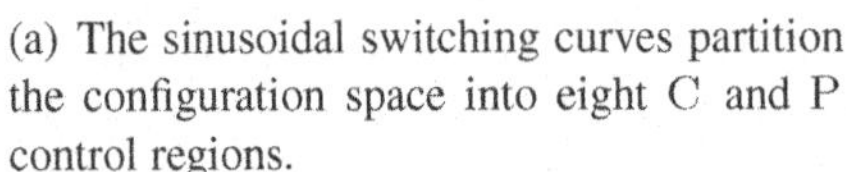
(a) The sinusoidal switching curves partition the configuration space into eight C and P control regions.

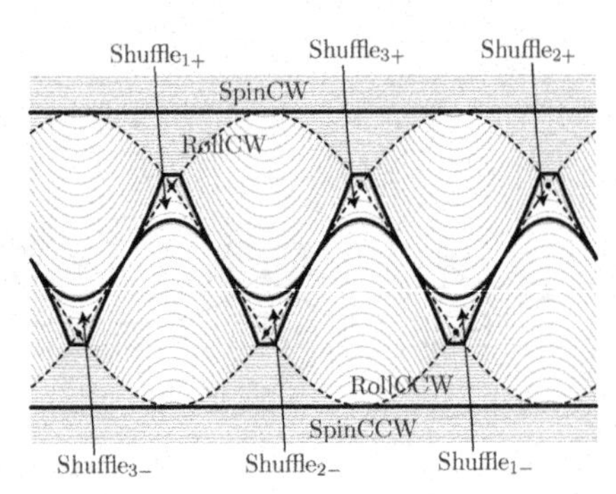

(b) Each trajectory corresponds to a level set (contour) of the Hamiltonian. The dashed lines represent control switches; the bold lines separate the trajectory classes.

Fig. 5. The configuration space of the robot relative to the switching line

robot relative to the switching line may be represented by (θ, y). Figure 5(a) shows the configuration space.

Each point on figure 5(a) corresponds to a configuration of the robot relative to the switching line. The sinusoidal curves defined by $\varphi_1 = 0$, $\varphi_2 = 0$, and $\varphi_3 = 0$ mark boundaries in configuration space; we call these curves the *switching curves*. The switching curves and their intersections divide the configuration space into cells, within each of which the controls are constant.

As an example, consider a point below switching curve 1, but above switching curves 2 and 3. The controls are $(-1, 1, 1)$, described by the symbol $C_{1}-$; the vehicle follows a circular arc around IC 1 in the clockwise direction. This trajectory is a sinusoidal curve in configuration space.

6.2 Level Sets of the Hamiltonian

The trajectory curves in configuration space can be drawn by considering each possible initial configuration, determining the constant control, and integrating to find the trajectory. When the trajectory crosses a switching curve, the control switches. However, the condition that the Hamiltonian remain constant over a trajectory provides an even simpler way to enumerate all trajectories in the configuration space.

Each extremal trajectory falls on a level set of the Hamiltonian (equation 12), and extremal trajectories may be classified by the value λ_0. Figure 5(b) shows the level sets of the Hamiltonian, or equivalently, extremal trajectories in configuration space.

- If $\lambda_0 > 6$, the level set is a pair of horizontal lines, one with $y = \lambda_0/3$, corresponding to a *spinCW* trajectory, and one with $y = -\lambda_0/3$, corresponding to a *spinCCW* trajectory.
- If $2\sqrt{3} \leq \lambda_0 \leq 6$, the level set is composed of two disjoint curves, one corresponding to *rollCW* trajectory and one corresponding to a *rollCCW* trajectory.

- If $\lambda_0 = 2\sqrt{3}$, the level set is the union of the bold curves shown in figure 5(b). Tangent trajectories follow these curves.
- If $3 < \lambda_0 < 2\sqrt{3}$, the level set is composed of six disjoint curves, one corresponding to each of the six symmetric *shuffle* trajectories.
- If $\lambda_0 = 3$, the level set is six isolated points, each corresponding to one of the six *slide* trajectories.

Table 2. Four of the five classes of extremal trajectories. Every optimal trajectory is composed of a sequence of controls that is a subsequence of one of the above types. (Doubly-singular *slide* trajectories are extremal, but never optimal; see section 7.) The structure of *tangent* trajectories is complicated, and shown explicitly in figure 6.

Class	Control sequence	Value of λ_0
SpinCW	P_-	$\lambda_0 \geq 6$
SpinCCW	P_+	
RollCW	$C_{3-}P_-C_{2-}P_-C_{1-}P_- \ldots$	$2\sqrt{3} \leq \lambda_0 < 6$
RollCCW	$C_{1+}P_+C_{2+}P_+C_{3+}P_+ \ldots$	
Tangent	CSCSCP ...	$\lambda_0 = 2\sqrt{3}$
Shuffle_{1-}	$C_{2+}C_{1-}C_{3+}P_+ \ldots$	$3 < \lambda_0 < 2\sqrt{3}$
Shuffle_{2-}	$C_{3+}C_{2-}C_{1+}P_+ \ldots$	
Shuffle_{3-}	$C_{1+}C_{3-}C_{2+}P_+ \ldots$	
Shuffle_{1+}	$C_{3-}C_{1+}C_{2-}P_- \ldots$	
Shuffle_{2+}	$C_{1-}C_{2+}C_{3-}P_- \ldots$	
Shuffle_{3+}	$C_{2-}C_{3+}C_{1-}P_- \ldots$	

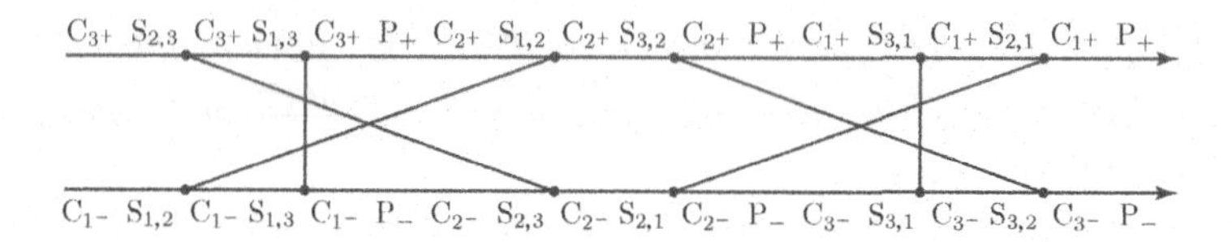

Fig. 6. The structure of *tangent* trajectories. The controls must occur in left-to-right order in the direction shown by either the top or the bottom arrows. However, after a singular control S, the trajectory may switch from one sequence to the other, as shown by the vertical and diagonal lines segments.

7 Optimal Trajectories

We have presented the five classes of extremal trajectory; every optimal trajectory must be extremal. However, not all extremal trajectories are optimal. In this section, we will present further conditions that optimal trajectories must satisfy. Specifically, we will show that doubly-singular slide trajectories are never optimal, and that the number of control switches in any optimal trajectory never exceeds 18. Finally, we show that the classification {*spin, roll, shuffle, tangent*} is *minimal*; for each trajectory class, there exists at least one pair of configurations for which a trajectory of that class is optimal.

Theorem 9. *Doubly-singular slide trajectories are not optimal for any pair of start and goal configurations.*

Proof: (sketch) First show that there exists a *shuffle* that connects any two configurations on a slide that are separated by less than $8\sqrt{6}/3$. Express the distance traveled and the time taken by a *shuffle* trajectory as a function of λ_0. Finally, show that the average forward velocity for such a trajectory is strictly less than -1 (the velocity for the doubly singular trajectory). ■

Theorem 10. *Optimal trajectories contain no more than 18 control switches. Specifically,*

(i) optimal spin trajectories contain zero control switches, and the maximum duration of an optimal spin trajectory is π;
(ii) optimal roll trajectories contain at most 8 control switches;
(iii) optimal shuffle trajectories contain at most 7 control switches;
(iv) optimal tangent trajectories contain at most 12 control switches if the trajectory is non-monotonic in θ, and at most 18 control switches if the trajectory is monotonic in θ;

Proof: (Sketch) The proof for *spin* trajectories is obvious. For each of *roll*, *shuffle*, and *tangent* trajectories we slice, reorder, or reflect segments to construct alternative trajectories that take the same time, but are not extremal. Since these equal-cost trajectories are not extremal, neither these nor the original *roll*, *shuffle*, and *tangent* are optimal. ■

Theorem 11. *There exist bounds on the displacements along the x and θ axis beyond which spin, roll, and shuffle trajectories are not optimal. In particular,*

(i) Roll trajectories with x-displacement more than $\frac{-40\sqrt{2}}{\sqrt{3}}$ are not optimal.
(ii) Shuffle trajectories with x-displacement more than $\frac{-16\sqrt{2}}{\sqrt{3}}$ and θ displacement more than $60°$ are not optimal.
(iii) Tangent trajectories are not optimal for configurations that are separated by more than $120°$, with distance between the configurations less than 4.

Proof: (Sketch) Theorem 10 gives the maximum number of segments that comprise optimal trajectories of each class. We compute the distance of each segment for each class. ■

Theorem 12. *Spin, Roll, Shuffle, and Tangent trajectories are each optimal for at least one pair of start and goal configurations of the omni-directional vehicle.*

Proof: (Sketch) For each class, we explicitly construct a pair of start and goal configurations for which no other trajectory class is optimal. For example, if the goal is sufficiently far away, no *roll*, *spin*, or *shuffle* trajectory can be optimal, since there are no more than nine segments, and each segment is of bounded length. Therefore a *tangent* trajectory is optimal. ■

8 Open Problems

We have presented a complete and minimal classification of optimal trajectories, and explicit descriptions of each trajectory. However, we have not addressed the problem of determining which of these trajectories is optimal for a particular pair of start and finish configurations. For the problem of determining the shortest trajectories for a steered car, Reeds and Shepp [10] suggest the simple approach of enumerating all possible structures that connect two configurations, and comparing the time of each. A similar approach should be possible for the omnidirectional vehicle.

Souères and Laumond [14] determined the complete *synthesis* of optimal trajectories for the steered car: an explicit mapping from pairs of configurations to trajectories. Balkcom and Mason [2] determined the synthesis for differential-drive vehicles. Such a result for the omnidirectional vehicle would remove the need for enumerating and comparing all trajectories between a pair of configurations, and would give the metric on the configuration space more explicitly.

The current work also does not consider the presence of obstacles. We expect that optimal trajectories among obstacles would consist of segments of obstacle-free trajectories, and segments that follow the boundary of the obstacles.

There are also broader questions. The shortest or fastest trajectories are now known for a few examples of specific systems: steered cars, the differential drive, and the omnidirectional vehicle considered here. The results share some features in common; each of the optimal trajectories can be described by motion of the robot relative to a switching line in the plane. The trajectories for steered cars include arcs of circles and straight lines; the trajectories for differential drives include spins in place and straight lines. The trajectories for the current system include straight lines, arcs of circles, and spins in place, and the system could in that sense be considered a hybrid of a steered car and a differential drive. What generalizations are possible, and can the optimal trajectories be determined for a generic mechanism whose design is described in terms of a set of variable parameters? Which mechanism should be chosen to be most efficient for a given distribution of start and goal configurations?

Acknowledgments

The authors would like to thank Steven LaValle, Hamid Chitsaz, Jean-Paul Laumond, Bruce Donald, the members of the CMU Center for the Foundations of Robotics, and the members of the Dartmouth Robotics Lab for invaluable advice and guidance in this work.

References

1. Balkcom, D.J., Kavathekar, P.A., Mason, M.T.: Time-optimal trajectories for an omnidirectional vehicle. International Journal of Robotics Research 25(10), 985–999 (2006)
2. Balkcom, D.J., Mason, M.T.: Time optimal trajectories for differential drive vehicles. International Journal of Robotics Research 21(3), 199–217 (2002)

3. Chitsaz, H., LaValle, S.M., Balkcom, D.J., Mason, M.T.: Minimum wheel-rotation paths for differential-drive mobile robots. In: IEEE International Conference on Robotics and Automation (2006)
4. Chyba, M., Haberkorn, T.: Designing efficient trajectories for underwater vehicles using geometric control theory. In: 24th International Conference on Offshore Mechanics and Artic Engineering, Halkidiki, Greece (2005)
5. Coombs, A.T., Lewis, A.D.: Optimal control for a simplified hovercraft model (preprint)
6. Desaulniers, G.: On shortest paths for a car-like robot maneuvering around obstacles. Robotics and Autonomous Systems 17, 139–148 (1996)
7. Dubins, L.E.: On curves of minimal length with a constraint on average curvature and with prescribed initial and terminal positions and tangents. American Journal of Mathematics 79, 497–516 (1957)
8. Kalmár-Nagy, T., D'Andrea, R., Ganguly, P.: Near-optimal dynamic trajectory generation and control of an omni directional vehicle. Robotics and Autonomous Systems 46, 47–64 (2004)
9. Pontryagin, L.S., Boltyanskii, V.G., Gamkrelidze, R.V., Mishchenko, E.F.: The Mathematical Theory of Optimal Processes. John Wiley, Chichester (1962)
10. Reeds, J.A., Shepp, L.A.: Optimal paths for a car that goes both forwards and backwards. Pacific Journal of Mathematics 145(2), 367–393 (1990)
11. Reister, D.B., Pin, F.G.: Time-optimal trajectories for mobile robots with two independently driven wheels. International Journal of Robotics Research 13(1), 38–54 (1994)
12. Renaud, M., Fourquet, J.-Y.: Minimum time motion of a mobile robot with two independent acceleration-driven wheels. In: Proceedings of the 1997 IEEE International Conference on Robotics and Automation, pp. 2608–2613 (1997)
13. Souères, P., Boissonnat, J.-D.: Optimal trajectories for nonholonomic mobile robots. In: Laumond, J.-P. (ed.) Robot Motion Planning and Control, pp. 93–170. Springer, Heidelberg (1998)
14. Souères, P., Laumond, J.-P.: Shortest paths synthesis for a car-like robot. IEEE Transactions on Automatic Control 41(5), 672–688 (1996)
15. Sussmann, H., Tang, G.: Shortest paths for the Reeds-Shepp car: a worked out example of the use of geometric techniques in nonlinear optimal control. SYCON 91-10, Department of Mathematics, Rutgers University, New Brunswick, NJ 08903 (1991)
16. Vendittelli, M., Laumond, J., Nissoux, C.: Obstacle distance for car-like robots. IEEE Transactions on Robotics and Automation 15(4), 678–691 (1999)

Mechanical Manipulation Using Reduced Models of Uncertainty

Todd D. Murphey

Department of Electrical and Computer Engineering,
University of Colorado at Boulder
murphey@colorado.edu

Abstract. In this paper we describe methods applicable to the modeling and control of mechanical manipulation problems, including those that experience uncertain stick/slip phenomena. Manipulation in unstructured environments often includes uncertainty arising from various environmental factors and intrinsic modeling uncertainty. This reality leads to the need for algorithms that are not sensitive to uncertainty, or at least not sensitive to the uncertainty we can neither model nor estimate. The particular contribution of this work is to point out that the use of an abstraction, in this case a kinematic reduction, not only reduces the computational complexity but additionally simplifies the representation of uncertainty in a system. Moreover, this simplified representation may be directly used in a stabilizing control law. The end result of this is two-fold. First, modeling for purposes of control is made more straight-forward by getting rid of some dependencies on low-level mechanics (in particular, the details of friction modeling). Second, the online estimation of the relevant uncertain variables is much more elegant and easily implementable than the online estimation of the full model and its associated uncertainties.

1 Introduction

It is traditional in robotics to view problems of manipulation, motion planning, and control in one of two extreme lights. First, if a system is kinematic (a word which for now we leave not specifically defined), we simplify the system description from a second-order system with forces and inertias to a first-order system that consists of velocities and constraints. Then motion plans and control laws (if necessary) are designed for this kinematic system. It is important to note that in order to *implement* this design based on kinematics, a backstepping algorithm is employed, either explicitly in an "inner-loop-outer-loop" control architecture, or implicitly by purchasing motor controllers (or other appropriate devices) that provide the inner loop control. In the end, the advantages of using kinematic structures include both lessened computational burden (due to the computation in a lower-dimensional space) and increased robustness to some classes of uncertainty (due to robustness properties of the backstepping, inner-loop controller).

S. Akella et al. (Eds.): Algorithmic Foundation of Robotics VII, STAR 47, pp. 359–374, 2008.
springerlink.com

If, however, there is some reason that a kinematic analysis is inappropriate, then we often revert to a more complex set of modeling choices. In particular, in multi-point contact many phenomena are introduced, including soft-contact models [2], elaborate models of frictional interfaces [17], and the inclusion of dynamic effects such as inertial terms and generalized forces. Nevertheless, it is not clear that the introduction of these additional modeling techniques helps for the purpose of control, motion planning, etcetera. In fact, it is often the case that this *hurts* our ability to successfully design control strategies. Not only does the introduction of these effects make problems computationally more complex, it also decreases robustness by introducing assumptions that are often not satisfied by the environment or, worse, may only sometimes be satisfied by the environment. Hence, we can be faced with a situation where our modeling assumptions are occasionally correct, but not reliably so.

From a design perspective (as opposed to a simulation perspective), it is thus desirable to, if necessary, introduce elements to a model that provide the full complexity of possible behavior of the system without introducing too much new information (thereby decreasing the applicability of the model). This is related to the idea of abstraction, which originated in the computer science community [10] and was then made formal in a control context in [18] and related works. In this paper, we focus on a particular type of abstraction that formalizes the idea of a system being kinematic, appropriately termed *kinematic reducibility*. However, this is merely the setting for the present work. Our main focus is to discuss what types of representation of uncertainty should be used in the abstracted setting.

Consider the conceptual block diagram in Fig.1. The blocks on the right-hand-side are familiar–these four blocks represent a traditional backstepping algorithm using "virtual inputs." In the case of a kinematic vehicle, we abstract the true dynamics of the vehicle to a kinematic representation where the abstracted inputs are now velocities and the input vector fields are vectors that satisfy the kinematic constraints. Then a backstepper is used that takes these velocities as reference signals for a lower-level controller. *It is useful to point out that doing so assumes that this low-level controller is robust to any uncertainties (coming from terrain, parametric uncertainty, etcetera).* In the context of Fig.1, this means that the reduction to the kinematic system induces a formal reduction of the uncertainty. In this case, there is no representation of uncertainty whatsoever in the abstract description of the system–all the robustness is built into the backstepping algorithm. What we will see is that this same abstraction in multi-point contact systems again reduces the representation of uncertainty, but not to the point that there is no uncertainty at all in the reduced equations. Instead, there is an *abstracted* uncertainty which corresponds to the hybrid, discrete-valued state that represents the contact state (whether any given contact is in contact or out of contact and, if in contact, whether it is slipping or sticking) of the system. It is this reduced representation of uncertainty we will use in designing controllers for these systems. The practical advantage of this approach is that the reduced representation of uncertainty makes some aspects of analysis more

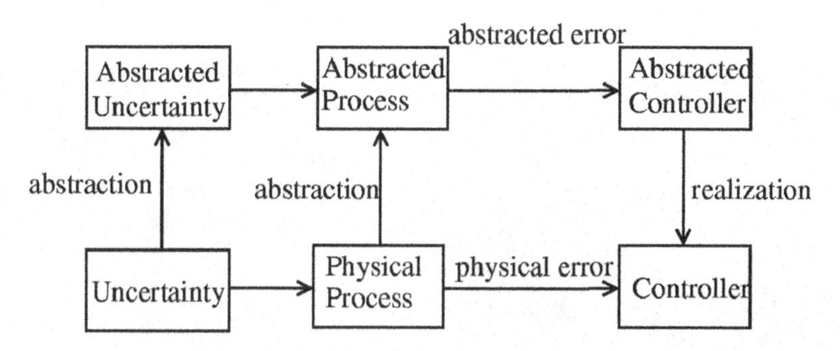

Fig. 1. Conceptual block diagram for reduction of both the physical model and uncertainty

simple and that it can make estimation less costly both in terms of computation and bandwidth requirements.

Our previous work [15] (first presented at WAFR 2002 [14]) in this area showed that some manipulation surfaces cannot stabilize an object without feedback, and then showed that by using the Power Dissipation Method (PDM) to model the system one can design a stabilizing controller that works surprisingly well experimentally. The weaknesses of this work were primarily that it was unclear *why* the power dissipation method would adequately capture the dynamics, and it was moreover unclear why a feedback controller could be designed in this context. The former issue was cleared up when we showed in [16] that the power dissipation method is actually a class of kinematic reductions, in the sense of the work by Lewis et al [3, 4]. The latter issue, that of understanding why we can do control design using a heuristic modeling technique, has only recently become clear, and this paper is intended to explain why and when controller design may occur in these traditionally heuristic settings.

The key contribution of this paper is that we present a methodology for combining kinematic reductions with stabilizing controllers that only use a reduced representation of uncertainty in their estimators. When possible, this allows one to use a reduced model for computational simplicity while not losing any of the behavior of which the system is capable. We present an example of multiple point manipulation as an example, but point out that the technique of uncertainty abstraction is potentially much more broadly applicable than just what is discussed here. This paper is organized as follows. Section 2 described an example system that motivates the present work. Section 3 discusses modeling of multi-point contact systems using Lagrangian mechanics and the constrained affine connection. Although the use of the constrained affine connection description of the mechanics is absolutely equivalent to the more traditional approach using a Lagrangian and generalized coordinates, we use it because it gives a precise statement of a test for kinematic reducibility, as discussed in Section 3.1. Section 4 discusses stability results relevant to these systems, and Section 5 gives, for purpose of

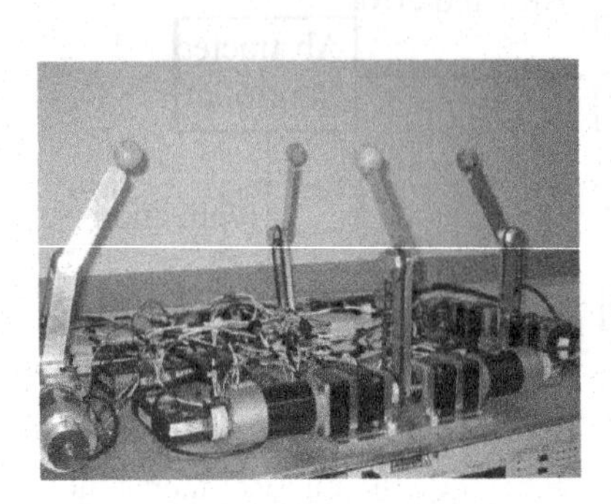

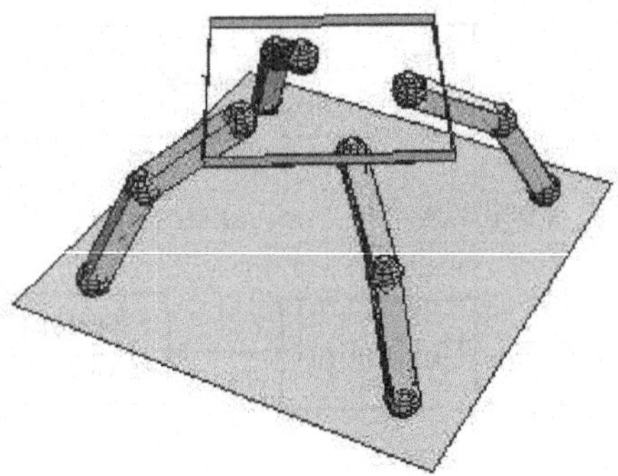

Fig. 2. An 8 degree-of-freedom four "finger" manipulator that manipulates objects it supports. The fingers are constrained, so stick/slip transitions between the actuator end-effectors and the manipulated object must occur when the actuators move. The present paper presents simulations of this device (the picture on the right with a "see-through" box supported and manipulated by four arms is the graphical representation we use in simulations).

illustration, a quick introduction to how one applies these results to the example in Section 2 .

2 Motivation: Multi-point Manipulation

A manipulation system consisting of many points of contact typically exhibits stick/slip phenomenon due to the point contacts moving in kinematically incompatible manners. We call this manner of manipulation *overconstrained manipulation* because not all of the constraints can be satisfied. Naturally, uncertainty due to overconstraint can sometimes be mitigated by having backdrivable actuators, soft contacts, and by other mechanical means [11], but these approaches avoid the difficulties associated with stick/slip phenomenon at the expense of losing information about the state of the mechanism. This, in turn, leads either to degraded performance or to requiring additional sensors. Consider the object in Fig. 2. It has eight degrees of freedom, all independently actuated by a DC brushless motor. The motion of the tips of the "fingers" can be constrained to be in a horizontal plane, so it can be used as a manipulation surface. However, the force any given finger exerts is constrained on a line–no finger can exert any "side-ways" force. In such cases friction forces and intermittent contact play an important role in the overall system dynamics, leading to non-smooth dynamical system behavior. The question is how to control the position and orientation of a supported object without being sensitive to the details of how the frictional stick/slip interactions adversely affect stability. This is addressed in Section 4 after discussing modeling issues in Section 3.

3 Modeling and Analysis of Multiple Point Contact

We assume that the systems we are interested in are finite-dimensional simple mechanical systems (as described for smooth systems in [3]). That is, their equations of motion may be found using a Lagrangian of the form kinetic energy minus potential energy ($L = K.E. - V$) along with a set of constraints on the system of the form $\omega(q)\dot{q} = 0$, where $\omega(q)$ is a matrix representing the configuration (q) dependent constraints. Moreover, there may be external forces acting on the system. If we ignore potential energy (as is appropriate for many planar systems including the one in Section 2), such a system's dynamics may be written down as: $\nabla_{\dot{q}}\dot{q} = u^{\alpha}Y_{\alpha}$, where the notation $u^{\alpha}Y_{\alpha}$ implies summation over the α. In this expression, ∇ is the constrained affine connection encoding the free kinetic energy and the constraints, in our case the nonslip constraints. Moreover, u represents external forces (not necessarily inputs) and Y represents the associated vector fields on the configuration manifold Q (i.e., $Y \in T_qQ$, the tangent space at $q \in Q$). If we wish to include potential energy, it will show up as a vector field on the right-hand side of the equation.

The systems of interest have two types of external forces–those that correspond to inputs and those that correspond to external disturbances. In the case of multiple point contact, the external disturbance forces generally correspond to reaction forces due to friction when a contact slips. Therefore, it will be useful to write the dynamic equations as: $\nabla_{\dot{q}}\dot{q} = u^{\alpha}Y_{\alpha} + d^{\beta}V_{\beta}$ so that we can distinguish between the different types of external forces. (Note that if a constraint is satisfied so that the contact is not slipping, there is still a reaction force. In that case the reaction force is incorporated into the definition of of the constrained affine connection ∇ in a manner identical to the constrained Euler-Lagrange equations).

Lastly, because the contact state changes over time (as the contacts transition between stick and slip), the constraints change over time. This implies that ∇ is not a single constrained affine connection, but rather comes from a set of constrained affine connections ∇^{σ}, each of which represents a different set of stick/slip states of the mechanism. The same holds true for Y^{σ} and V^{σ}. Hence, if we index the set of possible stick/slip states by σ, we get second-order equations of motion of the following form:

$$\nabla^{\sigma}_{\dot{q}}\dot{q} = u^{\alpha}Y^{\sigma}_{\alpha} + d^{\beta}V^{\sigma}_{\beta} \tag{1}$$

where u are input forces and d are external forces. Equation (1) represents the equations of motion for any multiple contact system or overconstrained system that experiences point contact with its environment. (Note that for this equation to make sense, one must assume that the switching signal σ is at least measurable, and often it is assumed that it is piecewise continuous.) Lastly, it is important to point out that the representation $\nabla_{\dot{q}}\dot{q} = u^{\alpha}Y_{\alpha}$ is neither more nor less than the Euler-Lagrange equations [3].

3.1 Kinematic Descriptions of Systems That Slip

We use the affine connection formalism to describe mechanical systems because it is in the context of this formalism that a useful technical connection between 2nd-order mechanical systems and 1st-order kinematic systems has been made (found for smooth systems in [9] and for nonsmooth systems in [16]). In particular, it would be useful to be able to write Eq. (1) in the form:

$$\dot{q} = \overline{u}^a X_a^\sigma, \tag{2}$$

where $\overline{u}$ are *velocity* inputs instead of force inputs. Roughly speaking, a system is kinematic if it can be written as a first order differential equation in q without losing any information about what trajectories the system is capable of producing. More precisely, this kinematic description is only useful if it satisfies two requirements. First, for every solution of the dynamic system in Eq. (1) there must exist a kinematic solution of the form in Eq. (2). In the case of a vehicle, this corresponds to requiring that for every *trajectory* of the vehicle there exists a corresponding *path* that can be obtained from kinematic considerations alone. Secondly, for every kinematic solution there must exist a dynamic solution that is equal to the kinematic solution coupled with its time derivative (so that it lies in TQ). This means that there must exist a dynamic solution for every feasible kinematic path. This way of viewing smooth kinematic systems has been studied extensively, including [9]. Motion planning has been studied using these concepts in [3, 4], but these works were all intended for smooth systems. However, it was shown in [16] that the kinematic reduction of a nonsmooth system of the form in Eq. (1) to one of the form in Eq. (2) is equivalent to the reduction of each smooth model of the multiple model system. The associated algebraic test of kinematic reducibility is that the *symmetric product* between two vector fields Y_i^σ and Y_j^σ (defined by $\langle Y_i^\sigma : Y_j^\sigma \rangle = \nabla^\sigma_{Y_i^\sigma} Y_j^\sigma + \nabla^\sigma_{Y_j^\sigma} Y_i^\sigma$ for given i, j, σ) lie within the distribution of the vector fields and that any reaction forces lie within the span of the input vector fields. That is,

$$\langle Y_i^\sigma : Y_j^\sigma \rangle \in span\{Y_i | i = 1, \ldots, m\} \;\; \forall \; i, j, \sigma \tag{3}$$

$$V_\beta^\sigma \in span\{Y_i | i = 1, \ldots, m\} \;\; \forall \; \beta, \sigma \tag{4}$$

Notice that this need only hold for each σ, so the calculation is a purely algebraic one, despite the fact that our system is nonsmooth. That is, even with the nonunique solutions these systems can have, one may test for each model independently (i.e., holding σ constant) whether a system is kinematic.

3.2 Uncertainty Representations

The main point of this paper comes from noting that the contact state enters solely in the σ dynamics in Eq. (1) and (2). Hence, the uncertainty for the full dynamic system depends on both σ and other uncertainties that drive σ, such as parametric uncertainties, choice of friction model governing the contact interaction, etcetera. However, the uncertainty in the kinematic model *only includes* σ, which means that the abstraction to the kinematic model reduces the representation of the uncertainty to a hybrid, discrete-valued structure (rather than a continuous one like that typically addressed in the robust control community). It is important to note that because of this the only assumption made regarding friction in a kinematic model is that it creates stick/slip effects. In the context of this paper, no other assumptions are necessary. However, when using such an abstraction, one must have confidence that the backstepping algorithm employed is robust with respect to the uncertainties that are left over, in our case model uncertainties and parametric uncertainties. Fortunately, motor controllers are known to be quite robust when following a desired reference velocity. Accordingly, we assume that we can track a desired velocity for the rest of this paper, ignoring transient behavior and coupling. If for some reason asymptotic tracking is not achieved, then an additional layer of analysis will be necessary. This reduced representation of uncertainty is what we will use in designing a stabilizing control for a mechanical manipulation system, and the online estimation of σ will in particular play a significant role in the stability results.

4 Stability Conditions

Now suppose we want to drive a (multiple-model, multi-point contact) mechanical system to a desired state. Then we have, for every choice of σ, a smooth system that must be stabilized (since a perfectly valid choice of σ is to have it be constant for all time). Moreover, because σ is uncertain, it must be thought of as an exogenous disturbance (albeit a discrete-valued one). Now, one could try to create a control law that is stable for all possible signals σ (in fact, one would have to do so if σ is not observable), but this is often impossible from a practical perspective. In fact, in the case of stabilizing the $SE(2)$ configuration of the object in Section 2, it is provably impossible [15]. Therefore, the question becomes one of estimation, the online estimation of the contact state σ (the abstracted uncertain variable) based on available outputs *and* the incorporation of this estimate into the controller. This latter part is important because the classical separation principle found in undergraduate controls textbooks is not valid for nonlinear or nonsmooth systems.

First, we need to know that σ is observable (i.e., different models can be distinguished based on available feedback). Although there are formal methods for determining this (see [19]), we will see in Section 5 that it is occasionally possible to see that σ is observable by inspection. If it is observable, there are generally two methods (and variations thereof) for determining the value of σ at any given time. The first is to directly compare predicted velocities (for every model indexed by σ) to the sensed velocity. This involves differentiating outputs, but may be an acceptable approach if only one derivative of the output is needed. (This is particularly true if the cardinality of values σ can take is small. That is, if the total number of models is small, so the models are relatively easy to distinguish from each other, then even with noisy data we should be able to distinguish them.) If differentiating outputs is not acceptable, then one may alternatively integrate the equations of motion for every model and compare these to the measured output. Either choice is an acceptable choice of estimator from a theoretical perspective because we are only interested in distinguishing different models from each other.

The stability results that are useful for the problems of interest here are from the adaptive control community, particularly multiple model adaptive control [6, 1, 5]. Suppose that we have a family of plants indexed by $p \in \mathbb{P}$, all of which are stabilized by a control law with Lyapunov function V_p. (For the moment, we ignore the design and implementation of these controllers. We will revisit this in Section 5.) Switching between plants is governed by the switching signal σ. In the case of a multiple contact system, σ_e encodes the externally determined *contact state* of the system, that is, which contacts are sticking and which are slipping. Moreover, σ_c encodes the current estimate of σ_e, and particularly tells us which controller is being used at any given time. Ideally, $\sigma_c = \sigma_e$, but there may be latencies that cause this not to be the case. Such systems can be written as:

$$\dot{x} = F_{\sigma(x,t)}(x,t) \qquad \sigma(x,t) \in \mathbb{P} \tag{5}$$

where $\mathbb{P}$ is an index over the set of all admissible plants. We assume that the F_p satisfy the following standard Lyapunov criteria; that there exist for all $p \in \mathbb{P}$ differentiable functions $V_p : \mathbb{R}^n \to \mathbb{R}$, positive constants λ_0, γ and class $\mathcal{K}_\infty$ [8] functions $\alpha, \overline{\alpha}$ satisfying:

$$\dot{V}_p = \frac{\partial V_p}{\partial x} F_q \leq -2\lambda_0 V_p \; for \; p = q, \tag{6}$$

$$\dot{V}_p = \frac{\partial V_p}{\partial x} F_q \leq 2\lambda_{F'} V_p \; for \; p \neq q, \tag{7}$$

$$\alpha(\|x\|) \leq V_p(x) \leq \overline{\alpha}(\|x\|), \tag{8}$$

$$V_p \leq \gamma V_q, \tag{9}$$

for all $x \in \mathbb{R}^n$ and $p, q \in \mathbb{P}$. These are relatively standard requirements for Lyapunov functions [8], except for the condition in Eq.(7) (which requires that whenever the plant and the controller are not matched the resulting instability is bounded by some growth rate $\lambda_{F'}$).

Switching signals σ are assumed to be a piecewise continuous (and therefore measurable) function coming from a family of functions S. We say that Eq.(5) is *uniformly exponentially stable over* S if there exist positive constants c and λ such that for any $\sigma \in S$ we have

$$\|\Phi_\sigma(t,\tau)\| \leq c e^{-\lambda(t-\tau)} \qquad \forall t \geq \tau \geq 0.$$

Here $\Phi_\sigma(t,\tau)$ denotes the flow (given σ) of Eq.(5). For such a system we say that λ is its *stability margin*.

To characterize and distinguish different families of functions S, we employ the following definitions (from [6]). Given $\sigma \in S$, we define $N_\sigma(t,\tau)$ to be the (integer) number of switches or discontinuities in σ in the interval (t,τ). Given two numbers τ_{AD} and N_0, called the *average dwell time* and *chatter bound* respectively, we say that $S_{\text{ave}}[\tau_{AD}, N_0]$ is the set of all switching signals satisfying

$N_\sigma(t,\tau) \leq N_0 + \frac{t-\tau}{\tau_{AD}}$. Lastly, let $S_{ave}[\tau_{AD}, N_0]$ be the set of all switching signals for which $N_\sigma(t,\tau) \leq N_0 + \frac{\tau - t}{\tau_{AD}}$. We will assume for the rest of the present work that switching signals σ_e (the external switching determining the contact state) can be characterized in this way.

Assumption 4.1. *Assume σ_e switching satisfies*

$$N_{\sigma_e}(t,\tau) \leq N_0^e + \frac{t-\tau}{\tau_{AD}^e}$$

for some $N_0^e > 0$ and τ_{AD}^e.

We can similarly require that the signal σ_c (the switching signal that dictates the current controller) also satisfy dwell-time requirements (i.e., $N_{\sigma_c}(t,\tau) \leq N_0^c + \frac{t-\tau}{\tau_{AD}^c}$) to ensure that the control switching does not destabilize the system.

It is well known that switching between a set of stable linear systems may well yield an unstable system [7]. This means that even in the most moderate case, where estimation of the contact state is perfect (i.e., $\sigma_c = \sigma_e$) and there are no latencies in sensing or actuation, our multiple contact system can in principle be destabilized by switching contact state. Our purpose in this section is to apply some results from the theory of switching systems to understand physically meaningful conditions that will guarantee stability for a multiple model system (even those without a common Lyapunov function). In particular, we will characterize such a condition in terms of the average dwell time as it was described above.

Due to space considerations, proofs of the following theorems are not presented here and the reader is directed to the conference proceeding [12, 13] where these technical results are presented in a theorem/proof format. First, the following result from [6] will be helpful. It states that for a collection of stable plants as Eq.(5) a bound on the average dwell time can be determined such that the hybrid system is stable with any desired stability margin.

Lemma 1 ([6]). *Given a system of the form in (10) such that all the F_p satisfy Eqs. (6), (8), and (9) hold, there is a finite constant τ_{AD}^* such that Eq.(5) is uniformly exponentially stable over $S_{ave}[\tau_D, N_0]$ with stability margin $\lambda < \lambda_0$ for any average dwell time $\tau_{AD} \geq \tau_{AD}^*$ and any chatter bound $0 < N_0$.*

In particular, the average dwell time must satisfy $\tau_{AD} > \frac{\log \gamma}{2(\lambda - \lambda_0)}$. Note that if we have a common Lyapunov function, then $\gamma = 1 \Rightarrow \log \gamma = 0 \Rightarrow \tau_{AD} = 0$ satisfies the stability requirements. Hence, common Lyapunov functions are highly desirable, if they can be found. A corollary of this result relevant to the multiple point contact example is Corollary 1 (proven in [12]).

Corollary 1. *If each contact state σ_e for a multi-point manipulation system is stabilized with a quadratic Lyapunov function V_p, if $\sigma_e \in Save[\frac{\log \gamma}{2(\lambda - \lambda_0)}, N_0]$ for some N_0, and if $\sigma_c = \sigma_e$ (i.e., the observer is perfect), then Eq. (1) or Eq. (2) (depending on whether the representation used is dynamic or kinematic) is exponentially stable with stability margin λ.*

Note that this result, and the results that follow, are equally applicable to both dynamic and kinematic systems. What does Corollary 1 mean for a multiple model system where there are external signals determining the switching, such as is the case in a multiple contact system? It means that so long as there are no latencies, no errors in estimation, and no noise in the sensors, the multiple model system is stable so long as the external switching signals σ_e are kept sufficiently slow on the average. How slow depends on how the controllers for each plant are designed and, more importantly, how they are related to each other. The closer γ can be kept to 1 (i.e., the closer we are to having a common Lyapunov function), the more quickly σ_e may switch without destabilizing the system.

What happens if there are noise sources, latencies, and time delays causing the controller switching σ_c to not coincide with the environmental switching σ_e? Most of these issues are adequately addressed in [1, 5]. However, if $\sigma_c \neq \sigma_e$, instabilities due to temporary mismatch between controllers and plants can occur. The basic consequence of this is roughly that the longer the mismatch, the slower the external switching must be in order to maintain stability. To address this issue, assume we have equations of motion of the following form:

$$\dot{x} = \begin{cases} F'_q x \text{ on } [t_i, t_i + d_\sigma) \\ F_p x \text{ on } [t_i + d_\sigma, t_{i+1}) \end{cases} \tag{10}$$

where (for each p) $\dot{x} = F_p(x)$ is asymptotically stable and (for each q) $\dot{x} = F'_q$ is potentially unstable but has a bound on the rate of growth $\lambda_{F'}$. Note that our example system in Section 2 satisfies these requirements because all the F_p are stable by design.

It is now useful to state an extension of Thm. 1 (also proven in [12]) to accommodate d_σ. The resulting trade-off is not surprising–the larger d_σ becomes, the more slowly σ_e is allowed to switch. In particular, if we can bound d_σ below by d_* then we find that choosing

$$\tau^e_{AD} > \frac{\frac{\log \gamma}{2} + 2\lambda_{F'} d_\sigma}{(\lambda_0 - \lambda)} \tag{11}$$

results in a stable system, as seen in the following Lemma.

Lemma 2. *Given a system of the form in (10) such that all the F_p satisfy Eqs. (6), (7), (8), and (9), there is a finite constant τ^*_{AD} and a finite constant d^*_σ such that Eq.(10) is uniformly exponentially stable over $S_{ave}[\tau_{AD}, N_0]$ with stability margin λ, for any average dwell time $\tau_{AD} \geq \tau^*_{AD}$, any chatter bound $0 < N_0$, and any $d_\sigma \leq d^*_\sigma$.*

With this, one may prove the following corollary.

Corollary 2. *If each contact state for a multi-point manipulation system is stabilized with a quadratic Lyapunov function V_p, and if τ^e_{AD} and d_σ satisfy Eq.(11), then for any N_0 the state output is exponentially stable.*

Proposition 2 indicates that if the contact states change slowly enough (i.e., τ^e_{AD} is large) and the estimator is fast enough (i.e., d_σ is small), then the system is stable. Among other things, this means that one does not have to concern oneself with the friction model to establish where switching occurs. Instead, the contact states can change arbitrarily, so long as they do so sufficiently slowly on the average and their effect is observable in the state output.

5 Example

Consider the eight degree of freedom manipulator in Fig. 2. This figure has four point contact actuators (corresponding to the inputs $u_1, \ldots, u_4$) located at $(1,1), (-1,1), (-1,-1), (1,-1)$ respectively (in the simulation), all oriented towards the origin. For each contact there are two independent constraints, a nonslip constraint in each direction tangent to the surface of the contact. Hence, there are $2^{2 \cdot 4} = 2^8 = 256$ possible combinations of stick and slip for the four point contact system. If one uses a symbolic software package such as *Mathematica* to compute the dynamic equations of motion for every possible contact state as in Eq. (1), one can exhaustively verify that all possible models are kinematic, so long as the contact interfaces are *dissipative* when slipping is occurring (i.e., the reaction force is nonzero and in the opposite direction of the slipping). This represents an extremely broad set of frictional interfaces, and the statement is proven in a non-exhaustive manner in [13]. Additionally, all the nontrivial, non-overconstrained kinematics are of one of the four forms in Table 5. There do exist σ_e with trivial kinematics (i.e., actuator velocities do not make the supported object move at all), however, and these correspond to constraints with no actuation. An example of this is a table with wheels that are all razor thin, so that spinning the wheels exerts very little force against an object, but sliding orthogonally to the wheel is very difficult. In such an example, no movement whatsoever occurs, and such a situation must be either be avoided through mechanical design or avoided online, but this is beyond the scope of what we discuss here.

For each of the four models in Table 5 a control law is calculated from the Lyapunov function $k(x^2 + y^2 + \theta^2)$ by solving $\dot{V} = -V$ for u_i, where k is some constant to be chosen during implementation. Moreover, by virtue of the design methodology, there is a common Lyapunov function (i.e., $\gamma = 1$ in Eq. (9)). Hence, chattering may occur (particularly near the planar origin), but will not affect stability. Things to note include the following.

1. The system is not smoothly locally controllable (since there are two constant input vector fields and three configuration variables to be controlled). However, all of the states are stabilizable to the the origin of $SE(2)$.
2. Note that these control laws are not only nonlinear, they are not even smooth. In fact, they have discontinuities at the origin.
3. The four models are distinguishable (e.g., based on state output, one can distinguish each model from the next) given nonzero inputs. This will be how we estimate σ_e in the simulations.
4. The four models are, in fact, distinguishable based entirely on θ output (i.e., one may construct a "reduced-order" hybrid observer).

Table 1. The four actuator manipulation surface shown in Fig 2 has all kinematic states, many of which are redundant. This figure shows the four distinct equations of motion that can occur in different contact states. Note that so long as u_1 $(= -u_3)$ and u_2 $(= -u_4)$ are nonzero, the four states can be distinguished from state output. In fact, just observation of θ is sufficient for distinguishing the states. Moreover, measurements of x and y are not helpful because the x and y dynamics are identical in all four models.

Equations of Motion	Control Law
$\dot{q} = \begin{bmatrix} -1 \\ -1 \\ 0 \end{bmatrix} u_1 + \begin{bmatrix} 1 \\ -1 \\ 1 \end{bmatrix} u_2$	$u_1 = \frac{-k\theta\,(\theta+x-y)+k\left(\theta^2+x^2+y^2\right)}{x+y}$ $u_2 = -k\theta$
$\dot{q} = \begin{bmatrix} -1 \\ -1 \\ -1 \end{bmatrix} u_1 + \begin{bmatrix} 1 \\ -1 \\ 0 \end{bmatrix} u_2$	$u_1 = k\theta$ $u_2 = \frac{k\theta\,(\theta+x+y)-k\left(\theta^2+x^2+y^2\right)}{x-y}$
$\dot{q} = \begin{bmatrix} -1 \\ -1 \\ 0 \end{bmatrix} u_1 + \begin{bmatrix} 1 \\ -1 \\ -1 \end{bmatrix} u_2$	$u_1 = \frac{k\theta\,(\theta-x+y)+k\left(\theta^2+x^2+y^2\right)}{x+y}$ $u_2 = k\theta$
$\dot{q} = \begin{bmatrix} -1 \\ -1 \\ 1 \end{bmatrix} u_1 + \begin{bmatrix} 1 \\ -1 \\ 0 \end{bmatrix} u_2$	$u_1 = -k\theta$ $u_2 = \frac{-k\theta\,(-\theta+x+y)-k\left(\theta^2+x^2+y^2\right)}{x-y}$

Figure 3 shows a simulation of the four actuator system using $k = 1$. We simulate the kinematic system rather than the dynamic one, but we are currently making a dynamic simulation to explicitly incorporate various modeling choices (particularly of friction) in the simulation. In either case, the control design should be done at the kinematic level to allow for the abstraction of uncertainty we are advocating. We use crossing from one quadrant to another as the way to drive σ_e in the simulation (which is motivated by minimizing the power dissipation, see [15]), but estimate σ_e online in the simulation. The actuators can only push in one direction for a short amount of space before reaching a kinematic singularity, so they are reset occasionally (this effect shows up in the estimation of σ_e). The object is indicated by a rectangle, but the reader should note that although the rectangle is illustrated as being small, the actual body it represents is in contact with all four actuators at all times, which are denoted in the figure by Nodes 1-4. Their range of motion is depicted by a dark line next to Node X. The initial condition is $\{x_0, y_0, \theta_0\} = \{.5, 2, \frac{\pi}{2}\}$, and progress in time is denoted by the lightening of the object. The three plots beneath the XY plot are X, Y, and θ versus time, respectively. This, and the other simulations, were all done in *Mathematica*, using Euler integration in order to avoid numerical singularities when crossing contact state boundaries. In Fig. 3, the object is stabilized to $(0, 0, 0)$ with no difficulty (and did so reliably over many simulation runs not depicted here). Moreover, this trajectory is qualitatively very similar to the trajectories found experimentally in [15].

The simulation is structured as follows. Depending on which quadrant the center of mass of the object is in, σ_e is chosen to be one of the four models in Table 5.

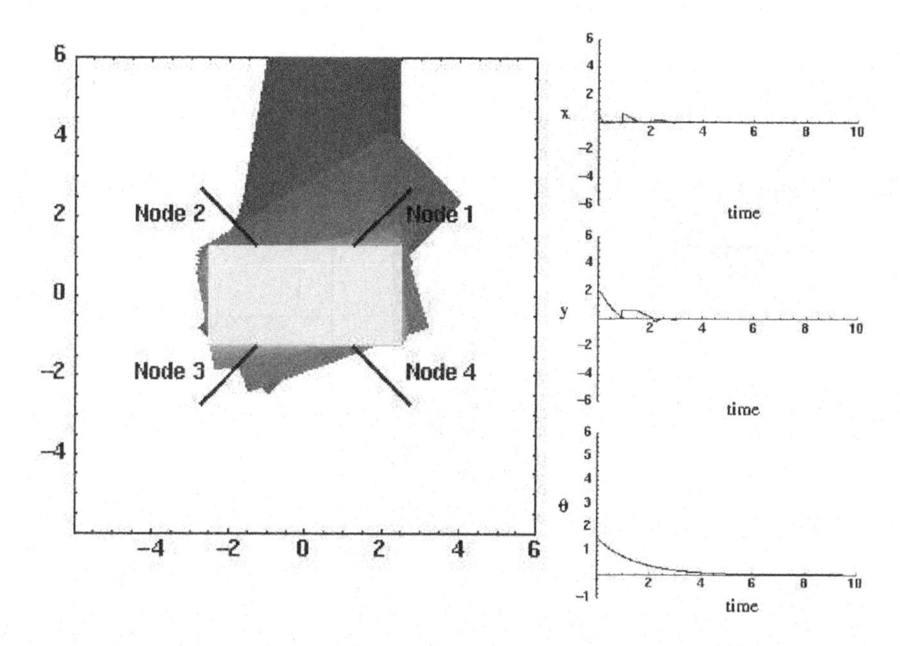

Fig. 3. Simulation of multi-point manipulation when $\sigma_c = \sigma_e$. The rectangle represents the center of the object which is actually in contact with all four of the actuators (Nodes 1-4). The time history progresses from dark rectangles at time 0 to the light rectangles at time 10. The three plots are plots of the X,Y, and θ coordinates against time.

Then, all four models are integrated with respect to time while applying an initial control value of $u_1 = 0, u_2 = 0.1$ (it doesn't matter what this initial control is, so long as it is nonzero). Based on this, σ_e can be immediately determined by looking at the evolution of θ. We did this rather than directly comparing velocities so that we are not differentiating the output. In any case, this then determines σ_c. Knowing σ_c, we ask the actuator tips to follow the velocities u_i based on the control laws in Table 5. These are implemented with an inverse Jacobian, except for when the actuator tip reaches one of its limits, in which case we reset it to the other end of its range (the dark lines in the figure). We can add noise to the sensed state variables and time delays to the estimated σ_c.

If σ_c is a bad estimate of σ_e, then performance degrades but stability is not lost, as seen in Fig. 4, where a time delay of one tenth of a second is introduced. (Note that the amount of time delay in a kinematic system is scalable by virtue of changing the gain on the controller.) Adding a small amount of noise to the sensed outputs has roughly the same effect as a small time delay, as we would expect. If the time delay for a given gain is made sufficiently large, the system becomes unstable. This indicates, at least in simulation, that the interpretation and application of the stability theorems in Section 4 are appropriate here, and that performance degrades reasonably gracefully as σ_c becomes less and less of a good estimate of σ_e until eventually the system destabilizes.

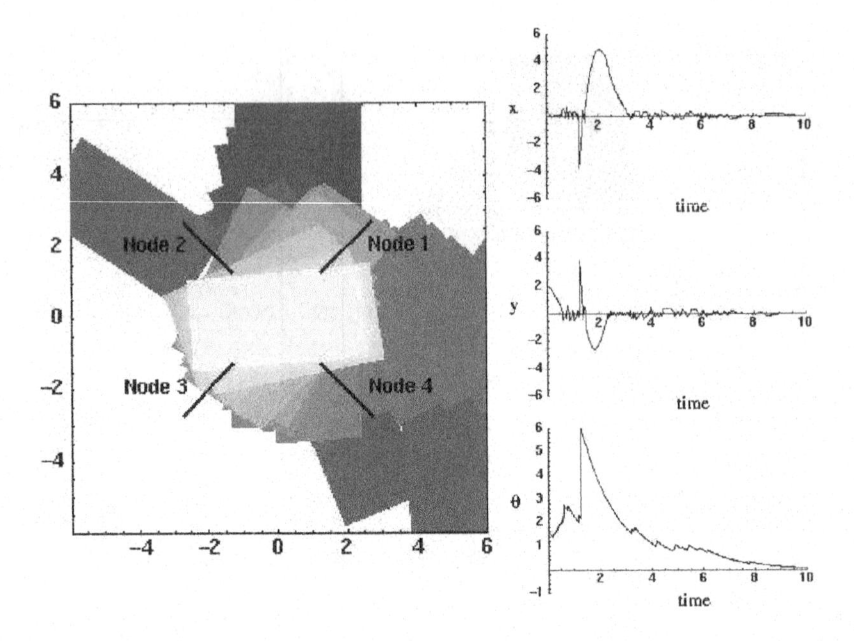

Fig. 4. Simulation of multi-point manipulation when $\sigma_c \neq \sigma_e$ (i.e., the estimated contact state is delayed). The object is only barely stabilized to the origin. (As in Fig. 3, the three plots are plots of the X,Y, and θ coordinates against time.)

6 Conclusions

In this paper we have described methods applicable to the modeling and control of mechanical manipulation problems, including those that experience uncertain stick/slip phenomena. The particular contribution of this work is to point out that the use of an abstraction, in this case a kinematic reduction, not only reduces the computational complexity but additionally simplifies the representation of uncertainty in a system. Moreover, this simplified representation may be directly used in a stabilizing control law. The end result of this is two-fold. First, modeling for purposes of control is made more straight-forward by getting rid of some dependencies on low-level mechanics (in particular, the details of friction modeling). Second, the online estimation of the relevant uncertain variables is much more elegant and easily implementable than the online estimation of the full model and its associated uncertainties. For instance, online friction system identification is quite complex and is not feasible for many applications. However, the presentation here assumes that all feasible states for the system are indeed kinematically reducible. If they are not, then one must switch back to a full analysis of the uncertainty. Moreover, enumeration of the kinematic states can be computationally challenging because in principle the number of kinematic states can go up exponentially in the number of contacts. We expect to be able to address this by more formally using the discrete symmetry properties that allowed us to reduce to only four states in Table 5.

We stabilize the system using techniques from multiple model adaptive control as developed in [1, 5, 7]. We demonstrate in simulation that this technique works well in the context of a simple example (based on experimental work seen in Fig. 2). Moreover, the model/controller presented in the context of this example does not include any explicit model of friction, making the proposed techniques applicable to cases where an unstructured environment makes it unlikely that one can model frictional interactions accurately. Instead, one moves some of the robustness requirements to the backstepping algorithm employed, hence reducing the uncertainty representation with which the high-level controller must contend.

Ultimately, the analytical techniques presented here should be extended to the more geometric setting of grasping and manipulation in the presence of gravitational forces. In particular, examples where a common Lyapunov does not exist should be examined in depth using the analytical techniques developed here. In the meantime, these results will be implemented both in a dynamic simulation environment we are developing and on a second generation version of the experiment discussed in Section 2 and seen in Fig. 2.

Acknowledgements. The author gratefully acknowledges the support for this work provided by the NSF under CAREER award CMS-0546430.

References

1. Anderson, B., Brinsmead, T., Bruyne, F.D., Hespanha, J., Liberzon, D., Morse, A.: Multiple model adaptive control. I. finite controller coverings. George Zames Special Issue of the Int. J. of Robust and Nonlinear Control 10(11-12), 909–929 (2000)
2. Bicchi, A., Kumar, V.: Robotic grasping and contact: a review. In: IEEE Int. Conf. on Robotics and Automation (ICRA), pp. 348–353 (2000)
3. Bullo, F., Lewis, A.: Geometric Control of Mechanical Systems. Number 49 in Texts in Applied Mathematics. Springer, Heidelberg (2004)
4. Bullo, F., Lewis, A., Lynch, K.: Controllable kinematic reductions for mechanical systems: Concepts, computational tools, and examples. In: Int. Symp. on Math. Theory of Networks and Systems (MTNS) (August 2002)
5. Hespanha, J., Liberzon, D., Morse, A., Anderson, B., Brinsmead, T., Bruyne, F.D.: Multiple model adaptive control, part 2: Switching. Int. J. of Robust and Nonlinear Control Special Issue on Hybrid Systems in Control 11(5), 479–496 (2001)
6. Hespanha, J., Morse, A.: Stability of switched systems with average dwell-time. Technical report, EE-Systems, University of Southern California (1999)
7. Liberzon, D.: S. in Systems and Control. Birkhäuser, Boston (2003)
8. Khalil, H.: Nonlinear Systems, 2nd edn. Prentice Hall, Englewood Cliffs (1996)
9. Lewis, A.: When is a mechanical control system kinematic? In: Proc. 38th IEEE Conf. on Decision and Control, pp. 1162–1167 (December 1999)
10. Loiseaux, C., Graf, S., Sifakis, J., Bouajjani, A., Bensalem, S.: Formal Methods in Systems Design. In: Property preserving abstractions for the verification of concurrent systems, vol. 6, pp. 1–35. Kluwer, Dordrecht (1995)

11. Luntz, J., Messner, W., Choset, H.: Distributed manipulation using discrete actuator arrays. Int. J. Robotics Research 20(7), 553–583 (2001)
12. Murphey, T.D.: Application of supervisory control methods to uncertain multiple model systems. In: Proc. American Controls Conference (ACC) (2005)
13. Murphey, T.D.: Modeling and control of multiple-contact manipulation without modeling friction. In: Proc. American Controls Conference (ACC) (2006)
14. Murphey, T.D., Burdick, J.W.: Algorithmic Foundations of Robotics V. In: Feedback Control for Distributed Manipulation, pp. 487–503. Springer, Heidelberg (2004)
15. Murphey, T.D., Burdick, J.W.: Feedback control for distributed manipulation with changing contacts. International Journal of Robotics Research 23(7/8), 763–782 (2004)
16. Murphey, T.D., Burdick, J.W.: The power dissipation method and kinematic reducibility of multiple model robotic systems. IEEE Transactions on Robotics 22(4), 694–710 (2006)
17. Olsson, H., Astrom, K., de Wit, C.C., Gafvert, M., Lischinsky, P.: Friction models and friction compensation. European Journal of Control 4(3), 176–195 (1998)
18. Pappas, G., Laffierier, G., Sastry, S.: Hierarchically consistent control sytems. IEEE Trans. Automatic Control 45(6), 1144–1160 (2000)
19. Vecchio, D.D., Murray, R.: Discrete state estimators for a class of nondeterministic hybrid systems on a lattice. In: Hybrid Systems: Computation and Control, Philadelphia, Pennsylvania (2004)

Motion Planning for Variable Inertia Mechanical Systems

Elie A. Shammas, Howie Choset, and Alfred A. Rizzi

Carnegie Mellon University
Pittsburgh, PA 15213, U.S.A.
eshammas@andrew.cmu.edu, choset@cs.cmu.edu, arizzi@cs.cmu.edu

Summary. In this paper, we generate gaits for mixed systems, that is, dynamic systems that are subject to a set of non-holonomic constraints. What is unique about mixed systems is that when we express their dynamics in body coordinates, the motion of these system can be attributed to two decoupled terms: the geometric and dynamic phase shifts. In our prior work, we analyzed systems whose dynamic phase shift was null by definition. Purely mechanical and principally kinematic systems are two classes of mechanical systems that have this property. We generated gaits for these two classes of systems by intuitively evaluating their geometric phase shift and relating it to a volume integral under well-defined *height* functions.

One of the contributions of this paper is to present a similar intuitive approach for computing the dynamic phase shift. We achieve this, by introducing a new scaled momentum variable that not only simplifies the momentum evolution equation but also allows us to introduce a new set of well-defined *gamma* functions which enable us to intuitively evaluate the dynamic phase shift. More specifically, by analyzing these novel gamma functions in a similar way to how we analyzed height functions, and by analyzing the sign-definiteness of the scaled momentum variable, we are able to ensure that the dynamic phase shift is non-zero solely along the desired fiber direction.

Finally, we also introduce a novel mechanical system, the *variable inertia snakeboard*, which is a generalization of the original snakeboard that was previously studied in the literature. Not only does this general system help us identify regions of the base space where we can not define a certain type of gaits, but also it helps us verify the generality and applicability of our gait generation approach.

1 Introduction

In our prior work, we generated *geometric* gaits for two classes of mechanical systems, *purely mechanical* in [16] and *principally kinematic* systems in [14]. These two systems seemingly belong at the two ends of a spectrum, that is, purely mechanical systems are systems whose motion is governed solely by the conservation of momentum while principally kinematic systems are systems whose motion is governed solely by the existence of a set of independent non-holonomic constraints that fully constrain the systems velocity. In this paper we generate gaits for the full range of this spectrum. Specifically, we generate gaits for *dynamic systems with non-holonomic constraints* (also known as *mixed systems*)

S. Akella et al. (Eds.): Algorithmic Foundation of Robotics VII, STAR 47, pp. 375–390, 2008.
springerlink.com

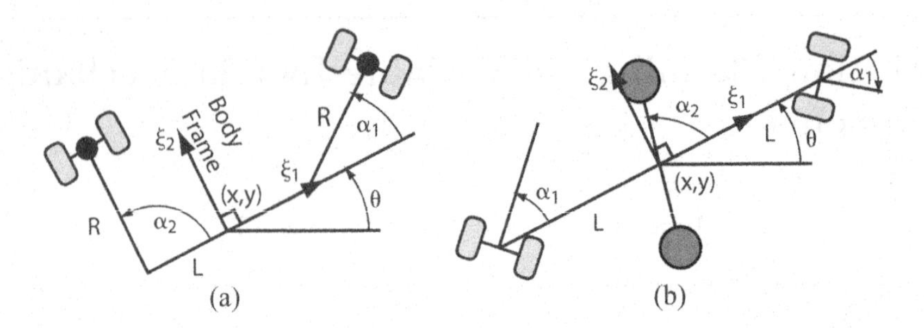

Fig. 1. A schematic of the variable inertia snakeboard in (a) and original snakeboard in (b) depicting their configuration variables

that is, systems whose motion is governed by *both* a non-holonomic set of constraints and generalized momentum being constrained by a set of differential equations.

In this paper, we generate gaits for a novel mechanical system, the *variable inertia snakeboard*, shown in Fig. 1(a). This system is a generalization of the *original snakeboard*, (Fig. 1(b)), which was extensively studied in the literature and which we analyzed in [15]. Both snakeboards belong to the mixed type systems, that is, the non-holonomic constraints do not fully span the fiber space. Thus, the generalized non-holonomic momentum must be instantaneously conserved along certain directions which for the above snakeboards are rotations about the wheel axes intersections. However, the inertia of the original snakeboard is independent of the base variables[1] which greatly simplifies the gait generation analysis. Thus, we consider the variable inertia snakeboard, which as its name suggests, has a non-constant inertia, to verify the generality and applicability of our gait analysis techniques.

Our gait generation techniques will allow us translate and rotate the variable inertia snakeboard in the plane by designing curves in the actuated *base space* which represents the internal degrees of freedom of the robot. In other words, we will generate gaits by using the actuated base variables to control the unactuated variables of the *fiber space* which denote the "position" of the system with respect to a fixed inertial frame. Thus, our goal is to design cyclic curves in the base space, which after a complete cycle, produce a desired motion along a specified fiber direction, hence, effectively moving the robot to a new position.

2 Prior Work

Gait generation has been extensively studied in the literature, next we present three approaches which are most relevant to our work.

[1] Changing the base variables, rotor and wheel axes angles, in Fig. 1(b) will not change the position of the system's center of mass.

Sinusoidal inputs: Ostrowski *et. al.* expressed the dynamics of a mechanical system in body coordinates and were able to represent it as an affine non-linear control system. Then by taking recourse to control theory, they were able to design *sinusoidal* gaits and specify the gait frequencies. Nonetheless, the gait amplitudes were empirically derived [12]. Ostrowski *et. al.* used their gait generation analysis to generate gaits for the original snakeboard (Fig. 1(b)) [11]. Moreover, Chitta *et. al.* developed several unconventional locomoting robots, such as the robo-trikke and the rollerblader, [4, 13], then used Ostrowski's techniques to generate sinusoidal gaits for these novel locomoting robots. Prior work related to Ostrowski's can also be found in [1, 8, 17].

Kinematic reduction: The work done by Bullo *et. al.* in [3] on kinematic reduction of simple mechanical systems is closely related to our work. They define a kinematic reduction for simple mechanical systems, or in other words, reduce the dynamics of a system so that it can be represented as a kinematic system. Then they study the controllability of these reduced systems and for certain examples, they were able to generate gaits for these systems. In fact, Bullo *et. al.* have designed gaits for the original snakeboard (Fig. 1(b)) which we analyzed in [15]. In this paper, we introduce one type of gait, *a purely kinematic gait*, which is structurally similar to gaits proposed by Bullo *et. al.* in [2]; however, we have a different way of generating these gaits.

Integration approach: We generated gaits for two classes of systems, purely mechanical and principally kinematic systems in [16] and [14]. For both systems, we were able to relate position change to a volume integral under a well-defined height function. This allowed us to define a variational problem to optimize gaits but more importantly it allowed us to generate gaits by intuitively designing curves in the base space. Finally, it is worth mentioning that Mukherjee *et. al.* in [7, 9, 10] and Yamada in [18] have done similar work where they generated gaits for specific systems. The fiber space of the systems they studied had an Abelian group structure which simplified their analysis.

3 Background Material

Here we present a rather abbreviated introduction to Lagrangian mechanics, introduce *mixed systems*, and finally we present several mechanics of locomotion results which we shall utiliize to generate gaits for mixed systems.

Lagrangian mechanics: The configuration space of a mechanical system, usually denoted by Q, is a trivial principal fiber bundle; that is, $Q = G \times M$ where G is the fiber space which has a Lie group structure and M is the base space. In this paper we assume the Lagrangian of a mechanical system to be its kinetic energy. Moreover, we assume that the non-holonomic constraints that are acting on the mechanical system can be written in a Pfaffian from, $\omega(q) \cdot \dot{q} = 0$, where $\omega(q)$ is a $k \times n$ matrix describing the constraints and $\dot{q}$ represents an element in the tangent space of the n-dimensional configuration manifold Q.

Associated with the Lie group structure of the fiber space, G, we can define the action, Φ_g, and the lifted action, $T_g\Phi_g$, which act on the entire configuration manifold, Q, and tangent bundle, TQ, respectively. Since we can verify that both the Lagrangian and non-holonomic constraints are invariant with respect to these action, we can express the system's dynamics at the Lie group identity[2] as was shown in [5]. In other words, we eliminate the dependence on the placement of the inertial frame. This invariance allows us to compute the *reduced Lagrangian*, $l(\xi, r, \dot{r})$, which according to [11] will have the form shown in (1) and the *reduced non-holonomic constraints* shown in (2) as we demonstrated in [14].

$$l(\xi, r, \dot{r}) = \frac{1}{2}\begin{pmatrix}\xi\\ \dot{r}\end{pmatrix}^T \tilde{M}\begin{pmatrix}\xi\\ \dot{r}\end{pmatrix} = \frac{1}{2}\begin{pmatrix}\xi\\ \dot{r}\end{pmatrix}^T \begin{pmatrix} I(r) & I(r)A(r)\\ A^T(r)I^T(r) & \tilde{m}(r)\end{pmatrix}\begin{pmatrix}\xi\\ \dot{r}\end{pmatrix} \tag{1}$$

$$\bar{\omega}(r)\begin{pmatrix}\xi\\ \dot{r}\end{pmatrix} = \begin{pmatrix}\bar{\omega}_\xi(r) & \bar{\omega}_r(r)\end{pmatrix}\begin{pmatrix}\xi\\ \dot{r}\end{pmatrix} = 0 \tag{2}$$

Here $\tilde{M}$ is the reduced mass matrix, $A(r)$ is the local form of the mechanical connection, $I(r)$ is the local form of the locked inertia tensor, that is, $I(r) = \mathbb{I}(e, r)$[3], and $m(r)$ is a matrix depending only on base variables. Finally, recall that ξ is an element of the Lie algebra and is given by $\xi = T_gL_{g^{-1}}\dot{g}$, where $T_gL_{g^{-1}}$ is the lifted action acting on a tangent space element $\dot{g}$.

Mixed systems: Such systems are a general type of dynamic mechanical system that are subject to a set of non-holonomic constraints which are invariant with respect to the Lie group action. Hence, a mechanical system whose configuration space has a trivial principal fiber structure, $Q = G \times M$, and is subjected to k non-holonomic constraints, $\omega(q) \cdot \dot{q} = 0$, is said to be *mixed* if

- $0 < k < l$ (number of constraints less than the dimension of fiber space),
- $\det(\omega(q)) \neq 0$ (linear independence), and
- $\omega(q) \cdot \dot{q} = \omega(\Phi_g(q)) \cdot T_g\Phi_g(\dot{q}) = 0$ (invariance).

Mechanics of locomotion: Now we borrow some well-known results from the mechanics of locomotion, [6], upon which we shall build our own gait generation techniques. For a mixed system, according to [11] the system's configuration velocity expressed in body coordinates, ξ, is given by the *reconstruction* equation shown in (3), where $\mathbf{A}(r)$ is an $l \times m$ matrix denoting the local form of the *mixed non-holonomic connection*, $\Gamma(r)$ is an $l \times (l-k)$ matrix, and p is the generalized non-holonomic momentum. We can compute this momentum variable by $p = \frac{\partial l}{\partial \xi}\bar{\Omega}^T$ where $\bar{\Omega}^T$ is a basis of $\mathcal{N}(\bar{\omega}_\xi)$, the null space of $\bar{\omega}_\xi$. Then using (1) we compute the expression for p as shown in (4).

$$\xi = -\mathbf{A}(r)\dot{r} + \Gamma(r)p^T \tag{3}$$

$$p^T = \bar{\Omega}\frac{\partial l}{\partial \xi} = \bar{\Omega}\,(I\xi + IA\dot{r}) = \begin{pmatrix}\bar{\Omega}I & \bar{\Omega}IA\end{pmatrix}\begin{pmatrix}\xi\\ \dot{r}\end{pmatrix} \tag{4}$$

[2] Note that the elements of the tangent space at the fiber space identity form a Lie algebra which is usually denoted by $\mathfrak{g}$.

[3] e is the Lie group identity element.

Moreover, for systems with a single generalized momentum variable[4], its evolution is governed by a first order differential equation[5] shown in (5), where the σ's are matrices of appropriate dimensions whose components depend solely on the base variables. Later in the paper, we will utilize both (3) and (5) and rewrite them in appropriate forms that will help us generate gaits.

$$\dot{p} = p^T \sigma_{pp}(r) p + p^T \sigma_{p\dot{r}}(r)\dot{r} + \dot{r}^T \sigma_{\dot{r}\dot{r}}(r)\dot{r} \tag{5}$$

Example: Now we introduce our example system, the variable inertia snakeboard, which is composed of three rigid links that are connected by two actuated revolute joints as shown in Fig. 1(a). The outer two links have mass, m, concentrated at the distal ends and an inertia, j, while the middle link is massless. Moreover, attached to the distal ends of the outer two links is a set of passive wheels whose axes are perpendicular to the robot's links. The no sideways slippage of these two sets of wheels provide the two non-holonomic constraints which act on the system.

We attach a body coordinate frame to the middle of the center link and align its first axis along that link. The location of the origin of this body attached frame is represented by the configuration variables (x, y) while its global orientation is represented by the variable θ. The two actuated internal degrees of freedom are represented by the relative angle between the links (α_1, α_2).

Hence, the variable inertia snakeboard has a five-dimensional, $(n = 5)$, configuration space $Q = G \times M$, where the associated Lie group fiber space denoting the robot's position and orientation in the plane is $G = SE(2)$, the special Euclidean group. The base space denoting the internal degrees of freedom is $M = \mathbb{S} \times \mathbb{S}$. The Lagrangian of the variable inertia snakeboard in the absence of gravity is computed using $L(q, \dot{q}) = \frac{1}{2}\sum_{i=1}^{3}(m_i \dot{x}_i^T \dot{x}_i + j_i \dot{\theta}_i^2)$. Let $2L$ and R be the length of the middle link and the outer links, respectively. Moreover, to simplify some expressions we assume that the mass and inertia of the two distal links are identical, that is, $m_i = m$ and $j_i = j = mR^2$. Given that the fiber space has an $SE(2)$ group structure, we can compute the group lifted action as shown in (6).

$$\xi = \begin{pmatrix} \xi^1 \\ \xi^2 \\ \xi^3 \end{pmatrix} = T_g L_{g^{-1}} \dot{g} = \begin{pmatrix} \cos(\theta) & -\sin(\theta) & 0 \\ \sin(\theta) & \cos(\theta) & 0 \\ 0 & 0 & 1 \end{pmatrix} \begin{pmatrix} \dot{x} \\ \dot{y} \\ \dot{\theta} \end{pmatrix} \tag{6}$$

$$I = mR \left(\begin{array}{c|c|c} \frac{2}{R} & 0 & -\sin(\alpha_1) + \sin(\alpha_2) \\ 0 & \frac{2}{R} & \cos(\alpha_1) - \cos(\alpha_2) \\ -\sin(\alpha_1) + \sin(\alpha_2) & \cos(\alpha_1) - \cos(\alpha_2) & \frac{L^2+2R^2}{R/2} + \frac{\cos(\alpha_1)+\cos(\alpha_2)}{1/2L} \end{array} \right) \tag{7}$$

$$IA = mR \begin{pmatrix} -\sin(\alpha_1) & \sin(\alpha_2) \\ \cos(\alpha_1) & -\cos(\alpha_2) \\ 2R + L\cos(\alpha_1) & 2R + L\cos(\alpha_2) \end{pmatrix} \text{ and } \tilde{m} = mR \begin{pmatrix} 2R & 0 \\ 0 & 2R \end{pmatrix} \tag{8}$$

[4] For systems with more than one momentum variables, (5) will be a systems of differential equations involving tensor operations as was shown in [1, 5].

[5] Recall that these equations are the dynamic equations of motion along the fiber variables expressed using the generalized momentum variables.

$$\bar{\omega}_\xi = \begin{pmatrix} -\sin(\alpha_1) & \cos(\alpha_1) & R + L\cos(\alpha_1) \\ \sin(\alpha_2) & -\cos(\alpha_2) & R + L\cos(\alpha_2) \end{pmatrix} \text{ and } \bar{\omega}_r = \begin{pmatrix} R & 0 \\ 0 & R \end{pmatrix} \tag{9}$$

$$p = \frac{m\left(p_{\xi^1}\xi^1 + p_{\xi^2}\xi^2 + p_{\xi^3}\xi^3 + p_\alpha(\dot{\alpha}_1 + \dot{\alpha}_2)\right)}{2\sin(\alpha_1 - \alpha_2)} \tag{10}$$

The lifted action allows us to verify the Lagrangian invariance and to compute the reduced Lagrangian. Thus, the components of the reduced mass matrix as is shown in (1) are given in (7) and (8). Note that the reduced mass matrix is not constant as was the case for the original snakeboard [15] but it depends solely on the base variables, α_1 and α_2. Similarly we can write the non-holonomic constraints in body coordinates for the variable inertia snakeboard where the components of (2) are given in (9). Moreover, note that the variable inertia snakeboard has a three-dimensional fiber space, $SE(2)$, and it has two non-holonomic constraints, one for each wheel set. We have verified that these constraints are invariant with respect to the group action and we know that these non-holonomic constraints are linearly independent away from singular configurations. Thus, we conclude that the variable inertia snakeboard a *mixed* type system.

Next, using (4) we compute the generalized non-holonomic momentum for the variable inertia snakeboard as shown in (10), where p_{ξ^i} and p_α are analytic functions of the base variables. For the sake of brevity, we will not present the explicit structure of these functions. As for the reconstruction equation, (3), we can easily compute it by solving for ξ from the system of equations in (2) and (4). Again, we will omit presenting the components of the mixed connection, but, it will be an $l \times m$ matrix.

4 Scaled Momentum

Now we manipulate (5) to a more manageable form which allows us to intuitively evaluate the dynamic phase shift. At this point we will limit ourselves to systems that have one less velocity constraint than the dimension of the fiber space, i.e., $l - k = 1$. This leaves us with only one generalized momentum variable and forces the term $\sigma_{pp}(r) = 0$ in (5) as was explained in [11]. Moreover, first order differential equation theory confirms that an integrating factor, $h(r)$, exists for (5). Thus, we define the scaled momentum as $\rho = h(r)p$ and rewrite (3) and (5) to arrive at

$$\xi = -\mathbf{A}(r)\dot{r} + \bar{\Gamma}(r)\rho, \text{ and} \tag{11}$$

$$\dot{\rho} = \dot{r}^T\bar{\Sigma}(r)\dot{r}, \tag{12}$$

where $\bar{\Gamma}(r) = \Gamma(r)/h(r)$ and $\bar{\Sigma}(r) = h(r)\sigma_{\dot{r}\dot{r}}(r)$. Now that we have written the reconstruction and momentum evolution equations in our simplified forms shown in (11) and (12), we are ready to generate gaits by studying and analyzing the three terms, $\mathbf{A}(r)$, $\bar{\Gamma}(r)$, and $\bar{\Sigma}(r)$. In fact, we will use $\mathbf{A}(r)$ and $\bar{\Gamma}(r)$ to respectively construct the *height* and *gamma* functions while we use $\bar{\Sigma}(r)$ to study the sign definiteness of the scaled momentum.

5 Gait Evaluation

In this section, we equate the position change due to any closed base-space curve to *two* decoupled terms. Then, we present how to design curves in such a way to *exclusively* ensure that any of these terms is non-zero along a specified fiber direction, that is, effectively synthesizing two types of gait associated with the each of the decoupled terms. In the next section, we will define a partition on the space of allowable gaits such that we can generate gaits by relating position change to either one of the decoupled terms or both. For the first case, we exclusively use the gait synthesis tools presented in this section while for the second case we define another type of gaits that simultaneously utilizes both gait synthesis tools.

We define a *gait* as a closed curve, ϕ, in the base space, M, of the robot. We require that our gaits be cyclic and continuous curves. Having written the body representation of a configuration velocity in a simplified manner as seen in (11), we solve for position change by integrating (11). Defining ζ as the integral of ξ and then integrating each row of (11) with respect to time we get

$$\Delta\zeta^i = \int_{t_0}^{t_1} \dot{\zeta}^i dt = \int_{t_0}^{t_1} \xi^i dt = \int_{t_0}^{t_1} \left(-\sum_{j=1}^{m} \mathbf{A}^i_j(r)\dot{r}^j + \sum_{j=1}^{l-k} \Gamma^i_j(r)\rho^j \right) dt$$

$$= \underbrace{\int\int_{\Phi} \sum_{o,j=1,o<j}^{m} \bar{\mathbf{A}}^i_{oj}(r)dr^o dr^j}_{I^{GEO}} + \underbrace{\int \sum_{j=1}^{l-k} \left(\bar{\Gamma}^i_j(r) \int \left(\dot{r}^T \bar{\Sigma}(r)\dot{r} \right)^j dt \right) dt}_{I^{DYN}} \quad (13)$$

Note that the first term can be written as a line integral and then by using Stokes' theorem we equate it to a volume integral. As for the second term we just substitute for the scaled momentum, ρ, using (12). Hence, we equated position change to two integrals, I^{GEO} which computes the geometric phase shift and I^{DYN} which computes the dynamic phase shift. Next, we analyze how to synthesize gaits using the two independent phase shifts.

5.1 Synthesizing Geometric Gaits

For simplicity, we limit ourselves to two-dimensional base spaces, that is, ($m = 2$). This allows us to equate the *geometric* position change contribution, I^{GEO}, due to any gait, ϕ, by computing the volume integral $\int\int_{\phi} F^i(r^1, r^2)dr^1 dr^2$, where $F^i = \frac{\partial A^i_2}{\partial r_1} - \frac{\partial A^i_1}{\partial r_2}$'s are the well-defined *height function* associated with the fiber velocity ξ^i. Then, we generate *geometric gaits* by studying certain properties of the height functions:

- *Symmetry:* to study smaller portions of the base space.
- *Signed regions:* to control the orientation of the designed curves as well as the magnitude of the the geometric phase shift.
- *Unboundedness:* to identify singular configurations of the robot.

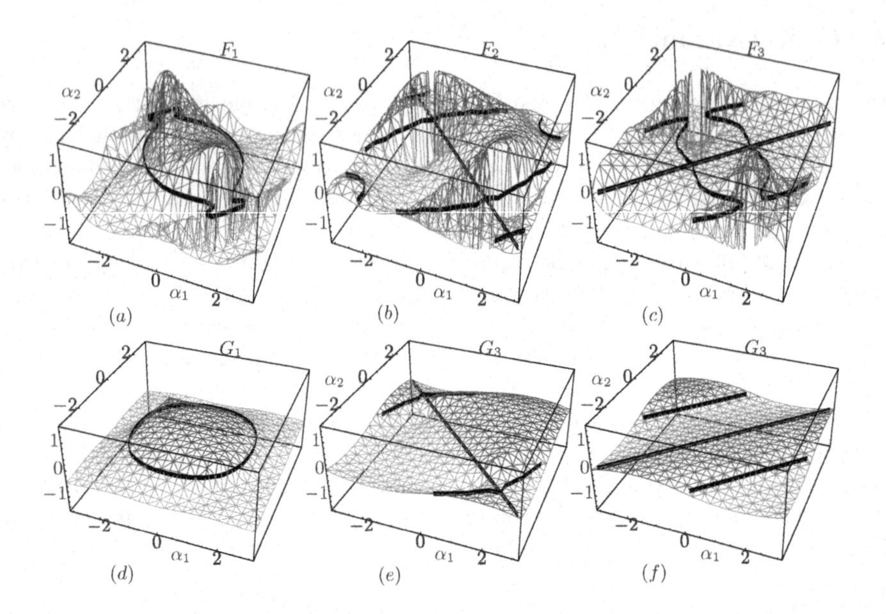

Fig. 2. The three height functions, (F_1, F_2, F_3), and three gamma functions, (G_1, G_2, G_3), corresponding to the three fiber directions in body representation, (ξ_1, ξ_2, ξ_3), for the variable inertia snakeboard are depicted in (a) through (f). The darker colors indicate the positive regions which are separated by solid lines from the lighter colored negative regions.

By inspecting the above properties of the height functions we are able to easily design curves that only envelope a non-zero volume under a desired height function while it encloses zero volume under the rest of the height functions. Again we remind the reader of some simple rules that are helpful in designing such curves:

- *Closed non-self-intersecting curves* that stay in a single signed region are guaranteed to enclose a non-zero volume.
- *Closed self-intersecting curves* that span two regions with opposite signs and that change orientation as they pass from one region to another are also guaranteed to enclose a non-zero volume.
- *Closed non-self-intersecting curves that are symmetric about odd points* are guaranteed to have zero volume.
- *Closed self-intersecting curves that are symmetric about even points* are guaranteed to have zero volume.

Note that these rules do not impose any additional constraints on the shape of the input curves.

5.2 Synthesizing Dynamic Gaits

Now, we will analyze the second term in (13), to propose gaits that ensure that I^{DYN} is non-zero along a desired fiber direction. Note that for each fiber direction the integrand of I^{DYN} in (13) is composed of the product of two terms, the *gamma function*, $\bar{\Gamma}^i(r)$, and the scaled momentum variable, ρ. Thus, by analyzing the $\bar{\Sigma}$ matrix in (13) we propose families gaits that ensure that the scaled momentum variable is sign-definite. Then, we analyze the the gamma functions in a similar way to how we analyze the height functions, that is, we study their symmetry, signed regions, and unbounded regions. Thus, by picking gaits that are located in a same signed region of $\bar{\Gamma}^i(r)$, we ensure the integrand of I^{DYN} is non-zero along a desired fiber direction.

Example: Now we compute the height and gamma functions for the variable inertia snakeboard. The expressions for this particular system are rather complicated and we will not present them here; however, we depict the graphs of the three height and gamma functions in Fig. 2(a) – (c) and (d) – (f), respectively. These functions have the following properties which we will utilize later to generate gaits.

- $F_2 = G_2 = 0$ for $\alpha_1 = -\alpha_2$,
- $F_3 = G_3 = 0$ for $\alpha_1 = \alpha_2$,
- F_1 and G_1 are even about both lines $\alpha_1 = \alpha_2$ and $\alpha_1 = -\alpha_2$,
- F_2 and G_2 are even about $\alpha_1 = \alpha_2$ and odd about $\alpha_1 = -\alpha_2$,
- F_3 and G_3 are odd about $\alpha_1 = \alpha_2$ and even about $\alpha_1 = -\alpha_2$.

6 Gait Generation for Mixed System

In this section, we utilize our geometric and dynamic gait synthesis to generate gaits for mixed systems. Next, we define a partition on the allowable gait space which allows us to independently analyze I^{GEO} and I^{DYN} and generate gaits using our synthesis tools. We respectively label the two families of gaits as *purely kinematic* and *purely dynamic* gaits. Moreover, we propose a third type of gait that simultaneously utilizes both shifts, I^{GEO} and I^{DYN}, to produce motions with relatively larger magnitudes. We label this family of gaits as *kino-dynamic* gaits.

6.1 Purely Kinematic Gaits

Purely kinematic gaits are gaits whose motions is solely due to I^{GEO}, that is, $I^{DYN} = 0$ for all time. A solution for such a family of gaits is to set $\rho = 0$ in (13) which sets the integrand of I^{DYN} to zero. Thus, we define purely kinematic gaits as gaits for which $\rho = 0$ for all time. Note that for purely mechanical systems $p = \rho = 0$ by definition and for principally kinematic systems $I^{DYN} = 0$ since $p = \varnothing$. Hence, any gait for these two types of systems is necessarily purely kinematic. However, for mixed systems, we generate purely kinematic gaits by the following two step process:

- Solving the scaled momentum evolution equation, (12), for which $\rho = \dot{\rho} = 0$. This step defines vector fields over the base space whose integral curves are candidate purely kinematic gaits.
- Using our geometric gait synthesis analysis on the above candidate gaits to concatenating parts of integral curves that enclose a non-zero volume under the desired height functions.

Sometimes, purely kinematic gaits are referred to as *geometric* gaits, since the produced motion is solely due to the generated geometric phase as defined in [1]. Moreover, purely kinematic gaits are structurally similar to gaits proposed by Bullo in his kinematic reduction of mechanical systems in [3]. The vector fields defined above essentially serve the same purpose of the de-coupling vector fields presented in Bullo's work.

Example: For the variable inertia snakeboard, we can easily design purely kinematic gaits by solving for the right hand side of (12) equal to zero. Since the right hand side of (12) is a quadratic in the base velocites[6], we ensure that the term $\Delta\rho(\alpha_1, \alpha_2) = \bar{\Sigma}_1^2\bar{\Sigma}_1^2 - \bar{\Sigma}_1^1\bar{\Sigma}_2^2 \geq 0$. A plot of a $\Delta\rho/\max(\Delta\rho)$ is shown in Fig. 3(a). The light colored regions indicate that $\Delta\rho(\alpha_1, \alpha_2) < 0$, that is, we can never compute any velocities for which $\dot{\rho} = 0$. In other words, we should avoid these regions of the base space while designing purely kinematic gaits.

Away from the negative regions of $\Delta\rho(\alpha_1, \alpha_2)$, we design purely kinematic gaits for the variable inertia snakeboard. The right hand side of (12) has four unknowns, $(\alpha_1, \alpha_2, \dot{\alpha}_1, \dot{\alpha}_2)$. Thus, at each point in the base space, that is, fixing (α_1, α_2), we need to solve the velocities $(\dot{\alpha}_1, \dot{\alpha}_2)$ for which the right hand side is zero. Since, we have two unknowns and one equation, we solve for the ratios, $\frac{\dot{\alpha}_1}{\dot{\alpha}_2}$ and $\frac{\dot{\alpha}_2}{\dot{\alpha}_1}$ for which the right hand side is zero. Thus, ignoring the magnitudes of the base velocities, the two ratios $\frac{\dot{\alpha}_1}{\dot{\alpha}_2}$ and $\frac{\dot{\alpha}_2}{\dot{\alpha}_1}$ define the slopes of vectors at each point in the base space which we use to define vector fields over the entire base space as depicted in Fig. 3(b). Hence, any part of an integral curve of the above vector fields is necessarily a purely kinematic gait. For example, the families of lines, $l_1 = \{\alpha_2 = \alpha_1 + k\pi, k \in \mathbb{Z}\}$ and $l_2 = \{\alpha_2 = -\alpha_1 + 2k\pi, k \in \mathbb{Z}\}$ are the simplest integral curves we could define whose velocities exactly match the above vector fields.

To design a purely kinematic gait that will move the variable inertia snakeboard along say the ξ^1 direction, we pick and closed integral curve that will enclose a non-zero volume solely under the first height function. Using the above lines, we know that the polygon given in the first row of the first column of Table 1 and depicted in Fig. 3(b) will move the snakeboard along the ξ_1 direction. Similarly we construct two other polygons shown in the second and third rows of the first column of Table 1 as depicted in Fig. 3(b) to respectively locomote the snakeboard along the ξ_2 and ξ_3 directions.

[6] For the original snakeboard, we verified in [15] that the right hand side of the scaled momentum evolution equation is not a quadratic. This simplified the generation of purely kinematic gaits and mislead us into believing that purely kinematic gaits could be defined everywhere on the base space.

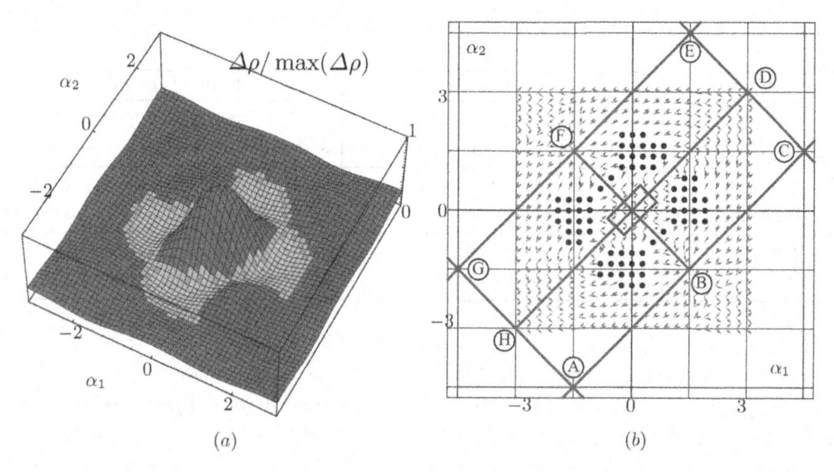

(a) (b)

Fig. 3. (a) Plot indicating the negative regions (lighter colored regions) of the base space where $\Delta\rho < 0$. (b) Two vector fields defined over the base space whose integral curves are purely kinematic gaits. The solid lines are integral curves of the vector fields which we will utilize to generate purely kinematic gaits. The solid dots indicate the negative regions of $\Delta\rho$ where the vector fields are not defined.

Inspecting the above polygonal gaits, we found out that they pass through the snakeboard's singular configurations, $(\alpha_1, \alpha_2) = \{(\frac{\pi}{2}, -\frac{\pi}{2}), (-\frac{\pi}{2}, \frac{\pi}{2})\}$, (Fig. 2$(a)$–$(c)$). So rather than solving numerically for other integral curves of the vector fields and solve for other possible gaits which is a tedious process, we simply shrunk the above proposed gaits around the center of the base space as shown in Fig. 3(b). These curves closely, but not exactly, match the vector fields. So we shall expect a change in the scaled momentum value as we traverse these gaits since they are an approximate solution. The motions of the variable inertia snakeboard due to these gaits are depicted in Fig. 4(a) – (c). Note, the small magnitudes of motion due to the small volumes under the height functions. However, we can clearly see that the gaits move the variable inertia snakeboard along the x and y directions in Fig. 4(a) and Fig. 4(b), respectively, and rotate the snakeboard in Fig. 4(c).

Finally, recall that, our analysis is done in body coordinates, ξ^i's, which are related by the map $T_gL_{g^{-1}}$ to the fiber variables, $\dot{g}^i$'s. Since, the fiber space for the variable inertia snakeboard is $SE(2)$ which is not Abelian, the map $T_gL_{g^{-1}}$ is non-trivial; hence, one should not expect a direct correspondence between say ξ^1 and $\dot{x}$. This explains the non-pure fiber motions in Fig. 4(a), where motion along ξ^1 transforms to major motion along x and minor motion along the y axis.

6.2 Purely Dynamic Gaits

As the name suggests, purely dynamic gaits are gaits that produce motion solely due to the dynamic phase shift, that is, $I^{GEO} = 0$ while $I^{DYN} \neq 0$. These gaits

Table 1. Three proposed gaits of each family for the variable inertia snakeboard

Purely Kinematic	Purely Dynamic	Kino-dynamic
Polygon ACEGA	$\alpha_1 = \frac{\pi}{4}(1 - \sin(t) - 2\sin^2(t))$ $\alpha_2 = \frac{\pi}{4}(1 + \sin(t) - 2\sin^2(t))$	$\alpha_1 = -\frac{1}{\sqrt{2}}\left(\frac{\pi}{2}\sin(t) + \frac{\pi}{4}\cos(t)\right)$ $\alpha_2 = \frac{1}{\sqrt{2}}\left(-\frac{\pi}{2}\sin(t) + \frac{\pi}{4}\cos(t)\right)$
Polygon ABFECBFGA	$\alpha_1 = \frac{\pi}{10}(2\sin(3t) - 5)$ $\alpha_2 = \frac{\pi}{6}(\sin(t) - 3)$	$\alpha_1 = \frac{1}{\sqrt{2}}\left(\frac{\pi}{2}\sin(2t) + \frac{\pi}{4}\sin(t)\right)$ $\alpha_2 = -\frac{1}{\sqrt{2}}\left(\frac{\pi}{2}\sin(2t) - \frac{\pi}{4}\sin(t)\right)$
Polygon ACDHGEDHA	$\alpha_1 = \frac{\pi}{4}(2\sin(t) + 1)$ $\alpha_2 = \frac{\pi}{4}(2\sin(t) - 1)$	$\alpha_1 = \frac{1}{\sqrt{2}}\left(\frac{\pi}{3}\sin(2t) + \frac{\pi}{3}\sin(t)\right)$ $\alpha_2 = \frac{1}{\sqrt{2}}\left(-\frac{\pi}{3}\sin(2t) + \frac{\pi}{3}\sin(t)\right)$

are relatively easy to design since these are gaits that enclose no "volume" in the base space. Note that all systems that have only *one* base variables have gaits that are necessarily purely dynamic, since setting $m = 1$ in (13) will yield $I^{GEO} = 0$. For example, all the gaits for robo-Trikke robot which was studied by Chitta *et. al.* in [4] are necessarily purely dynamic since there exists only one base variable. As for systems with more than one base space variable, it is still relatively easy to construct purely dynamic gaits. Such gaits should not enclose any area in the base space. A simple solution would be to ensure that a gait retraces the same curve in the second half cycle of the gait but in the opposite direction.

Thus, we propose the following purely dynamic families of gaits: $\{r_1, r_2\} = \{\sum_{i=0}^{n} a_i (f(t))^i, f(t)\}$, where $f(t) = f(t + \tau)$ is a periodic real function and a_i's are real numbers. We can verify that these gaits will have zero area in the base space (r_1, r_2). Moreover, we can verify that for the above family of gaits, the scaled momentum variable is sign-definite, that is, $\rho \leq 0$ or $\rho \geq 0$ for all time. Then, generating purely dynamic gaits reduces to the following simple procedure:

- Select gaits from the above described family and check the sign of the scaled momentum variable ρ.
- Analyze the gamma functions depicted in (11) and (13) to pick the gait that ensures that the integrand of I^{DYN} is non-zero for the desired fiber direction.

Example: For the variable inertia snakeboard, we construct three purely dynamic gaits depicted in the second column of Table 1. The motion due to these gaits are respectively shown in Fig. 4(d) – (f). For instance, we designed the first gait in the second column of Table 1 such that $\rho \leq 0$ for all time. The gait is located close to the center of the base space and is symmetric about the line $\alpha_1 = -\alpha_2$; moreover, only the first gamma function is non-zero and even about the line $\alpha_1 = -\alpha_2$ while the second and third gamma functions is odd about this line. Thus we expect a non-zero I^{DYN} only along the ξ^1 direction. This motion, is largely transformed to motion along the x direction as shown in Fig. 4(d). Similarly, we designed the other two gaits to move the variable inertia snakeboard along the y direction, (Fig. 4(e)) and to rotate it along the θ direction, (Fig. 4(f)).

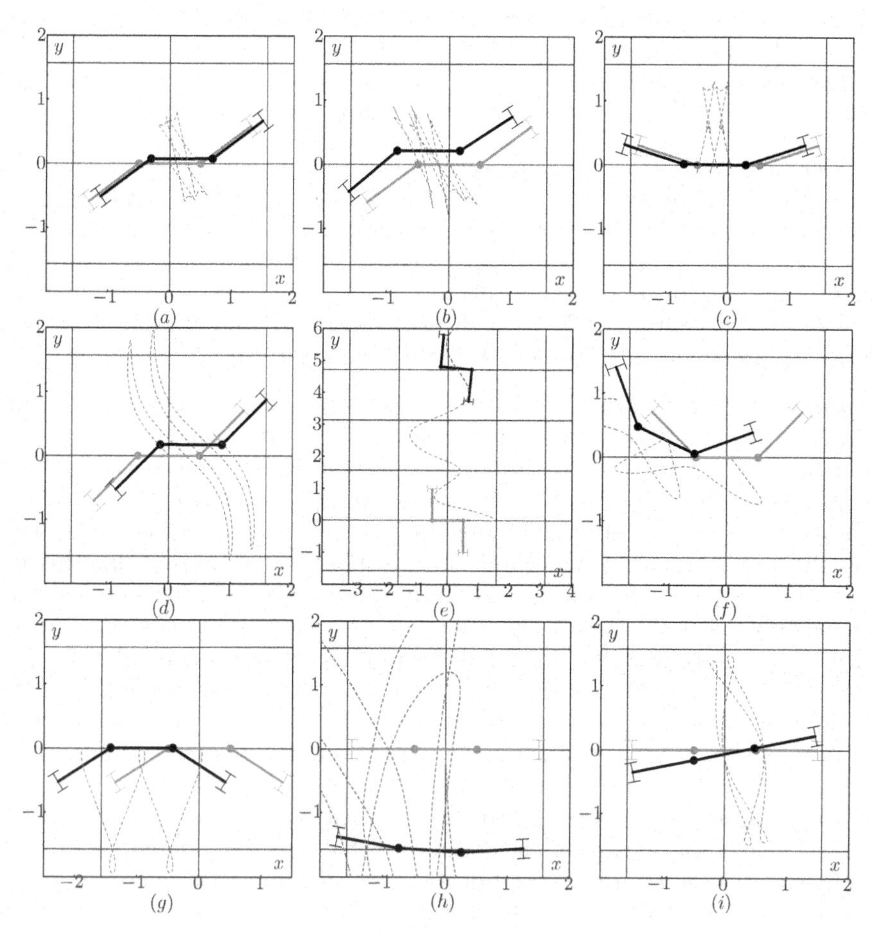

Fig. 4. The actual motion that the variable inertia snakeboard will follow as the base variables follow the three *purely kinematic* depicted in the first column of Table 1 shown respectively in a, b, and c; three *purely dynamic* gaits depicted in the second column of Table 1 shown respectively in d, e, and f; and three *kino-dynamic* gaits depicted in the third column of Table 1 shown respectively in g, h, and i. The initial and final configurations for each gait are shown in gray and black colors, respectively, while the dotted line depicts the trace of the origin of the body-attached coordinate frame.

6.3 Kino-Dynamic Gaits

Finally, we have the third type of gaits which we term as kino-dynamic gaits. These gaits have both I^{GEO} and I^{DYN} not equal to zero, that is, the motion of the system is due to both the geometric phase shift as well as the dynamic phase shift which are associated with I^{GEO} and I^{DYN}, respectively. We design kino-dynamic gaits in a two step process.

- First we do the volume integration analysis on I^{GEO} to find a set of candidate gaits that move the robot in the desired direction.
- The second step it to compute I^{DYN} for the candidate gaits and verify that the effect of I^{DYN} actually enhances the desired motion.

Essentially, kino-dynamic gaits are variations of purely kinematic gaits. In a sense, we start by generating a purely kinematic gait but by neglecting the constraints that the gaits has to be an integral curve of the vector fields that prescribes the purely kinematic gaits. Thus, we know that scaled momentum is not necessarily zero for all time, that is, $I^{DYN} \neq 0$. Then, we pick the gaits for which the magnitude of I^{DYN} additively contribute to that of I^{GEO}, hence, effectively producing fiber motions with bigger magnitudes.

Example: For the variable inertia snakeboard, we can generate kino-dynamic gaits by using the volume integration analysis to produce candidate gaits. For example, to generate a gait that rotates the variable inertia snakeboard in place, we start by designing a curve in the base space that envelopes a non-zero volume only under the third height function of the variable inertia snakeboard (Fig. 2(c)). A figure-eight type curve with each of its loops having opposite orientation and lying on the opposite side of the line $\alpha_1 = \alpha_2$, will envelope non-zero volume only under the third height function. This curve is the last curve in the third column of Table 1. We simulated this proposed gaits and indeed it does rotate the variable inertia snakeboard along the θ direction as shown in (Fig. 4(i)). Similarly, we designed two other curves depicted in the first and second rows of the last column of Table 1 to move the variable inertia snakeboard along the x direction, (Fig. 4(g)), and the y direction ,(Fig. 4(h)).

In this section we have generated three of each type of gaits that moved the variable inertia snakeboard in any specified global direction. Moreover, we have the freedom to choose from several of the types of gaits that we have proposed earlier. It is worth noting that the purely kinematic and purely dynamic gaits were the easiest to design since we are exclusively analyzing either I^{KIN} or I^{DYN} and not both at the same time as is the case for kin-dynamic gaits.

7 Conclusion

In this paper, we studied mixed non-holonomic systems and designed three families of gaits, purely kinematic, purely dynamic, and kino-dynamic gait, to move such systems along specified fiber directions. This work is a generalization over our prior work where we used one type of the gaits defined here to analyze two other systems, purely mechanical and principally kinematic. Moreover, our technique has better control over all parameters of the suggested gaits, which reduces the need for deeper intuition to manually set these parameters and our technique eliminated sinusoidal restrictions.

One of the contributions of this paper is the introduction of the scaled momentum variable which greatly simplified our gait generation analysis. This new variable allowed us to rewrite both the reconstruction as well as the momentum evolution equation in simpler forms that are suitable for our gait generation techniques.

Another contribution is the introduction of the novel mechanical system, the variable inertia snakeboard. This system is similar enough to the original snakeboard that we can relate our results to this well known system, but at the same time it did not over simplify the gait generation problem. In fact, through analyzing the variable inertia snakeboard, we identified regions in the base space where purely kinematic gaits are not possible. There are no such regions for the original snakeboard.

This paper constitutes a first step towards developing an algorithmic gait synthesis technique. Ideally, we would like to develop an algorithm whose inputs are the system's configuration space structure, its Lagrangian, and the set of nonholonomic constraint acting on the system. The algorithm would automatically generate gaits that will move the system along a desired global direction with a desired magnitude. However, we still need to develop several additional tools to complete this gait generating algorithm.

References

1. Bloch, A.: Nonholonomic Mechanics and Control. Springer, Heidelberg (2003)
2. Bullo, F., Lewis, A.D.: Kinematic Controllability and Motion Planning for the Snakeboard. IEEE Transactions on Robotics and Automation 19(3), 494–498 (2003)
3. Bullo, F., Lewis, A.D.: Geometric Control of Mechanical Systems: Modeling, Analysis, and Design for Simple Mechanical Control Systems. Springer, Heidelberg (2004)
4. Chitta, S., Cheng, P., Frazzoli, E., Kumar, V.: RoboTrikke: A Novel Undulatory Locomotion System. In: IEEE International Conference on Robotics and Automation (2005)
5. Marsden, J.: Introduction to Mechanics and Symmetry. Springer, Heidelberg (1994)
6. Marsden, J., Montgomery, R., Ratiu, T.: Reduction, Symmetry and Phases in Mechanics. Memoirs of the American Mathematical Society 436 (1990)
7. Mukherjee, R., Anderson, D.: Nonholonomic Motion Planning Using Stokes' Theorem. In: IEEE International Conference on Robotics and Automation (1993)
8. Murray, R.M., Sastry, S.S.: Nonholonomic Motion Planning: Steering Using Sinusoids. IEEE T. Automatic Control 38(5), 700 (1993)
9. Nakamura, Y., Mukherjee, R.: Nonholonomic Path Planning of Space Robots. In: IEEE International Conference on Robotics and Automation (1989)
10. Nakamura, Y., Mukherjee, R.: Nonholonomic Path Planning of Space Robots via a Bidirectional Approach. In: IEEE Transactions on Robotics and Autmation, vol. 7, pp. 500–514 (1991)
11. Ostrowski, J.: The Mechanics of Control of Undulatory Robotic Locomotion. Ph.D. dissertation, California Institute of Technology (1995)
12. Ostrowski, J., Burdick, J.: The Mechanics and Control of Undulatory Locomotion. International Journal of Robotics Research 17(7), 683 (1998)
13. Chitta, V.K.S., Heger, F.: Dynamics and Gait Control of a Rollerblading Robot. In: IEEE International Conference on Robotics and Automation (2004)

14. Shammas, E., Choset, H., Rizzi, A.: Natural Gait Generation Techniques for Principally Kinematic Mechanical Systems. In: Proceedings of Robotics: Science and Systems, Cambridge, USA (June 2005)
15. Shammas, E., Choset, H., Rizzi, A.A.: Towards Automated Gait Generation for Dynamic Systems with Non-holonomic Constraints. In: IEEE International Conference on Robotics and Automation (2006)
16. Shammas, E., Schmidt, K., Choset, H.: Natural Gait Generation Techniques for Multi-bodied Isolated Mechanical Systems. In: IEEE International Conference on Robotics and Automation (2005)
17. Walsh, G., Sastry, S.: On reorienting linked rigid bodies using internal motions. Robotics and Automation, IEEE Transactions on 11(1), 139–146 (1995)
18. Yamada, K.: Arm Path Planning for a Space Robot. In: IEEE/RSJ International Conference on Intelligent Robots and Systems (1993)

Sampling-Based Falsification and Verification of Controllers for Continuous Dynamic Systems

Peng Cheng and Vijay Kumar

GRASP Lab, University of Pennsylvania
{chpeng,kumar}@grasp.upenn.edu

Abstract. In this paper, we present a sampling-based verification algorithm for continuous dynamic systems with uncertainty due to adversaries, unmodeled disturbance inputs, unknown parameters, or initial conditions. The algorithm attempts to find inputs (and resulting trajectories) that falsify the specifications of the system thus providing examples of bad inputs to the system. The system is said to be verified if the algorithm cannot find falsifying inputs.

The main contribution of the paper is the analysis of the effects of discretization of the state and input spaces that are inherent to sampling-based techniques. We derive conditions that guarantee resolution completeness. These provide sufficient, although conservative, conditions for verifying Lipschitz continuous (but possibly non smooth) dynamic systems without known analytical solutions. We analyze the effects of transformations of the input and state space on these conditions. The main results of this paper are illustrated with several simple examples.

1 Introduction

Software-enabled control of dynamical systems finds applications not only in robotics, but also in manufacturing, fly-by-wire systems, air-traffic control, medical instrumentation, and biotechnology. There is currently no systematic approach to verifying controllers for systems with continuous input and state spaces except for a very special class of simple systems for which analytical solutions are readily available. Indeed, if we exclude this special class of systems, the verification problem is generally undecidable [1].

The *falsification* problem is similar to the *motion planning* problem. In the former, one tries to find the disturbance or adversarial inputs that result in trajectories which violate system specifications, for example, safety. In the later, we find inputs that guide the system to a state that satisfies specifications for the goal set. In our approach, the *verification* problem is solved by showing the absence of falsifying inputs or trajectories. Thus, a system is said to be verified if there are no falsifying inputs. Analogously, in motion planning, one can try to prove no motion plans exist to reach the goal set.

Because general verification problems are undecidable, semi-decidable approximation algorithms have been designed. Most of these algorithms [2, 8, 18] over-approximate the reachable set to check the safety. However such algorithms are

S. Akella et al. (Eds.): Algorithmic Foundation of Robotics VII, STAR 47, pp. 391–406, 2008.
springerlink.com

limited in their ability to handle complex dynamics in high dimensions. Recently, motivated by the successful application of sampling-based techniques in motion planning [6, 11, 17, 15, 14, 16] and the strong similarity between motion planning and falsification, researchers have developed algorithms [4, 9, 12] that use sampled controls to under-approximate the continuous search space to quickly find counter examples to show that the system is not safe. However, there is no principled way to verify system properties.

The paper presents a sampling-based verification algorithm for Lipschitz continuous but possibly non smooth systems. The verification is achieved by using sampling-based falsification algorithms, which iteratively construct solutions with sampled controls to falsify the given safety specification. Similar approaches have been proposed for linear systems [10] and for hybrid systems [5]. Because sampling-based control algorithms discretize the input and state spaces and approximate the set of trajectories (and therefore the reachable space), it is necessary to establish a relationship between the discretization of these spaces and the approximation of the reachable set, and quantify the confidence level associated with the falsification or verification result. The main goal of this paper is a set of conditions that establishes this connection. The basic result is that a proper choice of sampling dispersion (in input and state spaces) and an appropriate sampling algorithm will ensure that every falsifying control with a finite time horizon will be approximated within a desired level of fidelity by sampled controls in finite time.

This work is closely connected to previous work in which conditions for *resolution completeness* of sampling-based motion planning with differential constraints were established for the first time [7]. It is showed [7] that solutions to motion planning problems for dynamic systems will always be approximated by sample controls in finite time. Of course, no guarantees are offered for problems for which no solutions exist. The key idea is to use Lipschitz conditions on motion equations to develop resolution-complete algorithms. The proof for resolution-completeness relies on establishing that the reachable state set is densely covered by the states reached by sample controls.

In the same spirit, we introduce a relaxed problem, in which the safety specification is relaxed with a given tolerance to enlarge the set of falsifying controls. A resolution-complete (RC) falsification algorithm is designed to approximate falsifying controls for the relaxed problem. If no solutions are found for the relaxed problem, then there exist no falsifying controls for the original problem and the system is verified. This is illustrated schematically in Fig. 1. The shaded region represents the unsafe set for the original problem. The unsafe set of the relaxed problem shown as the set inside the dashed line includes all points which are in the ϵ neighborhood of the unsafe set of the original problem. All falsifying controls for the original problem turn into falsifying controls with violation ϵ for the relaxed problem. If all trajectories $\{\tilde{x}_i\}$ constructed from the RC falsification algorithm are outside of unsafe set of the relaxed problem, then the system is said to be verified.

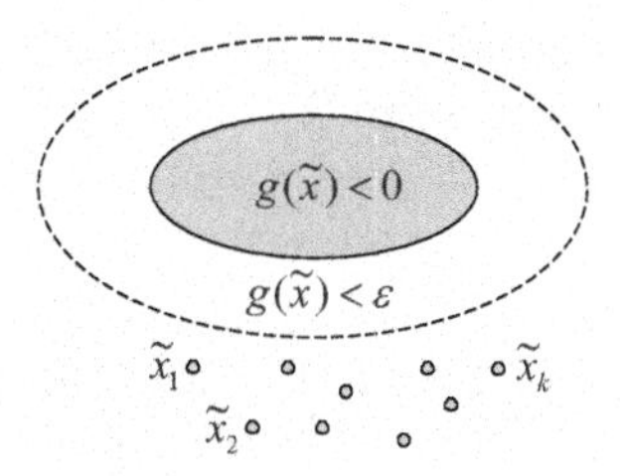

Fig. 1. Verification by falsification. $\tilde{x}$ denotes the system trajectory and the function $g(\tilde{x})$ defines the specification set or the unsafe set. Tolerance ϵ defines the relaxed problem. A trajectory is said to be falsifying for the relaxed problem if $g(\tilde{x}) < \epsilon$.

The organization of this paper is as follows. First, we formally define the dynamic system and the falsification and verification problems in Section 2. Section 3 provides a framework for sampling-based falsification and discusses the complexity of the algorithms. In Section 4, we present the verification algorithm through RC falsification and analyze the effects of scaling and transformation on RC conditions. Several examples are used to illustrate the application of the proposed algorithm in Section 5.

2 Falsification and Verification Problems

In this section, we formally define the dynamic systems of interest, the basic assumptions, and the falsification and verification problems. We use standard notation found in most books on systems theory (see, e.g., [13]).

The dynamic system is described as follows:

$$\dot{x} = \frac{dx}{dt} = f(x, u), x \in X, u \in U, \tag{1}$$

in which $X \subset \Re^n$ is the *state space* and $U \subset \Re^m$ is the *input space*. We assume that x and u are nondimensionalized. X and U are given the structure of a metric space using the infinity norm. We will assume that these sets are bounded and there exist D_u and D_x such that $\|u - u'\| < D_u$ for any $u, u' \in U$ and $\|x - x'\| < D_x$ for any $x, x' \in X$. We assume that the motion equation satisfies the Lipschitz condition with respect to state and input. There exist positive constants L_x and L_u such that:

$$\|f(x, u) - f(x', u')\| \leq L_x\|x - x'\| + L_u\|u - u'\| \tag{2}$$

for any $x, x' \in X$ and $u, u' \in U$. There also exists real constant $D_f > 0$ such that $\|f(x, u)\| < D_f$ for any $x \in X$ and $u \in U$.

The control space $\mathcal{U}$ (a function space) is assumed to include all piecewise constant controls $\tilde{u} : [0, t_f] \to U$. We will also assume that each input is only applied over a constant interval, δt, and there is a positive integer, k, such that

$t_f = k\delta t$. Both these assumptions are for simplicity. The extension to more general function spaces is described in [7].

Given a control $\tilde{u} : [0, t_f] \to U$ and a state $x_0 \in X$, the trajectory of the control from x_0 is

$$\tilde{x}(\tilde{u}, x_0, t) = x_0 + \int_0^t f(\tilde{x}(\tau), \tilde{u}(\tau))d\tau. \tag{3}$$

$\tilde{x}(\tilde{u}, x_0)$ is also used to denote the trajectory from x_0 as a function of time. The set X_{init} includes all possible initial states of the system. The trajectory space $\tilde{\mathcal{X}}$ for the problem is a function space defined by:

$$\tilde{\mathcal{X}} = \{\tilde{x}(\tilde{u}, x) \mid \tilde{u} \in \mathcal{U}, x \in X_{\text{init}}\}, \tag{4}$$

which could be generalized to the trajectory space for the system by replacing X_{init} with X. Assume that $\tilde{x} : [0, t_1] \to X$ and $\tilde{x}' : [0, t_2] \to X$ are two trajectories in $\tilde{\mathcal{X}}$ and $t_2 \geq t_1$, the metric for the trajectory space is

$$\rho_x(\tilde{x}, \tilde{x}') = |t_1 - t_2| + \max\left(\max_{t\in[0,t_1]} \|\tilde{x}(t) - \tilde{x}'(t)\|, \max_{t\in[t_1,t_2]} \|\tilde{x}'(t) - \tilde{x}(t_1)\|\right). \tag{5}$$

This metric can be easily shown to satisfy the standard metric axioms.

The unsafe set or the *specification set* is characterized by a continuous function $g : \tilde{\mathcal{X}} \to R$. If there exists $\tilde{x}(\tilde{u}, x) \in \tilde{\mathcal{X}}$ such that $g(\tilde{x}) < 0$, then the system is unsafe. Note that both spatial and temporal constraints can be incorporated in such functions. The function $g(\tilde{x})$ is assumed to be Lipschitz continuous with respect to $\tilde{x}$. For any $\tilde{x}, \tilde{x}' \in \tilde{\mathcal{X}}$

$$|g(\tilde{x}) - g(\tilde{x}')| \leq L_b \rho_x(\tilde{x}, \tilde{x}'). \tag{6}$$

Finally, we will only consider problems with finite time horizons. Further we require this time horizon, D_T, to be a integer multiple of δt. In other words, $D_T = K\delta t$ for some positive integer K. We are now in a position to define the verification and falsification problems.

Definition 1. Falsification problem: *Find a falsifying control $\tilde{u} \in \mathcal{U}$ and a state $x_0 \in X_{\text{init}}$ such that $g(\tilde{x}(\tilde{u}, x_0)) < 0$.*

Definition 2. Verification problem: *Verify that there does not exist any falsifying controls $\tilde{u} \in \mathcal{U}$ with a state $x_0 \in X_{\text{init}}$ such that $g(\tilde{x}(\tilde{u}, x_0)) < 0$.*

Definition 3. Falsifying control with *violation* **ϵ:** *A falsifying control $\tilde{u} \in \mathcal{U}$ and a state $x_0 \in X_{\text{init}}$ such that $g(\tilde{x}(\tilde{u}, x_0)) < -\epsilon$ for some $\epsilon > 0$.*

To facilitate the proof in Section 4, we will define relaxed version of the falsification problem below.

Definition 4. ϵ-relaxed falsification problem *Find a falsifying control $\tilde{u} \in \mathcal{U}$ and state $x_0 \in X_{\text{init}}$ such that its trajectory $\tilde{x}(\tilde{u}, x_0)$ satisfies $g(\tilde{x}(\tilde{u}, x_0)) < \epsilon$.*

3 Sampling-Based Falsification Algorithm

Because our verification algorithm is achieved through falsification, we will first describe a sampling-based falsification algorithm, which will be converted into a verification algorithm by RC conditions in Section 4.1. There are many sampling-based motion planning algorithms that can be used to design falsification algorithms. However, because our goal is to use the falsification algorithm for verification, we will only use the most basic algorithm and focus instead on its use for verification and not describe the different variants and heuristics of sampling-based algorithms.

3.1 The Basic Falsification Algorithm

To solve the falsification problem described in Section 2, we will assume that we are given a state sampling dispersion bound α_x, and an input sampling *dispersion* bound α_u. Dispersion is the radius of the largest empty ball in a given sample point set [19]. A finite sample state set $\mathcal{S}_x \subset X_{\text{init}}$ is chosen with dispersion less than the given α_x and a finite sample input set $\mathcal{S}_u \subset U$ is determined with dispersion less than α_u. These bounds are illustrated in Figure 2, in which dashed lines show the largest empty balls for the infinity norms, and small dots in (a) and (b) represent sample states in $\mathcal{S}_x$ and sample inputs in $\mathcal{S}_u$ respectively. The algorithm iteratively constructs a search graph using sample inputs in $\mathcal{S}_u$

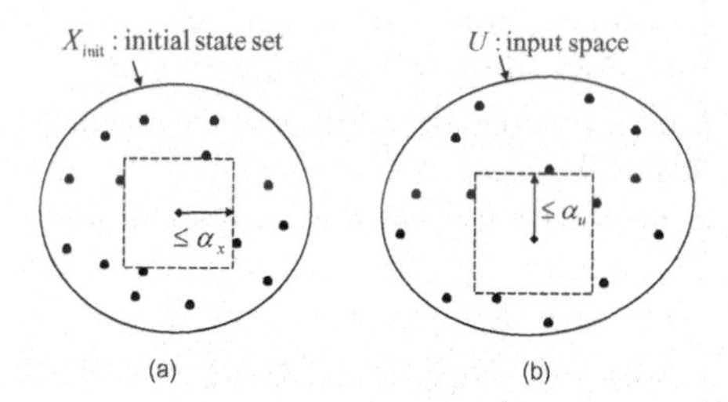

Fig. 2. The given dispersion bounds α_x and α_u are used to determine $\mathcal{S}_x$ and $\mathcal{S}_u$

from sample states in $\mathcal{S}_x$. The search graph is a directed graph. Every node n corresponds to a state $x(n) \in X$. If a sample input $u_l \in \mathcal{S}_u$ is applied for a duration δt from a node, n_k, to generate a control $\tilde{u} \in \mathcal{U}$, resulting in a trajectory segment $\tilde{x}(\tilde{u}, x(n_k))$, then this input u_l is said to have been *applied* for the node n_k. If all inputs in $\mathcal{S}_u$ have been applied for a node, then that node is called *expanded.*

For a problem with the unsafe set described by $g(\tilde{x}) < 0$, the sampling-based falsification algorithm is as follows.

1. Initialize the algorithm: Initialize the search graph by associating each state in $\mathcal{S}_x$ with a new node. There are no edges in the graph.

2. Select an unexpanded node in the search graph: If every node in the search graph is expanded, then the algorithm returns.
3. Generate a trajectory segment with an unapplied sampled input: Choose an unapplied input from $\mathcal{S}_u$ for the selected node. Apply the sample input on the selected node to generate a trajectory segment. Evaluate the function $g(\tilde{x})$ with respect to the current trajectory.
4. Update the search graph: If $g(\tilde{x}) \geq 0$ and the search depth is no larger than K (described in Section 2), then the final state is associated with a new node in the search graph and a new edge is inserted from the selected node to the new node; otherwise, a falsifying control is returned.
5. Iterate from Step 2 until no node is selected.

3.2 The Falsification Algorithm with State Space Discretization

In many algorithms, such as [3], state space discretization is used to decrease the computational complexity of the algorithm by restricting the maximal number of nodes in the search graph.

The discretization is governed by the dispersion bound α_x. The state space X is discretized into a finite number of non overlapping sets so that the maximal distance between any two states in a set is less than α_x. Every set allows at most one node in the search graph. If it contains one node, it is called *occupied*; otherwise, it is called *empty*. Thus before inserting a node for a new state at the end of the trajectory segment of duration δt, a check is performed to see if the discrete set in which the new state is in, is occupied or not. If it is occupied by an existing node, then no new nodes are added. However, a new edge must still be inserted from the selected node to the existing node.

State space discretization directly affects the computations that need to be performed for falsification. Because one input is applied on one unexpanded node in each iteration, the upper bound on the number of computations will be the product of the maximal number of nodes in the search graph and the number of sample inputs in $\mathcal{S}_u$. The size of $\mathcal{S}_u$ is $|\mathcal{S}_u| = O\left([D_u/\alpha_u]^m\right)$.

If we do not discretize the state space, every sample input from a node in the search graph can potentially generate a new node. Therefore, the number of nodes in a search graph starting from a node in K steps is bounded by summing a geometric series: $O\left((|\mathcal{S}_u|^{K+1} - 1)/(|\mathcal{S}_u| - 1)\right)$. Potentially we can have $|\mathcal{S}_x| = O([\frac{D_x}{\alpha_x}]^n)$ disjointed search graphs. Thus the number of iterations of the basic algorithm without discretizing the state space is:

$$O\left([D_x/\alpha_x]^n \, [D_u/\alpha_u]^{m(K+1)}\right). \tag{7}$$

If we do discretize the state space, the number of nodes in the search graph is bounded by the number of non overlapping sets in the partition. The number of sets is $O([\frac{D_x}{\alpha_x}]^n)$ and the number of iterations of the algorithm is

$$O\left([D_u/\alpha_u]^m \, [D_x/\alpha_x]^n \right). \tag{8}$$

Thus state space discretization greatly reduces the upper bound on the number of iterations.

4 A Resolution Complete Algorithm for Verification

In this section, the falsification algorithm in Section 3 is first converted into a verification algorithm by adapting RC conditions for motion planning with differential constraints in Section 4.1. The choice of dispersion bounds with respect to the computation budget, the effects of state and input space transformation on algorithm parameters are respectively provided in Sections 4.2 and 4.3.

We will first define ϵ-resolution completeness.

Definition 5. ϵ-Resolution Complete (ϵ-RC) falsification algorithm *Given a falsification problem, if there exists a falsifying control $\tilde{u}$ with violation $\epsilon > 0$, then an ϵ-RC falsification algorithm will find a falsifying control $\tilde{u}'$ in finite time.*

In other words, if there exists $\tilde{u}$ and $x_0 \in X_{\text{init}}$ such that $g(\tilde{x}(\tilde{u}, x_0)) < -\epsilon$, then an ϵ-RC falsification algorithm will find a control $\tilde{u}'$ and $x_0' \in X_{\text{init}}$ such that $g(\tilde{x}(\tilde{u}', x_0')) < 0$.

4.1 Verification through RC Falsification

To solve the verification problem, we simply run the algorithms in Sections 3.1 and 3.2 on the ϵ-relaxed falsification problem. It will be shown in the following that if α_x and α_u are appropriately chosen, then all falsifying controls for the original falsification problem will be approximated and returned as solutions of the relaxed problem. If no solution is returned, then the system is verified.

Note: The function describing the unsafe set for the ϵ-relaxed falsification problem is $g'(\tilde{x}) = g(\tilde{x}) - \epsilon < 0$. Therefore, if $g'(\tilde{x}) = g(\tilde{x}) - \epsilon < 0$ in Step 4 of the algorithm in Section 3.1, a falsifying control will be returned.

Theorem 1. *For a given $\epsilon > 0$, if an ϵ-RC algorithm does not find a solution with respect to the ϵ-relaxed falsification problem in finite time, then the system in the original problem is verified.*

Proof. Every falsifying control for the original problem is a falsifying control with violation ϵ for the ϵ-relaxed problem, which will be approximated and returned as a solution to the relaxed problem in finite time by an ϵ-RC algorithm. Conversely, if no solution is returned for ϵ-relaxed problem, then the system is verified. $\diamond$

Recall α_x and α_u are the dispersion bounds for sampling in X and U. The following theorem provides the choice of algorithm parameters to ensure that the falsification algorithm in Section 3 is ϵ-resolution complete.

Theorem 2. *If the dispersion bounds satisfy the RC inequality $\lambda\alpha_x + \gamma\alpha_u < \sigma$ with*

$$\sigma = \frac{\epsilon}{L_b}, \lambda = \frac{e^{L_x \delta t(K+1)} - 1}{e^{L_x \delta t} - 1}, \gamma = L_u \delta t e^{L_x \delta t} \frac{e^{L_x \delta t K} - 1}{e^{L_x \delta t} - 1}, \tag{9}$$

then the falsification algorithms in Section 3 are ϵ-RC falsification algorithms.

Proof. The proof will show that under the conditions in the above theorem, every falsifying control $\tilde{u}$ with violation ϵ will be approximated and returned by the algorithm. The proof follows a similar reasoning as in [7]. Instead of presenting the proof, we present the main intuition behind the idea.

As shown in Fig. 3, sampling in the control space means only an approximate solution $\tilde{u}'$ of a falsifying control $\tilde{u}$ can be returned from our sampling-based algorithm (see (a)). Furthermore, the state space discretization and state sampling in X_{init} result in discontinuities in the trajectories $\hat{x}(\tilde{u}', x_0')$ in our search graph. There is a discontinuity in (c) because x_{new} and $x(n_e)$ are not the same point and the initial state x_0 is approximated by x_0' in (b). The main observation is that the dispersion bounds α_x and α_u bound the variation of initial states, the trajectory discontinuities, and the control mismatches. Because the system is Lipschitz continuous and the time horizon is finite, for any $\tilde{u}$ and $x_0 \in X_{\text{init}}$ there always exist (adapted from Theorem 2.5 in [13]) $\tilde{u}'$ and $x_0' \in X_{\text{init}}$ such that

$$\rho_x(\tilde{x}(\tilde{u}, x_0), \hat{x}(\tilde{u}', x_0')) < \lambda\alpha_x + \gamma\alpha_u, \tag{10}$$

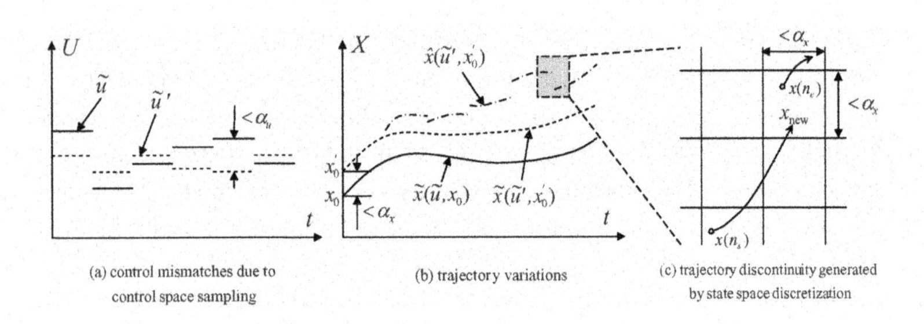

Fig. 3. The intuition of RC conditions

in which λ and γ are given as above. Recall from (6) the function g has a Lipschitz constant L_b. Therefore, if there exists a falsifying trajectory $\tilde{x}(\tilde{u}, x_0)$ with violation ϵ, an approximation $\tilde{x}(\tilde{u}', x_0')$ that satisfies

$$\rho_x(\tilde{x}, \tilde{x}') < \epsilon / L_b, \tag{11}$$

will be a falsifying control. The conditions in the theorem immediately follow by requiring the right side of (10) be less than the right side of (11). ◇

4.2 Choice of Dispersion Bounds

The computational burden is determined by first determining an upper bound T_{iter} on running time for each iteration and the upper bound on the number of iterations. Since the later directly depends on the dispersion bounds α_x and α_u (see (8)), the RC inequality in Theorem 2 indirectly determines the computations required for the ϵ-relaxed falsification problem.

This is illustrated in Fig. 4 (a) for a simple example, in which $D_u = D_x = 1$, $m = 1$, $n = 2$, $\lambda = 0.278$, and $\gamma = 1.43$. For a given T_{iter}, the solid lines are iso-cost curves representing a fixed computational cost for different choices of dispersion bounds. The closer the iso-cost lines are to the origin, the higher the required computational cost. The straight dashed lines represents the RC inequality in Theorem 2 for different choices of relaxation ϵ. The closer the lines are to the origin, the smaller the relaxation ϵ. For a given computational

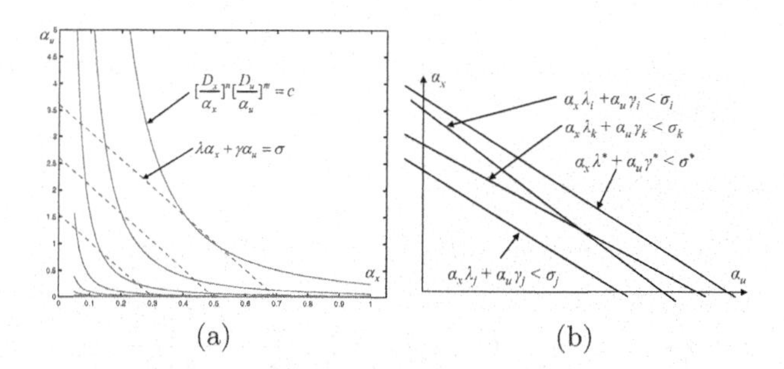

Fig. 4. (a) Selection of algorithm parameters with respect to computational resources (b) Comparison of different RC inequalities

budget we can, in principle, find the minimally relaxed falsification problem, for which the RC inequality line will be tangent to the iso-cost curve with the given computational budget allowing us to determine the dispersion bounds.

4.3 Transformations on X and U

We have assumed that the underlying spaces are metric spaces. Often it is necessary to introduce scaling transformations to non dimensionalize the system so that we are not affected by using non homogeneous coordinates (for example, Cartesian coordinates and angles) and inputs (for example, torques and forces). But one can also imagine transforming the underlying spaces to take advantage of dimensions along which the dynamic system may evolve slowly (slow time scale) and focus instead on dimensions along which changes happen more rapidly (fast time scale). In what follows, we will explore the effects of transformations allowing for general transformation of X and U.

Consider the following transformation: $x = T_x\bar{x}$, $u = T_u\bar{u}$, and $t = T_c\bar{t}$, in which T_x and T_u are full rank square matrices, and T_c is a positive real number. The state space X and input space U are respectively transformed into $\bar{X}$ and $\bar{U}$. If $T_x = \beta I$, $T_u = \beta I$, and $T_c = \beta$ for some real constant $\beta > 0$, then the transformation is called a *uniform scaling* transformation.

The transformed motion equation is

$$\dot{\bar{x}} = \frac{d\bar{x}}{d\bar{t}} = \bar{f}(\bar{x}, \bar{u}) = T_c T_x^{-1} f(T_x\bar{x}, T_u\bar{u}). \tag{12}$$

The Lipschitz constants with respect to the state and input for (1) are

$$L_x = \sup_{x,u} \left\| \frac{\partial f}{\partial x}(x,u) \right\|, \; L_u = \sup_{x,u} \left\| \frac{\partial f}{\partial u}(x,u) \right\|, \tag{13}$$

and for (12) are:

$$L_{\bar{x}} = T_c \sup_{x,u} \left\| T_x^{-1} \frac{\partial f}{\partial x}(x,u) T_x \right\|, L_{\bar{u}} = T_c \sup_{x,u} \left\| T_x^{-1} \frac{\partial f}{\partial u}(x,u) T_u \right\|. \tag{14}$$

Because matrix multiplication does not commute, L_x is different from $L_{\bar{x}}$ for general transformations.

RC inequalities under transformation.

Theorem 3. *The RC inequality after the transformation has*

$$\sigma = \frac{\epsilon}{L_b}, \; \lambda = \max(T_c, \|T_x\|_\infty) \|T_x^{-1}\|_\infty \frac{e^{L_{\bar{x}} \bar{\delta t}(\bar{K}+1)} - 1}{e^{L_{\bar{x}} \bar{\delta t}} - 1}, \tag{15}$$

and

$$\gamma = \max(T_c, \|T_x\|_\infty) \|T_u^{-1}\|_\infty L_{\bar{u}} \bar{\delta t} e^{L_{\bar{x}} \bar{\delta t}} \frac{e^{L_{\bar{x}} \bar{\delta t} \bar{K}} - 1}{e^{L_{\bar{x}} \bar{\delta t}} - 1}. \tag{16}$$

Proof. With the given transformation, we have

$$\rho_x(\tilde{x}, \tilde{x}') \leq \max(T_c, \|T_x\|_\infty) \rho_x(\tilde{\bar{x}}, \tilde{\bar{x}}'). \tag{17}$$

The given algorithm parameters $\epsilon_{\bar{x}}$ and $\epsilon_{\bar{u}}$ are described with respect to the new spaces. With these algorithm parameters, for any falsifying control $\tilde{\bar{u}}$ and initial state $\bar{x}_0$, there exists an approximation $\tilde{\bar{u}}$ and state $\bar{x}_0'$ such that

$$\rho_x(\tilde{\bar{x}}(\tilde{\bar{u}}, \bar{x}_0), \hat{\bar{x}}(\tilde{\bar{u}}', \bar{x}_0')) < \epsilon_{\bar{x}} \frac{e^{L_{\bar{x}} \bar{\delta t}(\bar{K}+1)} - 1}{e^{L_{\bar{x}} \bar{\delta t}} - 1} + \epsilon_{\bar{u}} L_{\bar{u}} \bar{\delta t} e^{L_{\bar{x}} \bar{\delta t}} \frac{e^{L_{\bar{x}} \bar{\delta t} \bar{K}} - 1}{e^{L_{\bar{x}} \bar{\delta t}} - 1} \tag{18}$$

With infinity norms on the state and input space, it can be verified that

$$\epsilon_{\bar{x}} \leq \|T_x^{-1}\|_\infty \epsilon_x, \epsilon_{\bar{u}} \leq \|T_u^{-1}\|_\infty \epsilon_u. \tag{19}$$

Substituting the above inequalities into (18) and requiring the right side of (17) be less than ϵ/L_b will complete the proof. ◇

Corollary 1. *The RC inequality is invariant for any uniform scaling.*

Proof. With any uniform scaling, it can be verified that $\max(T_c, \|T_x\|_\infty) = 1/\|T_x^{-1}\|_\infty = 1/\|T_u^{-1}\|_\infty$, $K = \bar{K}$, $\delta t = \bar{\delta t}$, $L_x = L_{\bar{x}}/\beta$, and $L_u = L_{\bar{u}}/\beta$. Therefore, the same RC inequality coefficients in Theorem 2 will always be derived by substituting these equalities into the RC inequality coefficients in Theorem 3. ◇

Comparison of RC conditions with different transformations

From the above description, we can see that the derived RC inequality might not be invariant for non-uniform scaling, such as transformation $T_x = \beta I$, $T_u = \beta I$, and $T_c = \xi > \beta$. Assume that λ_i, γ_i, and σ_i are coefficients of the derived RC inequality, which are obtained from Transformation i. Let

$$E_i = \{(\alpha_x, \alpha_u) \mid \alpha_x \lambda_i + \alpha_u \gamma_i < \sigma_i, \alpha_x > 0, \alpha_u > 0\}. \tag{20}$$

The inequalities and set E_i are shown in Fig. 4 (b). For Transformations i, j, and k, Transformation i is said to be superior to Transformation j if $E_j \subset E_i$. If $E_i \not\subset E_k$ and $E_k \not\subset E_i$, then Transformation i is neither better nor worse than Transformation k. A transformation can be said to be "optimal" from the standpoint of resolution completeness if the set defined by $\alpha_x \lambda^* + \alpha_u \gamma^* < \sigma^*$ is not the subset of E_i for all other transformations. Again this "optimal" transformation will generate dispersion bounds that are larger so that the maximal number of nodes, the size of $\mathcal{S}_u$, and therefore the computational cost will be smaller.

5 Examples

In this section we illustrate the sampling-based falsification and verification methodology, the use of ϵ-relaxation, and transformation of state and input spaces. We choose several simple verification problems that allow easy interpretation. The first problem has parametric uncertainty in inputs while the second problem has parametric uncertainty in the initial state. The third problem incorporates uncertainty in the form of disturbance input functions. The final problem presents an analysis of control policies for pursuit evasion.

5.1 Verification Problems

Problem 1: Verification of a system with an uncertain parameter Consider a point mass which moves freely on a plane with constant but unknown external force, u, along the y-axis (see Fig. 5 (a)). The state x of the system includes (p_x, v_x, p_y, v_y) in $X = [0, 15] \times [1, 3] \times [-1, 1] \times [-1, 2]$, which denote the position and velocity along x and y axes respectively. Its motion equation is $\dot{p}_x = v_x$, $\dot{v}_x = 0$, $\dot{p}_y = v_y$, and $\dot{v}_y = u$, in which $u \in U = [5, 15]$ is the system parameter determining the magnitude of the constant input. The system has initial state $x_0 = (0.0, 2.0, 0.0, 1.0)$. The system is safe if the trajectory of the point mass from initial state x_0 always stays outside of an unsafe region (shown shaded in Fig. 5 (a)), which is a square of width $d = 0.5$ with its center at point $(10, 0)$. The function defining the unsafe set[1] is

$$g(\tilde{x}(\tilde{u}, x_0)) = \min_t(\|\tilde{x}(\tilde{u}, x_0, t) - [10, 0]^T\|) - 0.5 < 0.$$

It can be verified that $g(\tilde{x})$ satisfies (6) with Lipschitz constant $L_b = 1$.

[1] Recall that we are considering X as a metric space with the infinity norm.

The verification problem is to check whether the system is safe for all inputs. There is a natural choice for the finite time horizon, D_T. For $t > 0.7$, p_y can be shown to be less than -0.5 and decreasing. Therefore, we choose $D_T = 0.7$. Because analytical solutions are available for this simple system, it is straightforward to show that the system is safe. We will verify this using a sampling-based algorithm in the next subsection.

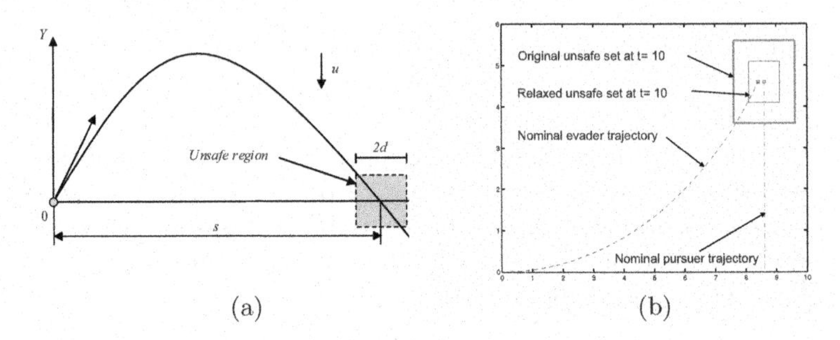

Fig. 5. Simple verification problems

Problem 2: Verification of a system with an uncertain initial state Consider the autonomous system with no control: $\dot{y} = 0.2y \sin t^2$ and $\dot{t} = 1$. We define the extended state $x = [y, t]^T \in X = [0, 2] \times [0, 10]$. The initial state is unknown, but restricted to lie in the set $X_{\text{init}} = [0, 0.1] \times 0$. The system is considered to be safe if at $t = 9$ seconds, $\|y(t) - 1.0\| > 0.5$. Again the Lipschitz constant L_b for the function g is 1. The time horizon is $D_T = 9$ seconds. We consider ϵ-relaxed problems with $\epsilon = 0.5$ first and then 0.05.

Problem 3: Verification of a system under input disturbances Consider the kinematic model of a UAV whose nominal inputs are constant but are subject to bounded disturbances. The dynamics is characterized as $\dot{x} = (v_0 + v)\cos\theta$, $\dot{y} = (v_0 + v)\sin\theta$, and $\dot{\theta} = w_0 + w$, in which $v_0 = 1$ and $w_0 = 0.1$ are the nominal inputs for the system, $x \in [0, 10]$, $y \in [0, 5]$, and $\theta \in [0, 2\pi]$ are position and orientation, $v \in [-0.01, 0.01]$ and $w \in [-0.001, 0.001]$ denote the disturbances. The system starts from the initial state $(0, 0, 0)$ at time 0. See Fig. 6 (a). The question is whether the system will stay in the 1.0-neighborhood of the goal position $[x_g, y_g]^T = [8.41, 4.60]^T$ under the input disturbance at time 10 seconds. Thus, the system is said to be unsafe if [2] $\|[x, y]^T - [x_g, y_g]^T\| > 1.0$ at $t = 10$. Again, the Lipschitz constant $L_b = 1$. The disturbance control space consists of piecewise-constant controls with $\delta t = 2$ seconds. We will consider ϵ relaxation with $\epsilon = 0.5$.

Problem 4: Verification of a control policy for the pursuer Consider the UAV in Problem 3 as an evader and a point mass model for a pursuer with position p_x and p_y. The pursuer captures the evader if

[2] The infinity norm is used here.

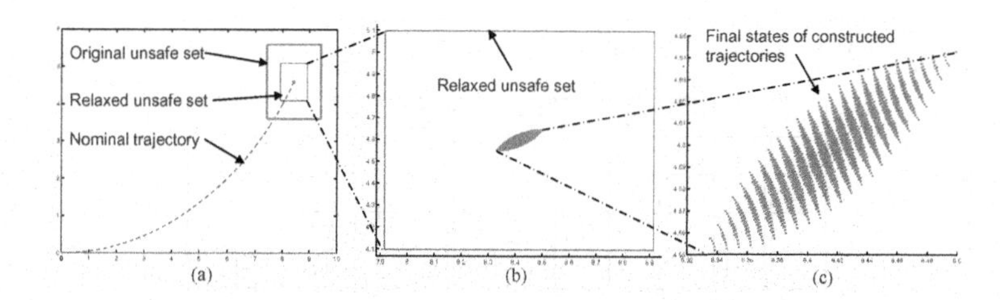

Fig. 6. Verification of the system under input disturbances

$$\|[x(t), y(t)]^T - [p_x(t), p_y(t)]^T\| < 1.0$$

for some t in a finite time horizon of $D_T = 10$ seconds. The UAV has a given nominal control input but with bounded disturbance in the input. An open-loop trajectory ($p_x(t) = 8.61$ and $p_y(t) = 0.46t$) is computed for the pursuer according to the nominal trajectory to achieve capture (see Fig. 5 (b)). The pursuer trajectory is verified if the pursuer can capture the evader over any disturbance from its nominal control. Again, the Lipschitz constant $L_b = 1$. The disturbance control space consists of piecewise-constant controls with $\delta t = 2$ seconds. We will consider ϵ relaxation with $\epsilon = 0.5$.

5.2 RC Inequalities under Scaling and Transformation

The transformation is achieved with following diagonal matrices

$$T_x = \text{Diag}(a_{11}, a_{22}, \cdots, a_{nn}), T_u = \text{Diag}(b_{11}, b_{22}, \cdots, b_{mm}). \tag{21}$$

Problem 1: Because the control is constant, the control space is $\mathcal{U} = \{\tilde{u} \mid \tilde{u}(t) = c, c \in U\}$. Because the algorithm without state space discretization is used and the initial state is a point, RC inequality will be in form $\gamma\alpha_u < \sigma$. We use the ϵ-relaxed falsification problem with $\epsilon = 0.2$. With this ϵ, the ϵ-RC inequalities after four different transformations are listed in Table 1. It can be seen that the RC inequalities are the same for the uniform scaling transformation between 1 and 2 (see Table 1).

Problem 2: Because input space sampling does not exist, RC inequality for this problem will be in form $\lambda\alpha_x < \sigma$. RC inequalities are calculated in Table 2 (a) for a fixed $T_c = 1$.

Problem 3: The ϵ-RC inequalities are calculated in Table 2 (b) with $T_c = 1$ and T_u equal to an identity matrix. The algorithm with state space discretization is used for falsification and verification.

Problem 4: The same ϵ-RC inequalities are obtained as for Problem 3.

Table 1. RC inequalities under different transformations

No.	a_{11}	a_{22}	a_{33}	a_{44}	b_{11}	T_c	$L_{\bar{x}}$	$L_{\bar{u}}$	γ_i	σ_i
1	1	1	1	1	1	1	1	1	1.41	0.2
2	10	10	10	10	10	10	10	10	1.41	0.2
3	10	1	10	1	1	1	0.1	1	7.51	0.2
4	10	100	10	100	1	100	1000	1	767.64	0.2

Table 2. RC inequalities under different transformations

No.	a_{11}	a_{22}	$L_{\bar{x}}$	λ_i	σ_i
1	1.0	1.0	8.2	1.12e32	0.5
2	10.0	1.0	1.0	8.10e4	0.5
3	100.0	1.0	0.28	1.34e3	0.5
4	1000.0	1.0	0.208	7.50e3	0.5

(a)

a_{11}	a_{22}	a_{33}	$L_{\bar{x}}$	$L_{\bar{u}}$	λ_i	γ_i	σ_i
1	1	1	1.01	1	2.81e4	5.61e4	0.5
10	10	1	1.01e-1	1	105.4	190.88	0.5
100	100	1	1.01e-2	1	631.5	1.06e3	0.5
1000	1000	1	1.01e-3	1	6.03e3	1.01e4	0.5

(b)

5.3 Simulation Results

Problem 1: Under Transformation 1 in Table 1, we choose $\alpha_u = 0.141$. A sample input set S_u with this dispersion bound is $\{5, 5.28, 5.56, \cdots, 15\}$. The system was verified because no solution was returned.

Problem 2: From Table 2 (a), we can see that Transformation 3 yields the best RC inequality in terms of the lowest λ (highest dispersion). The dispersion bound α_x is chosen to be 3.7×10^{-4} to satisfy this inequality. Sample states from X_{init} are $\{0, 7.0 \times 10^{-4}, 1.4 \times 10^{-3}, \cdots, 0.0994, 0.1\}$ are used for simulation. As shown in Fig. 7 (a), the final state of the trajectory from $y = 0.05$ and $t = 0$ is returned by the ϵ-RC falsification problem with $\epsilon = 0.5$, and therefore, the system is not verified.

In order to investigate this problem further, the relaxation tolerance ϵ is reduced to 0.05. For the same transformation, the state sampling dispersion bound α_x is calculated to be 2.5×10^{-5}. Now all the final states of the approximated trajectories are outside of the 0.05-relaxed unsafe set. Therefore, the system is verified. Three sample trajectories are illustrated in Fig. 7 (b).

Problem 3: From Table 2(b), we can see that Transformation 2 has the best RC inequality. We chose dispersion bounds $\alpha_x = 1.1 \times 10^{-3}$ and $\alpha_u = 2.01 \times 10^{-3}$ which satisfy this inequality. The chosen sample input (v, w) with the specified input dispersion is in $\{-0.01, -0.006, -0.002, 0.002, 0.006, 0.01\} \times \{0.0\}$. As shown in Fig. 6 (b) and (c), since the final states of all constructed trajectories do not enter the unsafe region, the system is verified.

Problem 4: The verification algorithm runs with the same choice of the sample input set as in Problem 3. The pursuer trajectory is verified because no disturbance input for the evader is a falsifying control for the relaxed problem. Note

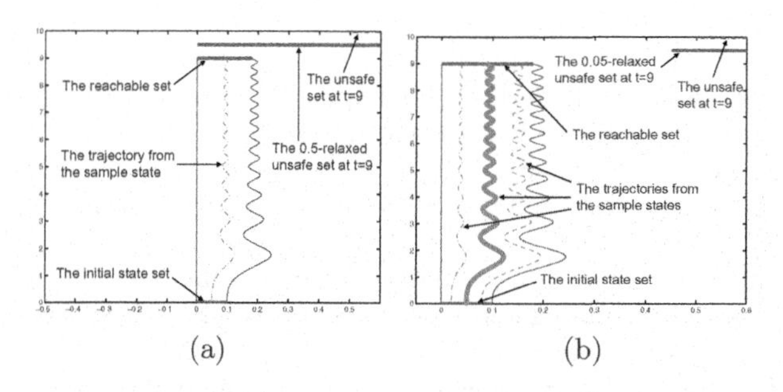

Fig. 7. Trajectories computed by the verification algorithm for ϵ-relaxed problems

that the complexity of the proposed verification for this problem depends only on the state and input space of the evader. Increasing the number of pursuers does not change the computational cost.

6 Conclusion

In this paper, we proposed a sampling-based verification algorithm based on resolution complete falsification, which involves the iterative construction of solutions that falsify the given safety specification with sampled controls. We derive sufficient conditions for the discretization of the state and input spaces to guarantee that we can find approximations to any falsifying control inputs, if they exist. Thus the paper provides a novel and systematic approach to verifying controllers for continuous dynamic systems.

While the paper presents sufficient conditions for resolution completeness, these conditions are conservative and require a high resolution sampling in state and input spaces for most practical problems. This is because the verification problem is extremely hard. (Recall that the path planning problem (without dynamics) is NP-hard.) We provide a partial solution to this problem by pursuing transformations of input and state spaces that might allow a lower resolution while guaranteeing resolution completeness. This continues to be an area of ongoing research.

Of course heuristics can improve performance by several orders. As shown in the RC inequality in Theorem 2, the complexity of the verification algorithm increases exponentially with the time horizon, and dimension of the state space and input spaces. Thus it is important to prune the search space based on domain knowledge. Our preliminary work in this direction is discussed in [9].

Acknowledgements. We gratefully acknowledge support from NSF grant CNS-0410514 and ONR grant FA8650-04-C-7133.

References

1. Alur, R., Henzinger, T., Lafferriere, G., Pappas, G.: Discrete abstractions of hybrid systems. Proccedings of the IEEE 88(2), 971–984 (2000)
2. Asarin, E., Bournez, O., Dang, T., Maler, O.: Reachability analysis of piecewise-linear dynamical systems. In: Hybrid Systems: Computation and Control, Springer, Heidelberg (2000)
3. Barraquand, J., Latombe, J.-C.: Nonholonomic multibody mobile robots: Controllability and motion planning in the presence of obstacles. Algorithmica 10, 121–155 (1993)
4. Bhatia, A., Frazzoli, E.: Incremental search methods for reachability analysis of continuous and hybrid systems. In: Hybrid Systems: Computation and Control, Philadelphia, USA (2004)
5. Branicky, M., Curtiss, M., Levine, J., Morgan, S.: Sampling-based planning, control, and verification of hybrid systems. IEEE Proc. Control Theory and Applications (accepted)
6. Burns, B., Brock, O.: Sampling-based motion planning using predictive models. In: IEEE Int. Conf. Robot. & Autom. (2005)
7. Cheng, P.: Sampling-based Motion Planning with Differential Constraints. PhD thesis, University of Illinois, Urbana (2005)
8. Chutinan, A., Krogh, B.: Verification of infinite-state dynamic systems using approximate quotient transition systems. IEEE Transactions on Automatic Control 46, 1401–1410 (2001)
9. Esposito, J.M., Kim, J., Kumar, V.: Adaptive RRTs for validating hybrid robotic control systems. In: Proc. Workshop on Algorithmic Foundation of Robotics (2004)
10. Girard, A., Pappas, G.: Verification using simulation. In: Hybrid Systems: Computation and Control, Springer, Heidelberg (2006)
11. Hsu, D., Latombe, J.-C., Motwani, R.: Path planning in expansive configuration spaces. Int. J. Comput. Geom. & Appl. 4, 495–512 (1999)
12. Kapinski, J., Krogh, B., Maler, O., Stursberg, O.: On systematic simulation of open continuous systems. In: Hybrid Systems: Computation and Control, Springer, Heidelberg (2003)
13. Khalil, H.: Nonlinear systems. Prentice-Hall, Upper Saddle River (1996)
14. Ladd, A., Kavraki, L.: Measure theoretic analysis of probabilistic path planning. IEEE Transactions on Robotics and Automation 20(2), 229–242 (2004)
15. LaValle, S., Branicky, M., Lindemann, S.: On the relationship between classical grid search and probabilistic roadmaps. International Journal of Robotics Research 24 (2004)
16. LaValle, S., Kuffner, J.J.: Randomized kinodynamic planning. International Journal of Robotics Research 20(5), 378–400 (2001)
17. Kavraki, L., Svestka, P., Latombe, J.-C., Overmars, M.: Probabilistic roadmaps for path planning in high-dimensional configuration spaces. IEEE Trans. Robot. & Autom. 12(4), 566–580 (1996)
18. Mitchell, I., Tomlin, C.: Overappoximating reachable sets by hamilton-jacobi projections. J. of Sci. Comput. 19 (December 2003)
19. Niederreiter, H.: Random Number Generation and Quasi-Monte-Carlo Methods. Society for Industrial and Applied Mathematics, Philadelphia, USA (1992)

Part VIII

Sensor Networks and Reconfiguration

Surrounding Nodes in Coordinate-Free Networks*

R. Ghrist[1], D. Lipsky[1], S. Poduri[2], and G. Sukhatme[2]

[1] Department of Mathematics, University of Illinois, Urbana, IL 61801, USA
[2] Department of Computer Science, University of Southern California, Los Angeles, CA, USA

Summary. Consider a network of nodes in the plane whose locations are unknown but which establish communication links based on proximity. We solve the following problems: given a node in the network, (1) determine if a given cycle surrounds the node; and (2) find some cycle that surrounds the node. The only localization capabilities assumed are unique IDs with binary proximity measure, and, in some cases, cyclic orientation of neighbors. We give complete algorithms for finding and verifying surrounding cycles when cyclic orientation data is available. We also provide an efficient but non-complete algorithm in the case where angular data is not available.

1 Introduction

It is increasingly important to analyze networked collections of sensors, robots, communication devices, or other local agents which coordinate to solve global problems. A similar problem arose in mathematics a century ago — how to extract global properties of a space built from local, combinatorially defined pieces, or *simplices*. It is not a coincidence that the techniques developed to solve such mathematical problems (algebraic topology) provide perspectives and tools applicable to this latest incarnation of the problem.

This paper considers a network version of a simple classical problem in algebraic/differential topology. Given a point x_0 in the plane $\mathbb{R}^2$ and a simple closed curve, determine whether or not the curve surrounds the point — that is, whether the *winding number* of the curve about x_0 is nonzero (Fig. 1[left]). In the topological setting, this problem is very easily solved using simple topological methods [10]. In a network-theoretic version of the problem, x_0 is a node in a network graph Γ whose vertices represent non-localized sensors in the plane and whose edges encode proximity; and $\mathcal{L}$ is an abstract cycle in this graph (Fig. 1[right]).

Several such problems about winding numbers have very natural motivations, and are especially challenging when the nodes are not localized. Consider as an example, a networked collection of sensors (e.g., accelerometers or acoustic

* Supported by DARPA HR0011-05-1-0008 and NSF PECASE DMS-0337713 [RG,DL] and by NSF CCR-0120778, CNS-0520305, and IIS-0133947 [SP,GS].

S. Akella et al. (Eds.): Algorithmic Foundation of Robotics VII, STAR 47, pp. 409–424, 2008.
springerlink.com

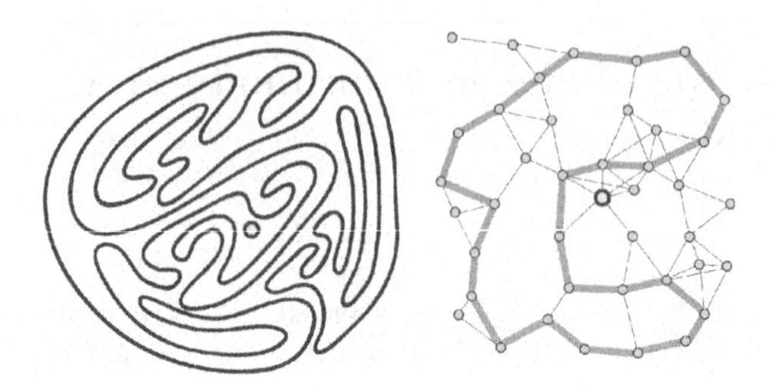

Fig. 1. [left] Is the node inside the curve or outside? For a network without localization [right], this can be challenging

sensors) which are distributed in a 2-d domain. Given a certain node x_0 which registers an important reading (an alarm), one problem relevant to security applications is to determine whether the detection has occurred within a region of particular importance whose perimeter is defined by a cycle in the network. Similarly, it an alarm goes off at a node, one might wish to find a small subcollection $\mathcal{L}$ of sensors whose sensing domains are guaranteed to 'surround' the node x_0 in the plane: thus, the embedding of the cycle $\mathcal{L}$ in the plane is a curve which surround x_0.

If one has sufficient data to localize nodes, then all such problems about winding numbers are trivial to solve and computationally efficient solutions exist in the computer graphics literature under various simplifying assumptions. The assumption of localized nodes is natural for any number of systems involving stationary nodes placed intentionally, e.g., video cameras. However, in the case of nodes which are distributed in an unpredictable and non-uniform manner, or in which the nodes are mobile, then localized nodes are no longer a priori natural. Robotics, in particular, presents a natural setting in which mobile devices communicating via an ad hoc wireless network can provide localization challenges.

1.1 Related Work

There is a substantial and growing literature on geometric properties of ad hoc networks in which localization is weakened or not assumed at all. The recent work on routing without localization initiated by [14] uses a heat-flow to determine virtual coordinates for a non-localized network for applications to weighted routing problems. In many cases [14, 7] a set of known landmarks is used to estimate system geometry. All these methods are effective, but, with a few exceptions [11] non-rigorous. Recent work of Fekete et al. [8] gives a distributed algorithm for rigorous topology exploration, boundary detection, and surrounding cycles: the algorithm is complete when the nodes are sufficiently dense.

There is a large body of work on coarse distance estimation in ad hoc networks augmented with angular data in the form of the angle of separation between a node's neighbors. This arises in the the paper [6], which uses a network graph along with exact angular measures of neighbors to detect holes in the physical network and perform routing. The work detailed in [13] gives criteria for ensuring coverage in a sensor network using bounds on separation angles among neighbors.

Very recently, algebraic topology has been recognized as a novel tool for problems in sensor and ad hoc networks. The papers [3, 4, 5] present algebraic topological criteria for coverage in sensor networks. Recent unpublished results of Y. Baryshnikov relating to hole-detection in networks with randomly distributed nodes uses Betti numbers to perform boundary detection.

As a problem in computational geometry, networks with no localization and proximity measurements arise in the literature on *unit disc graphs*: abstract graphs whose vertices correspond to a set of nodes in the plane and whose edges are determined by nodes within unit distance. Clearly, not all graphs are realizable as a unit disc graph. Recognizing whether a graph is a realizable unit disc graph is NP-complete [1]. It follows that finding some embedding of an abstract unit disc graph into the plane for which the graph is the unit disc proximity network is also NP hard. Even finding an 'approximate' embedding which realizes a unit disc graph up to local errors is NP hard [12]. But, using angular data, [2] gives an algorithm for finding a realization of a spanner of the unit disc graph, which enables one to compute virtual coordinates and approximate some locations.

1.2 Innovations

The perspective that guides our techniques is that of topology, more specifically, winding numbers [10]. Recall that the winding number of a planar cycle $\mathcal{L}$ about a point $x \in \mathbb{R}^2$ is, roughly speaking, the number of times the cycle wraps around the point. It can be computed in a number of ways: analytically, via integrating a tangent vector about $\mathcal{L}$; topologically, via computing the homology class of $\mathcal{L}$ in the complement of $x_0 \in \mathbb{R}^2$; or combinatorially, via computing the intersection number of $\mathcal{L}$ with a ray based at x in $\mathbb{R}^2$. We develop an approach to computing winding numbers which is adapted to networks and differs from all three above.

In §3, we solve a **separation** problem: to compute whether a given node x_0 is surrounded by the image of a given cycle $\mathcal{L}$. In §4, we solve an **isolation** problem of determining whether a given node x_0 is surrounded by some cycle $\mathcal{L}$ and constructing an explicit cycle. The algorithms we present can be implemented in systems as a distributed local computation, in which nodes are assumed to have a limited amount of memory and simple processing abilities. We present results of simulations in §7. In §5 we deal with managing uncertainty in cyclic orientation data, and in §6 we present an algorithm for systems which possess no cyclic orientation data whatsoever. In both these cases, as in [8], non-complete algorithms exist which provide certificates.

2 Problem Formulation

2.1 Assumptions

Throughout this paper, we consider networks which satisfy some or all of the following assumptions.

P (Planar) Nodes with unique labels lie in the Euclidean plane $\mathbb{R}^2$.
N (Network) Nodes form the vertices of a connected unit disc network graph Γ of sufficiently large diameter.
O (Ordering type) Each node can determine the clockwise cyclic ordering of its neighbors in the plane.

There are no coordinates, no node localization, and no assumptions about node density or distribution other than sufficient extent. Assumptions **P** and **N** will be in force for the remainder of this paper. Assumption **O** will sometimes not be imposed.

Assumption **P** and **N** imply that nodes can broadcast their unique IDs and these can be detected by any neighboring nodes within unit distance. This creates a network graph whose vertices correspond to the labeled nodes and whose edges correspond to communication links. There is no metric information encoded in an edge beyond the coarse datum that the distance between the nodes in the plane is no more than one.

Assumption **O** means that each node can perform a clockwise "sweep" of its neighborhood and determine the order in which neighbors appear. More specifically, there is a cyclic total ordering $\lhd$ on the neighbors of a node x_0 which defines the counterclockwise (CCW) order in which they appear. There is no "compass" and thus ordering is known only up to a cyclic permutation. There is also no angular data assigned to the ordering: an oriented pair of neighbors may form an arbitrary (nonzero) angle with x_0 without changing the angular ordering. This type of coarse angular data is not too uncommon in robotics contexts. Cyclic orientation data is natural in, e.g., primitive landmark vision systems, radar networks, and robots with gap sensors.

Definition: Let x_0 be a node and $\{x_i\}_1^3$ be a triple of distinct neighbors of x_0. Define the index

$$\mathcal{I}_{x_0}(x_1; x_2, x_3) := \begin{cases} +1 : x_1 \lhd x_2 \lhd x_3 \lhd x_1 \\ -1 : x_1 \rhd x_2 \rhd x_3 \rhd x_1 \end{cases} \tag{1}$$

Geometric interpretation: The pair of rays in $\mathbb{R}^2$ from x_0 passing through x_2 and x_3 join at x_0 to form an bent line that divides the plane. Orient this bent line using the ordering $(x_2 \to x_0 \to x_3)$. The index $\mathcal{I}_{x_0}(x_1; x_2, x_3) = -1$ iff x_1 lies to the left of this line, and $\mathcal{I}_{x_0}(x_1; x_2, x_3) = +1$ iff x_1 lies to the right of it.

The limit in which two neighbors have angle zero with x_0 leads naturally to the problem of uncertainty in angular ordering. When the angle between neighbors is so small as to interfere with the orientation type, the data is weaker and the problems more subtle. For the present, we assume that nodes are in a 'general position' so as to possess a positive lower bound on angles. This is, of course, completely unrealistic in practice. Later, in §6 we consider this more carefully and allow for nodes to be unable to distinguish the angular ordering of certain neighbors. For most (but not all) networks, there is a surprisingly large tolerance for angular ordering blindness. A complete analysis of this situation is presented in [9].

The input data for the problem is the network graph Γ. When Assumption **O** is in place, the graph has vertices augmented with the cyclic ordering type of its immediate neighbors.

2.2 Problem Statements

Definition: The projection map $\Gamma \mapsto \overline{\Gamma} \subset \mathbb{R}^2$ maps vertices of Γ to the position of the corresponding node in the Euclidean plane and edges of Γ to the line segment connecting the nodes. These line segments all have length bounded above by one. We solve two problems concerning winding numbers of cycles:

> **Separation:** Given a cycle $\mathcal{L}$ in the network graph Γ and a node $x_0 \in V(\Gamma)$ which is disjoint from the nodes of $\mathcal{L}$, determine whether the projected cycle $\overline{\mathcal{L}}$ surrounds x_0.

The image of the cycle $\overline{\mathcal{L}}$ in the plane is a closed piecewise-linear curve. If the curve is *simple* (that is, non-self-intersecting), then the Jordan Curve Theorem implies that the cycle separates the plane into two connected components, only one of which is bounded. We will restrict attention to simple cycles, using network criteria to satisfy this condition (Corollary 1).

Our second problem is a constructive version of the previous.

> **Isolation:** Given a node $x_0 \in \mathcal{X}$, find a cycle $\mathcal{L}$ in the network graph Γ whose projection $\overline{\mathcal{L}}$ surrounds x_0 or determine that no such cycle exists.

For reasons of robustness with respect to error, we desire a cycle $\mathcal{L}$ which is not too close to x_0. An interesting generalization of this problem relevant to security applications is to construct a sequence of concentric cycles which isolate the target node x_0 and form 'moats' in $\mathbb{R}^2$ to ring x_0.

3 Separation

3.1 Restrictions

We consider the separation problem for a network satisfying Assumptions **P**, **N**, and **O**. Namely, given a node x_0 and an oriented cycle $\mathcal{L} = (x_i)_1^N$ of cyclically connected nodes distinct from x_0, determine whether the projected cycle $\overline{\mathcal{L}}$ surrounds the node x_0. Let d denote the "hop" distance function on Γ. We assume that $d(x_0, \mathcal{L}) > 1$, meaning that the shortest path from x_0 to a vertex of $\mathcal{L}$ in Γ requires more than one "hop" or edge.

Both Assumption **O** and the bound on $d(x_0, \mathcal{L})$ are necessary. A critical example is illustrated in Fig. 2, which gives two labeled graphs in $\mathbb{R}^2$ with isomorphic network graphs and identical cyclic orientation data. The target node x_0 lies on opposite sides of the cycles illustrated. It is much easier to construct examples of unit disc graphs which can be realized in ways which change winding numbers of cycles.

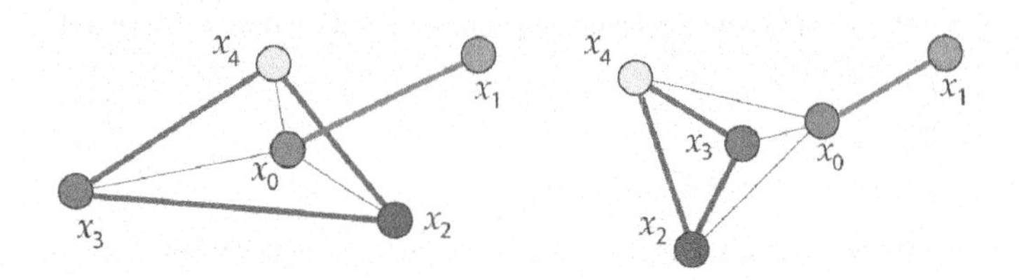

Fig. 2. Two examples of embedded network graphs with identical network and cyclic orientation data

The problem makes the most sense when the projected cycle $\overline{\mathcal{L}}$ is a simple closed curve in $\mathbb{R}^2$. The easiest way to guarantee such a cycle is to choose a cycle which is 'minimal' with respect to communication between nodes.

Definition: For any subgraph $\Delta \subset \Gamma$, let $\langle \Delta \rangle$ denote the maximal subgraph of Γ spanned by the vertices of Δ. Say that Δ is **chord-free** if $\langle \Delta \rangle = \Delta$. The simplest criterion for a cycle $\mathcal{L}$ to have a simple projection to the plane is that $\langle \mathcal{L} \rangle = \mathcal{L}$. The following lemma is both trivial and well-known [8, 3, 9].

Lemma 1. *If the projections of two edges of a unit disc graph Γ intersect in $\mathbb{R}^2$, then these span a subgraph of Γ containing a cycle of three edges.*

Corollary 1. *Any path (or cycle) $\mathcal{P}$ in a unit disc graph Γ satisfying $\langle \mathcal{P} \rangle = \mathcal{P}$ has image $\overline{\mathcal{P}}$ a non-intersecting (closed) curve in $\mathbb{R}^2$.*

3.2 Algorithm

In differential topology, the way one decides whether a loop in the plane encloses a point is to choose a path from the point which terminates sufficiently far from the starting point as to be definitely outside the loop. For a 'generic' choice of such a path, the path and the loop intersect without tangencies, and the number of intersection points counted mod 2 is zero if and only if the loop does not surround the point [10].

The obvious generalization of this strategy is to choose any path $\mathcal{P}$ in Γ from x_0 to a terminal point which is sufficiently far away from $\mathcal{L}$ to guarantee that it is outside the cycle in $\mathbb{R}^2$, and then court intersections. However, this counting is not always easy or even possible. The inspiration for our method is nearly opposite to that coming from differential topology. Instead of trying to force intersections to be a discrete set of points, one thinks of manipulating the path so as to *maximize* the amount of intersection with the cycle with the result of having a single connected component in the intersection. Then, one could compute whether the endpoints of $\mathcal{P}$ lie on the same side of $\overline{\mathcal{L}}$. This last step is what we do, using the orientation data in a crucial manner.

Fix an orientation for the cycle $\mathcal{L}$ in Γ and order the nodes (ℓ_i) of $\mathcal{L}$ cyclically. Choose a node x_∞ sufficiently far from $\mathcal{L}$ in Γ. Generate chord-free paths $\mathcal{P}_0$ and $\mathcal{P}_\infty$ from x_0 and x_∞ respectively to points on $\mathcal{L}$. The projection of these paths to $\mathbb{R}^2$ are not self-intersecting, and can only intersect $\overline{\mathcal{L}}$ at most once at the last segment of the path.

The crucial step is to determine whether the paths $\overline{\mathcal{P}}_0$ and $\overline{\mathcal{P}}_\infty$ lie on the same side of $\overline{\mathcal{L}}$ or different sides. In the simplest configuration, the final point on a path is connected to $\mathcal{L}$ in Γ by only one edge, as in Fig. 3[left]. The angular orientation data then suffices to determine on which side of $\overline{\mathcal{L}}$ the path lies. However, the situation may be complicated, as in Fig. 3[right]. A more subtle analysis is required in this case: see Algorithm IndexCheck.

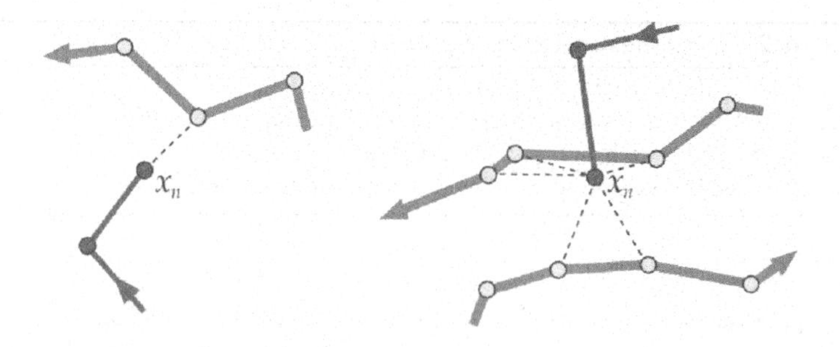

Fig. 3. Determining whether an oriented path with terminal node x_n approaches the projected oriented cycle $\overline{\mathcal{L}}$ from the left or from the right can be simple [left] or complicated [right] depending on the number of communication links between x_n and $\mathcal{L}$

3.3 Proofs

Lemma 2. *If $\mathcal{L}$ is a cycle in Γ with $\langle\mathcal{L}\rangle = \mathcal{L}$ and $x \in V(\Gamma)$ with $d(x, \mathcal{L}) > 2|\mathcal{L}|^2/\pi^2$, then x is not in the region of $\mathbb{R}^2$ bounded by $\overline{\mathcal{L}}$.*

Proof. The Isoperimetric Inequality says that the area A enclosed by the simple closed curve $\overline{\mathcal{L}}$ in $\mathbb{R}^2$ is bounded above by $1/(4\pi)$ times the square of the perimeter of $\overline{\mathcal{L}}$. This perimeter is bounded above by $|\mathcal{L}|$. Let $\mathcal{P}$ be a chord-free path. By placing a ball of radius $\frac{1}{2}$ about every other vertex of $\mathcal{P}$, one obtains disjoint balls of total area $\frac{1}{8}\pi|\mathcal{P}|$. Such a path $\mathcal{P}$ from x to $\mathcal{L}$ of length at least $2|\mathcal{L}|^2/\pi^2$ violates the area constraint: the endpoint is thus not surrounded by $\overline{\mathcal{L}}$.

Better constants are possible, but the lower bound must be quadratic in $|\mathcal{L}|$, since a chord-free path in the interior of $\overline{\mathcal{L}}$ can fill up the area bound by $\overline{\mathcal{L}}$, which is quadratic in the perimeter. The following lemmas are critical for turning local

Algorithm 1. $\mathcal{I}$ = IndexVertexLoop($x, \mathcal{L}, \Gamma$)

Require: $\Gamma = (V, E)$ is a graph satisfying **P**, **N**, and **O**
Require: $x \in V(\Gamma)$, $\mathcal{L} = (\ell_i)$ is an oriented cycle of Γ, $\langle \mathcal{L} \rangle = \mathcal{L}$, and $d(x, \mathcal{L}) > 1$.
1: choose a path $\mathcal{P} = (x_i)_0^n$ in G with $x_0 = x$, $\langle \mathcal{P} \rangle = \mathcal{P}$, and $d(x_i, \mathcal{L}) = 1$ iff $i = n$.
2: **if** at x_n, for some j, either $\ell_j \lhd x_{n-1} \lhd \ell_{j+1}$ or $\ell_{j+1} \lhd x_{n-1} \lhd \ell_j$ and no other ℓ_i separates x_{n-1} from these neighbors in $\lhd$ **then**
3: return $\mathcal{I} \Leftarrow \mathcal{I}_{\ell_j}(\ell_{j-1}; \ell_{j+1}, x_n) \cdot \mathcal{I}_{\ell_{j+1}}(\ell_{j+2}; x_n, \ell_j) \cdot \mathcal{I}_{\ell_{x_n}}(x_{n-1}; \ell_j, \ell_{j+1})$
4: **else**
5: **if** for some j, $d(x_n, \ell_j) = d(x_n, \ell_{j+1}) = 1$ **then**
6: return $\mathcal{I} \Leftarrow \mathcal{I}_{\ell_j}(\ell_{j-1}; \ell_{j+1}, x_n) \cdot \mathcal{I}_{\ell_{j+1}}(\ell_{j+2}; x_n, \ell_j) \cdot \mathcal{I}_{\ell_{x_n}}(x_{n-1}; \ell_j, \ell_{j+1})$
7: **else**
8: choose any ℓ_j with $d(\ell_j, x_n) = 1$.
9: return $\mathcal{I} \Leftarrow \mathcal{I}_{\ell_j}(x_n; \ell_{j-1}, \ell_{j+1})$
10: **end if**
11: **end if**

Algorithm 2. $\mathcal{I}$ = IndexCheck($x_0, \mathcal{L}, \Gamma$)

Require: $\Gamma = (V, E)$ is a graph satisfying **P**, **N**, and **O**
Require: $x_0 \in V(\Gamma)$, $\mathcal{L}$ is an oriented cycle of Γ, $\langle \mathcal{L} \rangle = \mathcal{L}$, and $d(x_0, \mathcal{L}) > 1$
1: choose $x_\infty \in V(\Gamma)$ with $d(x_\infty, \mathcal{L}) > 2|\mathcal{L}|^2/\pi^2$
2: return $\mathcal{I} \Leftarrow$ IndexVertexLoop($x_0, \mathcal{L}, \Gamma$) − IndexVertexLoop($x_\infty, \mathcal{L}, \Gamma$)

cyclic orientation data into global cyclic orientation data. These can be proved by direct enumeration, but an approach which is both more elegant and more easily generalized is to use simple algebraic topology (homology theory): see [9] for detailed proofs.

Lemma 3. *Consider a graph Δ having one oriented cycle (x_1, x_2, x_3), with each inner node x_i connected to an outer node y_i. If this graph has planar image $\overline{\Delta}$ as in Fig. 4, then the cyclic orientation of the outer nodes $(y_i)_1^3$ with respect to any point in their convex hull is equal to*

$$\prod_{i=1}^{3} \mathcal{I}_{x_i}(y_i; x_{i-1}, x_{i+1}) \tag{2}$$

under the convention −1=CW and +1=CCW, and where the subscript indices are cyclic (computed mod 3).

Lemma 4. *Consider a graph Y having one central node x_0 attached to an ordered triple of non-colinear points $(y_i)_1^3$. Then, for any i, $\mathcal{I}_{x_0}(y_i; y_{i-1}, y_{i+1})$ is equal to the cyclic orientation of the ordered triple $(y_j)_1^3$ in $\mathbb{R}^2$ about x_0, under the convention −1=CW and +1=CCW.*

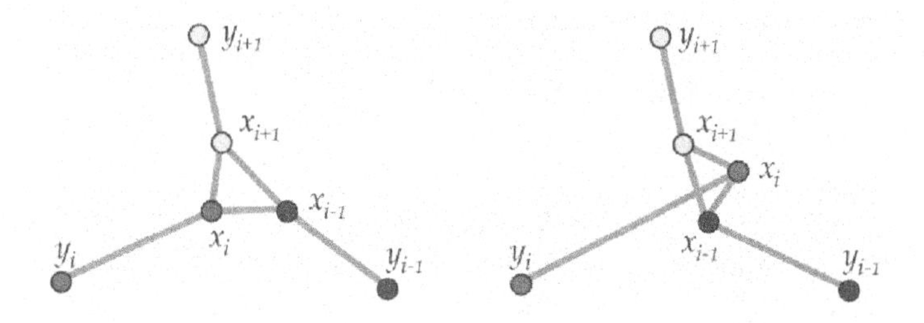

Fig. 4. The cyclic orientation of the outer nodes can be derived from the indices of the inner nodes in the above cases by computing the product of indices

Theorem 1. *In any network satisfying Assumptions* **P**, **N**, *and* **O**, *let $\mathcal{L}$ be a cycle satisfying $\langle\mathcal{L}\rangle = \mathcal{L}$ and x_0 a node with $d(x_0, \mathcal{L}) > 1$. Algorithm* IndexCheck *returns $\mathcal{I} = 0$ iff the winding number of $\overline{\mathcal{L}}$ about the node $x_0 \in \mathbb{R}^2$ vanishes.*

Proof. Corollary 1 implies that $\overline{\mathcal{L}}$ is embedded in $\mathbb{R}^2$. This simple closed curve separates the plane in two connected components, thanks to the Jordan Curve Theorem. Fixing an orientation on $\mathcal{L}$ induces an (unknown) orientation on $\overline{\mathcal{L}}$.

Choose a chord-free path $\mathcal{P}_0 = \{x_i\}_0^n$ from x_0 to x_n with $d(x_i, \mathcal{L}) = 1$ iff $i = n$. Via Corollary 1, the image of this path, $\overline{\mathcal{P}}_0$, is simple and the restriction of this path to the subpath between nodes x_0 and x_{n-1} lies entirely on one side of $\overline{\mathcal{L}}$ in $\mathbb{R}^2$. The edge from x_{n-1} to x_n may or may not cross $\overline{\mathcal{L}}$.

If there do not exist consecutive cycle nodes ℓ_j, ℓ_{j+1} incident to x_n, then choose any ℓ_j incident to x_n. In this case, the 'Y' graph connecting ℓ_j to x_n, ℓ_{j-1}, and ℓ_{j+1} has no additional connections between outer nodes, and the index $\mathcal{I}_{\ell_j}(x_n; \ell_{j-1}, \ell_{j+1})$ shows on which side of $\overline{\mathcal{L}}$ the node x_n (hence x_0) lies.

If, however, consecutive cycle neighbors exist, one argues that the subgraph Δ consisting of the cycle $(x_n, \ell_j, \ell_{j+1})$ and the connections of these inner nodes to respective outer nodes $(x_{n-1}, \ell_{j-1}, \ell_{j+2})$ has image $\overline{\Delta}$ as in Fig. 4. A more complicated embedding cannot appear thanks to repeated application of Lemma 1. Thanks to Lemma 3, the product of the three indices $\mathcal{I}_{\ell_j}(\ell_{j-1}; \ell_{j+1}, x_n)$, $\mathcal{I}_{\ell_{j+1}}(\ell_{j+2}; x_n, \ell_j)$, and $\mathcal{I}_{\ell_{x_n}}(x_{n-1}; \ell_j, \ell_{j+1})$ gives the cyclic orientation of the ordered triple $(x_{n-1}, \ell_{j+2}, \ell_{j-1})$ of outer nodes. Via Lemma 4, this tells whether x_{n-1} (and thus x_0) lies to the 'left' or to the 'right' of the embedded segment $(\ell_i)_{j-1}^{j+2}$ of $\overline{\mathcal{L}}$. However, it is possible that $\overline{\mathcal{L}}$ doubles back and crosses the segment between x_{n-1} and x_n, as in Fig. 3[right]. In this case, one needs to be sure to use the subgraph Δ generated by consecutive nodes $(\ell_i)_{j-1}^{j+2}$ where, from the vantage of x_n, $\ell_j \lhd x_{n-1} \lhd \ell_{j+1}$ or $\ell_{j+1} \lhd x_{n-1} \lhd \ell_j$, and no other ℓ_i separates.

There is an ambiguity in $\mathcal{I}$ resulting from the fact that we do not know if the orientation on $\overline{\mathcal{L}}$ is clockwise or counterclockwise; thus we do not know which sign for $\mathcal{I}$ (i.e., the 'left' or the 'right' side of $\overline{\mathcal{L}}$) corresponds to the bounded component of $\mathbb{R}^2 - \overline{\mathcal{L}}$. To determine this, choose a node x_∞ with $d(x_\infty, \mathcal{L}) > 2|\mathcal{L}|^2/\pi^2$. From Lemma 2, x_∞ lies within the unbounded component of $\mathbb{R}^2 - \overline{\mathcal{L}}$. That one can choose such a node and a chord-free path $\mathcal{P}_\infty$ from $\mathcal{L}$ to x_∞ is possible thanks to Assumption **N**. Computing the index of x_∞ with respect to $\mathcal{L}$ and comparing it to that of x_0 as in IndexCheck determines whether x_0 and x_∞ are on the same or different sides of $\overline{\mathcal{L}}$.

4 Isolation

We consider the isolation problem for a network satisfying Assumptions **P**, **N**, and **O**. Given a node x_0, determine whether there exists a cycle $\mathcal{L}$ which surrounds x_0 and construct one if it exists. We restrict the location of the cycle we search for by specifying a lower R_α and upper R_ω bound on the hop distance to x_0. We search for surrounding cycles within the subgraph whose vertices satisfy both bounds.

4.1 Algorithm

The first algorithm we give to solve this problem is similar is spirit to the Bound-Hole algorithm of [6], in that it relies on angular ordering to perform a depth-first search with constraints. The algorithm of [6] was intended to find holes in a network at a known boundary (or 'stuck') point assuming known exact pairwise angles between neighbors.

To solve the Isolation problem, we choose a chord-free path $\mathcal{P}$ in Γ from x_0 to some terminal point x_∞ which is more than R_ω hops from x_0. Truncate the graph Γ to Γ', the subgraph generated by nodes within R_α and R_ω hops of x_0. The path $\mathcal{P}$ restricts to a path $\mathcal{P}' = \{p_i\}_1^N$ in Γ'.

Beginning with the first node p_1 of $\mathcal{P}'$, construct a path $\mathcal{L}$ by performing a depth-first search of Γ' with the following conditions (Algorithm SweepCycle).

1. The depth-first search augments the path $\mathcal{L}$ by choosing the next available node of Γ' which is clockwise (CW) from the prior edge of $\mathcal{L}$.
2. The search backtracks whenever there is no available CW node, or when the path $\mathcal{L}$ has endpoint in the 1-hop neighborhood of $\mathcal{P}$ approaching from the 'right' side.
3. If the search backtracks all the way to the starting point of $\mathcal{L}$, this starting point is changed to be the next point on $\mathcal{P}$.
4. If the path $\mathcal{L}$ has endpoint in the 1-hop neighborhood of $\mathcal{P}$ on the 'left' side, then $\mathcal{L}$ is completed to a cycle in Γ' by connecting the ends along $\mathcal{P}$.

The justification for this algorithm is that any cycle in Γ' whose image in $\mathbb{R}^2$ has winding number ± 1 about x_0 must intersect $\overline{\mathcal{P}}'$ (since the path is chord-free and thus embedded). Thus, constructing any path which approaches $\mathcal{P}'$ from each side once is automatically a path which encircles x_0.

4.2 Proofs

Theorem 2. *For any system satisfying Assumptions* **P**, **N**, *and* **O**, *Algorithm* SweepCycle *returns a cycle $\mathcal{L}$ in Γ' whose image $\overline{\mathcal{L}}$ encloses x_0 in $\mathbb{R}^2$ if and only if such a cycle exists.*

Algorithm 3. $\mathcal{L}$ = SweepCycle($x_0, R_\alpha, R_\omega, \Gamma$)

Require: Γ satisfies **P**, **N**, and **O**
Require: $R_\omega > R_\alpha > 1$
1: let $\mathcal{P}$ be a chord-free path from x_0 to x_∞ with $d(x_0, x_\infty) > R_\omega$
2: truncate $\mathcal{P} \to \mathcal{P}' = (p_i)$, $\Gamma \to \Gamma'$ with distance to x_0 between R_α and R_ω
3: split 1-hop neighborhood of $\mathcal{P}'$ into two sides $\mathcal{P}'_+$, $\mathcal{P}'_-$ using orientation data
4: **while** (ℓ_j) has endpoint not in $\mathcal{P}'_-$ **do**
5: **while** (ℓ_j) does not have endpoint in $\mathcal{P}'_+$ and interior point not in $\mathcal{P}'_+$ **do**
6: augment (ℓ_j) via CW depth-first search of $\Gamma' - (\mathcal{P}' \cup \mathcal{P}'_+)$
7: **end while**
8: $\ell_1 \Leftarrow p_i$, the next available node of $\mathcal{P}'$
9: **end while**
10: return $\mathcal{L} \Leftarrow \emptyset$ if search is exhausted, else return $\mathcal{L} = (\ell_j)$ union segment of $\mathcal{P}'$ connecting ends of (ℓ_j)

Proof. The image of the truncated graph Γ' lies in a topological annulus $A \subset \mathbb{R}^2$, whose boundary components are connected by the embedded path $\overline{\mathcal{P}}$. Any simple closed curve in A which consists of a segment of $\overline{\mathcal{P}}$ and a segment in $A - \overline{\mathcal{P}}$ which approaches $\overline{\mathcal{P}}$ from both sides surrounds x_0 in $\mathbb{R}^2$ (this is proved using, e.g., homology theory). Thus, if Algorithm SweepCycle returns a non-empty cycle $\mathcal{L}$, then its image $\overline{\mathcal{L}}$ surrounds x_0.

The Jordan Curve Theorem applied to A implies that any simple closed curve in A which surrounds x_0 must intersect $\overline{\mathcal{P}}$. If $\mathcal{L}$ is any cycle of Γ' whose image surrounds x_0, then there are at least two nodes of $\mathcal{L}$ within the 1-hop neighborhood of $\mathcal{P}'$, thanks to Lemma 1. Choose two such nodes on opposite sides of $\mathcal{P}'$ and such that no further nodes of $\mathcal{L}$ are within the 1-hop neighborhood of $\mathcal{P}'$ In a search of Γ', Algorithm SweepCycle will eventually hit one of these two nodes. The depth first search the algorithm performs cannot exhaust Γ' without sweeping through $\mathcal{L}$.

5 Angles and Uncertainty

As a step toward removing angular orientation data, we modify Assumption **O** to account for uncertainty in angular orientation types.

U (Ordering type with Uncertainty) Each node can determine the clockwise cyclic ordering of any neighbors separated by angles of at least α_0.

That is, any triple of neighbors can be cyclically ordered if none of the three pairwise angles is below the threshold α_0. We do not assume that the cyclic ordering measurement always fails whenever an angle is below α_0, but rather that the reading either returns a true angular reading or an empty (i.e., uncertain) reading.

The surprising fact is that for α_0 as large as $\pi/3$, it is often possible to rigorously determine winding numbers. Some choices of Γ and $\mathcal{L}$ present too much uncertainty, but criteria for knowing when you can compute winding numbers are possible. We outline this procedure in the setting of the Separation Problem as an example.

Consider a network satisfying Assumptions **P**, **N**, and **U**, with α_0. Let $\mathcal{L}$ be a cycle satisfying $\langle \mathcal{L} \rangle = \mathcal{L}$ and x_0 a node with $d(x_0, \mathcal{L}) > 1$. In the simplest case where x_n is not within 1-hop of a consecutive pair of cycle nodes, choose any isolated incident cycle node ℓ_j of $\mathcal{L}$. That $\mathcal{L}$ is chord-free implies all three angles at ℓ_i to x_n, ℓ_{i-1}, and ℓ_{i+1} are greater than $\pi/3$ and thus α_0. Therefore $\mathcal{I}$ is well-defined here and yields winding information.

In the more complicated case where there is a subgraph of the form in Fig. 4, then some of the indices at the three inner nodes may be undefined. Since the angles of a triangle sum to π, at least one angle in the interior triangle is no less that $\pi/3$. We consider cases based on how many of these three interior nodes admit a well-defined cyclic orientation of neighbors.

Case 1: If all three indices exist, we are obviously done.

Case 2: If only one index exists, we claim that this single index is equal to the full index of $\mathcal{P}$ with respect to $\mathcal{L}$. This breaks into two cases, according to Fig. 4. In the case on the left, each vertex has the same index, and choosing any one yields the same as their product. In the case on the right, if only one index exists, a brief argument involving plane geometry shows that the angle out of which the bisector of the inner triangle emanates is the largest of the three angles; thus, if only one index is well-defined, it is this one. This node always has index equal to the product of the three inner node indices.

Case 3: In the case where only two indices exist, they either have the same sign or different sign. If they have the same sign, then, using the 'largest angle' result of Case 2, we know that Δ has no self-intersections in $\mathbb{R}^2$. Thus, the index of this path with respect to $\mathcal{L}$ is equal to the index of either of the two well-defined nodes.

Case 4: If two indices only are defined and the two computed indices differ, then we are certainly in the case where Δ is not embedded in the plane. However, if it is not possible to compute the third index and it is not possible to determine which of the two nodes has the larger subtended angle, then there is no information by which the index of this path can be determined. One may attempt to modify either $\mathcal{P}$ or $\mathcal{L}$ locally to remove the ambiguity, but there is no guarantee that this is possible for every Γ.

Indeed, with this model of uncertainty, it is not possible to solve the separation problem. See [9] for examples, along with an alternate uncertainty model which does allow for a complete algorithm.

6 Isolation without Cyclic Orientation Data

For systems which do not satisfy Assumption **O**, no solutions are possible which apply to arbitrary networks: very sparse graphs can be embedded in the plane as a unit disc graph in inequivalent ways. However, there are non-complete algorithms which, upon successful termination, return rigorous winding number information. The 'flower' graphs of [8] provide one example of rigorous containment certificate. We briefly present a different approach which uses a modification of Assumption **P** as follows.

P' There is a simply-connected domain $\mathcal{D} \subset \mathbb{R}^2$ which partitions the nodes by membership in $\mathcal{D}$ and yields an 'interior' graph Γ^o of all nodes in $\mathcal{D}$ which satisfies $\overline{\Gamma^o} \subset \mathcal{D}$.

It is not necessary to know the precise geometry of $\mathcal{D}$ (cf. [3]). Algorithm TriPath performs the following operations. From x_0, choose three chord-free paths $\{\mathcal{P}_i\}_1^3$ from x_0 to the 'exterior' graph $\Gamma - \Gamma^o$ such that the 1-hop neighborhood of each $\mathcal{P}_i$ is disjoint from $\mathcal{P}_{i-1} \cup \mathcal{P}_{i+1}$ outside an R_α-hop neighborhood of x_0. (The existence of such paths is of course not guaranteed for all networks.) The algorithm searches within Γ^o for a sequence of arcs connecting $\mathcal{P}_i$ to $\mathcal{P}_{i+1}$ avoiding $\mathcal{P}_{i-1}$ for each i. Chaining these arcs together yields a cycle in Γ^o: see Fig. 6[left].

A proof analogous to that of Theorem 2 implies that this cycle surrounds x_0 in $\mathbb{R}^2$. Assumption **P'** implies that there is a topological annulus $A \subset \mathbb{R}^2$ whose outer boundary is $\partial\mathcal{D}$ and whose inner boundary is a simple closed curve surrounding x_0. The three paths $\{\overline{\mathcal{P}}_i\}$ intersect A in three pairwise-disjoint arcs, each connecting the inner boundary of A to the outer boundary of A. A simple homological argument reveals that any loop in A whose intersections with the $\overline{\mathcal{P}}_i$ are cyclically ordered must surround the inner hole of A and thus surround x_0.

We repeat that should Algorithm TriPath fail to construct a surrounding cycle, it does not indicate the non-existence of such a cycle.

Algorithm 4. $\mathcal{L} = \mathsf{TriPath}(x_0, R_\alpha, \Gamma, \Gamma^o)$

Require: Γ satisfies **P'** and **N**
Require: $x_0 \in \Gamma^o$
1: search for chord-free paths $\{\mathcal{P}_i\}_1^3$ in Γ from x_0 to $\Gamma - \Gamma^o$ with $d(\mathcal{P}_i, \mathcal{P}_{j\neq i}) > 1$ outside an R_α neighborhood of x_0.
2: search for paths $\{\mathcal{L}_i\}_1^3$ in Γ^o from $\mathcal{P}_i$ to $\mathcal{P}_{i+1}$ with $d(\mathcal{L}_i, \mathcal{P}_{i-1}) > 1$.
3: **if** either search fails **then**
4: $\quad \mathcal{L} \Leftarrow \emptyset$
5: **else**
6: $\quad \mathcal{L} \Leftarrow$cycle in $\cup_i \mathcal{P}_i \cup_i \mathcal{L}_i$
7: **end if**
8: return $\mathcal{L}$

---

7 Simulations

Algorithm IndexCheck is easily implemented, However, as it produces no output except for a winding number that is easily seen when the graph is illustrated, we waste no space illustrating test runs of this algorithm.

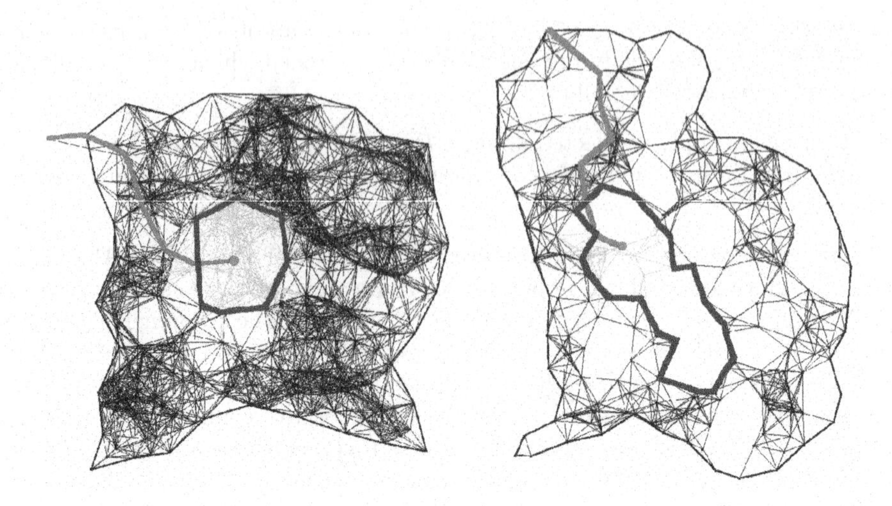

Fig. 5. Results of Algorithm SweepCycle on a dense [left] and sparse [right] randomly generated network. In both cases, the algorithm successfully produces a chord-free surrounding cycle outside of the 2-hop neighborhood of the encircled node.

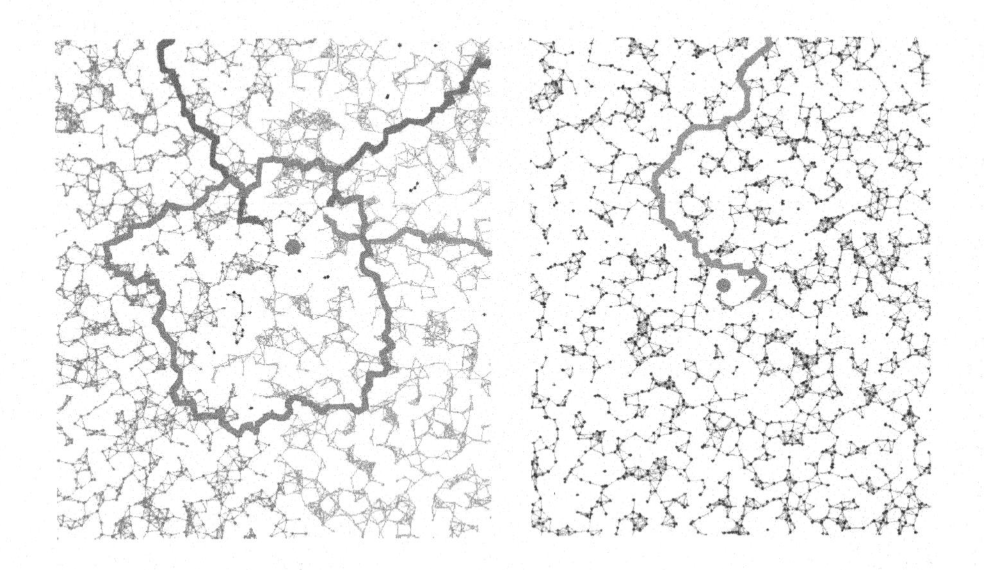

Fig. 6. Algorithm TriPath applied to a randomly generated network: [left] a network for which TriPath is successful. Three disjoint paths exist to 'infinity' and a loop which cyclically connects these paths generates a surrounding cycle. [right] The algorithm fails to find a surrounding cycle about the highlighted node, even though a surrounding cycle exists. The algorithm cannot find three disjoint paths from the node to the boundary.

Algorithm SweepCycle has been implemented [in C] for randomly generated sets of nodes with node density, input node, and the radius R_α as user-defined parameters. Examples of networks which are relatively dense and sparse are illustrated in Fig. 5. This algorithm inherits the time- and space-complexity of a depth-first search on the truncated graph.

Algorithm Tripath has likewise been implemented [in Java] with randomly generated sets of nodes. Fig. 6[left] displays a typical output, with the three paths $(\mathcal{P}_i)_1^3$ in bold and the cycle $\mathcal{L}$ marked. Depending on the exact form of Γ, it may be impossible to find three such paths $(\mathcal{P}_i)_1^3$ which are properly separated. Fig. 6[right] illustrates such an example, and reinforces the result that this algorithm does not always find a surrounding cycle. This algorithm is distributed and local: both the operation of finding $(\mathcal{P}_i)$ and the connecting segments between them are distributed in this software.

8 Concluding Remarks

The challenge of localization in an unknown environment is significant across many areas of robotics and sensor networks, and has generated an impressive array of techniques and perspectives. We demonstrate that localization is not a prerequisite to solving problems about winding numbers in a planar network. For many systems, the unit disc graph possesses sufficient information to find separating cycles about a node. We also demonstrate that an angular ordering of neighbors suffices to solve winding number problems for all possible networks. Absolute angles are not needed, and large uncertainty in the angular ordering data may be tolerated. As with many problems in manipulation, localization, mapping, etc., the amount of sensory information needed to solve the problem is sometimes far below what one would expect.

References

1. Breu, H., Kirkpatrick, D.G.: Unit Disk Graph Recognition is NP-hard. Comput. Geom. Theory Appl. 9(1-2), 3–24 (1998)
2. Bruck, J., Gao, J., Jiang, A.: Localization and routing in sensor networks by local angle information. In: Proc. MobiHoc, pp. 181–192 (2005)
3. de Silva, V., Ghrist, R.: Coordinate-free coverage in sensor networks with controlled boundaries via homology. Intl. J. Robotics Research 25(12), 1205–1222 (2006)
4. de Silva, V., Ghrist, R.: Coverage in sensor networks via persistent homology. Alg. & Geom. Topology 7, 339–358 (2007)
5. de Silva, V., Ghrist, R., Muhammad, A.: Blind swarms for coverage in 2-d. In: Proc. Robotics: Systems & Science (2005)
6. Fang, Q., Gao, J., Guibas, L.: Locating and Bypassing Routing Holes in Sensor Networks. In: Proc. 23rd Conference of the IEEE Communications Society (InfoCom) (2004)
7. Fang, Q., Gao, J., Guibas, L.J., de Silva, V., Zhang, L.: GLIDER: Gradient landmark-based distributed routing for sensor networks. In: IEEE INFOCOM 2005 (March 2005)

8. Fekete, S., Kröller, A., Pfisterer, D., Fischer, S.: Deterministic boundary recongnition and topology extraction for large sensor networks. In: Algorithmic Aspects of Large and Complex Networks (2006)
9. Ghrist, R.: Winding numbers for networks with weak angular data. Contemp. Math., American Mathematical Society (to appear, 2007)
10. Guillemin, V., Pollack, A.: Differential Topology. Prentice Hall, Englewood Cliffs (1974)
11. Jadbabaie, A.: On geographic routing without location information. In: Proc. IEEE Conf. on Decision and Control (2004)
12. Kuhn, F., Moscibroda, T., Wattenhofer, R.: Unit Disk Graph Approximation. In: Workshop on Discrete Algorithms and Methods for Mobile Computing and Communications (DIAL-M) (2004)
13. Poduri, S., Pattem, S., Krishnamachari, B., Sukhatme, G.: A unifying framework for tunable topology control in sensor networks. USC CRES Technical Report CRES-05-004 (2005)
14. Rao, A., Papadimitriou, C., Shenker, S., Stoica, I.: Geographic Routing without Location Information. In: Proceedings of 9th Annual International Conference on Mobile Computing and Networking (Mobicom 2003) (September 2003)

Passive Mobile Robot Localization within a Fixed Beacon Field

Carrick Detweiler, John Leonard, Daniela Rus, and Seth Teller

Computer Science and Artificial Intelligence Laboratory,
Massachusetts Institute of Technology
{carrick,jleondard,rus,teller}@csail.mit.edu

Abstract. This paper describes an intuitive geometric algorithm for the localization of mobile nodes in networks of sensors and robots using range-only or angle-only measurements. The algorithm is a minimalistic approach to localization and tracking when dead reckoning is too inaccurate to be useful. The only knowledge required about the mobile node is its maximum speed. Geometric regions are formed and grown to account for the motion of the mobile node. New measurements introduce new constraints which are propagated back in time to refine previous localization regions. The mobile robots are passive listeners while the sensor nodes actively broadcast making the algorithm scalable to many mobile nodes while maintaining the privacy of individual nodes. We prove that the localization regions found are optimal–that is, they are the smallest regions which must contain the mobile node at that time. We prove that each new measurement requires quadratic time in the number of measurements to update the system, however, we demonstrate experimentally that this can be reduced to constant time.

1 Introduction

Localization is a critical issue for many field robotics applications. In open outdoor environments, differential GPS systems can provide precise positioning information. There are many applications, however, in which GPS cannot be used, such as indoor, underwater, extraterrestrial, or urban environments. For situations when GPS is unavailable, dead reckoning may provide an alternative. Dead reckoning, however, is subject to accumulated error over time and is insufficient for many tasks. Most current localization methods make use of range or angle measurements to other nodes (pre-deployed beacons or other robots) to constrain dead reckoning error growth [4, 7, 14, 15, 19, 20].

In this paper, we present a localization algorithm for mobile agents for situations in which dead reckoning capabilities are poor, or simply unavailable. This includes the important case of passively tracking a non-cooperative target. The method is also applicable to low cost underwater robots, such as AMOUR [25], and other non-robotic mobile agents, such as animals [2] and people [10].

Our approach is based on a field of statically fixed nodes that communicate within a limited distance and are capable of estimating either ranges (in one

S. Akella et al. (Eds.): Algorithmic Foundation of Robotics VII, STAR 47, pp. 425–440, 2008.
springerlink.com

case) or angles (in the other case) to neighbors. These agents are assumed to have been previously localized by a static localization algorithm (e.g. [18]). A mobile node moves through this field, passively obtaining ranges (respectively angles) to nearby fixed nodes and listens to broadcasts from the static nodes. Based on this information, and an upper bound on the speed of the mobile node, our method recovers an estimate of the path traversed. As additional measurements are obtained, this new information is propagated backwards to refine previous location estimates. We prove that our algorithm finds optimal localization regions–that is, the smallest regions that must contain the mobile node.

Our algorithm allows for significant delays between measurements which makes traditional trilateration or triangulation approaches impossible. The algorithm is scalable to any number of mobile nodes as the mobile nodes are passive. The passivity of listeners also maintains the privacy of the mobile nodes.

This paper is organized as follows. We first introduce the intuition behind the range-only and angle-only versions of the algorithm. We then present the general algorithm, which can be instantiated with either range or angle information, and prove that it is optimal. Finally, we discuss an implementation of the range-only and angle-only algorithms, present experimental results of the range-only algorithm, and discuss extensions to the algorithm.

2 Related Work

A wide variety of strategies have been pursued for representing uncertainty in robot localization. Many approaches employ probabilistic state estimation to compute a posterior for the robot trajectory, based on assumed measurement and motion models. A variety of filtering methods have been employed for localization of mobile robots, including EKFs [7, 14, 15], Markov methods [3, 14], Monte Carlo methods [6, 7, 14] and batch techniques [4, 14, 15, 16]. Much of this recent work falls into the broad area of simultaneous localization and mapping (SLAM), in which the goal is to concurrently build a map of an unknown environment while localizing the robot.

Our approach assumes that the robot operates in a static field of nodes whose positions are known *a priori*. We assume only a known velocity bound for the vehicle, in contrast to the more detailed motion models assumed by modern SLAM methods. Most current localization algorithms assume an accurate model of the vehicle's uncertain motion. If a detailed probabilistic model is available, then a state estimation approach will likely produce a more accurate trajectory estimate for the robot. There are numerous real-world situations, however, that require localization algorithms that can operate with minimal proprioceptive sensing, poor motion models, and/or highly nonlinear measurement constraints. In these situations, the adoption of a bounded error representation, instead of a representation based on Gaussians or particles, is justified.

In previous work, Smith *et al.* also explored the problem of localization of mobile nodes without dead reckoning [22]. They compare the case where the mobile node is a passive listener verses actively pinging to obtain range estimates. In the passive listening case an EKF is used. However, the inherent difficulty of (re)initializing the filter leads them to conclude a hybrid approach is necessary. The mobile node remains passive until it detects a bad state. At this point it becomes active. In our work we maintain a passive state. Additionally, our approach introduces a geometric approach to localization which can stand alone, as we demonstrate experimentally in Section 6, or be post-processed by a Kalman filter or other filtering methods.

Our approach represents uncertainty using bounded regions that are computed based on worst-case assumptions of dead-reckoning and measurement errors. This can be contrasted with the conventional assumption of Gaussian errors in EKF approaches, or the representation of uncertainty with sets of particles in Markov Chain Monte Carlo state estimation [24]. Previous work adopting a bounded region representation of uncertainty includes Meizel *et al.* [17], Briechle and Hanebeck [1], Spletzer and Taylor [23], and Isler and Bajcsy [12]. Meizel *et al.* investigated the initial location estimation problem for a single robot given a prior geometric model based on noisy sonar range measurements. Briechle and Hanebeck [1] formulated a bounded uncertainty pose estimation algorithm given noisy relative angle measurements to point features. Doherty *et al.* [9] investigated localizations methods based only on wireless connectivity, with no range estimation. Spletzer and Taylor developed an algorithm for multi-robot localization based on a bounded uncertainty model [23]. Finally, Isler and Bajscy examine the sensor selection problem based on bounded uncertainty models [12].

The robot localization problem bears similarities with the classical problem of robot motion planning with uncertainty. In the seminal work of Erdmann bounded sets are used for forward projection of possible robot configurations. These are restricted by the control uncertainty cone [11]. In our work the step to compute the set of feasible poses for a robot moving through time is similar to Erdmann's forward projection step.

Our approach differs from all these approaches by incorporating a dynamic motion component for the robot. By assuming a worst-case model for robot motion, in terms of a maximum allowable speed, we are able to develop a bounded region localization algorithm that can handle the trajectory estimation problem given non-simultaneous measurements. This scenario is particularly important for underwater acoustic tracking applications, where significant delays between measurements are common due to the speed of sound.

3 Algorithm Intuition

This section formally describes the localization problem we solve and the intuition behind the range-only version and angle-only version of the localization algorithms. The generic algorithm is presented in section 4.

3.1 Problem Formulation

We will now define a generic formulation for the localization problem. This setup can be used to evaluate range-only, angle-only, and other localization problems. We start by defining a *localization region.*

Definition 1. *A localization region at some time t is the set of points in which a node is assumed to be at time t.*

We will often refer to a localization region simply as a region. It is useful to formulate the localization problem in terms of regions as the problem is typically under-constrained, so exact solutions are not possible. Probabilistic regions can also be used, however, we will use a discrete formulation. In this framework the localization problem can be stated in terms of finding *optimal* localization regions.

Definition 2. *We say that a localization region is optimal with respect to a set of measurements at time t if at that time it is the* smallest *region that must contain the true location of the mobile node, given the measurements and the known velocity bound. A region is individually optimal if it is optimal with respect to a single measurement.*

For example, for a range measurement the individually optimal region is an annulus and for the angle case it is a cone. Another way to phrase optimality is if a region is optimal at some time t, then the region contains the true location of the mobile node and all points in the region are reachable by the mobile node.

Suppose that from time $1 \cdots t$ we are given regions $A_1 \cdots A_t$ each of which is individually optimal. The times need not be uniformly distributed, however, we will assume that they are in sorted order. By definition, at time k region A_k must contain the true location of the mobile node and furthermore, if this is the only information we have about the mobile node, it is the smallest region that must contain the true location. We now want to form regions, $I_1 \cdots I_t$, which are optimal given all the regions $A_1 \cdots A_t$ and an upper bound on the speed of the mobile node which we will call s. We will refer to these regions as *intersection regions* as they will be formed by intersecting regions.

3.2 Range-Only Localization and Tracking

Figure 1 shows critical steps in the range-only localization of mobile Node **m**. Node **m** is moving through a field of localized static nodes (Nodes **a**, **b**, **c**) along the trajectory indicated by the dotted line.

At time t Node **m** passively obtains a range to Node **a**. This allows Node **m** to localize itself to the circle indicated in Figure 1(a). At time $t+1$ Node **m** has moved along the trajectory as shown in Figure 1(b). It expands its localization estimation to the annulus in Figure 1(b). Node **m** then enters the communication range of Node **b** and obtains a ranging to Node **b** (see Figure 1(c)). Next, Node **m** intersects the circle and annulus to obtain a localization region for time $t+1$ as indicated by the bold red arcs in Figure 1(d).

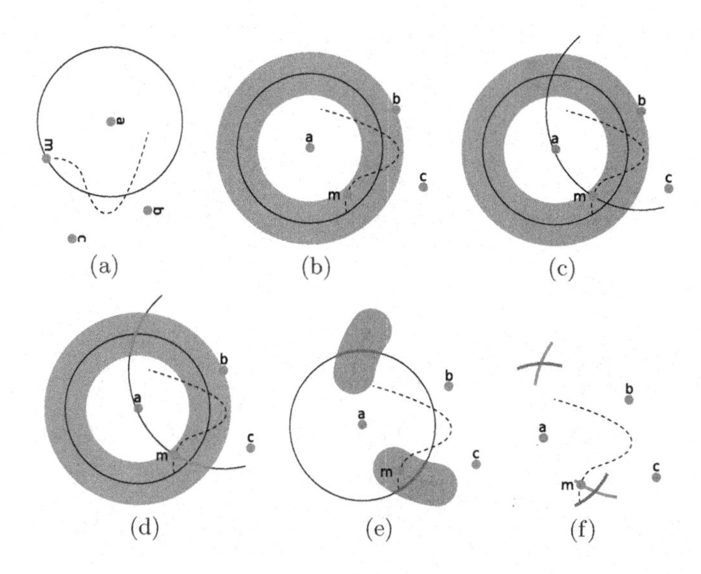

Fig. 1. Example of the range-only localization algorithm

The range taken at time $t+1$ can be used to improve the localization at time t as shown in Figure 1(e). The arcs from time $t+1$ are expanded to account for all locations the mobile node could have come from. This is then intersected with the range taken at time t to obtain the refined location region illustrated by the bold blue arcs. Figure 1(f) shows the final result. Note that for times t and $t+1$ there are two possible location regions. This is because two range measurements do not provide sufficient information to fully constrain the system. Range measurements from other nodes will quickly eliminate this.

3.3 Angle-Only Localization and Tracking

Consider Figure 2. Each snapshot shows three static nodes that have self-localized (Nodes **a**, **b**, **c**). Node **m** is a mobile node moving through the field of static nodes along the trajectory indicated by the dotted line. Each snapshot shows a critical point in the angle-only location and trajectory estimation for Node **m**.

At time t Node **m** enters the communication range of Node **a** and passively computes the angle to Node **a**. This allows Node **m** to estimate its position to be along the line shown in Figure 2(a). At time $t+1$ Node **m** has moved as shown in Figure 2(b). Based on its maximum possible speed, Node **m** expands its location estimate to that shown in Figure 2(b). Node **m** now obtains an angle measurement to Node **b** as shown in Figure 2(c). Based on the intersection of the region and the angle measurement Node **m** can constrain its location at time $t+1$ to be the bold red line indicated in Figure 2(d).

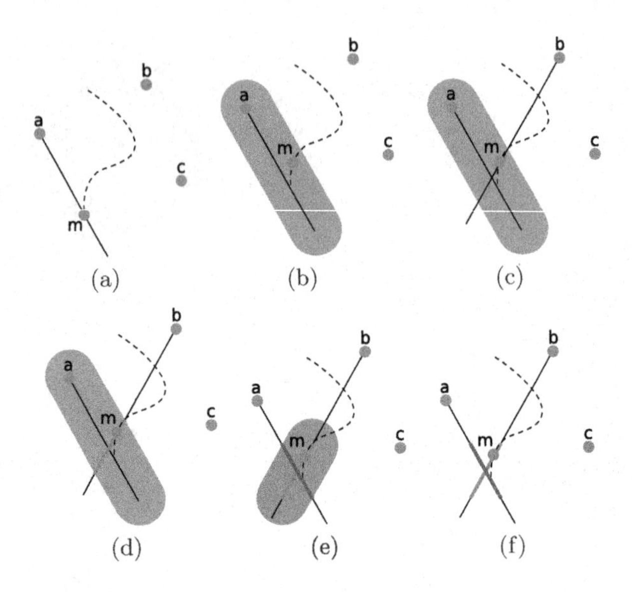

Fig. 2. Example of the angle-only localization algorithm

The angle measurement at time $t+1$ can be used to further refine the position estimate of the mobile node at time t as shown in Figure 2(e). The line that Node **m** was localized to at time $t+1$ is expanded to represent all possible locations Node **m** could have come from. This is region is then intersected with the original angle measurement from Node **a** to obtain the bold blue line which is the refined localization estimate of Node **m** at time t. Figure 2(f) shows the two resulting location regions. New angle measurements will further refine these regions.

4 The Localization Algorithm

4.1 Generic Algorithm

The localization algorithm follows the same idea as in Section 3. Each new region computed will be intersected with the grown version of the previous region and the information gained from the new region will be propagated backwards. Algorithm 1 shows the details.

Algorithm 1 can be run online by omitting the outer loop (lines 4-6 and 11) and executing the inner loop whenever a new region/measurement is obtained.

The first step in Algorithm 1 (line 3), is to initialize the first intersection region to be the first region. Then we iterate through each successive region.

The new region is intersected with the previous intersection region grown to account for any motion (line 6). Finally, the information gained from the new region is propagated back by successively intersecting each optimal region grown backwards with the previous region, as shown in line 9.

Algorithm 1. Localization Algorithm

```
1: procedure LOCALIZE(A_1 ··· A_t)
2:     s ← max speed
3:     I_1 = A_1                                   ▷ Initialize the first intersection region
4:     for k = 2 to t do
5:         Δt ← k − (k − 1)
6:         I_k = Grow(I_{k−1}, sΔt) ∩ A_k          ▷ Create the new intersection region
7:         for j = k − 1 to 1 do                   ▷ Propagate measurements back
8:             Δt ← j − (j − 1)
9:             I_j = Grow(I_{j+1}, sΔt) ∩ A_j
10:        end for
11:    end for
12: end procedure
```

4.2 Algorithm Details

Two key operations in the algorithm which we will now examine in detail are `Grow` and `Intersect`. `Grow` accounts for the motion of the mobile node over time. `Intersect` produces a region that contains only those points found in both localization regions being intersected.

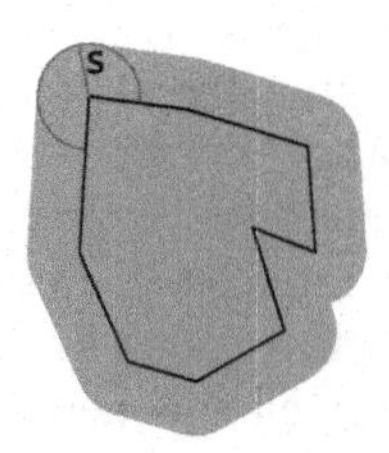

Fig. 3. Growing a region by s. Acute angles, when grown, turn into circles as illustrated. Obtuse angles, on the other hand, are eventually consumed by the growth of the surroundings.

Figure 3 illustrates how a region grows. Let the region bounded by the black lines contain the mobile node at time t. To determine the smallest possible region that must contain the mobile node at time $t+1$ we `Grow` the region by s, where s is the maximum speed of the mobile node. The `Grow` operation is the Minkowski sum [8] (frequently used in motion planning) of the region and a circle with diameter s.

Notice that obtuse corners become circle arcs when grown, while everything else "expands." If a region is convex, it will remain convex. Let the complexity of a region be the number of simple geometric features (lines and circles) needed to describe it. Growing convex regions will never increase the complexity of a region by more than a constant factor. This is true as everything just expands except for obtuse angles which are turned into circles and there are never more than a linear number of obtuse angles. Thus, growing can be done in time proportional to the complexity of the region.

A simple algorithm for `Intersect` is to check each feature of one region for intersection with all features of the other region. This can be done in time proportional to the product of the complexities of the two regions. While better algorithms exist for this, for our purposes this is sufficient as we will always ensure that one of the regions we are intersecting has constant complexity as shown in Sections 5.2 and 5.3. Additionally, if both regions are convex, the intersection will also be convex.

4.3 Correctness and Optimality

We now prove the correctness and optimality of Algorithm 1. We will show that the algorithm finds the location region of the node, and that this computed location region is the smallest region that can be determined using only maximum speed. We assume that the individual regions given as input are optimal, which is trivially true for both the range and angle-only cases.

Theorem 1. *Given the maximum speed of a mobile node and t individually optimal regions, $A_1 \cdots A_t$, Algorithm 1 will produce optimal intersection regions $I_1 \cdots I_t$.*

Without loss of generality assume that $A_1 \cdots A_t$ are in time order. We will prove this theorem inductively on the number of range measurements for the online version of the localization algorithm. The base case is when there is only a single range measurement. Line 3 implies $I_1 = A_1$ and we already know A_1 is optimal.

Now inductively assume that intersection regions $I_1 \cdots I_{t-1}$ are optimal. We must now show that when we add region A_t, I_t is optimal and the update of $I_1 \cdots I_{t-1}$ maintains optimality given this new information. Call these updated intersection regions $I'_1 \cdots I'_{t-1}$.

First we will show that the new intersection region, I'_t, is optimal. Line 6 of the localization algorithm is

$$I'_t = \mathbf{Grow}(I_{t-1}, s\triangle t) \cap A_t. \tag{1}$$

The region $\mathbf{Grow}(I_{t-1})$ contains all possible locations of the mobile node at time t ignoring the measurement A_t. The intersection region I'_t must contain all possible locations of the mobile node as it is the intersection of two regions that constrain the location of the mobile node. If this were not the case, then there would be some point p which was not in the intersection. This would imply that p was neither in I_{t-1} nor A_t, a contradiction as this would mean p was not reachable. Additionally, all points in I'_t are reachable as it is the intersection of a reachable region with another region. Therefore, I'_t is optimal.

Finally we will show that the propagation backwards, line 9, produces optimal regions. The propagation is given by

$$I'_j = \mathbf{Grow}(I'_{j+1}, s\triangle t) \cap A_j \tag{2}$$

for all $1 \leq j \leq t-1$. The algorithm starts with $j = t-1$. We just showed that I'_t is optimal, so using the same argument as above I_{t-1} is optimal. Applying this recursively, all $I_{t-2} \cdots I_1$ are optimal. Q.E.D.

5 Complexity

5.1 General Complexity

Algorithm 1 has both an inner and outer loop over all regions which suggests an $O(n^2)$ runtime, where n is the number of input regions. However, `Grow` and `Intersect` also take $O(n)$ time as proven later by Theorem 2 and 3 for the range and angle only cases. Thus, overall, we have an algorithm which runs in $O(n^3)$ time. We show, however, that we expect the cost of `Grow` and `Intersect` will be $O(1)$, which suggests $O(n^2)$ runtime overall.

The runtime can be further improved by noting that the correlation of the current measurement with the past will typically decrease rapidly as time increases. This implies that information need only be propagated back a fixed number of steps, eliminating the inner loop of Algorithm 1. Thus, we can reduce the complexity of the algorithm to $O(n)$.

5.2 Range-Only Complexity

The range-only instantiation of Algorithm 1 is obtained by taking range measurements to the nodes in the sensor fields. Let $A_1 \cdots A_n$ be the circular regions formed by n range measurements. $A_1 \cdots A_n$ are individually optimal and as such can be used as input to Algorithm 1. We now prove the complexity of localization regions is worst case $O(n)$. Experimentally we find they are actually $O(1)$ leading to an $O(n^2)$ runtime.

Theorem 2. *The complexity of the regions formed by the range-only version of the localization algorithm is $O(n)$, where n is the number of regions.*

The algorithm intersects a grown intersection region with a regular region. This will be the intersection of some grown segments of an annulus with a circle (as shown in the Figure 4). Let regions that contain multiple disjoint sub-regions be called *compound regions*. Since one of the regions is always composed of simple arcs, the result of an intersection will be a collection of circle segments. We will show that each intersection can at most increase the number of circle segments by two, implying linear complexity in the worst case.

Consider Figure 4. At most the circular region can cross the inner circle of the annulus that contains the compound region twice. Similarly, the circle can cross the outer circle of the annulus at most twice. The only way the number of sub-regions formed can increase is if the circle enters a sub-region, exits that sub-region, and then reenters it as illustrated in the figure. If any of the entering or exiting involves crossing the inner or outer circles of the annulus, then it must cross at least twice. This means that at most two regions could be split using

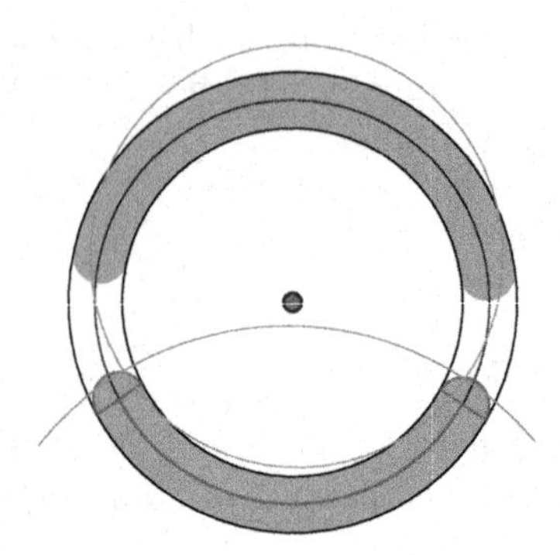

Fig. 4. An example compound intersection region (in blue) and some new range measurement in red. With each iteration it is possible to increase the number of regions in the compound intersection region by at most two.

this method, implying a maximum increase in the number of regions of at most two.

If the circle does not cut the interior or exterior of the annulus within a sub-region then it must enter and exit though the ends of the sub-region. But notice that the ends of the sub-regions are grown such that they are circular, so the circle being intersected can only cross each end twice. Furthermore, the only way to split the subregion is to cross both ends. To do this the annulus must be entered and exited on each end, implying all of the crosses of the annulus have been used up. Therefore, the number of regions can only be increased by two with each intersection proving Theorem 2.

In practice it is unlikely that the regions will have linear complexity. As seen in Figure 4 the circle which is intersecting the compound region must be very precisely aligned to increase the number of regions (note that the bottom circle does not increase the number of regions). In the experiments described in Section 6 we found some regions divided in two or three (e.g. when there are only two range measurements, recall Figure 1). However, there were none with more than three sub-regions. Thus, in practice the complexity of the regions is constant leading to $O(n^2)$ runtime.

5.3 Angle-Only Region Complexity

Algorithm 1 is instantiated in the angle-only case by taking angle measurements $\theta_1 \cdots \theta_t$ with corresponding bounded errors $e_1 \cdots e_t$ to the field of sensors. These form the individually optimal regions $A_1 \cdots A_t$ used as input to Algorithm 1. These regions appear as triangular wedges as shown in Figure 5(a). We now show the complexity of the localization regions is at worst $O(n)$ letting us conclude an $O(n^3)$ algorithm (in practice, the complexity is $O(1)$ implying a runtime of $O(n^2)$ see below).

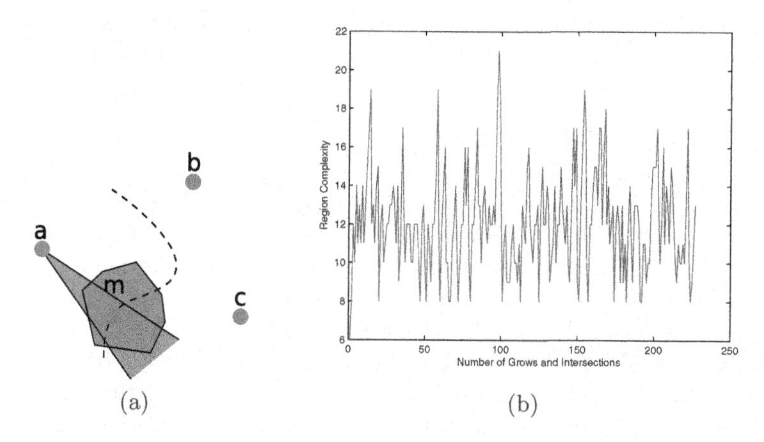

Fig. 5. (a)An example of the intersecting an angle measurement region with another region. Notice that the bottom line increases the complexity of the region by one, while the top line decreases the complexity of the region by two. (b)Growth in complexity of a localization region as a function of the number of *Grow* and *Intersect* operations.

Theorem 3. *The complexity of the regions formed by the angle-only version of the localization algorithm is $O(n)$, where n is the number of regions.*

Examining the Algorithm 1, each of the intersection regions $I_1 \cdots I_n$ are formed by intersecting a region with a grown intersection region n times. We know growing a region never increases the complexity of the region by more than a constant factor. We will now show that intersecting a region with a grown region will never cause the complexity of the new region to increase by more than a constant factor.

Each of these regions is convex as we start with convex regions and `Grow` and `Intersect` preserve convexity. Assume we are intersecting `Grow`(I_k) with A_{k-1}. Note that A_{k-1} is composed of two lines. As `Grow`(I_k) is convex, each line of A_{k-1} can only enter and exit `Grow`(I_k) once. Most of the time this will actually decrease the complexity as it will "cut" away part of the region. The only time that it will increase the complexity is when the line enters a simple feature and exits on the adjacent one as shown in Figure 5(a). This increases the complexity by one. Thus, with the two lines from the region A_{k-1} the complexity is increased by at most two.

In practice the complexity of the regions is not linear, rather it is constant. When a region is intersected with another region there is a high probability that some of the complex parts of the region will actually be removed, simplifying the region. Figure 5(b) illustrates the results of a simulation where the complexity of a region was tracked over time. In the simulation a single region was repeatedly grown and then intersected with a region formed by an angle measurement from a randomly chosen node from a field of 30 static nodes. Figure 5(b) shows that increasing the number of intersections does not increase the complexity of the region. The average complexity of the region was 12.0. Thus, the complexity of the regions is constant so the angle-only localization algorithm runs in $O(n^2)$.

6 Experimental Results

We implemented the range-only version of Algorithm 1 and tested it on a dataset created by Moore *et al.* [18]. Range measurements were obtained from a static network of six MIT Crickets [21] to one mounted on a mobile robot. The robot moved within a 2.0 by 1.5 meter space. Ground truth was collected by a video capture system with sub-centimeter accuracy [18].

Figure 6(a) shows the static nodes (green), the arcs recovered by the localization algorithm (orange), and a path generated by connecting the mid-points of the arcs (green). Figure 6(b) shows the same recovered trajectory (green) along with ground truth data (red). Inset in Figure 6(b) is an enlarged view showing a location where an error occurred due to a lack of measurements while the robot was making a turn.

The mean absolute error from the ground truth was 7.0cm with a maximum of 15.6cm. This was computed by taking the absolute error from the ground truth data at each point in the recovered path. This compares well with the 5cm

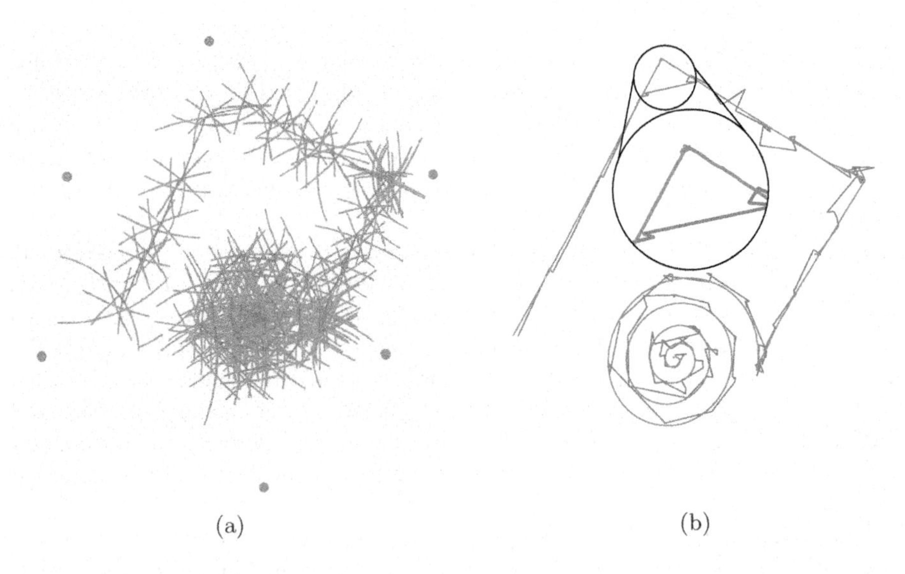

(a) (b)

Fig. 6. (a)The arcs found by the range-only implementation of Algorithm 1 and the path reconstructed from these. Orange arcs represent the raw regions, while the green line connects the midpoints of these arcs. (b)Ground truth (red) and recovered path (green). Inset is an enlarged portion illustrating error caused by a lack of measurements during a turn of the robot.

of measurement error inherent to the Crickets [18]. Figure 7 shows the absolute error as a function of time.

The algorithm handles the error intrinsic to the sensors well. One reason for this is the robot did not always travel at its maximum speed in a straight line. This meant the regions were larger than if the robot were traveling at its maximum speed in a straight line, making up for measurement error. In most applications this will be the case, however, to be conservative the maximum speed could be increased slightly to account for error in the sensors. Most sensors also have occasional large non-Gaussian errors. To account for these, measurements which have no intersection with the current localization region are rejected as outliers.

This implementation only propagates information back to 5-8 localization regions. We did not find it sensitive to the exact number. We also never encountered a region with a complexity higher than three. These two facts gives an overall runtime of $O(n)$, or $O(1)$ for each new range measurement.

7 Discussion

The propagation of the information gained from new measurements back to previous measurements can take a significant amount of time. In the experiments presented in Section 6 we found that we only had to propagate the information back through a small, fixed, period of time. This led to a significant reduction in the runtime. There are, however, cases where further propagation may be needed. For instance, the localization of a mobile node traveling in a straight line at maximum speed could be improved by propagating the measurements back far. An adaptive system that decides at runtime how far to propagate information back may improve localization results while maintaining the constant time update.

Additional knowledge about the motion characteristics of the mobile node can also be added to the system to further refine localization. Maximum acceleration is likely to be known in most physical systems. With a bound on acceleration the regions need only be grown in the current estimate of the direction of travel and directions that could be achieved by the bounded acceleration. In many systems this would significantly reduce the size of the localization regions.

With real sensors there are some pathological cases which will cause the algorithm to fail. If the maximum speed has not been increased sufficiently to account for sensor errors then a number of consecutive erroneous measurements could produce a region which does not contain the true location of the mobile node. This situation can be detected as subsequent measurements will have no intersection with the current localization region. At this point the algorithm could be restarted or the previous regions could be modified to allow for a larger error.

Our algorithm can be used with a variety of different sensors. Many are passive in nature which allows for the scaling of our algorithm to any number of mobile nodes. On land we have used MIT Crickets [21] to obtain range measurements.

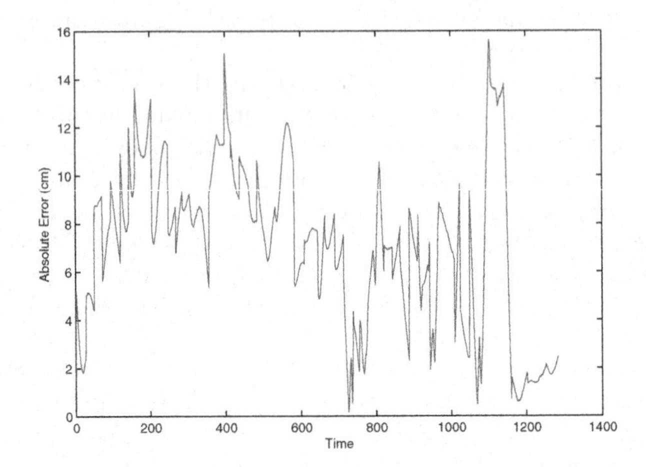

Fig. 7. Absolute error in the recovered position estimates in centimeters as a function of time

Range measurements are obtained passively by taking the difference in time of flight of a radio and ultrasonic pulse. Angle measurements can be obtained passively using an omnidirectional camera [4]. Underwater, acoustic ranging can be done passively using synchronized clocks [5]. Angle measurements can be obtained by listening to sounds using an acoustic array [13].

The accuracy of the produced localization regions depends on a number of factors in addition to the properties of the sensors. One of the most important is the time between measurements. The more frequent the measurements the more precise the localization regions will be. Additionally, the selection of the static nodes to be used in measurements is important. For instance, taking consecutive measurements to nodes that are close together will yield poor results. Thus, a selection algorithm, such as that presented by Isler et al. [12], will improve results by choosing the best nodes for measurements.

We have implemented this algorithm on our underwater robot AMOUR and underwater sensor network [25], and plan to collect data at the Gump research station in Moorea in June 2006. This system and many other underwater systems have poor dead-reckoning. We expect to enhance the localization and tracking of our robot using the algorithm presented in this paper.

References

1. Briechle, K., Hanebeck, U.D.: Localization of a mobile robot using relative bearing measurements. IEEE Transactions on Robotics and Automation 20(1), 36–44 (2004)
2. Butler, Z., Corke, P., Peterson, R., Rus, D.: From animals to robots: virtual fences for controlling cattle. In: International Symposium on Experimental Robotics, New Orleans, pp. 4429–4436 (2004)

3. Castelnovi, M., Sgorbissa, A., Zaccaria, R.: Markov-localization through color features comparison. In: Proceedings of the 2004 IEEE International Symposium on Intelligent Control, pp. 437–442 (2004)
4. Deans, M., Hebert, M.: Experimental comparison of techniques for localization and mapping using a bearings only sensor. In: Proc. of the ISER 2000 Seventh International Symposium on Experimental Robotics, pp. 395–404. Springer, Heidelberg (2000)
5. Deffenbaugh, M., Bellingham, J.G., Schmidt, H.: The relationship between spherical and hyperbolic positioning. In: OCEANS 1996. MTS/IEEE. Prospects for the 21st Century. Conference Proceedings, vol. 2, pp. 590–595 (1996)
6. Dellaert, F., Fox, D., Burgard, W., Thrun, S.: Monte carlo localization for mobile robots. In: IEEE International Conference on Robotics and Automation (ICRA 1999), vol. 2, pp. 1322–1328 (May 1999)
7. Djugash, J., Singh, S., Corke, P.I.: Further results with localization and mapping using range from radio. In: International Conference on Field and Service Robotics (July 2005)
8. Dobkin, D.P., Hershberger, J., Kirkpatrick, D.G., Suri, S.: Computing the intersection-depth of polyhedra. Algorithmica 9(6), 518–533 (1993)
9. Doherty, L., Pister, K.S.J., Ghaoui, L.E.: Convex optimization methods for sensor node position estimation. In: INFOCOM, pp. 1655–1663 (2001)
10. Eagle, N., Pentland, A.: Reality mining: Sensing complex social systems. Personal and Ubiquitous Computing, 1–14 (September 2005)
11. Erdmann, M.: Using backprojections for fine motion planning with uncertainty. IJRR 5(1), 19–45 (1986)
12. Isler, V., Bajcsy, R.: The sensor selection problem for bounded uncertainty sensing models. In: Fourth International Symposium on nformation Processing in Sensor Networks, 2005. IPSN 2005, pp. 151–158 (April 2005)
13. Johnson, D., Dudgeon, D.: Array Signal Processing. Prentice Hall, Englewood Cliffs (1993)
14. Kantor, G.A., Singh, S.: Preliminary results in range-only localization and mapping. In: Proceedings of the IEEE Conference on Robotics and Automation (ICRA 2002), vol. 2, pp. 1818–1823 (May 2002)
15. Kurth, D.: Range-only robot localization and SLAM with radio. Master's thesis, Robotics Institute Carnegie Mellon University, Pittsburgh PA (May 2004)
16. McLauchlan, P.F.: A batch/recursive algorithm for 3d scene reconstruction. In: Computer Vision and Pattern Recognition, vol. 2, pp. 738–743 (2000)
17. Meizel, D., Leveque, O., Jaulin, L., Walter, E.: Initial localization by set inversion. IEEE Transactions on Robotics and Automation 18(3), 966–971 (2002)
18. Moore, D., Leonard, J., Rus, D., Teller, S.: Robust distributed network localization with noisy range measurements. In: Proc. 2nd ACM Sen.Sys., Baltimore, pp. 50–61 (November 2004)
19. Olson, E., Leonard, J., Teller, S.: Robust range-only beacon localization. In: Proceedings of Autonomous Underwater Vehicles, pp. 66–75 (2004)
20. Olson, E., Walter, M., Leonard, J., Teller, S.: Single cluster graph partitioning for robotics applications. In: Robotics Science and Systems, pp. 265–272 (2005)
21. Priyantha, N.B., Chakraborty, A., Balakrishnan, H.: The cricket location-support system. In: MobiCom 2000: Proceedings of the 6th annual international conference on Mobile computing and networking, pp. 32–43. ACM Press, New York (2000)

22. Smith, A., Balakrishnan, H., Goraczko, M., Priyantha, N.: Tracking moving devices with the cricket location system. In: MobiSys 2004: Proceedings of the 2nd international conference on Mobile systems applications and services, pp. 190–202. ACM Press, New York (2004)
23. Spletzer, J.R., Taylor, C.J.: A bounded uncertainty approach to multi-robot localization. In: IEEE/RSJ International Conference on Intelligent Robots and Systems (IROS 2003), vol. 2, pp. 1258–1265 (2003)
24. Thrun, S., Burgard, W., Fox, D.: Probabilistic Robotics. MIT Press, Cambridge (2005)
25. Vasilescu, I., Kotay, K., Rus, D., Dunbabin, M., Corke, P.: Data collection, storage, and retrieval with an underwater sensor network. In: SenSys 2005: Proceedings of the 3rd international conference on Embedded networked sensor systems, pp. 154–165. ACM Press, New York (2005)

Efficient Motion Planning Strategies for Large-Scale Sensor Networks

Jason C. Derenick, Christopher R. Mansley, and John R. Spletzer

Department of Computer Science and Engineering, Lehigh University
{jcd6,crm5,josa}@lehigh.edu

Abstract. In this paper, we develop a suite of motion planning strategies suitable for large-scale sensor networks. These solve the problem of reconfiguring the network to a new shape while minimizing either the total distance traveled by the nodes or the maximum distance traveled by any node. Three network paradigms are investigated: centralized, computationally distributed, and decentralized. For the centralized case, optimal solutions are obtained in $O(m)$ time in practice using a logarithmic-barrier method. Key to this complexity is transforming the Karush-Kuhn-Tucker (KKT) matrix associated with the Newton step sub-problem into a mono-banded system solvable in $O(m)$ time. These results are then extended to a distributed approach that allows the computation to be evenly partitioned across the m nodes in exchange for $O(m)$ messages in the overlay network. Finally, we offer a decentralized, hierarchical approach whereby follower nodes are able to solve for their objective positions in $O(1)$ time from observing the headings of a small number (2-4) of leader nodes. This is akin to biological systems (*e.g.* schools of fish, flocks of birds, *etc.*) capable of complex formation changes using only local sensor feedback. We expect these results will prove useful in extending the mission lives of large-scale mobile sensor networks.

1 Introduction

Consider the initial deployment of a wireless sensor network (WSN). Ideally, the WSN is fully connected with a topology to facilitate coverage, sensing, localization, and data routing. Unfortunately, since deployment methods can vary from aerial to manual, the initial configuration could be far from ideal. As a result, the WSN may be congested, disconnected, and incapable of localizing itself in the environment. Node failures in established networks could have similar effects. Such limitations in static networks have lead to an increased research interest into improving network efficiency via nodes that support at least limited mobility [2].

Also of fundamental importance to WSN research is resource management, and (perhaps most importantly) power management. Energy consumption is the most limiting factor in the use of wireless sensor networks, as service life is limited by onboard battery capacity. This constraint has driven research into power sensitive routing protocols, sleeping protocols, and even network architectures for minimizing data traffic [1, 13]. It would seem only natural to develop motion planning strategies with similar performance objectives.

S. Akella et al. (Eds.): Algorithmic Foundation of Robotics VII, STAR 47, pp. 441–456, 2008.
www.springerlink.com

In this vein, we propose a set of *motion planning strategies* that allow a mobile network to reconfigure to a new geometry while minimizing the total distance the nodes must travel, or the maximum distance that any node must travel. We believe a suite of strategies is critical due to the proliferation of non-standard sensor network architectures which are often implementation specific. As such, we provide centralized, computationally distributed, and decentralized approaches suitable for use with large-scale sensor network architectures. Each is computationally efficient, and without onerous communication overhead.

2 Related Work

Changes to the environment, mission objectives, and node failures are all factors that can contribute to need for reconfiguring a sensor network. However, topology changes can also be driven by performance objectives. For example, Cortes *et al* applied optimization based techniques to motion planning for improving network coverage [7]. Similarly, Zhang and Sukhatme investigated using motion to control node density [20]. The work of Hidaka *et al* investigated deployment strategies for optimizing localization performance [16], while the work of Butler and Rus was motivated by event monitoring using constrained resources [6]. Also worth noting is work in the areas of formation control [21], conflict resolution [17], and cooperative control [3]. A recent survey/tutorial outlining additional relevant work within each of these areas can be found in [11].

In contrast to these efforts, the focus of our work is efficient motion planning strategies suitable for large-scale networks. Given initial and objective network geometries, we determine how to optimally reposition each node in order to achieve the objective configuration while minimizing the distances that the nodes must travel. The objective positions can then be fed to appropriate controllers to drive the nodes to their desired destinations. When servo/actuator costs dominate the power budget, such approaches can dramatically improve the network mission life. We also emphasize applicability to large-scale systems. Our methods scale well in terms of both computational and message complexity to ensure that advantages gained through efficient motion planning are not compromised by excessive computation or routing requirements. Finally, we provide centralized, computationally distributed, and decentralized models to support the diverse array of WSN architectures.

3 The Motion Planning Problem

In developing our motion planning strategies, we consider the problem of having a multi-agent team transition to a new shape formation while minimizing either the total distance or maximum distance metric. For our purposes, we adopt the traditional definition of shape that is often employed in statistical shape analysis [10]:

Definition 1. *The shape of a formation is the geometrical information that remains when location, scale, and rotational effects are removed.*

Thus, formation shape is invariant under the Euclidean similarity transformations of translation, rotation and scale [10].

For brevity, in this paper we only consider operations in $SE(2)$ and refer the reader to [9] for details on obtaining optimal solutions in $\mathbb{R}^3$. Letting $Q = [q_1, \ldots, q_m]^T \in \mathbb{R}^{m\times 2}$ denote the concatenated coordinates of the objective shape formation with respect to some world frame $\mathcal{W}$ and letting $S = [s_1, \ldots, s_m]^T \in \mathbb{R}^{m\times 2}$ denote an instance (or an *icon*) of our objective shape with respect to some local frame $\mathcal{F}$, the shape of a robot formation can be represented as the set of equality constraints:

$$\begin{aligned} q_i^x - q_1^x &= \alpha\left(s_i^x \cos\theta - s_i^y \sin\theta\right) \\ q_i^y - q_1^y &= \alpha\left(s_i^x \sin\theta + s_i^y \cos\theta\right) \end{aligned} \tag{1}$$

for $i = 2, \ldots, m$. In this formulation, $\alpha \in \mathbb{R}_+$ and θ respectively denote the scale and orientation of the formation, while the (x, y) superscripts denote the specific Euclidean coordinate.

Without loss of generality, we can define the objective formation scale and orientation respectively as:

$$\alpha = \frac{\| q_2 - q_1 \|}{\| s_2 - s_1 \|} = \frac{\| q_2 - q_1 \|}{\| s_2 \|} \qquad \theta = \arctan\frac{q_2^y - q_1^y}{q_2^x - q_1^x} \tag{2}$$

The former equalities hold as we choose $s_1 \stackrel{\triangle}{=} O_{\mathcal{F}}$. From the latter, we obtain:

$$\cos\theta = \frac{q_2^x - q_1^x}{\| q_2 - q_1 \|} \qquad \sin\theta = \frac{q_2^y - q_1^y}{\| q_2 - q_1 \|} \tag{3}$$

Given these definitions, the non-convex constraints in (1) can be restated as the following set of linear equalities:

$$\begin{aligned} \| s_2 \| \, (q_i^x - q_1^x) - (s_i^x, s_i^y)^T (q_2 - q_1) &= 0, \qquad i = 3, \ldots, m \\ \| s_2 \| \, (q_i^y - q_1^y) - (s_i^y, s_i^x)^T (q_2 - q_1) &= 0, \qquad i = 3, \ldots, m \end{aligned} \tag{4}$$

These constraints are now convex, and they define the equivalence class of the full set of similarity transformations of the formation. Thus, if an objective shape Q and an icon S satisfy these constraints, the two shapes are equivalent under the Euclidean similarity transformations of translation, rotation and scaling. So, given an initial formation position $P = [p_1, \ldots, p_m]^T \in \mathbb{R}^{m\times 2}$, and an objective shape icon S, the problem becomes finding the set of objective positions $Q \sim S$ such that

1. $\max \| q_i - p_i \|$ is minimized for $i = 1, \ldots, m$ OR
2. $\sum_{i=1}^{k} \| q_i - p_i \|$ is minimized.

In other words, if the network were given an objective icon S, it must determine the objective positions for each node that minimizes the chosen metric, while ensuring the final shape formation $Q \subset \mathcal{W}$ is equivalent to S.

As the constraints are linear in Q, the problems can be modeled as the respective second-order cone programs (SOCPs)

$$\begin{array}{ll} \min\limits_{q,t_1} t_1 & \min\limits_{q,t} \sum\limits_{i=1}^{m} t_i \\ \text{s.t.} \ \| q_i - p_i \|_2 \leq t_1 & \text{s.t.} \ \| q_i - p_i \|_2 \leq t_i \\ \quad Aq = 0 & \quad Aq = 0 \end{array} \tag{5}$$

for $i = 1, \ldots, m$. Since the SOCPs are convex, a local minimum corresponds to a global minimum. This allows optimal solutions to be obtained through a variety of methods such as descent techniques or (more efficiently) by interior point methods (IPMs). While primal-dual IPMs represent perhaps the most efficient algorithms, we employ a simpler barrier IPM. It provides good computational complexity in practice, and as we shall see lends itself to a computationally distributed implementation.

Finally, in the interest of brevity, the results presented in this paper largely focus on the *mini-max* distance problem as defined in Equation 5 (left). It should be noted that similar results have been obtained for the total distance variation [9].

4 A Centralized Approach

Centralized approaches are appropriate for hierarchical network architectures such as the TENET [13]. For the motion planning problem, "master" nodes acting as cluster-heads would calculate the objective positions for the cluster and communicate these to supporting nodes in the network. While simple in design, the hierarchy requires that algorithms scale well computationally with the size of the network.

To address this, we solve the motion planning problem by adapting the logarithmic penalty-barrier approach outlined in [4]. Like other IPMs, the complexity is largely defined by solving a linear system of equations. In this case, Equality-constrained Newton's method (ENM) is used for internal minimization and the linear system is in KKT form. As solving this system provides a solution to the Newton step sub-problem, we accordingly refer to it as the "Newton KKT system." We show that by reformulating the SOCP, we can band the coefficient matrix to solve the system in $O(m)$ time via algorithms that exploit knowledge of matrix bandwidth. Furthermore, we show empirically that the total number of iterations required to reduce the duality gap to a desired tolerance is $O(1)$. The result is a simple IPM that in practice solves the motion planning (and similar) problems in $O(m)$ time.

4.1 Reformulating the Motion Planning Problem

The original *mini-max* motion planning problem can be restated in a relaxed form suitable for solving via the barrier approach. Conversion requires augmenting the objective function given in (5) with log-barrier terms corresponding to the problem's conic constraints as follows:

$$\begin{array}{l} \min\limits_{q,t_1} \tau_k t_1 - \sum\limits_{i=1}^{m} \log\left(t_1^2 - (q_i - p_i)^T (q_i - p_i)\right) \\ \text{s. t. } Aq = 0 \end{array} \tag{6}$$

where τ_k is the inverse log-barrier scaler for the kth iteration. Essentially, solving our SOCPs reduces to solving a sequence of convex optimization problems of this form, where after each iteration τ_{k+1} is chosen such that $\tau_{k+1} > \tau_k$.

4.2 Banding the Newton KKT System

During each iteration of the log-barrier approach, we aim to minimize the second-order Taylor approximation of our objective function as a function of the Newton step, δx, subject to $A\delta x = 0$. As a result, obtaining δx is equivalent to analytically solving the KKT conditions associated with this equality-constrained sub-problem. In other words, we must solve the following linear system of equations [4]:

$$\begin{bmatrix} H & A^T \\ A & 0 \end{bmatrix} \begin{bmatrix} \delta x \\ w \end{bmatrix} = \begin{bmatrix} -g \\ 0 \end{bmatrix} \tag{7}$$

where H and g respectively denote the evaluated Hessian and gradient of the objective function given in (6) at x, w is the corresponding dual variable for δx, and A is as previously defined. Solving (7) is the bottleneck of the algorithm; however, we will show that it can be solved very efficiently (*i.e.* in $O(m)$ time) by simply reposing the problem given in (6).

Noting that the coefficient matrix of (7) is symmetric indefinite, we employ Gaussian elimination with non-symmetric partial pivoting. The performance of this technique suffers significantly when the linear system in question features dense rows and/or columns due to fill-in [19]. In particular, the algorithm could yield a worst-case performance of $O(m^3)$ when solving an instance of (7) associated with the nominal problem formulation given in (6). To illustrate this point, we include Figure 1 (left) which shows the corresponding non-zero sparsity structure (*a.k.a.* the dot-plot) of the Newton KKT system. As the rows of system are permuted during reduction, the dense rows and columns respectively located in the upper-right and lower-left quadrants of (7) could introduce a solid sub-block of order $m \times m$, which itself would require $O(m^3)$ basic operations to reduce. Such a workload is highly impractical, especially when considering large-scale configurations that inherently feature 1000's of decision variables.

To address this issue, we present the following auxiliary formulation of (6) that facilitates transforming the Newton KKT system into a mono-banded form:

$$\begin{aligned}
\min_{q,t} \; & \tfrac{\tau_k}{m} \sum_{i=1}^{m} t_i - \sum_{i=1}^{m} \log\left(t_i^2 - (q_i - p_i)^T (q_i - p_i)\right) \\
\text{s. t.} \; & \| s_2 \| \, (q_i^x - d_j^x) - (s_i^x, s_i^y)^T (d_{j+1} - d_j) = 0, \quad i = 3, \ldots, m \\
& \| s_2 \| \, (q_i^y - d_j^y) - (s_i^y, s_i^x)^T (d_{j+1} - d_j) = 0, \quad i = 3, \ldots, m \\
& t_{i+1} = t_i, \quad i = 1, \ldots, m-1 \\
& d_{2i+1} = d_{2i-1}, \quad i = 1, \ldots, m-3 \\
& d_{2(i+1)} = d_{2i}, \quad i = 1, \ldots, m-3 \\
& d_i = q_i, \quad i \in \{1, 2\}
\end{aligned} \tag{8}$$

where $j = 2(i-3) + 1$.

Notice that the objective has changed from (6); however, we see that both forms are equivalent since:

$$\frac{\tau_k}{m} \sum_{i=1}^{m} t_i = \frac{\tau_k}{m} \sum_{i=1}^{m} t_1 = \left(\frac{\tau_k}{m}\right) m t_1 = \tau_k t_1 \tag{9}$$

where the first equality holds due to the equality constraints placed on t_i.

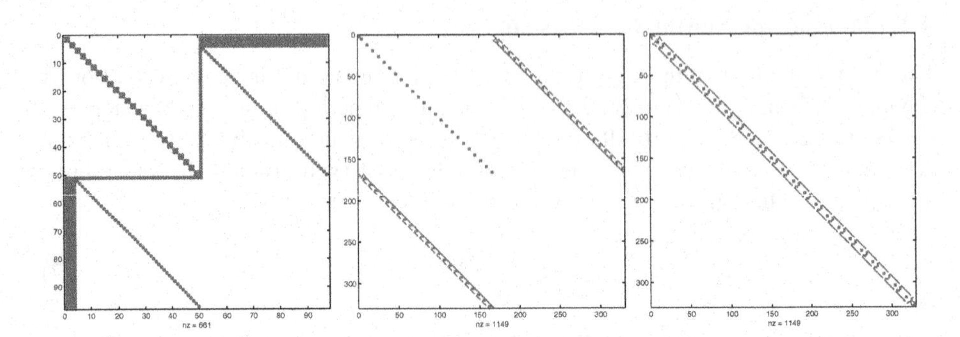

Fig. 1. (left) The nominal Newton KKT system sparsity structure for the *mini-max* motion planning problem in $SE(2)$. (center) Augmented Newton KKT system sparsity structure. (right) The banded system with lower and upper bandwidths of 8.

Given this augmented formulation, our claim is that the system can be made mono-banded. To show this, we begin by defining the nominal solution vector for the coefficient structure of (7) as follows:

$$\left[\delta\eta_1^T, \delta\eta_2^T, \delta\kappa_1^T, \ldots, \delta\kappa_{(m-2)}^T, \mu^T\right]^T \tag{10}$$

$$\delta\eta_i = \begin{bmatrix} \delta q_i \\ \delta t_i \end{bmatrix} \quad \delta\kappa_i = \begin{bmatrix} \delta d_{2(i-1)+1} \\ \delta d_{2(i-1)+2} \\ \delta\eta_{(i+2)} \end{bmatrix} \quad \mu = \begin{bmatrix} w_1 \\ \vdots \\ w_{7m-13} \end{bmatrix}$$

where the δ variables correspond to the primal Newton step components associated with each of the respective system variables.

Given the new objective, (9), and assuming the shape problem's solution vector permutation corresponds with (10), the Hessian for our problem is now

$$H = \begin{bmatrix} \psi_1 & \cdots & \cdots & 0 \\ \vdots & \psi_2 & & \vdots \\ \vdots & & \ddots & \vdots \\ 0 & \cdots & \cdots & \psi_m \end{bmatrix} \qquad \begin{array}{l} \psi_i = \nabla^2\phi(u_i, t_i),\ i \in \{1, 2\} \\ \\ \psi_i = \begin{bmatrix} 0_{4\times 4} & 0_{4\times 3} \\ 0_{3\times 4} & \nabla^2\phi(u_i, t_i) \end{bmatrix},\ i \in \{3, \ldots, m\} \end{array} \tag{11}$$

where $\nabla^2\phi(u_i, t_i)$ is defined as in [14] with $u_i = q_i - p_i$. Notice that this Hessian is block-diagonal and separable. This differs from the its nominal form, which features a dense row and column corresponding to the variable, t_1. This is evident by observing the upper-left quadrant (defined by H) of the KKT matrix given in Figure 1 (left).

Similarly, we can eliminate the dense columns and rows in A (and A^T) by introducing $2(m-2)$ auxiliary d_j variables along with their associated $4(m-3)$ equality constraints. Doing so allows us to rewrite (4) as the first two constraint sets given in (8). By reformulating the linear shape constraints in this fashion, we are now able to construct A as a *pseudo-banded* system. We say *pseudo-banded*, because the matrix is non-square and exhibits a band-like structure.

To show this, we begin by stating the constraint/row permutation that yields A in *pseudo-banded* form. We define the constraints associated with q_1 and q_2 as:

$$\begin{array}{ll} \varrho_1 \triangleq q_1^x = d_1^x & \\ \varrho_2 \triangleq q_1^y = d_1^y & \varrho_4 \triangleq q_2^x = d_2^x \\ & \varrho_5 \triangleq q_2^y = d_2^y \\ \varrho_3 \triangleq t_1 = t_2 & \end{array} \tag{12}$$

Similarly, for $3 \leq i \leq (m-1)$, we define the constraints associated with q_i as:

$$\begin{array}{ll} \varphi_{i_1} \triangleq \parallel s_2 \parallel (q_i^x - d_j^x) = (s_i^x, s_i^y)^T (d_{j+1} - d_j) & \varphi_{i_4} \triangleq d_{j+2}^x = d_j^x \\ \varphi_{i_2} \triangleq \parallel s_2 \parallel (q_i^y - d_j^y) = (s_i^y, s_i^x)^T (d_{j+1} - d_j) & \varphi_{i_5} \triangleq d_{j+2}^y = d_j^y \\ \varphi_{i_3} \triangleq t_i = t_{i-1} & \varphi_{i_6} \triangleq d_{j+3}^x = d_{j+1}^x \\ & \varphi_{i_7} \triangleq d_{j+3}^y = d_{j+1}^y \end{array} \tag{13}$$

where j is as previously defined.

With q_m, we associate the remaining three constraints:

$$\begin{array}{l} \varphi_{m_1} \triangleq \parallel s_2 \parallel (q_m^x - d_j^x) = (s_m^x, s_m^y)^T (d_{j+1} - d_j) \\ \varphi_{m_2} \triangleq \parallel s_2 \parallel (q_m^y - d_j^y) = (s_m^y, s_m^x)^T (d_{j+1} - d_j) \\ \varphi_{m_3} \triangleq t_m = t_{m-1} \end{array} \tag{14}$$

where j is as previously stated with $i = m$.

Given these definitions, we provide the following row permutation for A, which yields the *pseudo-banded* form that appears in the lower-left (and upper-right) quadrant of Figure 1 (center):

$$\left[\vartheta^T, \varkappa_1^T, \ldots, \varkappa_{(m-1)}^T, \varsigma^T\right]^T \tag{15}$$

$$\vartheta = \begin{bmatrix} \varrho_1 \\ \varrho_2 \\ \vdots \\ \varrho_5 \end{bmatrix} \quad \varkappa_i = \begin{bmatrix} \varphi_{(i+2)_1} \\ \varphi_{(i+2)_2} \\ \vdots \\ \varphi_{(i+2)_7} \end{bmatrix} \quad \varsigma = \begin{bmatrix} \varphi_{m_1} \\ \varphi_{m_2} \\ \varphi_{m_3} \end{bmatrix}$$

Notice that all of the primal constraints defined in (8) have been included.

Given the definitions of A and H, the mono-banded form of (7) can now be constructed. Symmetrically applying the permutation that yields the following Newton KKT system solution vector ordering:

$$\left[\lambda^T, \xi_1^T, \ldots, \xi_{(m-3)}^T, \chi^T\right]^T \tag{16}$$

$$\lambda = \begin{bmatrix} \delta q_1 \\ \delta t_1 \\ w_1 \\ \vdots \\ w_5 \\ \delta q_2 \\ \delta t_2 \end{bmatrix} \quad \xi_i = \begin{bmatrix} \delta d_{2(i-1)+1} \\ \delta d_{2(i-1)+2} \\ w_{6+7(i-1)} \\ \vdots \\ w_{12+7(i-1)} \\ \delta q_{(i+2)} \\ \delta t_{(i+2)} \end{bmatrix} \quad \chi = \begin{bmatrix} \delta d_{2m-5} \\ \delta d_{2m-4} \\ w_{7m-15} \\ w_{7m-14} \\ w_{7m-13} \\ \delta q_m \\ \delta t_m \end{bmatrix}$$

produces a mono-banded coefficient structure having a total bandwidth of 17. Simply using the standard Reverse Cuthill-McKee (RCM) reordering algorithm [8] would yield a bandwidth 47% larger than that obtained with our approach.

In Figure 1 (center), we show the "augmented" Newton KKT system constructed from the Hessian given by (11) and the linear constraint set given in (8). The latter is permuted according to (15). Taking the coefficient structure of (7) in this form and symmetrically permuting its rows and columns according to (16) yields the mono-banded system appearing in Figure 1 (right). The system corresponds to a team of 25 agents dispersed in $SE(2)$. It can now be solved in $O(m)$ using a band-diagonal LU-based solver [18].

4.3 Total Complexity

Assuming a fixed duality gap reduction, the iteration complexity of the barrier approach grows as $O(\sqrt{m})$ [4]. Noting that the per-iteration complexity is defined by solving the mono-banded Newton KKT system as well as computing banded matrix-vector products, the total number of basic operations required to achieve optimality grows only as $O(m^{1.5})$. The generality of this result should not go overlooked as the bound applies to any SOCP that yields a mono-banded Newton KKT system. This includes regulated cases of the motion planning problem in $\mathbb{R}^3$ [9].

4.4 Performance in Practice

To gauge the performance of the framework in practice, we assumed both a fixed duality gap reduction and barrier parameter μ as suggested in [4]. A total of 5,000 random instances of the motion planning problem were solved using an implementation of the barrier algorithm. The objective was to minimize the total distance traveled by the team. Values of m were considered between 10 and 1000 at intervals of 10, where m denotes configuration size. Our implementation was validated by comparing obtained solutions against those of the MOSEK industrial solver [15]. All problems were solved using a standard desktop PC having a 3.0 GHz Pentium 4 processor and 2.0 GB of RAM.

Figure 2 (left) shows the results of these trials. Each data point corresponds to the mean of 50 samples with the error bars corresponding to a single standard deviation. The trend indicates that the total number of Newton iterations remains constant (for $m \gtrapprox 170$). This result in tandem with the linear per-iteration complexity established earlier shows that in practice the motion planning problem is solvable in $O(m)$ time.

These results are associated with the simple barrier method outlined in [4]; however, it should be noted that empirical results show that similar performance can be achieved using more sophisticated solvers. Figure 2 (right) shows the CPU time required by the MOSEK industrial solver for configuration sizes having up to 2000 nodes. Each point corresponds to the mean obtained from solving 50 randomly generated motion planning SOCPs (total distance metric). This data shows that for a configuration of 2000 nodes in $SE(2)$ an optimal solution can be obtained in only 0.28 seconds. Furthermore, the CPU time clearly scales as $O(m)$ with linear regression analysis revealing $r^2 = 0.9869$, where r is the associated correlation coefficient. Results obtained using the solver also

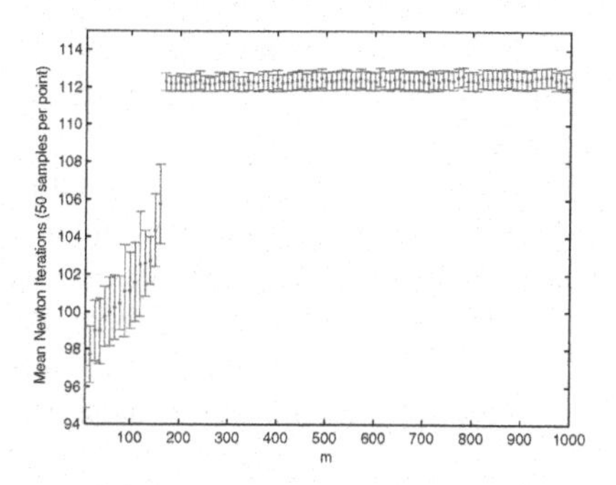

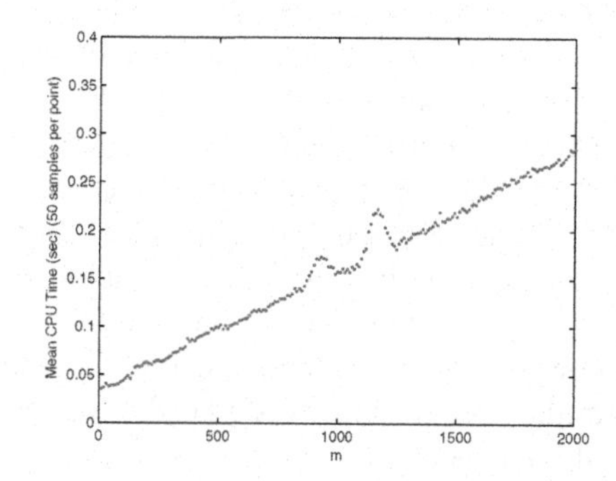

Fig. 2. (left) The mean number of Newton iterations required to solve the motion planning problem (total distance metric) as a function of configuration size, m. For $m \gtrapprox 170$ the number of iterations appears constant using the log-barrier approach. Error bars signify a single standard deviation. (right) The mean CPU-utilization time required to solve the motion planning problem using the MOSEK solver. The trend is strongly linear, with $r^2 = 0.9869$.

indicate that an optimal solution can typically be found in less than 12 iterations - regardless of configuration size.

5 A Computationally Distributed Approach

Our centralized solution features both a band-diagonal linear system as well as a separable objective function (w.r.t. the variables each node introduces). We shall leverage these characteristics to distribute the computational workload evenly across the network. The resulting $O(1)$ expected per-node workload will enable our approach to be employed by a significantly less sophisticated class of processors, or to significantly larger-scale networks. We now define a hierarchical, cluster-based architecture for achieving this objective.

5.1 Architectural Overview

Our paradigm solves convex optimization problems in the context of a cluster-based network architecture under the direction of some *root* node(s). The *root* is responsible for orchestrating the solve process; thus, it maintains a global state reflecting the status of the distributed computation. It is responsible for performing such tasks as initializing the network and determining when the solve is complete. Although the *root* maintains a "global perspective", its data view is primarily limited to that which affects the computation of its associated decision variables. The only exception is when it requests data needed to manage the IPM solve process. For instance, when it requests Newton decrement data.

At the *root*'s disposal are the remaining nodes in the network, which we term the *secondary* peers. These nodes are considered *secondary*, because they serve only as a distributed memory pool and a computational engine for the *root* during the solve process; individually, they lack a global view of the solver and only manage data relevant to their computations. They wait until a data request is received originating from the *root* before transitioning into a new state of computation.

To reduce the communication overhead, we define the architecture to have a hierarchical scheme based upon network clusters. The role of clusterheads is to ensure that each request of the *root* is satisfied at the lowest level. Sub-nodes treat their clusterhead as a local accumulator and forward the requested information to that node where it is aggregated before being passed up the hierarchy, ultimately to the *root*. The result is that the *root* (and all clusterheads) only need to send a constant number of messages with each data request.

5.2 Distributing and Solving the Newton KKT System

Given the objective function and Hessian are separable, implementing a distributed Newton decrement and line search computation (see [4]) reduces to having each node pass its contribution to the greater value up the cluster hierarchy at request. For this reason, along with the fact that the per-iteration complexity is defined by solving the Newton KKT system, we focus our discussion on distributing the LU solver. As will be seen, we can effectively distribute the process while providing per-node computation, storage, and overlay message complexities of $O(1)$.

To distribute the Newton KKT system, $K \in \mathbb{R}^{y \times y}$, among m nodes, we make the assumption that the system is band-diagonal with respective upper and lower bandwidths of b_u and b_l. Additionally, we assume the matrix is represented in its equivalent compact form, K_c, where $K_c \in \mathbb{R}^{y \times (b_l + b_u + 1)}$ [18]. We respectively denote the corresponding right-hand-side and permutation vectors as $b \in \mathbb{R}^y$ and $p \in \mathbb{Z}^y_+$.

Adopting this representation for K, we adapt the LU-based solver with partial pivoting outlined in [18]. Distributing this algorithm, we begin by assigning the ith node, n_i, a sub-block $K^i_c \subset K_c$. Additionally, each n_i manages the corresponding sub-vectors $b_i \subset b$ and $p_i \subset p$. To illustrate the decomposition, we provide Figure 3, which shows the distribution of K_c for a team of 5 nodes in $SE(2)$ solving the total distance problem. Given the dependencies between the equations in the linear system, devising a completely concurrent solution is not feasible. Thus, we assume the decomposition and subsequent solves are done one node at a time in a "pass-the-bucket" fashion, where node n_i decomposes K^i_c and then hands the process off to node n_{i+1}. This process continues iteratively until decomposition is complete.

Decomposition

During decomposition, the algorithm employs partial pivoting by searching at most b_l sub-diagonal elements in order to identify one with greater magnitude. This implies that a node in our WSN that is performing its respective decomposition may only need information pertaining to at most b_l rows, which can be buffered at one or more peers. In the worst case scenario, where each node only manages a single row, node n_i may have to query up to b_l of its peers. With this result in mind, we offer the following theorem:

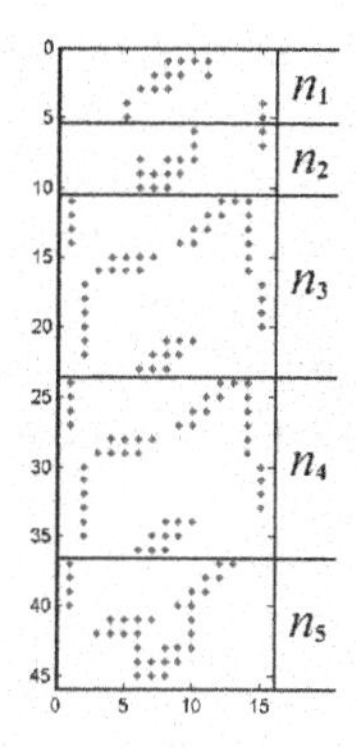

Fig. 3. A non-zero dot-plot illustrating the decomposition of the compact Newton KKT system (i.e. K_c) for a configuration of 5 nodes in $SE(2)$ minimizing the total distance metric. For this problem, $b_l = b_u = 7$. Notice that the middle $(m-3)$ nodes (i.e. n_3 and n_4) are assigned sub-blocks with identical structure.

Theorem 1. *Let $i \in \mathcal{I} = \{1, \ldots, m\}$ and let $K_c^j \in \mathbb{R}^{u_j \times (b_l + b_u + 1)}$, $u_j \in \mathbb{Z}_+$ for $j = 1, \ldots, m$. Define $\psi(i) : \mathcal{I} \to \mathbb{Z}_+$ as a mapping to the number of nodes that have to be contacted by n_i during the decomposition of K_c^i. The following holds:*

$$\psi(i) \leq \phi(b_l, u_1, \ldots, u_m) = \left\lceil \frac{b_l}{\left(\min\limits_{i \in \{1, \ldots, m\}} u_i \right)} \right\rceil$$

Proof. By contradiction.
Assume $\psi(i) > \phi(b_l, u_1, \ldots, u_m)$. Choosing $u_i = 1, \forall i \in \{1, \ldots, m\}$, we see:

$$\psi(i) > \left\lceil \frac{b_l}{1} \right\rceil = b_l$$

However, it must hold that $\psi(i) \leq b_l$, since n_i will only ever require data about b_l rows during the decomposition of K_c^i. $\rightarrow\leftarrow$

Using the data it acquired from its $\psi(i) \leq \phi$ supporting peers (in particular, $n_{i+1}, \ldots,$ $n_{i+\psi(i)}$), n_i performs standard LU decomposition on K_c^i. Upon completion, it sends an update to each of the supporting nodes. The update contains the modified row(s) information and the adjusted permutation vectors corresponding to the changes it made with respect to the data the recipient provided. This allows each supporting node to update its cache before the process is handed off to n_{i+1}.

Forward and Backward Substitution

Similar to decomposition, both forward and backward substitution are done in an iterative manner. In both cases, the active node, n_i, will have to communicate with a small (bounded) number of its peers. During forward substitution, it will have to acquire information from each of the $\psi(i)$ nodes that provided it with data during the decomposition. This differs from the backward substitution phase, which may require n_i to communicate with up to 2ϕ nodes. The additional messaging is introduced via the upper triangular factor, U, having a bandwidth now constrained by $(b_u + b_l + 1)$, which is latent to the use of partial pivoting [12]. Since n_m is the last node to perform both decomposition and forward substitution, it is responsible for signaling the start of a new phase in the LU-solver process.

Message and Storage Complexity

For simplicity, the assumption is made that whenever n_i requests information from any node, the data is received in a single message. This assumption is reasonable, because the amount of information (including row data) that has to be shared between any two nodes is a function of b_u and b_l, which are both independent of configuration size. As such, the number of messages required to transmit said data is also constant. Noting that information is delivered upon request, the total number of messages sent by n_i is:

$$O(2\psi(i) + 2\psi(i) + 4\psi(i) + \gamma(i)) \equiv O(8\phi + 3) \equiv O(1) \tag{17}$$

where $\gamma(i) \leq 3$ is a mapping to the number of hand-off/signal messages sent by n_i.

As all nodes send $O(1)$ messages during the solve, the aggregate message complexity for the distributed LU process is $O(m)$. Recalling that the number of Newton iterations will be $O(1)$ in practice, we expect that no more than $O(m)$ messages will be generated in the overlay network. Furthermore, since n_i manages some fixed-size K_c^i, b_i, p_i, and row data received by as many as 2ϕ peers, per-node storage is $O(1)$.

5.3 Experimental Results

To demonstrate our approach, we implemented the distributed framework on a team of six Sony Aibos and charged the team with transitioning to a delta formation. The

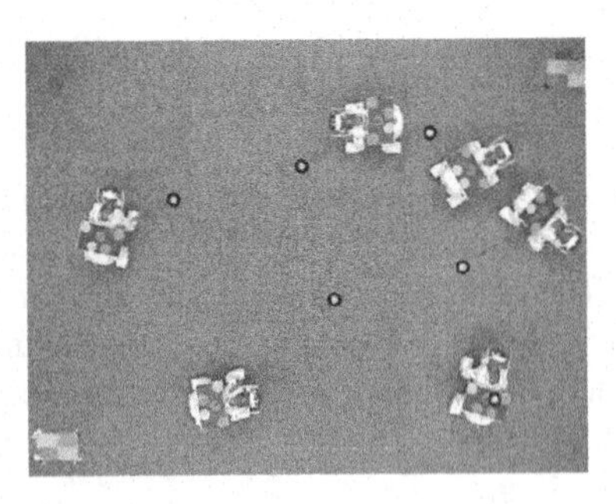

Fig. 4. (left) An initial dispersion of 6 Aibos, along with overlaid lines/points mapping each to its computed optimal position. (right) The Aibos after reconfiguring to the desired delta shape formation. All computations were done in a distributed fashion, with each dog being responsible for computing its optimal position and local control inputs.

objective was to minimize the total distance traveled by team members. Each Aibo was outfitted with a unique butterfly pattern [5] that was tracked via an overhead camera system serving as an indoor "GPS". Figure 4 (left) shows the initial configuration, along with lines mapping each to its computed optimal position, while Figure 4 (right) shows the Aibos after transitioning to the optimal shape configuration.

6 A Decentralized Hierarchical Approach

For our decentralized approach, we assume a hierarchical model whereby a small number of *leader* nodes acting as exemplars solve the motion planning problem. This allows the remaining *follower* nodes to infer their objective positions through local observations. Such a model is attractive to not only hierarchical network architectures [13], but also models where minimizing data communication is a primary objective [1]. For our decentralized approach, we make the following assumptions.

1. Each node knows the objective shape icon S for the network.
2. Leader nodes (individually or collectively) know the current network shape.
3. Follower nodes have *no knowledge* of the current network shape.
4. Follower nodes can identify their neighbors and measure their *relative* position.
5. Follower nodes can observe the *relative* heading of their immediate neighbors.

6.1 An $O(1)$ Decentralized Solution

Key to this approach is the realization that although the optimization problem in (5) includes $2m$ decision variables (corresponding to the m robot positions), the feasible set is constrained to the equivalence class of the full set of similarity transformations for the objective formation shape. More concisely: there are only 4 degrees of freedom in determining a node's objective position on the plane which correspond to the translation, rotation, and scale of the objective shape.

As the leader nodes have knowledge of the current and objective shapes, they can solve for their objective positions using either of the approaches outlined in Sections 4-5. Follower nodes have more constrained knowledge, and as a result are incapable of estimating their objective positions. However, an observation of the heading ω_l of leader l introduces an additional constraint on the objective shape of the form $(q_l - p_l)^T(\sin\omega_l, -\cos\omega_l) = 0$ where all measurements are relative to the follower's coordinate frame $\mathcal{F}$. If the headings of 4 leader nodes can be observed, the motion planning problem becomes fully constrained via the equality constraints in (4). Perhaps more significant is that the problem can now be solved by the follower nodes in a decentralized fashion, and in $O(1)$ time regardless of formation size.

To see this, recall that in addition to this heading constraint, each robot imposes two additional equality constraints on the objective network shape as shown in (4). With 4 leader nodes and 1 follower node, this corresponds to a total of 4 bearing and 10 shape constraints over 14 decision variables. However, noting that the shape index (*not* coordinate) assignments are arbitrary, the follower node can designate itself as the first index corresponding to the 3-tuple $\{p_1, q_1, s_1\}$ and associate one of the observed leaders with $\{p_2, q_2, s_2\}$. This eliminates the associated shape constraints for these two nodes, and reduces the set to

$$\begin{aligned} (q_l - p_l)^T(\sin\omega_l, -\cos\omega_l) &= 0, \quad & l = 2,\ldots,5 \\ \| s_2 - s_1 \| (q_l^x - q_1^x) - (s_l^x, s_l^y)^T(q_2 - q_1) &= 0, \quad & l = 3,\ldots,5 \\ \| s_2 - s_1 \| (q_l^y - q_1^y) - (s_l^y, s_l^x)^T(q_2 - q_1) &= 0, \quad & l = 3,\ldots,5 \end{aligned} \tag{18}$$

where $l \in \{2\ldots5\}$ now corresponds to the set of observed leaders. The constraint set is linear in q, and can be written in the form $A\hat{q} = b$, where the solution vector $\hat{q} \subset q$ is the objective positions of follower and 4 observed leader nodes. It is a linear system of 10 equations in 10 unknowns, and is readily solvable via Gaussian elimination techniques.

Thus, each follower node can solve for its objective position (as well as its neighbors) so long as the *relative* position and headings of 4 neighbors can be observed. This is akin to biological systems (*e.g.* schools of fish, flocks of birds, *etc.*) capable of complex formation changes using only local sensor feedback. Furthermore, the solution is obtained from solving an $O(1)$ sized (10×10) linear system of equations - regardless of the number of nodes in the network. The assumption of knowledge of the objective shape does however require $O(m)$ storage for each node.

It should also be noted that after solving for its objective position, each follower is "promoted" to leader status. As it migrates to its objective position, its heading can be observed by other follower nodes to solve their own decentralized problem. So, while in practice the actual number of leader nodes will be a function of the sensor network topology, in theory only 4 are *necessary*. This is illustrated in Figure 5.

6.2 Simulation Results

Figure 5 models the initial deployment of a sensor network. The objective configuration was a {4,4} tessellation on the plane with a tiling size of 10 meters. Unfortunately, positional errors introduced during deployment - modeled as Gaussian noise $\sim N(0, \sigma_x = \sigma_y = 7.5)$ - result in a significantly different geometry (Figure 5a). To compensate for these errors, four leader nodes (red circles) solve the motion planning problem, and begin migrating to their objective positions. Relative sensor measurements allow the remaining follower nodes (blue triangles) to solve for their objective positions in decentralized fashion. The propagation of decentralized solutions through the network is reflected in Figure 5b. The decentralized trajectories that minimize the maximum distance that any node must travel, and the optimal network configuration achieving the desired shape are shown in Figures 5c-d. It was assumed that the sensing range of each node was 25 meters.

Note that in this case, the orientation of the shape was not constrained. If a fixed orientation was desired (*e.g.*, orthogonal to the $x-y$ axes), the number of degrees of freedom would be reduced to 3 - as would the number of observations required to solve the decentralized problem. Fixing the scale would simplify the problem even further, requiring only 2 observations for each decentralized node solution. We should also emphasize that although in this example the decentralized solution was able to propagate through the entire network using the minimum number of leader nodes, this will *not*

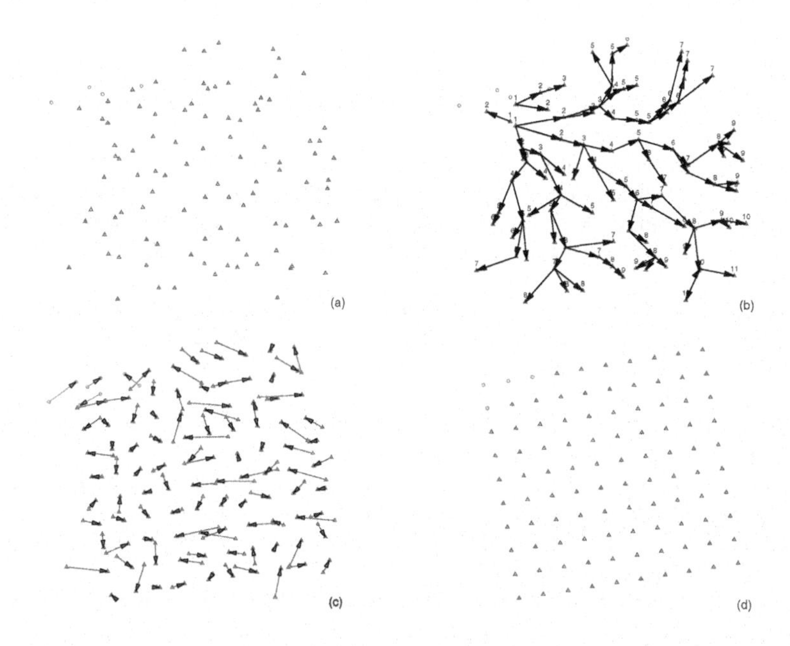

Fig. 5. Decentralized Motion Planning: (a) The initial network configuration with leader (red circle) and follower (blue triangle) nodes. (b) Evolution of the decentralized solution. (c) Node trajectories (d) Final network configuration achieving the desired $\{4,4\}$ tessellation.

typically be the case. More than likely, a small number of leader nodes will be associated with disjoint clusters in the network.

7 Discussion

In this paper, we developed a set of motion planning strategies suitable for large-scale sensor networks. These solve the problem of reconfiguring the network to a new shape while minimizing either the total distance traveled by the nodes or the maximum distance traveled by any node. The centralized approach runs in $O(m)$ time in practice through banding the Newton KKT system. The distributed approach reduces the expected per-node workload to $O(1)$ in exchange for $O(1)$ messages per-node in the overlay network. Finally, we derived a decentralized, hierarchical approach whereby follower nodes are able to solve for their objective positions in $O(1)$ time from observing the headings of a small number of leader nodes.

We are currently extending these results to a more general motion planning framework. To achieve this, issues such as collision/obstacle avoidance will have to be addressed. The latter is a particularly challenging task, as the presence of obstacles introduces concave constraints on the feasible set, and the resulting problem is no longer solvable as a SOCP. We hope that randomization and convex restriction techniques will still allow the problem to be solved for real-time applications.

References

1. Compass: Collaborative multiscale processing and architecture for sensornetworks, http://compass.cs.rice.edu/
2. Networking technology and systems (NeTS). NSF Solicitation 06-516 (December 2005)
3. Bachmayer, R., Leonard, N.E.: Vehicle networks for gradient descent in a sampled environment. In: Proc. IEEE Conf. on Decision and Control, Las Vegas (December 2002)
4. Boyd, S., Vandenberghe, L.: Convex Optimization. Cambridge Unviersity Press, Cambridge (2004)
5. Bruce, J., Veloso, M.: Fast and accurate vision-based pattern detection and identification. In: IEEE International Conference on Robotics and Automation (May 2003)
6. Butler, Z., Rus, D.: Event-based control for mobile sensor networks. IEEE Pervasive Computing 2(4), 10–18 (2003)
7. Cortés, J., Martínez, S., Karatas, T., Bullo, F.: Coverage control for mobile sensing networks. IEEE Trans. on Robotics and Automation 20(2), 243–255 (2004)
8. Cuthill, E., McKee, J.: Reducing the bandwidth of sparse symmetric matrices. In: Proceedings of the 1969 24th national conference, New York, USA, pp. 157–172 (1969)
9. Derenick, J., Spletzer, J.: TR LU-CSE-05-029: Optimal shape changes for robot teams. Technical report, Lehigh University (2005)
10. Dryden, I.L., Mardia, K.V.: Statistical Shape Analysis. John Wiley and Sons, Chichester (1998)
11. Ganguli, A., Susca, S., Martínez, S., Bullo, F., Cortés, J.: On collective motion in sensor networks: sample problems and distributed algorithms. In: Proc. IEEE Conf. on Decision and Control, Seville, Spain, pp. 4239–4244 (December 2005)
12. Golub, G.H., Loan, C.F.V.: Matrix computations, 3rd edn. Johns Hopkins University Press, Baltimore (1996)
13. Govindan, R., et al.: Tenet: An architecture for tiered embedded networks. Technical report, Center for Embedded Networked Sensing (CENS) (November 2005)
14. Lobo, M., Vandenberghe, L., Boyd, S., Lebret, H.: Applications of second-order cone programming. Linear Algebra and Applications, Special Issue on Linear Algebra in Control, Signals and Image Processing (1998)
15. MOSEK ApS. The MOSEK Optimization Tools Version 3.2 (Revision 8) User's Manual and Reference, http://www.mosek.com
16. Mourikis, A., Roumeliotis, S.: Optimal sensing strategies for mobile robot formations: Resource constrained localization. In: Robotics: Science & Sys., pp. 281–288 (June 2005)
17. Ögren, P., Leonard, N.: A tractable convergent dynamic window approach to obstacle avoidance. In: IEEE/RSJ IROS, Lausanne, Switzerland, vol. 1 (October 2002)
18. Press, W., et al.: Numerical Recipes in C. Cambridge University Press, Cambridge (1993)
19. Saad, Y.: Iterative Methods for Sparse Linear Systems. Society for Industrial and Applied Mathematics. Philadelphia, PA, USA (2003)
20. Zhang, B., Sukhatme, G.S.: Controlling sensor density using mobility. In: The Second IEEE Workshop on Embedded Networked Sensors, pp. 141–149 (May 2005)
21. Zhang, F., Goldgeier, M., Krishnaprasad, P.S.: Control of small formations using shape coordinates. In: Proc. IEEE Int. Conf. Robot. Automat., Taipei, vol. 2 (September 2003)

Asymptotically Optimal Kinodynamic Motion Planning for Self-reconfigurable Robots

John H. Reif[1] and Sam Slee[1]

Department of Computer Science, Duke University, Durham, NC, USA
{reif,sgs}@cs.duke.edu

Abstract. Self-reconfigurable robots are composed of many individual modules that can autonomously move to transform the shape and structure of the robot. In this paper we present a kinodynamically optimal algorithm for the following "x-axis to y-axis" reconfiguration problem: given a horizontal row of n modules, reconfigure that collection into a vertical column of n modules. The goal is to determine the sequence of movements of the modules that minimizes the movement time needed to achieve the desired reconfiguration of the modules. Prior work on self-reconfigurable (SR) robots assumed a constant velocity bound on module movement and so required time linear in n to solve this problem.

In this paper we define an abstract model that assumes unit bounds on various physical properties of modules such as shape, aspect ratio, mass, and the maximum magnitude of force that an individual module can exert. We also define concrete instances of our abstract model similar to those found in the prior literature on reconfigurable robots, including various examples where the modules are cubes that are attached and can apply forces to neighboring cubes. In one of these concrete models, the cube's sides can contract and expand with controllable force, and in another the cubes can apply rotational torque to their neighbors. Our main result is a proof of tight $\Theta(\sqrt{n})$ upper and lower bounds on the movement time for the above reconguration problem for concrete instances of our abstract model.

This paper's analysis characterizes optimal reconfiguration movements in terms of basic laws of physics relating force, mass, acceleration, distance traveled, and movement time. A key property resulting from this is that through the simultaneous application of constant-bounded forces by a system of modules, certain modules in the system can achieve velocities exceeding any constant bounds. This delays modules with the least distance to travel when reconfiguring in order to accelerate modules that have the farthest to travel. We utilize this tradeoff in our algorithm for the x-axis to y-axis problem to achieve an $O(\sqrt{n})$ movement time.

1 Introduction

This paper develops efficient algorithms by treating reconfiguration as a kinodynamic planning problem. Kinodynamic planning refers to motion planning problems subject to simultaneous kinematic and dynamics constraints [2]. To that end, in this paper we define an abstract model for modules in SR robots.

S. Akella et al. (Eds.): Algorithmic Foundation of Robotics VII, STAR 47, pp. 457–472, 2008.
springerlink.com

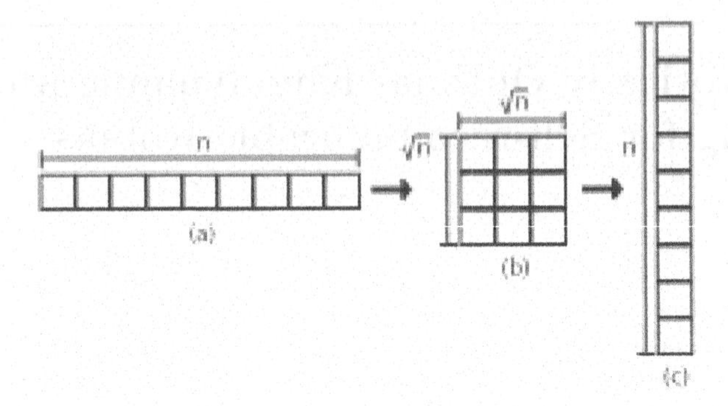

Fig. 1. The x-axis to y-axis problem: transforming a row of modules into a column. Our algorithm uses an intermediate step of forming a square.

This extends the definitions of previous models [3, 4] by setting fixed unit bounds on a module's shape, aspect ratio, mass, and the force it can apply, among other requirements.

To exhibit the significance of these bounds we consider the following "x-axis to y-axis" reconfiguration problem. Given a horizontal row of n modules, reconfigure that collection into a vertical column of n modules. This simple problem, illustrated in Figure 1, provides a worst-case example in that $\Omega(n)$ modules must move $\Omega(n)$ module lengths to reach any position in the goal configuration regardless of that goal column's horizontal placement. Any horizontal positioning of the vertical column along the initial row configuration is deemed acceptable in our treatment of the problem in this paper.

We define the *movement time* of a reconfiguration problem to be the time taken for a system of n modules to reconfigure from an initial configuration to a desired goal configuration. Modules are assumed to be interchangeable so the exact placement of a given module in the initial or goal configuration is irrelevant. An implicit assumption in various prior papers on reconfiguring robotic motion planning [3, 5, 6] is that the modules are permitted only a fixed unit velocity. This assumption is not essential to the physics of these systems and has constrained prior work in the area.

For the x-axis to y-axis problem, if only one module in the SR robot is permitted to move at a given time and with only a fixed unit velocity, a lower bound of $\Omega(n^2)$-time is clear [5]. Allowing concurrent movement of modules, a movement time of $O(n)$ is possible while still keeping all modules connected to the system and keeping unit velocities. However, we observe that faster reconfiguration is possible if we do not restrict modules to move at a fixed, uniform pace.

Another major principle that we use, and that has seen use in prior reconfiguration algorithms [1, 7], is what we refer to as the principle of *time reversal.* This principle is simply that executing reconfiguration movements in reverse is always possible, and they take precisely the same movement time as in the forward direction. This is, of course, ignoring concerns such as gravity. Otherwise an example of rolling a ball down a hill would require less time or less force than moving that ball back up the hill. We ignore gravity and use this principle extensively in work of this paper.

Before continuing further, it will be useful to define some notation that we will use throughout the remainder of this paper. As given above, let n denote the number of modules in the SR robot undergoing reconfiguration. In the initial row configuration of the x-axis to y-axis problem, let the modules be numbered $1, \ldots, n$ from left to right. Let $x_i(t)$ be the x-axis location of module i at time t. Similarly, let $v_i(t)$ and $a_i(t)$ be the velocity and acceleration, respectively, of module i at time t. For simplicity, the analysis in our examples will ignore aspects such as friction or gravity. The effect is similar to having the reconfiguration take place with ideal modules while lying flat on a frictionless planar surface.

In the following Section 2, we differentiate between different styles of self-reconfigurable robots and survey related work in the field. We begin introducing the work of this paper in Section 3 by giving our abstract model for SR robot modules. Given the bounds set by this model, Section 4 references physics equations that govern the movement of modules and define what reconfiguration performance is possible. To explain our algorithm, in Section 4 we also begin with a 1-dimensional case of n masses represented as a row of n points with a separation of 1 unit between adjacent points, for some unit of distance measure. The points will then contract so that they have only 1/2 unit separation. After showing that this contraction takes $O(\sqrt{n})$ movement time while still satisfying the bounds of our abstract model, we then immediately extend this result to a matching example using a known physical architecture for modules.

Section 5 will then extend the result to a contraction/expansion case in 2 dimensions while maintaining the same time bound. This will then lead to a $O(\sqrt{n})$ movement time algorithm for the x-axis to y-axis reconfiguration problem. This algorithm recursively uses the 1-dimensional contraction operation and uses the reversible process of transforming the initial n module row into an intermediate stage $\sqrt{n} \times \sqrt{n}$ cube. The process is then reversed to go from the cube to the goal column configuration. Section 6 then matches this with a $\Omega(\sqrt{n})$ lower bound for the 1-dimensional example and the x-axis to y-axis problem, showing both cases to be $\Theta(\sqrt{n})$. Finally, the conclusion in Section 7 summarizes the results of this paper. Some proofs are omitted due to length requirements. A version of this paper including all proofs can be found at: `www.cs.duke.edu/~sgs/publications/reifsleeWAFR06.pdf` .

2 Related Work

When developing models and algorithmic bounds for self-reconfigurable (SR) robots (also known as metamorphic robots [8, 4, 5]) it is important to note the style of SR robot we are dealing with. Two of the main types of SR robots are closed-chain style robots and lattice style robots. Closed-chain SR robots are composed of open or closed kinematic chains of modules. Their topology is described by one-dimensional combinatorial topology. [1] To reconfigure these modules are required to swing chains of other modules to new locations. One such implementation of this design by Yim et al. has had several demonstrations of locomotion [9].

For the other major style, lattice or substrate SR robots attach together only at discrete locations to form lattice-like structures. Individual modules typically move by walking along the surfaces formed by other modules. The hardware requirements for this style of module are more relaxed than those for closed-chain style systems. Here individual modules need only be strong enough to move themselves or one or two neighbors. Closed-chain style modules typically must be strong enough to swing long chains of other modules. The models and algorithms presented in this paper are meant for lattice style modules.

Previous work by several research groups has developed abstract models for lattice style SR robots. One of the most recent is the sliding cube model proposed by Rus et al. [3]. As the name implies, this model represents modules as identical cubes that can slide along the flat surfaces created by lattices of other modules. In addition, modules have the ability to make convex or concave transitions to other orthogonal surfaces. In this abstraction, a single step action for a module would be to detach from its current location and then either transition to a neighboring location on the lattice surface, or make a convex or concave transition to another orthogonal surface to which it is next. Transitions are only made in the cardinal directions (no diagonal movements) and for a module to transition to a neighboring location that location must first be unoccupied. Most architectures for lattice style SR robots satisfy the requirements of this model.

One such physical implementation is the compressible unit or expanding cube design [6, 7]. Here an individual module can expand from its original size to double its length in any given dimension, or alternatively compress to half its original length. Neighbor modules are then pushed or pulled by this action to generate movement and allow reconfiguration. This particular module design is relevant because it provides a good visual aide for the algorithms we present in this paper. For this purpose we also add the ability for expanding cubes to slide relative to each other. Otherwise a single row of these modules can only exert an expanding or contracting force along the length of that row and is unable to reconfigure in 2 dimensions. Prior work has noted that while individual expanding cube modules do not have this ability, it can be approximated when groups of modules are treated as atomic units [6, 7].

3 Abstract Model

We now begin introducing our results for this paper by defining our abstract model for SR robot modules. It generalizes many of the requirements and properties of reconfigurable robots and in particular, the properties of the sliding cube model and the expanding cube hardware design described in the previous section. Our abstract model explicitly states bounds on physical properties such as the size, mass, and force exerted for each module. These properties will be utilized by the algorithms and complexity bounds that later follow. The requirements of our model are described by the following set of axioms.

- Each module is assumed to be an object in 3D with either (i) fixed shape and size or (ii) a limited set of geometric shapes that it can acquire. In either case, the module also has a constant bound on its total volume and the aspect ratio of its shape.
- Each module can latch onto or grip adjacent modules and apply to these neighboring modules a force or torque.
- Generally, modules are all connected either directly or indirectly at any time.
- For each module there is a constant bound on its mass.
- There is a constant bound on the magnitude of the force and/or torque that a module may apply to those modules with which it is in contact.
- The motion of modules is such that they never collide with a velocity above a fixed constant magnitude.
- For modules in direct contact with each other, the magnitude of the difference in velocity between these contacting modules is always bounded by a constant.
- Each module, when attached (or latched) to other modules, can dynamically set the stiffness of the attachment. This in addition to the ability of the module to apply contraction/expansion or rotational forces at its specified attachments.

Note: The axiom on the stiffness of the attachments to neighboring modules is required to ensure that forces (external to the module) are transmitted through it to neighboring modules. We add this since contraction and/or rotational forces along a chain of modules can accumulate to amounts more than the unit maximum applied by any one module, and may need to be transfered from neighbor to neighbor. Note also that we assume idealized modules that act in synchronized movements under centralized control. This reduces analysis difficulties for cases when many sets of modules operate in parallel.

While the axioms given above state important module requirements, more concrete models are necessary for algorithm design and analysis. The sliding cube model and the expanding cube hardware design described in the previous section satisfy the axioms of our abstract model so long as bounds on the physical abilities of modules are implied. In particular, for these modules the "stiffness" requirement means that the transmitted forces are translational. The exact use of this will become apparent in our first 1-dimensional example with expanding cube modules.

Finally, while not used in this paper, we note two other useful abilities for physical modules: the ability to *push* and the ability to *tunnel*. The ability to push requires one module exerting enough force to push a second module in front of it while sliding along a surface in a straight line. The tunneling ability would allow modules to transition through the interior of a lattice structure in addition to moving along external surfaces.

4 1D Force Analysis

All analysis of movement planning in this paper is based on the physical limitations of individual modules stated in the last section. Given this abstract model, we can now make use of the elementary equations of Newtonian physics governing the modules' movement in space and time. For these equations we use the notation first mention in Section 1. Let $x_i(t)$ be the distance traveled by object i during movement time t. Similarly, $v_i(t)$ is its velocity after time t. Finally, given an object i with mass m_i, let F_i be the net force applied to that object and a_i be the resulting constant acceleration. Although the relevant equations are basic, we state them here so that they can be referred to again later in the paper:

$$F_i = m_i a_i \tag{1}$$

$$x_i(t) = x_i(0) + v_i(0)t + \frac{1}{2}a_i t^2 \tag{2}$$

$$v_i(t) = v_i(0) + a_i t \ . \tag{3}$$

In the first equation we get that a net force of F_i is required to move an object with mass m_i at a constant acceleration with magnitude a_i. In the second equation, an object's location $x_i(t)$ after traveling for a time t is given by it's initial position $x_i(0)$, it's initial velocity $v_i(0)$ multiplied by the time, and a function of a constant acceleration a_i and the time traveled squared. Similarly, the third equation gives the velocity after time t, $v_i(t)$, to be the initial velocity $v_i(0)$ plus the constant acceleration a_i multiplied by the time traveled.

All modules in the examples that follow are assumed to have unit mass $m = 1$ and sides with unit length 1. Our algorithms all require 2 stages of motion: one stage to begin motion and a second stage to slow it and ensure zero final velocity. So, we assume that each module is capable of producing a unit amount of force 1 once for each stage. This still keeps within the constant-bounded force requirement of our abstract model. For an expanding cube module this unit force is capable of contracting or expanding it in unit time while pulling or pushing one neighbor module. Also, since the module has unit mass, by equation (1) we get a bound on its acceleration $a \leq 1$ given the force applied in just a single motion stage. Again, all concerns for friction or gravity (or a detailed description of physical materials to ensure the proper stiffness of modules) are ignored to simplify calculations.

Before tackling the x-axis to y-axis reconfiguration problem, we first begin by analyzing a simpler 1-dimensional case which we will refer to as the *Point Mass Contraction* problem. Here a row of n point masses will contract from a total length of n units, to a length of $n/2$. Although these point masses are not connected and do not grip each other, they otherwise satisfy the axioms of our model. After showing this reconfiguration to take $O(\sqrt{n})$ movement time, we will show that same result holds for a case with expanding-cube style modules instead of point masses. This 1-dimensional contraction case, which we refer to as the *Squeeze* problem, will be a recursive step in our algorithm for the x-axis to y-axis reconfiguration problem.

We assume the initial configuration consists of an even number n of point masses arranged in a row on the x-axis, each initially having 0 velocity. For $i = 1, \ldots, n$ the ith point initially at time $t = 0$ has x-coordinate $x_i(0) = i - (n+1)/2$. We assume each point has unit mass and can move in the x-direction with acceleration magnitude upper bounded by 1. For simplicity, we assume no friction nor any gravitational forces. Our goal configuration at the final time T is the point masses arranged in a row on the x-axis with 0 final velocity, each distance 1/2 from the next in the x-axis direction, so that for $i = 1, \ldots, n$ the ith point at the final time T has x-coordinate

$$x_i(T) = \frac{x_i(0)}{2} = \frac{i}{2} - \frac{n+1}{4} .$$

To differentiate notation, we'll use parentheses to denote a function of time, such as $x_i(t)$, and square brackets to denote order of operations, such as $2 * [3 - 1] = 4$. Furthermore, we require that the velocity difference between consecutive points is at most 1, and that consecutive points never get closer than a distance of 1/2 from each other. Given this, our goal is to find the minimal possible movement time duration from the initial configuration at time 0 to final configuration at time T.

Lemma 1. *The Point Mass Contraction problem requires at most total movement time $T = \sqrt{n-1}$.*

Proof: Fix $T = \sqrt{n-1}$. For each point mass $i = 1, \ldots, n$ at time t, for $0 \leq t \leq T$, let $x_i(t)$ be the x-coordinate of the ith point mass at time t and let $v_i(t), a_i(t)$ be its velocity and acceleration, respectively, in the x-direction (our algorithm for this Point Mass Contraction Problem will provide velocity and acceleration only in the x-direction).

For $i = 1, \ldots, n$ the ith point mass needs to move from initial x-coordinate $x_i(0) = i - (n+1)/2$ starting with initial velocity $v_i(0) = 0$ to final x-coordinate $x_i(T) = x_i(0)/2 = i/2 - (n+1)/4$ ending with final velocity $v_i(T) = 0$. To ensure the velocity at the final time is 0, during the first half of the movement time the ith point mass will be accelerated by an amount α_i in the intended direction, and then during the second half of the movement time the ith point mass will accelerate by $-\alpha_i$ (in the reverse of the intended direction). For $i = 1, \ldots, n$, set that acceleration as:

$$\alpha_i = \frac{n+1-2i}{n-1} .$$

Note that the maximum acceleration bound is unit, since $|\alpha_i| \leq 1$, satisfying the requirement of our abstract model. During the first half of the movement time t, for $0 \leq t \leq T/2$, by equations (3) and (2) we get that the velocity at time t is $v_i(t) = v_i(0) + \alpha_i t$ which implies $v_i(t) = \alpha_i t$ and the x-coordinate is given by:

$$x_i(t) = x_i(0) + v_i(0)t + \frac{\alpha_i}{2}t^2$$

$$x_i(t) = i - \frac{n+1}{2} + \frac{\alpha_i}{2}t^2 .$$

At the midway point $T/2$, this gives the ith point mass velocity $v_i(T/2) = \alpha_i T/2$ and x-coordinate

$$x_i(T/2) = i - \frac{n+1}{2} + \frac{\alpha_i}{2}\left[\frac{T}{2}\right]^2$$

$$x_i(T/2) = i - \frac{n+1}{2} + \frac{a_i T^2}{8} .$$

During the second half of the movement time t, for $T/2 < t \leq T$, we will set the ith point mass acceleration at time t to be $a_i(t) = -\alpha_i$. By equation (3) we get that the velocity at time t during the second half of the movement time is

$$v_i(t) = v_i\left(T/2\right) - \alpha_i\left[t - \frac{T}{2}\right]$$

$$v_i(t) = \alpha_i\left[\frac{T}{2}\right] - \alpha_i\left[t - \frac{T}{2}\right]$$

$$v_i(t) = \alpha_i[T - t] .$$

and the x-coordinate given by equation (2) is

$$x_i(t) = x_i(T/2) + v_i(T/2)\left[t - \frac{T}{2}\right] - \frac{\alpha_i}{2}\left[t - \frac{T}{2}\right]^2$$

$$x_i(t) = \left[i - \frac{n+1}{2} + \frac{\alpha_i T^2}{8}\right] + \frac{\alpha_i T}{2}\left[t - \frac{T}{2}\right] - \frac{\alpha_i}{2}\left[t - \frac{T}{2}\right]^2 .$$

This implies that at the final time T, the ith point mass acceleration has velocity $v_i(T) = \alpha_i(T - T) = 0$ and x-coordinate

$$x_i(T) = \left[i - \frac{n+1}{2} + \frac{\alpha_i T^2}{8}\right] + \frac{\alpha_i T}{2}\left[T - \frac{T}{2}\right] - \frac{\alpha_i}{2}\left[T - \frac{T}{2}\right]^2$$

$$x_i(T) = \left[i - \frac{n+1}{2} + \frac{\alpha_i T^2}{8}\right] + \frac{\alpha_i T^2}{4} - \frac{\alpha_i T^2}{8}$$

$$x_i(T) = i - \frac{n+1}{2} + \frac{\alpha_i T^2}{4} .$$

Recalling that we initially set $\alpha_i = \frac{n+1-2i}{n-1}$ and $T = \sqrt{n-1}$ we get:

$$x_i(T) = i - \frac{n+1}{2} + \frac{n+1-2i}{4}$$

$$x_i(T) = \frac{i}{2} - \frac{n+1}{4}$$

$$x_i(T) = \frac{x_i(0)}{2} .$$

So we get that $x_i(T) = x_i(0)/2$ as required. Moreover, the time any consecutive points are closest is the final time T, and at that time they are distance $1/2$ from each other, as required in the specification of the problem.

It is easy to verify that the velocity difference between consecutive points masses i and $i+1$ is maximized at time $T/2$, and at that time the the magnitude of the velocity difference is

$$\begin{aligned}|v_i(T/2) - v_{i+1}(T/2)| &= |\alpha_i - \alpha_{i+1}|T/2 \\ &\leq \frac{2}{n-1}\frac{\sqrt{n-1}}{2} \\ &\leq 1\end{aligned}$$

as required in the specification of the problem. Thus, we have shown it possible to complete this problem in time $T = \sqrt{n-1}$ while still satisfying the axioms of our abstract model (excluding connectivity and gripping). □

The *Squeeze* Problem. With this result proved for the Point Mass Contraction problem, the same reconfiguration using expanding cube modules can be solved by directly applying Lemma 1. We refer to this reconfiguration task as the *Squeeze* problem and define it as follows. Assume an even number of n modules, numbered $i = 1, \ldots, n$ from left to right as before. Keeping the modules connected as a single row, our goal is to contract the modules from each having length 1 to each having length $1/2$ for some unit length in the x-axis direction. This reconfiguration task is shown in Figure 2. Given the bounds on the module's physical properties required by our abstract model, the goal is to perform this reconfiguration in the minimum possible movement time T.

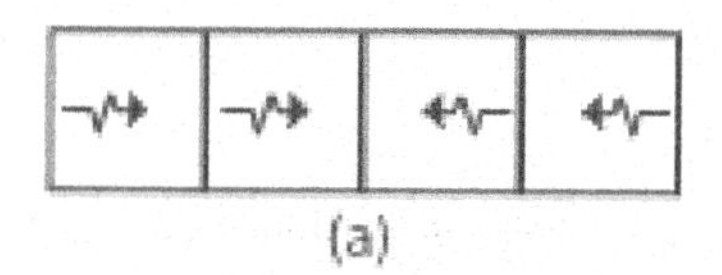

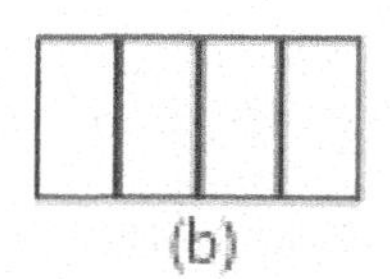

Fig. 2. The *Squeeze* problem: Expanding cube modules in (a), each of length 1, contract to each have length 1/2 and form the configuration in (b)

Similarly, we define the *Reverse Squeeze* problem as the operation that undoes the first reconfiguration. Given a connected row of n contracted modules, each with length $1/2$ in the x-axis direction, expand those modules to each have length 1 while keeping the system connected as a single row. Since this problem is exactly the reverse of the original *Squeeze* problem, steps for an algorithm solving the *Squeeze* problem can be run in reverse order to solve the Reverse *Squeeze* problem in the same movement time with the same amount of force.

We first perform the force analysis for the case of solving the *Squeeze* problem. At the initial time $t = 0$ cube i's center x-coordinate has location

$x_i(0) = i - (n+1)/2$. Recall that each cube is assumed to have unit mass 1, can grip its adjoining cubes, and can exert expanding or contracting forces against those neighbors. Furthermore, a unit upper bound is assumed on the magnitude of this force, which as stated earlier also causes an acceleration bound $a_i \leq 1$ for all modules. We wish to contract the center x-coordinates of the modules from location $x_i(0)$ to $x_i(0)/2 = i/2 - [n+1]/4$ as before in the contraction example with point masses. Again, we wish to find the minimum possible movement time for contraction from the initial configuration at time 0 to the goal configuration at time T.

Lemma 2. *The Squeeze problem requires at most total movement time* $T = \sqrt{n-1}$.

The proof is omitted here to meet space requirements.

Given that the above problem requires at most $T = \sqrt{n-1}$ time to be solved, we get the same result for the *Reverse Squeeze* problem.

Corollary 1. *The Reverse Squeeze problem requires at most movement time* $T = \sqrt{n-1}$.

Note that all of the operations performed in solving the *Squeeze* problem can be done in reverse. The forces and resulting accelerations used to contract modules can be performed in reverse to expand modules that were previously contracted. Thus, we can reverse the above *Squeeze* algorithm in order to solve the *Reverse Squeeze* problem in exactly the same movement time as the original *Squeeze* problem. (Note: Although the forces are reversed in the Reverse Squeeze problem, the stiffness settings remain the same. It is important to observe that without these stiff attachments between neighboring modules, the accumulated expansion forces would make the entire assembly fly apart.)

5 2D Reconfiguration

The above analysis of the 1-dimensional *Squeeze* problem has laid the groundwork for reconfiguration in 2 dimensions. This includes the x-axis to y-axis reconfiguration problem which we will provide an algorithm for at the end of this section. We build to that result by first looking at a simpler example of 2-dimensional reconfiguration that will serve as an intermediate step in our final algorithm. We continue to use the same expanding cube model here as was used in the previous section.

In that previous section a connected row of an even number of n expanding cube modules was contracted from each module having length 1 in the x-axis direction to each having length 1/2. We now begin with that contracted row and reconfigure it into two stacked rows of $n/2$ contracted modules. Modules begin with $\frac{1}{2} \times 1$ width by height dimensions and then finish with $1 \times \frac{1}{2}$ dimensions. This allows pairs of adjacent modules to rotate positions within a bounded 1×1 unit dimension square.We denote this reconfiguration task as the *confined cubes swapping* problem.

The process that we use to achieve reconfiguration is shown in Figure 3. We begin with a row of modules, each with dimension $\frac{1}{2} \times 1$. Here motion occurs in two parts. First, fix the bottom edge of odd numbered modules so that edge does not move. Do the same for the top edge of even numbered modules. Then contract all modules in the y direction from length 1 to length 1/2. Note that to achieve this two counterbalancing forces are required: (1) a force within each module to contract it, and (2) sliding forces between adjacent modules in the row to keep the required top/bottom edges in fixed locations as described earlier. This process creates the "checkered" configuration in part (c) of Figure 3.

We can then reverse the process, but execute it in the x direction instead, to expand the modules in the x direction and create 2 stacked rows of modules, each with dimension $1 \times \frac{1}{2}$. Note that requiring certain edges to stay in fixed locations had an important byproduct. This results in pairs of adjacent modules moving within the same 1×1 square at all times during reconfiguration. This also means that the bounding box of the entire row does not change as it reconfigures into 2 rows. This trait will be of great significance when this reconfiguration is executed in parallel on an initial configuration of several stacked rows.

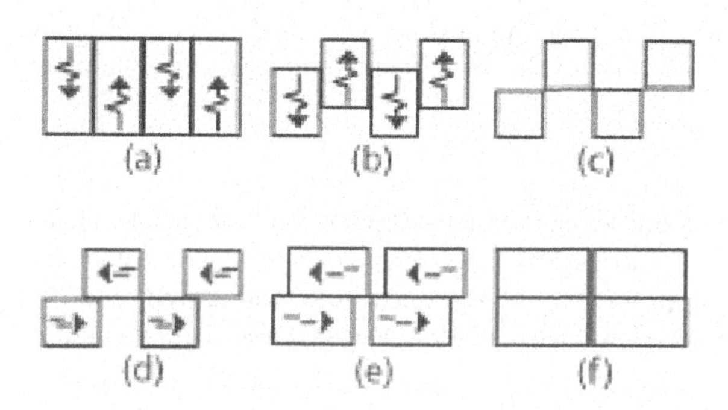

Fig. 3. Expanding cube modules contracting vertically (a - c), then expanding horizontally (d - f). Here the scrunched arrow represents contraction and the 3 piece arrow denotes expansion.

Again, number modules $i = 1, \ldots, n$ from left to right and assume modules have unit mass 1, can grip each other, and can apply contraction, expansion and sliding forces. In this problem we will only use a unit force 1 total per module for both stages of motion. This means a force of 1/2 applied in each motion stage and, by the physics equations in Section 4, an acceleration upper bounded by 1/2 in each stage as well. Finally, concerns for gravity or friction are again ignored for the sake of simplicity. Given these bounds, our goal is to find the minimum reconfiguration time for this confined cubes swapping problem.

Lemma 3. *The confined cubes swapping problem described above requires $T = O(1)$ movement time for reconfiguration.*

The proof is omitted here to meet space requirements.

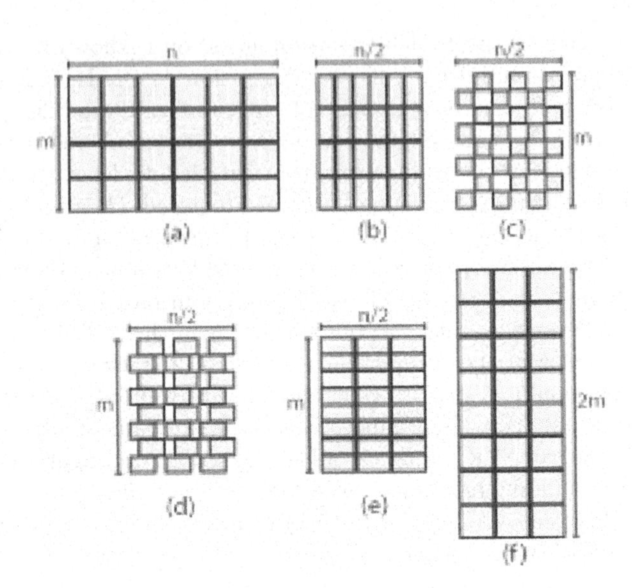

Fig. 4. Horizontal contraction from stage a to b. Vertical expansion from state e to f. Note that the dimensions of the array remain unchanged through stages b - e.

Extending this analysis, we now consider the case of an $m \times n$ array of normal, unit-dimension modules. That is, we have m rows with n modules each. We wish to transform this into a $2m \times \frac{n}{2}$ array configuration. All of the same bounds on the physical properties of modules hold and gravity and friction are still ignored. Once more we wish to find the minimum movement time T for this reconfiguration.

Lemma 4. *Reconfiguring from an $m \times n$ array of unit-dimension modules to an array of $2m \times \frac{n}{2}$ unit-dimension modules takes $O(\sqrt{m} + \sqrt{n})$ movement time.*

The proof is omitted here to meet space requirements.

The problem just analyzed may now be used iteratively to solve the x-axis to y-axis reconfiguration problem. By repeatedly applying the above array reconfiguration step, we double the height and halve the width of the array each time until we progress from a $1 \times n$ to an $n \times 1$ array of modules.

This process will take $O(\lg_2 n)$ such steps, and so there would seem to be a danger of the reconfiguration problem requiring an extra $\lg_2 n$ factor in its movement time. This is avoided because far less time is required to reconfigure arrays in intermediate steps. Note that the $O(\sqrt{m} + \sqrt{n})$ time bound will be dominated by the larger of the two values m and n. For the first $\lfloor (\lg_2 n)/2 \rfloor$ steps the larger value will be the number of columns n, until an array of $\lfloor \sqrt{n} \rfloor \times \lceil \sqrt{n} \rceil$ dimensions is reached. In the next step a $\lceil \sqrt{n} \rceil \times \lfloor \sqrt{n} \rfloor$ array of modules is created, and from that point on we have more rows than columns and the time bound is dominated by m.

The key aspect is that the movement time for each reconfiguration step is decreased by half from the time we begin until we reach an intermediate array of dimensions about $\sqrt{n} \times \sqrt{n}$. By the principle of time reversal it should take us the same amount of movement time to go from a single row of n modules to an $\sqrt{n} \times \sqrt{n}$ cube as it does to go from that cube to a single column of n modules. This tactic is now used in our analysis to find the minimum reconfiguration time for the x-axis to y-axis problem.

Lemma 5. *The x-axis to y-axis reconfiguration problem only requires movement time $O(\sqrt{n})$.*

Proof: For simplicity, let n be even and let $n = p^2$ for some integer $p > 0$. Let $r(i)$ and $c(i)$ be the number of rows and columns, respectively, in the module system after i reconfiguration steps. From the previous Lemma 4 in this section we have that a single step of reconfiguring an $m \times n$ array of modules into a $2m \times \frac{n}{2}$ array requires time $O(\sqrt{m} + \sqrt{n})$. Initially, assuming a large initial row length, then $c(0) = n$, $n \gg 1$, and the reconfiguration step takes $O(\sqrt{n})$ time. For subsequent steps we still have $c(i) \gg r(i)$, but $c(1) = n/2, c(2) = n/4$, etc. while $r(1) = 2, r(2) = 4$, etc. In general $c(i) = n/2^i$ and $r(i) = 2^i$. Eventually, we get $c(i') = r(i') = \sqrt{n}$ at $i' = (\lg_2 n)/2$. Up until that point the time for each reconfiguration stage $i+1$ is $O(\sqrt{c(i)})$. So, the total reconfiguration time to that point is given by the following summation.

$$
\begin{aligned}
\sum_{i=0}^{(\lg_2 n)/2} \sqrt{c(i)} &= \sum_{i=0}^{(\lg_2 n)/2} \sqrt{\frac{n}{2^i}} \\
&\leq \sqrt{n} \sum_{i=0}^{\infty} \left(\frac{1}{\sqrt{2}}\right)^i \\
&= \frac{\sqrt{n}}{1 - (1/\sqrt{2})} \\
&= O(\sqrt{n}) \, .
\end{aligned}
$$

Thus we have that reconfiguration from the initial row of n modules to the intermediate $\sqrt{n} \times \sqrt{n}$ square configuration takes $O(\sqrt{n})$ movement time. Reconfiguring from this cube to the goal configuration is just the reverse operation: we are simply creating a "vertical row" now instead of a horizontal one. This reverse operation will then take the exact same movement time using the same amounts of force as the original operation, so it too takes $O(\sqrt{n})$ movement time. Thus, we have that while satisfying the requirements of our abstract model the x-axis to y-axis problem takes $O(\sqrt{n})$ movement time. □

6 Lower Bounds

In Section 4 we showed that the 1-dimensional Point Mass Contraction Problem, reconfiguring a row of n point masses with unit separation to have 1/2 distance separation, could be solved in time $T = \sqrt{n-1}$. We now show a matching lower bound for this problem. Again, the same assumptions about the physical properties of the point masses are held and concerns for friction or gravity are ignored.

Lemma 6. *The Point Mass Contraction Problem requires at least total movement time $T = \sqrt{n-1}$.*

Proof: Consider the movement of the 1st point mass which needs to move from initial x-coordinate $x_1(0) = 1 - [n+1]/2 = 1/2 - n/2$ starting with initial velocity $v_1(0) = 0$ to final x-coordinate $x_1(T) = x_1(0)/2 = 1/4 - n/4$ and ending with final velocity $v_1(T) = 0$. The total distance this point mass needs to travel is $x_1(T) - x_1(0) = [n-1]/4$.

Taking into consideration the constraint that the final velocity is to be 0, it is easy to verify that the time-optimal trajectory for point mass $i = 1$ is an acceleration of 1 in the positive x-direction from time 0 to time $T/2$, followed by a reverse acceleration of the same magnitude in the negative x-direction. The total distance traversed in each of the two stages is at most $[a_1/2]t^2 = 1 * [T/2]^2/2$. So, the total distance traversed by the 1st point mass is at most $[T/2]^2 = T^2/4$ which needs to be $[n-1]/4$. Hence, $T^2/4 \geq [n-1]/4$ and so $T \geq \sqrt{n-1}$. □

Hence, we have shown:

Theorem 1. *The lower and upper bound for the total movement time for the Point Mass Contraction Problem is exactly $T = \sqrt{n-1}$.*

As in Section 4, we can use an extension of this lower bound argument to prove the following lemma.

Lemma 7. *The Squeeze problem requires total movement time $\Omega(\sqrt{n})$.*

The proof is omitted here to meet space requirements.

Hence, we have also shown:

Theorem 2. *The total movement time for the Squeeze problem is both upper and lower bounded by $\Theta(\sqrt{n})$.*

By the same argument, we can also get a bound on the x-axis to y-axis problem.

Corollary 2. *The x-axis to y-axis reconfiguration problem requires total movement time $\Omega(\sqrt{n})$.*

Proof: Given the initial row configuration and the goal column configuration, pick the end of the row farthest away from the goal column configuration's horizontal placement. Select the n/c cubes at this end of the row, for some real number greater than 2. Among these n/c cubes we can again find some ith cube that must have an acceleration at most $a_i \approx c/2$ and that it must travel a distance $\geq n[1/4 - 1/[2c]] + 1/4$. This again leads to the movement time being bounded as $T = \Omega(\sqrt{n})$. □

7 Conclusion

In this paper we have presented a novel abstract model for self-recongurable (SR) robots that provides a basis for kinodynamic motion planning for these robots. Our model explicitly requires that SR robot modules have unit bounds on their size, mass, magnitude of force or torque they can apply, and the relative velocity between directly connected modules. The model allows for feasible physical implementations and permits the use of basic laws of physics to derive improved reconguration algorithms and lower bounds.

In this paper we have focused on a simple and basic reconguration problem. Our main results were tight upper and lower bounds for the movement time for this problem. Our recursive *Squeeze* algorithm recongures a horizontal row of n modules into a vertical column in $O(\sqrt{n})$-time. This result significantly improves on the running time of previous reconguration algorithms. Our algorithm satisfies the restrictions imposed by our abstract model and we also show that it is kinodynamically optimal given the assumptions of that model.

While carefully using the forces produced by modules, our analysis ignored forces caused by gravity and friction. Addressing these concerns is a topic for future work as the algorithm is brought closer to physical implementation. Also, the algorithm given was a centralized planner and only solved a simple example to demonstrate how faster reconfiguration algorithms were possible. Yet the multipart nature of SR robots makes distributed algorithms a necessity. Extending our lower-bound analysis to more complex analysis, and developing distributed algorithms to match those bounds, is a topic of future work.

References

1. Casal, A., Yim, M.: Self-reconfiguration planning for a class of modular robots. In: Proceedings of SPIE, Sensor Fusion and Decentralized Control in Robotic Systems II, vol. 3839, pp. 246–255 (1999)
2. Donald, B.R., Xavier, P.G., Canny, J.F., Reif, J.H.: Kinodynamic motion planning. Journal of the ACM 40(5), 1048–1066 (1993)
3. Kotay, K., Rus, D.: Generic distributed assembly and repair algorithms for self-reconfiguring robots. In: Proc. of IEEE Intl. Conf. on Intelligent Robots and Systems (2004)
4. Pamecha, A., Chiang, C., Stein, D., Chirikjian, G.: Design and implementation of metamorphic robots. In: Proceedings of the 1996 ASME Design Engineering Technical Conference and Computers in Engineering Conference (1996)
5. Pamecha, A., Ebert-Uphoff, I., Chirikjian, G.: Useful metrics for modular robot motion planning. In: IEEE Trans. Robot. Automat., pp. 531–545 (1997)
6. Vassilvitskii, S., Kubica, J., Rieffel, E., Suh, J., Yim, M.: On the general reconfiguration problem for expanding cube style modular robots. In: Proceedings of the 2002 IEEE Int. Conference on Robotics and Automation, May 11-15, pp. 801–808 (2002)

7. Vona, M., Rus, D.: Self-reconfiguration planning with compressible unit modules. In: 1999 IEEE International Conference on Robotics and Automation (1999)
8. Walter, J.E., Welch, J.L., Amato, N.M.: Distributed reconfiguration of metamorphic robot chains. In: PODC 2000, pp. 171–180 (2000)
9. Yim, M., Duff, D., Roufas, K.: Polybot: A modular reconfigurable robot. In: ICRA, pp. 514–520 (2000)

Part IX

Planning for Games, VR, and Humanoid Motion

Visibility-Based Pursuit-Evasion with Bounded Speed

Benjamín Tovar and Steven M. LaValle

Dept. of Computer Science
University of Illinois at Urbana-Champaign
61801, USA
{btovar,lavalle}@uiuc.edu

Summary. This paper presents an algorithm for a visibility-based pursuit-evasion problem in which bounds on the speeds of the pursuer and evader are given. The pursuer tries to find the evader inside of a simply-connected polygonal environment, and the evader in turn tries actively to avoid detection. The algorithm is at least as powerful as the complete algorithm for the unbounded speed case, and with the knowledge of speed bounds, generates solutions for environments that were previously unsolvable. Furthermore, the paper develops a characterization of the set of possible evader positions as a function of time. This characterization is more complex than in the unbound-speed case, because it no longer depends only on the combinatorial changes in the visibility region of the pursuer.

1 Introduction

Consider a robot in a search-and-rescue operation, such as firefighting inside a building. Victims have to be located before they can receive proper aid. The objective of this robot, called here the *pursuer*, is to find each person inside the building. In the worst-case, the robot should plan as if a person, called an *evader*, is actively hiding. However, the pursuer can make some safe assumptions about each evader. For example, a person does not move at more than $12m/s$. This paper studies the search taking into consideration such speed bounds.

We consider a version of the visibility-based pursuit-evasion problem, in which bounds on the speed of the pursuer, and the evader are given. This yields a major complication for describing the set of possible positions where the evader might be. Perhaps surprisingly, describing the possible positions of the evader with unbounded speed is much easier; they depend only on the combinatorial changes in the visibility region of the pursuer. This is no longer true in the bounded speed case, because the set of possible evader positions is also a function of *time*.

Determining the set of possible evader's positions as a function of time, called the *reachable set of the evader* has been previously studied in [2, 7, 14]. Even in the absence of obstacles, the exact computation of the reachable set is computational intractable, since it involves finding a solution to the Hamilton-Jacobi-Bellman equation [6]. The present paper is the first attempt to describe the set of evader's position inside a polygonal environment. Whereas this description is

S. Akella et al. (Eds.): Algorithmic Foundation of Robotics VII, STAR 47, pp. 475–489, 2008.
springerlink.com

made exact, it is rather used to prove that an approximation to the reachable set that is easier to compute is conservative. This approximation is then used together with the combinatorial changes in the visibility of the robot, enlarging the class of environments that can be searched by a single pursuer. Thus, even though knowing speed bounds makes the problem *easier* to the pursuer, since evader capabilities are decreased, the design of a complete algorithm becomes much more complicated. This is one of the reasons of why the speed bounds have been ignored for visibility-based pursuit-evasion.

The visibility-based pursuit-evasion problem was proposed in [15]. The unbounded speed case has been discussed extensively in the literature. A complete algorithm for a pursuer with an omnidirectional field of view was presented in [4]. A solution for a limited pursuer's field of view was presented in [3]. For pursuers moving on the boundary of the environment, having a single ray of visibility, a complete algorithm was presented in [12]. For the same problem, an finite state automaton was designed in [10]. A randomized solution for a pursuer moving under polyhedral kinematic constraints was described in [8], based on a randomized strategy presented in [9]. The randomized algorithm gives an arbitrarily high probability of evader detection, even when the environment is not searchable with one pursuer by the complete algorithm in [4]. Minimal sensing solutions, in which the environment is unknown to the pursuer, have been presented in [5, 13].

This paper formalizes the problem of pursuit-evasion with bounded speed. We give a description of the set of evader possible positions, *contaminated regions*, in the form of an information state. This information state takes advantage of the combinatorial structure studied in previous approaches to compute the worst-case contamination of a region. Contaminated regions are not kept explicitly, but are computed selectively as the pursuer needs them. Assuming a pursuer that moves in piecewise-linear paths, we present a search algorithm that uses the description of the contaminated regions as a function of the evader speed. This algorithm is as powerful as any complete algorithm for the unbounded speed case. The movement of the pursuer presents a challenging optimization problem [7, 17, 18, 19]; thus, moving the pursuer in piecewise-linear paths may not lead to a complete algorithm. However, by taking into account the speed bounds defined in the problem, this algorithm solves many instances of pursuit-evasion tasks in environments for which no solution exists in the unbounded speed case.

2 Problem Formulation

The pursuer and the evader are modeled as points moving in an open set $\mathcal{R} \subset \mathbb{R}^2$. It is assumed that $\mathcal{R}$ is simply-connected, with a polygonal boundary $\partial\mathcal{R}$. Let $e(t) \in \mathcal{R}$ denote the position of the evader at time $t \geq 0$. It is assumed that $e : [0, \infty) \to \mathcal{R}$ is a continuous function. Let $\mathcal{V}_e(t)$ be the speed of the pursuer at time t. The mapping $\mathcal{V}_e : [0, \infty) \to [0, v_e]$ may not be continuous, but sets a maximum speed for the evader at v_e. Similarly, let $p(t) \in \mathcal{R}$ denote the position of the pursuer at time $t \geq 0$. It is assumed that $p : [0, \infty) \to \mathcal{R}$ is continuous and piecewise-differentiable. The pursuer moves with a maximum speed of v_p according to the speed map $\mathcal{V}_p : [0, \infty) \to [0, v_p]$, which may not be continuous. Since dynamics is disregarded, the implication of the speed bounds are better understood if the position mappings are parametrized as a function of their arclength s. For example, if the length of the path from $e(t_0)$ to $e(t_f)$ is s, then $s \leq v_e(t_f - t_0)$, for any $t_0, t_f \in [0, \infty)$, $t_0 \leq t_f$.

For a point $q \in \mathcal{R}$, let $V(q)$ denote the set of all points in $\mathcal{R}$ that are visible from q (i.e., the line segment joining q and any point in $V(q)$ lies in $\mathcal{R}$). The set $V(q)$ is called the visibility region at q. A mapping $p(t)$ is called a solution strategy if for every continuous path $e : [0, \infty) \to \mathcal{R}$ subject to arclength$(e(t_0), e(t)) \leq v_e(t - t_0)$, $\forall t_0, t \in [0, \infty)$, there exists a time $t_c \in [0, \infty)$ such that $e(t_c) \in V(p(t_c))$. The time t_c is called the time of capture for the strategy $p(t)$. Thus, the position of the evader remains unknown to the pursuer until t_c. The pursuer's task is to find a $p(t)$ solution strategy with a finite time of capture. A complete algorithm reports such a solution strategy if it exits, or reports that evader remains undetected for the given speed bounds.

It is clear that the particular values of speeds of the pursuer and evader are not as important as the ratio v_e/v_p between them. For each simple polygon and each evader speed, a pursuer speed can be found such that a solution strategy exists:

Proposition 1. *Given a simply-connected polygonal environment $\mathcal{R}$ and a maximum evader speed v_e, a speed of a pursuer v_p can be found such that a solution strategy exists.*

Proof. Compute the visibility graph of $\mathcal{R}$ and find the edge with the smallest length l_{min}. Set $v_p = l_{\mathcal{R}} v_e / l_{min}$, in which $l_{\mathcal{R}}$ is the length of $\partial\mathcal{R}$. If the pursuer transverses $\partial\mathcal{R}$ at such speed, any evader is detected. This is because the evader can only hide from reflex vertex to reflex vertex (bitangents), but the pursuer sees all such paths before the evader can transverse them. □

Proposition 1 motivates the study of visibility pursuit-evasion with bounds on the speed. Rather than declaring the problem unsolvable, bounds on the speed may be found such that a strategy exists. An interesting question considers finding the maximum v_e/v_p for which there exists a solution. If the maximum v_e/v_p approaches infinity, then $\mathcal{R}$ can be searched assuming an evader with unbounded speed. Likewise, if it approaches 0, then the search looks like a visibility coverage problem. An upper bound for this ratio can be found with a binary search using the strategy presented in this paper.

2.1 The Model

The pursuer has perfect information about its position and orientation with respect to $\mathcal{R}$. It has two sensors, a clock and a visibility sensor. The clock reports a positive real number that indicates the time elapsed from the beginning of the pursuing task. The visibility sensor $V : \mathcal{R} \to \text{pow}(\mathcal{R})$ reports the visibility region from the current position of the pursuer. An observation space is defined as $Y = [0, \infty) \times \text{pow}(\mathcal{R})$. It can be interpreted as visibility regions together with a timestamp. Let $\tilde{y}_t$ be the history (sequence) of observations up to time t, and let $\tilde{p}_t$ be the history of all the pursuer positions up to time t. Also, let $x = (p(t), e(t))$ be the *state* of the pursuit-evasion task. This leads to a state space $X = \mathbb{R}^2 \times \mathbb{R}^2 = \mathbb{R}^4$. Consider the history information state $(\eta_0, \tilde{p}_t, \tilde{y}_t)$, in which the initial condition $\eta_0 = (p(0), \mathcal{R})$ reflects the fact that at $t = 0$ the position of the pursuer is known, but the evader can be anywhere in $\mathcal{R}$. Let $X(\eta_0, \tilde{p}_t, \tilde{y}_t) \subseteq X$ be the smallest set of states in which the pursuit-evasion task might be, as it is deduced from $(\eta_0, \tilde{p}_t, \tilde{y}_t)$. Each $\eta_t = X(\eta_0, \tilde{p}_t, \tilde{y}_t)$ is an information state of the nondeterministic information space $\mathcal{I}_{ndet} = \text{pow}(X)$ (see [11]). The information state η_t is represented as $\eta_t = (p(t), E(t))$, in which the set $E(t) \subset \mathcal{R}$ is the set of all positions in which the evader might be at time t. Consider the maximal connected sets of points in $E(t)$. Each of these sets is referred to as a *contaminated* region of $\mathcal{R}$. Each of the maximal connected sets of points in $\mathcal{R} \backslash E(t)$ is referred to as a *cleared* region of $\mathcal{R}$. When a contaminated region becomes cleared and later it becomes contaminated again, such region is referred to as *recontaminated.* An equivalent way to describe a solution strategy is to find the mapping $p(t)$ such that the state (p, e) is known. In other words, the task is completed when $E(t)$ contains a single point (the location of the evader), or it is the null set (no evader is in $\mathcal{R}$).

Let $y_{t'} \in Y$ be the observation made at time t'. The information transition equation is defined as $\eta_{t'} = f_{\mathcal{I}}(\eta_t, p(t'), y_{t'})$. The main complication of determining $f_{\mathcal{I}}$ lies in describing how $E(t)$ changes. The next section describes the changes in $E(t)$ as a function of time and the pursuer movements. This description is used later to find a solution strategy.

3 Describing Contaminations

The edges of a visibility polygon $V(p)$ alternate between being part of $\partial\mathcal{R}$ or crossing the interior of $\mathcal{R}$. The latter ones are collinear with p, and are referred to as *gaps.* A label of *contaminated* or of *cleared* is assigned to each gap. The label indicates whether the maximal connected region in $\mathcal{R} \backslash V(p)$ for which the gap is an edge might contain the evader. As the pursuer moves, $V(p)$ changes combinatorially. For the unbounded speed case, $E(t)$ depends uniquely on the gaps changes. If gap α appears, the region behind α is cleared, and α is labeled accordingly. If a cleared gap α merges with a contaminated gap β to form gap γ, then the whole region behind α gets recontaminated, and γ is labeled as contaminated. Gap changes occur when the pursuer crosses inflection rays (appearances and disappearances), or when it crosses bitangent complements (merges and splits). For a deeper discussion of gap changes the reader is referred to [16].

These conditions no longer hold when bounds on the speed are present. A gap does not have to disappear and appear again to mark the whole region behind it as cleared. Likewise, when two gaps merge, a cleared region does not contaminate immediately. Before modeling contaminations, we introduce some gap terminology. A gap that can disappear is called *primitive*. Otherwise, if it can split, it is called *nonprimitive*. For a gap α starting at reflex vertex a_α, let l_v and l_n be the edges of $\partial\mathcal{R}$ that intersect at a_α. As shown in Figure 1, l_v is the edge that is (perhaps partially) visible from p, while l_n is completely hidden. Let $\theta(t)$ be the angle between α and l_n at time t. If $\theta(t)$ increases, α is said to move in the *positive direction*. The length of α at time t is denoted by $\lambda(t)$. Let i_α and i_n be the ray extensions of α and l_n respectively, until an edge of $\partial\mathcal{R}$ is hit. If α is primitive, i_n is the inflection ray that, if crossed, forces α to disappear. Also, let $\alpha(r, \theta(t))$ be the point on α at r distance from a_α when α is at the angular position $\theta(t)$. Finally, let $b(t) = \alpha(\lambda(t), \theta(t))$ be the intersection point of α with $\partial\mathcal{R}$. Thus, α is the line segment $[a_\alpha, b(t)]$.

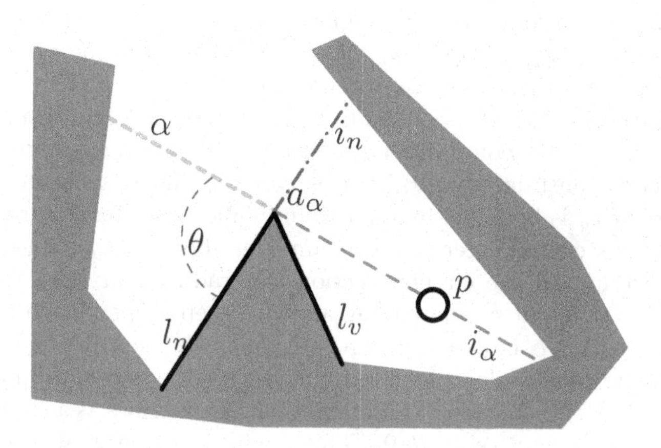

Fig. 1. Gap description. The gap α starts at reflex vertex a_α. It has an angular position of θ, as measured from the edge l_n. The rays i_α, and i_n, extend from a_α in the direction of α and l_n respectively, until they hit an edge of the polygon.

3.1 A Recontamination Fan

Let α be a gap currently visible for which the region behind it is completely contaminated at time $t = 0$. Assume that from $t = 0$ to $t = t_f$, α does not disappear, split or merge. The region between α at $\theta(0)$, and α at $\theta(t_f)$ is called the *recontamination fan* of α, or *α-fan*. Recontamination inside the α-fan is described next. If α moved in the negative direction, that is $\theta(0) > \theta(t_f)$, the whole region behind α is still contaminated and no more computations are needed. For the positive direction the angular velocity $\omega(t)$ of α is needed. Assume that the vertex a_α is placed at the origin of the plane, and that the pursuer position is given by $p = (p_x(t), p_y(t))$. Let $\mathbf{p}$ be the position vector of p, and let $\mathbf{n}$ be its unit normal vector. Also, let $\mathbf{v_p} = [\mathrm{d}p_x/\mathrm{d}t, \mathrm{d}p_y/\mathrm{d}t]$ be the velocity of the pursuer. We have:

$$\omega(t) = \frac{\mathbf{v_p} \cdot \mathbf{n}}{|\mathbf{p}|} = \frac{p_x \frac{\mathrm{d}y}{\mathrm{d}t} - p_y \frac{\mathrm{d}x}{\mathrm{d}t}}{p_x^2 + p_y^2} \tag{1}$$

Assume that $\omega(t)$ has a single maximal value at $\omega(t_{max}) = \omega_{max}$. The general case, when $\omega(t)$ has several critical points is discussed later in this section.

Recontamination from $t = 0$ to $t = t_{max}$

Consider an evader arbitrarily close to $\alpha(r, \theta(0))$. If the evader is to remain arbitrarily close to $\alpha(r, \theta(t))$ during $t \in [0, t_{max}]$, then $\omega_{max} \leq v_e/r$. Let $r_n = v_e/\omega_{max}$. Any evader at distance less than or equal to r_n from a_α can follow exactly the angular motion described by α. Thus, the evader could be anywhere along this arc and such region remains contaminated (see Figure 2.a). Consider now $r_b = v_e/\omega(0)$. Any evader arbitrarily close to $\alpha(r_b, \theta(0))$ cannot follow the angular motion described by α. This is true for positions on the gap with $r \geq r_b$. Their recontamination regions can be described with a circular section with radius $v_e t_{max}$, centered at the original position of the evader, as shown in Figure 2.b. The circular sections do not grow towards a_α. This is because either the positions reachable are already considered by a radius of smaller length, or because the evader would otherwise cross α. The latter is better exemplify if $w(t)$ remains constant at $w(0)$. In this case, r_n and r_b coincide. The region after the arc at r_n is bounded by the involute of a circle (see Figure 2.b). These circular sections cannot intersect α at any time before t_{max}, since $w(t)$ never decreases. Adding all the circular sections for radii larger than r_b makes the portion of α starting at r_b and ending at $b(0)$ sweep perpendicularly by $v_e t_{max}$ (see Figure 2.b). As α moves, its length $\lambda(t)$ may change. If $\lambda(t)$ decreases, the $v_e t_{max}$ sweep is done as described, but eliminating the part that intersects with the polygon. If $\lambda(t)$ increases, a contamination region grows as a circular section, with radius $v_e t_{max}$ and center $b(0)$. This circular section is also present when $r_n > \lambda(0)$. In this case, the region of the circular section crossing (above) α is eliminated.

For $r \in (r_n, r_b)$, let t_r be the smallest time for which $v_e \geq w(t)r$ does not hold. This time is referred to as the *breaking time* of r. Before t_r, an evader placed arbitrarily close to $\alpha(r, \theta(t))$ follows the arc described by α. From t_r to $t_{\max}$, the evader could be anywhere in a growing circular section centered at the last point of contact with α at $\theta(t_r)$, and bounded by α at $\theta(t)$. When adding the effect of such contamination regions, a sweep similar to the one described before takes place. The difference is that, per radius r, the sweep length is $v_e(t_{max} - t_r)$ and the evader travels perpendicularly to α at $\theta(t_r)$.

Recontamination from $t = t_{max}$ to $t = t_f$

Since $w(t)$ is now decreasing, positions for which a breaking time existed may be able to intersect α in the sweep $v_e(t_f - t_{max})$. Suppose the sweep intersects α at $\alpha(r_m, \theta(t_f))$, and r_m is the maximum radius for all such intersections. Then the circular section with radius r_m and center a_α, bounded by α at $\theta(0)$ and $\theta(t_f)$, is completely contaminated. This circular section is added to the contamination sweep that did not intersect α, namely the sweep of the radii after r_m. In general, the contamination boundary consists of:

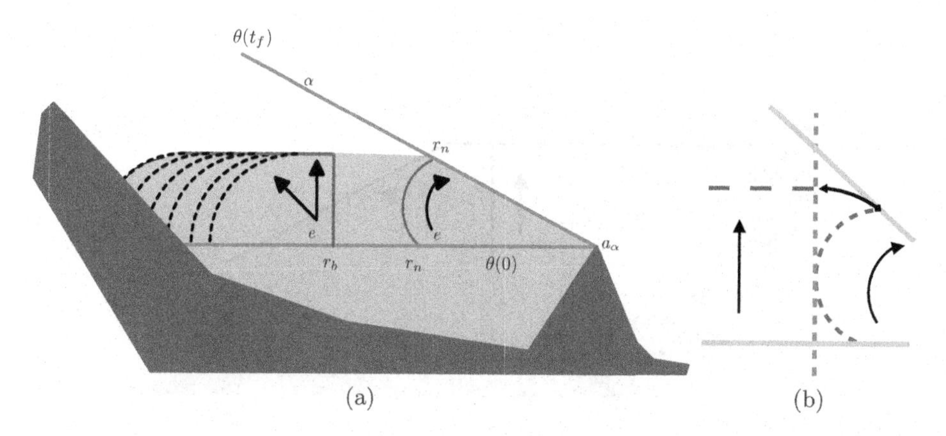

Fig. 2. Recontamination fan. (a) Gap α moved from $\theta(0)$ to $\theta(t_f)$. Any evader before r_n follows exactly the angular motion of α. Any evader after r_b can be anywhere inside a circular section of radius $v_e t_f$. The net effect is a line segment sweeping perpendicularly to the original position. (b) For $w(t)$ constant, the contamination region between the vertical dashed line and the dashed arc is bounded by the involute of a circle.

1. A line segment from a_α to $\alpha(r_m, \theta(t_f))$
2. A curve function of $p(t)$ from $\alpha(r_m, \theta(t_f))$ to the point $(r_b, v_e t_f)$,
3. A line segment $[(r_b, v_e t_f), (b(0), v_e t_f)]$, parallel to α at $\theta(0)$.
4. A circular section, with center $b(0)$ and radius $v_e t_f$.

For some values of r_m some elements may not be present. Note that the region between α at $\theta(t_f)$ and the contamination boundary is cleared, In the unbounded speed case, the whole region behind α would be marked as contaminated. When $w(t)$ has more than one maximum, the concepts before described are applied as follows. Find all the local minima of $\omega(t)$ in $[0, t_f]$. Assume the local minima occur at $t_1, t_2, ..., t_n$. The recontamination fan is computed by parts, from $t = 0$ to t_1, from the gap at $\theta(0)$ to $\theta(t_1)$, and so on. At each step, new breaking times should be computed, and the sweep of $v_e t$ should be checked for intersection. Note that the contamination boundary may contain two or more elements of the same type. This is because the breaking times would change for each radius at each period of time. Particularly, we can consider the line segment up to r_m of the previous time period as a gap, for which a recontamination fan is computed.

Piecewise-Linear Approximation

While a complete algorithm should compute the contamination boundary exactly, a piecewise linear approximation is easily computed. Let $r_{n'} = v_e t_f / \tan(\theta(t_f) - \theta(0))$. An evader traveling $v_e t_f$ from $\alpha(r_{n'}, \theta(0))$ and perpendicularly to α at $\theta(0)$, intersects α at $\alpha(r_{m'}, \theta(t_f))$, with $r_{m'}^2 = r_{n'}^2 + (v_e t)^2$. The approximation is a line segment from a_α to $\alpha(r_{m'}, \theta(t_f))$, and a line segment $\alpha(r_{m'}, \theta(t_f))$ parallel to α at $\theta(0)$ that extends until an edge of $\partial\mathcal{R}$ is hit (see Figure 3). Note

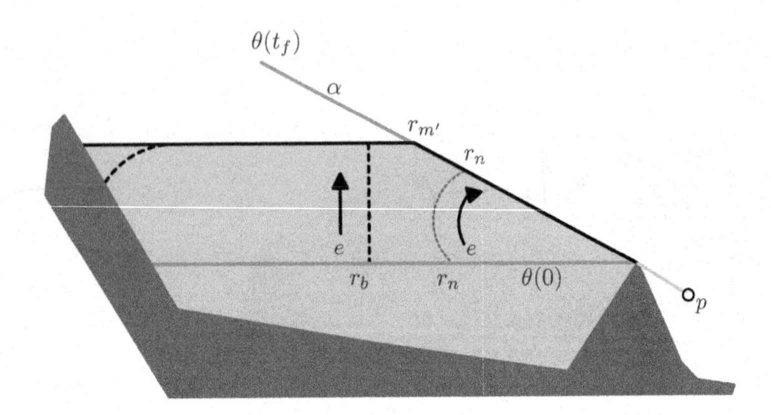

Fig. 3. Piecewise-linear approximation. Instead of computing the exact contamination boundary, an approximation is easily generated by ignoring the angular velocity of the gap, considering only its final angular position.

that $r_{m'} > r_m$, because traveling from $r_{m'}$ did not have to wait for a breaking time to intersect α. Thus, this approximation is conservative and may be preferred over the exact one in real robotic implementations given its simplicity.

Fan Contamination for a Pursuer Moving in a Piecewise-Linear Path

As an example of a fan recontamination, consider a pursuer that moves in a piecewise-linear path with constant velocity v_p. To simplify the example, assume that the pursuer moves in a vertical line at a distance x_0 of the y-axis. From Equation 1:

$$\omega(t) = \frac{x_0 v_p}{x_0^2 + p_y^2} = \frac{x_0 v_p}{x_0^2 + (t v_p - y_0)^2} \tag{2}$$

Equation 2 has a maximum at $\omega(y_0/v_p) = v_p/x_0$. Thus, $r_n = v_e x_0/v_p$, the last radius for which an evader can follow exactly the angular motion of α. Based on Equation 2, an expression for the breaking time of each radius can be found using $\omega(t)r = v_e$:

$$t(r) = \frac{1}{v_p}\left(y_0 + \sqrt{\frac{r x_o v_p}{v_e} - x_o^2}\right) \tag{3}$$

Note that $t(r)$ for $r < r_n$ generates complex solutions. This means, as expected, that such breaking times do not exist. Figure 4 shows a computed example of the fan recontamination for two different speeds of the evader.

3.2 Merges Spreading Contamination

When two gaps merge, contamination spreads in the regions behind them. For two or more consecutive reflex vertices in $\partial\mathcal{R}$, a merge is also considered when one of them occludes the others. While this is not entirely true, since a bitangent

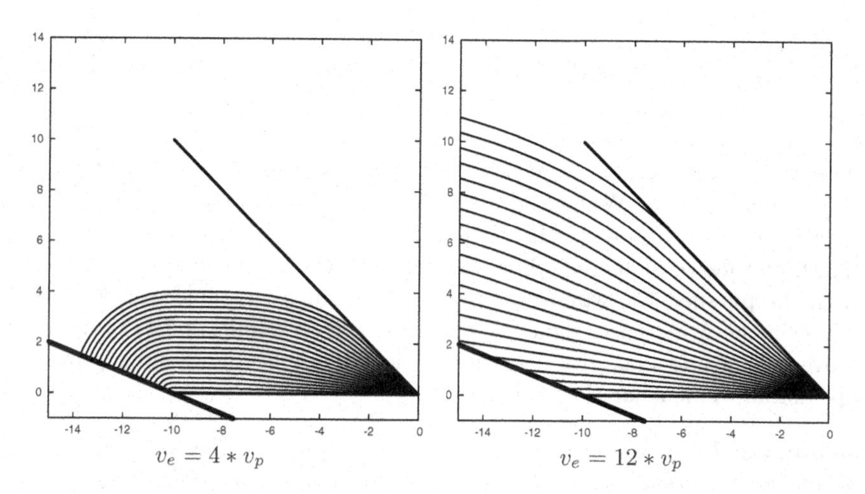

Fig. 4. Recontamination fan computed example for a pursuer linear motion. The gap start at $\theta(0) = 0°$, and ends at $\theta(1) = 45°$. The thick line on the bottom-right represents an edge of $\mathcal{R}$. The evolution of the contamination is shown for times between $[0, 1]$.

does not exists, it simplifies the description of contamination. Particularly, a primitive gap is assigned to the first occluded vertex, while a nonprimitive gap is assigned to each of the remaining consecutive reflex vertices. The *split* only generates one gap. To describe the recontamination between gaps, the following lemma is proposed:

Lemma 1. *Let α and β be two gaps that merge into gap γ. When γ splits, α and β appear at the same angular position at the time of the merge, independently from the pursuer motion.*

Proof. Merges and splits occur when the pursuer crosses a bitangent complement of $\partial\mathcal{R}$. Thus α, β, and γ are aligned with the bitangent at the split or the merge. This is independent from where the bitangent complement is crossed. □

Lemma 1 provides a tool to encode the contamination of cleared regions. If α is a gap for which the region behind is completely cleared, the angular position of α before a merge (i.e., when it was last seen) is recorded. The merge may allow a path between α and a contaminated gap β that the evader can transverse without being detected. Assume that the merge occurs at time $t = 0$, and contamination should be determined for time t_f. Further, assume that γ splits at time $t_s \in [t_0, t_f]$. If the evader does not cross γ by the time t_s, it cannot contaminate α anymore. The worst-case contamination of α has two general cases, function of whether α is visible from the position of the evader or not.

First, assume that α is completely visible from the current position of the evader, which is at h distance from a_α. For clarity, we disregard the effect of the recontamination fans for now. The evader should move as to maximize the contamination of the region behind α. If $v_e t_f > h$, then the worst-case appears when the evader moves to a_α. This forces the pursuer to cross the inflection ray i_n to completely clear the region again. If $v_e t_f \leq h$, the evader cannot reach a_α, but recontamination may still exist. Imagine the evader moves towards some point in α. When the evader reaches this point and keeps moving, allow α and its extension ray i_α to move with the evader around a_α. We said that the evader is *pushing* the gap. When the evader reaches a point in l_n, the rays i_n and i_α coincide (see Figure 5). If the pursuer crosses i_α before the evader gets to l_n, the evader is detected. Since at the moment of the merge i_α is collinear with the pursuer, the worst-case pushes i_α as far as possible from the pursuer. Thus, in the worst-case, the evader pushes α as to minimize $\theta(t_f)$, as any other movement would make its detection easier. The following lemma provides the optimal movement for the evader:

Lemma 2. *Let α be a cleared gap, let $l = v_e t_f$, and consider an evader standing at distance h from a_α, with $t = 0$ and α completely visible from the evader position. If $l > h$, then the optimal strategy for the evader is to move to a_α. If $l \leq h$, then it should move in a straight line, with an angle of $\phi = \arccos(l/h)$ as measured from a parallel line to α, passing through the position of the evader.*

Proof. When $l \geq h$, moving to a_α makes i_α coincide with i_n. When $l < h$, assume the optimal strategy is not straight line. Such strategy has an endpoint b, which can be joined with the position of the evader by a line segment, reaching b faster, a contradiction. To find the angle ϕ, consider the angle $\sigma = \theta(0) - \theta(t_f)$, in which $\theta(0)$ and $\theta(t_f)$ are the positions of the gap before the evader pushed it. To maximize σ, consider the triangle (see Figure 5) with angles ϕ, σ and $\pi - \phi - \sigma$. Now, σ is maximized when $\pi - \phi - \sigma = \pi/2$, from which $\sigma = \arctan\left(l/\sqrt{h^2 - l^2}\right)$, and $\phi = \arccos(l/h)$ follows.

If α is partially visible from the evader position, and the path found in Lemma 2 intersects $\partial\mathcal{R}$, then the evader should move to the last reflex vertex obstructing the path. Once reached, a new path from Lemma 2 should be computed, taking into account the time elapsed. This can be extended for the case when α is completely hidden. The evader moves in a shortest-path, until α is visible. Consider now that the evader is presented with two choices: either, to start at time t_{s1} at distance h_1 from a_α, or to start at time t_{s2} at distance h_2 from a_α, with the inequalities $t_{s1} < t_{s2}$, and $h_1 > h_2$. Given the expression for σ in Lemma 2, t_{s1} is the better choice when $l_1^2/l_2^2 > (h_1^2 - l_1^2)/(h_2^2 - l_2^2)$, for $l_i = v_e t_{si}$. Thus, given the time t_f, the best path for the evader when α is not visible can be determined.

The recontamination fans alter the computation of the paths, since a shortest-path may cross a gap visible to the pursuer. A recontamination fan is computed for every gap that merges, assuming that the region behind the original position of the gap is completely contaminated. If a path is completely contained inside the contamination regions defined by the fans, no further modification to the path is required. Otherwise, the amount of time spent crossing the region between a contamination boundary and the next merge is added to the path.

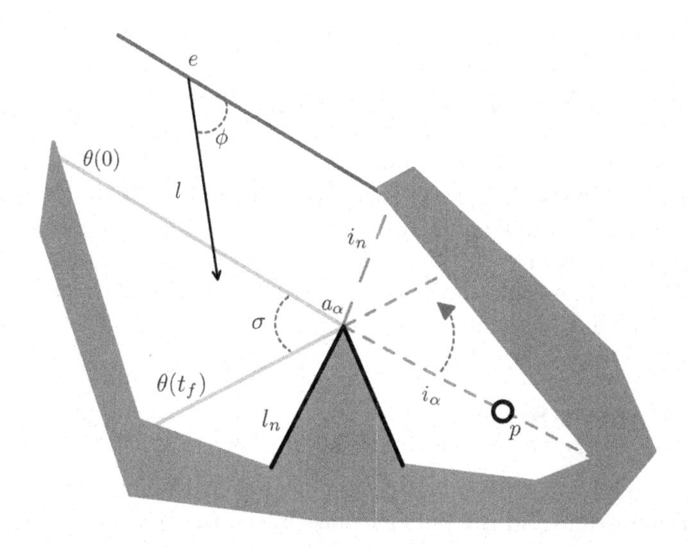

Fig. 5. Pushing a gap. The evader travels on l at angle ϕ as to maximize the angle σ, thus maximizing contamination. As the evader moves, the gap moves with it, and so does the gap extension ray i_α. In the extreme case, i_α coincides with i_n. If the pursuer crosses i_α, the evader is detected.

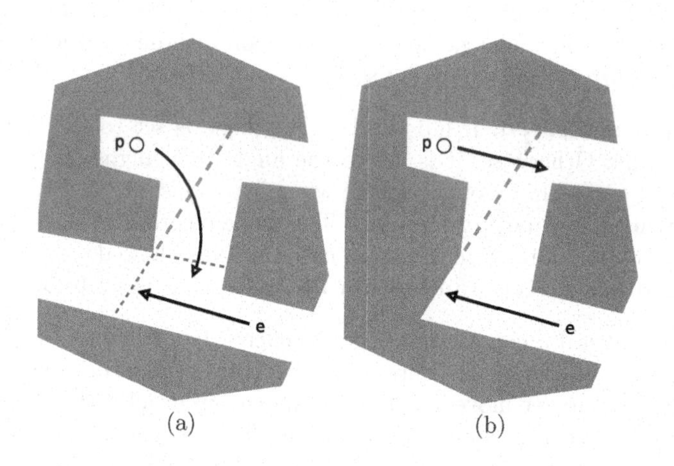

Fig. 6. Clearing equivalent an polygon. The pursuer considers the gap extension ray in (a) as the actual inflection ray. This is equivalent as clearing the polygon in (b).

Clearing an Equivalent Polygon

From the pursuer perspective, the gap extension ray i_α is a real inflection ray. When the pursuer crosses i_α, it clears an equivalent polygon in which α is one of the edges (see Figure 6). At the extreme case, α will coincide with l_n and the pursuer clears the original polygon. This holds also when the gap γ is a nonprimitive gap. As γ moves, i_γ aligns with a bitangent complement. From the pursuer perspective, crossing i_γ before the alignment is like clearing a polygon in which γ is one of the edges, and for a pursuer strategy this polygon is equivalent to the original one. Since contamination paths are computed in a pairwise manner, this simplifies the contamination computation when the evader force a gap to split. When γ splits, say in gaps α and β, the evader could have reached α and β already, by the triangle inequality. In fact, γ will coincide with one of α or β, the one originated by the same reflex vertex. Thus the pursuer should consider, at the same time, the worst positions for α, β, and γ, since the evader may be pushing any of them.

4 The Pursuit Status

Up to now, we have described how the set $E(t)$ of contaminated regions changes as a function of time and the pursuer movements. In this section we provide an appropriate representation for the information state $\eta_t = (p(t), E(t))$. As seen in the previous section, a structure that provides the gap relations is needed. Namely, we need to know which gap will split in which other gaps. This gap hierarchy is represented with a shortest-path tree T, rooted at the pursuer position. Except the root, there is a one-to-one correspondence between the reflex vertices of $\partial\mathcal{R}$, and the nodes in T. Thus the gaps of the corresponding vertex are assigned to each node. The gap's angular position recorded at the node depends on the gap's current contamination status:

- **The gap is cleared.** If it is visible, the angle is set to the current angular position. Otherwise, it is set to the angle that aligns the gap with the bitangent complement of the merge when it was last seen.
- **The evader is pushing the gap.** The angle recorded is the one that minimizes $\theta(t)$, for the period the contamination was allowed.
- **The gap has a recontamination fan.** The angle is set to 0.

Note that even if a contaminated gap is currently visible, the angle recorded depends on the status of its contamination. The combinatorial structure of T changes as the pursuer moves. In fact, it changes in exactly the same places as gaps do [1, 16]. When T changes combinatorially, together with each merge and split the time t of the event is recorded. The value of t is necessary to compute how far the evader could have moved since the last state (i.e., in fan recontamination and pushing of gaps). The tree is modified every time a bitangent complement is crossed, an inflection ray is crossed, and when the gap extension ray of a gap being pushed is crossed. Note that if there exists a path in T between a clear gap and a contaminated one, and this path does not visit the root, then there is a contamination path between the two gaps.

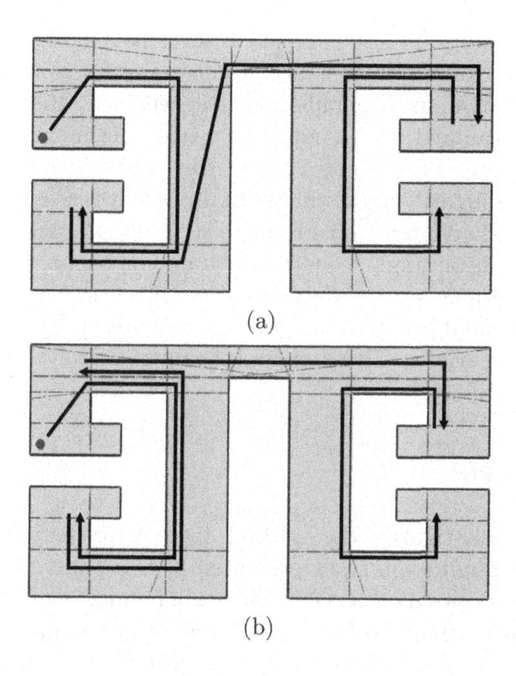

Fig. 7. Example. The path of the pursuer is shown, from the initial position marked as a black circle. (a) $v_e \ll v_p$. Note that the room at the upper-left does not get contaminated. (b) $v_e = v_p$. The upper-left room gets recontaminated, and the clearing path is longer. This example cannot be solved without bounding the speed of the evader. The algorithm presented here also finds a solution for each problem solvable without bounding the speed of the evader.

5 An Improved Pursuit Strategy

Once the representation of an information state has been defined, a search in the information space $\mathcal{I}_{ndet}$ can be performed to find a pursuit strategy. The search starting node is $\eta_0 \in \mathcal{I}_{ndet}$, which has all gaps labeled as contaminated. An information state $\eta_{t_c} \in \mathcal{I}_{ndet}$ is a search goal if it has all its gaps labeled as cleared, and no gap is being pushed. The search strategy is similar to the unbounded speed case in [4]. The visibility-cell decomposition is computed as in the unbounded speed case. The center of each cell is computed. For $\eta_t \in \mathcal{I}_{ndet}$, a set of actions U_{η_t} is defined. An action $u_\beta \in U_{\eta_t}$ takes the pursuer to a neighboring cell through a straight line. Thus, the paths are restricted to be piece-wise linear. A state is a candidate to add to the search queue when T is modified combinatorially, or when a gap extension ray is crossed, as the states are expanded with U_{η_t}. The candidate is accepted if it has at least one gap in which progress to clear it is better than in any previous state. Progress here is defined as: either the gap is cleared, or if the gap is being pushed, the angular position of the gap is bigger than in any other state.

The pursuer strategy is not complete. Nevertheless, there is an important guarantee to its performance. It is at least as powerful as any strategy for the unbounded speed case. In the unbounded speed case, there are no breaking times in the recontamination fans and the evader can move arbitrarily close to any point in the gap. When merges occur, the evader is able to transverse any shortest-path in arbitrarily small time. Finally, when it sees a gap, it can travel arbitrarily fast to the vertex that produces it. Thus, the information state correctly encodes the recontaminations for the unbounded speed case. Crossing gap extension rays becomes immediately crossing inflection rays, and the search is performed as presented originally in [4]. Figure 7 presents examples for two different speeds of the evader. These examples cannot be solved without bounding the speed of the evader.

6 Future Work

We are currently investigating the optimal strategy for the pursuer based on the description of contamination state presented in this paper. The main practical difficulty is the description of the contamination boundary of the fans. The piecewise linear approximation may be a useful tool for providing better paths for the pursuer since it is simpler to analyze. Finding the pursuers movements presents interesting challenges in optimization. For example, since the pursuer has some control in the contamination inside a fan, it can control to some extent the optimal positions for the evader once it reached a cleared gap. It may be possible to model such scenario as a zero-sum game in which the evader tries to maximize recontamination. Other interesting questions remain to be explored. For example, given that contamination travels in shortest paths, we conjecture that environments with the same shortest-path graph will require the same pursuer speed. Our future work considers the study of these questions.

References

1. Aronov, B., Guibas, L., Teichmann, M., Zhang, L.: Visibility queries in simple polygons and applications. In: Chwa, K.-Y., H. Ibarra, O. (eds.) ISAAC 1998. LNCS, vol. 1533, Springer, Heidelberg (1998)
2. Crandall, M.G., Evans, L.C., Lions, P.L.: Some properties of viscosity solutions of hamilton-jacobi equations. Trans. Amer. Math. Soc. 282 (1984)
3. Gerkey, B., Thrun, S., Gordon, G.: Clear the building: Pursuit-evasion with teams of robots. In: Proceedings of the AAAI National Conference on Artificial Intelligence (2004)
4. Guibas, L.J., Latombe, J.-C., LaValle, S.M., Lin, D., Motwani, R.: Visibility-based pursuit-evasion in a polygonal environment. In: Rau-Chaplin, A., Dehne, F., Sack, J.-R., Tamassia, R. (eds.) WADS 1997. LNCS, vol. 1272, pp. 17–30. Springer, Heidelberg (1997)
5. Guilamo, L., Tovar, B., LaValle, S.M.: Pursuit-evasion in an unknown environment using gap navigation graphs. In: IEEE/RSJ Int. Conf. on Intelligent Robots & Systems (2004)

6. Hwang, I., Stipanovic, D., Tomlin, C.J.: Polytopic approximations of reachable sets applied to linear dynamic games and a class of nonlinear systems. In: Advances in Control, Communication Networks, and Transportation Systems In Honor of Pravin Varaiya (2005)
7. Isaacs, R.: Differential Games. Wiley, New York (1965)
8. Isler, V., Daniilidis, K., Pappas, G.J., Belta, C.: Hybrid control for visibility-based pursuit-evasion games. In: IEEE/RSJ Int. Conf. on Intelligent Robots & Systems (2004)
9. Isler, V., Kannan, S., Khanna, S.: Locating and capturing an evader in a polygonal environment. In: Workshop on the Algorithmic Foundations of Robotics (2004)
10. Kameda, T., Yamashita, M., Suzuki, I.: On-line polygon search by a six-state boundary 1-searcher. Technical Report CMPT-TR 2003-07, School of Computing Science, SFU (2003)
11. LaValle, S.M.: Planning Algorithms. Cambridge University Press, Cambridge (2006), `http://msl.cs.uiuc.edu/planning/`
12. LaValle, S.M., Simov, B., Slutzki, G.: An algorithm for searching a polygonal region with a flashlight. International Journal of Computational Geometry and Applications 12(1-2), 87–113 (2002)
13. Sachs, S., Rajko, S., LaValle, S.M.: Visibility-based pursuit-evasion in an unknown planar environment. International Journal of Robotics Research (to appear, 2003)
14. Stipanovic, D.M., Hwang, U., Tomlin, C.J.: Computation of an over-approximation of the backward reachable set using subsystem level set functions. Dynamics Of Continuous Discrete And Impulsive Systems 11, 397–412 (2004)
15. Suzuki, I., Yamashita, M.: Searching for a mobile intruder in a polygonal region. SIAM J. Computing 21(5), 863–888 (1992)
16. Tovar, B., Guilamo, L., LaValle, S.M.: Gap navigation trees: Minimal representation for visibility-based tasks. In: Proc. Workshop on the Algorithmic Foundations of Robotics (2004)
17. Yavin, Y., Pachter, M.: Pursuit-Evasion Differential Games. Pergamon Press, Oxford (1987)
18. Yong, J.: On differential evasion games. SIAM J. Control & Optimization 26(1), 1–22 (1988)
19. Zaremba, L.S.: Differential games reducible to optimal control problems. In: IEEE Conf. Decision & Control, Tampa, pp. 2449–2450 (December 1989)

Planning Near-Optimal Corridors Amidst Obstacles*

Ron Wein[1], Jur van den Berg[2], and Dan Halperin[1]

[1] School of Computer Science, Tel-Aviv University
{wein,danha}@tau.ac.il
[2] Institute of Information and Computing Sciences, Utrecht University
berg@cs.uu.nl

Abstract. Planning corridors among obstacles has arisen as a central problem in game design. Instead of devising a one-dimensional motion path for a moving entity, it is possible to let it move in a corridor, where the exact motion path is determined by a local planner. In this paper we introduce a quantitative measure for the quality of such corridors. We analyze the structure of optimal corridors amidst point obstacles and polygonal obstacles in the plane, and propose an algorithm to compute approximations for optimal corridors according to our measure.

1 Introduction

The task of planning a natural path for a moving entity that avoids obstacles plays an important role in robotics, as well as in game design. The problem is often solved by constructing a graph that discretizes the environment, and extracting a collision-free path from this graph. The nodes of such a graph may be the cells of a uniform grid (see, e.g., [17]), or — according to Probabilistic Roadmap (PRM) paradigm [1, 5] — free configurations that are randomly chosen, attempting to capture the connectivity of the free configuration space.

A common drawback of the above methods is that they output a fixed path in response to a query. This is often not the ideal solution for motion planning, as it lacks flexibility to avoid local hazards (such as small obstacles, other moving entities, etc.) that are encountered during the motion. It also leads to predictable, and possibly unrealistic motions, which are not suitable for some applications, such as computer games. One approach for tackling these problems is a potential-field planner, in which the moving entity is attracted to its goal configuration, and repelled by obstacles, or other moving entities (see, e.g., [6]). However, this approach is prone to get stuck in local minima of the potential field; while there are methods that help in resolving such situations (see, e.g., [7]), they may still not yield valid motions at all.

* This work has been supported in part by the IST Programme of the EU as Shared-cost RTD (FET Open) Projects under Contract No IST-2001-39250 (MOVIE — Motion Planning in Virtual Environments) and IST-006413 (ACS - Algorithms for Complex Shapes) , and by the Hermann Minkowski–Minerva Center for Geometry at Tel Aviv University.

S. Akella et al. (Eds.): Algorithmic Foundation of Robotics VII, STAR 47, pp. 491–506, 2008.
springerlink.com

We would therefore like to indicate the global direction of movement for the moving entity, while leaving enough flexibility for some *local planner* to avoid local hazards. An ideal solution for this is to use *corridors*, which have recently been introduced in the game design field [15]. Corridors are defined as a union of balls whose center points lie along a backbone path. The radius of the balls is determined by the *clearance* (i.e., the distance to the nearest obstacle) along the backbone path. The more restricted task of locally planning the motion around the backbone path can be successfully performed by potential-field methods. In order to guarantee that the local planner operates on a restricted environment, the radii of the balls are upper bounded by some predetermined value.[1] As a result, rather than moving along a fixed path, the moving entity moves within a corridor around the backbone path. This gives a strict global direction of movement, yet provides the local flexibility we look for.

Planning within corridors has many applications. It has been used to plan motions for coherent groups of entities, where the backbone path provides the global motion of the group [3]. The interactions between entities of the group are locally controlled by a social potential-field method [16]. Corridors have also been used to plan the motion of a camera that follows a moving character (a *guide*) [12]. If the guide moves along the backbone path, the corridor gives the flexibility for the camera to swerve if necessary. Another advantage of corridors is that they allow for non-holonomic and kinodynamic planning, if the motion of a single entity (or multiple entities) is planned using a potential field method within the corridor [4]. This is very difficult to achieve and incorporate into a fixed path. A common property of the applications of corridors is that the moving entity is small compared to the scale of the environment. In many fields (open field robotic navigation, games, etc.) this is indeed the case.

The problem we consider in this paper is how to plan a good corridor. A good corridor is short, avoiding unnecessary detours, and at the same time it should be wide (up to some prescribed maximum) to provide local maneuvering space. These requirements often contradict. Given start and goal configurations and a set of obstacles, the shortest collision-free path is contained in the *visibility graph* of the obstacles; see, e.g., [9]. However, such a path is incident to obstacle boundaries and cannot serve as a backbone path of a valid corridor. If one is only concerned with clearance, allowing paths that are as long as needed, then such paths are easily found using the *Voronoi diagram* of the given obstacles [14]. It is also possible to consider interpolates of these two structures, named *visibility–Voronoi diagrams*, as suggested in [19]. Indeed, a good corridor makes a good trade-off between length and clearance.

In this paper we introduce a measure for the quality of corridors, and present methods to plan corridors that are (nearly) optimal with respect to this measure amidst point obstacles or polygonal obstacles in the plane.

The rest of this paper is organized as follows. In Section 2 we formally define corridors and introduce the quality measure. Section 3 discusses properties of

---

[1] The fact that the radii of the balls are bounded is also a major difference between a corridor and the *medial axis transform* of the free workspace.

optimal corridors amidst point obstacles in the plane, and in Section 4 we generalize our results to polygonal obstacles. We give some concluding remarks and future-work directions in Section 5.

2 Measuring Corridors

A *corridor* $C = \langle \gamma(t), w(t), w_{\max} \rangle$ in a d-dimensional workspace (typically $d = 2$ or $d = 3$) is defined as the union of a set of d-dimensional balls whose center points lie along the *backbone path* of the corridor, which is given by the continuous function $\gamma : [0, L] \longrightarrow \mathbb{R}^d$, where L is the length of γ. The radii of the balls along the backbone path are given by the function $w : [0, L] \longrightarrow (0, w_{\max}]$. Both γ and w are parameterized by the length of the backbone path. In the following, we will refer to $w(t)$ as the *width* of the corridor at point t. The width is positive at any point along the corridor, and does not exceed $w_{\max}$, a prescribed *desired width* of the corridor.

Given a corridor $C = \langle \gamma(t), w(t), w_{\max} \rangle$ of length L in $\mathbb{R}^d$, the interior of the corridor is thus defined by $\bigcup_{t \in [0,L]} B(\gamma(t); w(t))$, where $B(p; r)$ is an open d-dimensional ball with radius r that is centered at p. In typical motion-planning applications we are given a set of obstacles $\mathcal{O}$ that the moving entities should avoid. The interior of the corridor should be disjoint from the interior of the given obstacles, otherwise it is an *invalid* corridor. In this paper we study the problem of computing *valid* corridors amidst obstacles in the plane.

2.1 The Weighted Length Measure

As we have already indicated, a good corridor must be short — namely its backbone path should avoid unnecessarily long detours — and its width should be as wide as some predefined maximum in order to allow maximal flexibility for the motion within the corridor. The corridor should contain narrow passages only if they allow considerable shortcuts.

If we examine the intersection of the corridor $C = \langle \gamma(t), w(t), w_{\max} \rangle$ with an orthogonal $(d-1)$-dimensional hyperplane at $\gamma(t)$, the volume of the cut is proportional to $w^{d-1}(t)$. Thus, in order to combine the two desired properties of the corridor as discussed above, we define the *weighted length* $L^*(C)$ of a corridor $C = \langle \gamma(t), w(t), w_{\max} \rangle$ to be:

$$L^*(C) = \int_{\gamma} \left(\frac{w_{\max}}{w(t)} \right)^{d-1} dt \; . \tag{1}$$

We wish to minimize the weighted length by either shortening the backbone path or by extending the corridor's width (up to $w_{\max}$). Given a start position $s \in \mathbb{R}^d$ and a goal position $g \in \mathbb{R}^d$, a corridor $C = \langle \gamma(t), w(t), w_{\max} \rangle$ satisfying $\gamma(0) = s$ and $\gamma(L) = g$ is *optimal* if for any other valid corridor C' connecting the two endpoints we have $L^*(C) \leq L^*(C')$.

Our weighting scheme can be directly applied for extracting backbone paths from PRMs that contain cycles [11, 13], where instead of considering the Euclidean length we try to minimize the weighted length of the backbone path we compute, in order to obtain a better corridor. However, for some sets of obstacles we can actually devise a complete scheme for computing an optimal corridor, as we show in the rest of this paper.

2.2 Properties of an Optimal Corridor

Observation 1. *If for some portion of the backbone path γ of a corridor C, we have $w(t) < \min\{c(\gamma(t)), w_{\max}\}$ for $t \in [t_0, t_0 + \tau]$ ($\tau > 0$), where $c(p)$ is the* clearance *of the point p, namely its distance to the nearest obstacle, we can improve the quality of the corridor by letting $w(t) \longleftarrow \min\{c(\gamma(t)), w_{\max}\}$ for each $t \in [t_0, t_0 + \tau]$.*

Given a set of obstacles and a $w_{\max}$ value, we can associate the *bounded clearance measure* $\hat{c}(p)$ with each point $p \in \mathbb{R}^d$, where $\hat{c}(p) = \min\{c(p), w_{\max}\}$. Using the observation above, it is clear that the width function of an optimal path $C = \langle \gamma(t), w(t), w_{\max} \rangle$ is simply $w(t) = \hat{c}(\gamma(t))$. Note that $\hat{c}(\gamma(t))$ is a continuous function along any path γ.

Lemma 2. *Given a set of obstacles and $w_{\max}$, the backbone path of the optimal corridor connecting any given start position s with any goal position g is* smooth.

Proof. We have already observed that the weight function of the optimal corridor connecting s and g is the bounded clearance function of the backbone path and it is a continuous function. Assume that γ contains a sharp turn (a $\mathcal{C}_1$-discontinuity). Let us shortcut the sharp turn using a circular arc of radius r (as r approaches 0 the approximation is tighter). Let ℓ_1 be the length of the original path segment we shortcut and let ℓ_2 be the length of the circular arc. It is easy to show that there exist $\hat{r} > 0$ and some constants $A_1 > A_2 > 0$ such that for each $0 < r < \hat{r}$ we have $\ell_1 \geq A_1 r$ and $\ell_2 = A_2 r$. If the maximal width w^* along the original path segment is obtained at some point p^*, then as the distance of any point p along the circular arc from p^* is bounded by Kr, where K is some constant, and as the weight function is continuous, we can write $w^* - w(p) < Mr$ for some positive constant M. Let L_1^* be the weighted length of the original path segment and let L_2^* be the weighted length of the circular arc. We can write:

$$\frac{L_1^*}{L_2^*} \geq \frac{\frac{w_{\max}}{w^*} l_1}{\frac{w_{\max}}{w^* - Mr} l_2} = \frac{w^* - Mr}{w^*} \cdot \frac{A_1}{A_2} .$$

As $A_1 > A_2$, we can choose $0 < r < \min\left\{\frac{w^*}{M}\left(1 - \frac{A_2}{A_1}\right), \hat{r}\right\}$ such that the entire expression above is greater than 1. We thus have $L_1^* > L_2^*$, and we managed to decrease the weighted length of the corridor, in contradiction to its optimality. We conclude that $\gamma(t)$ must be a smooth function. □

At several places in this paper we apply infinitesimal analysis, where we assume that the bounded clearance measure (hence the weight function) is not continuous. Assume that we have some hyperplane $\mathcal{H}$ in $\mathbb{R}^d$ that separates two regions, such that in one region the bounded clearance is w_1 and in the other it is w_2.

Minimizing the weighted length between two endpoints that are separated by $\mathcal{H}$ is equivalent to applying Fermat's principle, stating that the actual path between two points taken by a beam of light is the one which is traversed in the least time. The optimal backbone thus crosses the separating hyperplane once, such that the angles α_1 and α_2 it forms with the normal to $\mathcal{H}$ obey Snell's Law of refraction,[2] with w_1 and w_2 playing the role of the "speed of light" in the respective regions:

$$w_2 \sin \alpha_1 = w_1 \sin \alpha_2 \ . \tag{2}$$

3 Optimal Corridors Amidst Point Obstacles

In this section we consider planar environments cluttered with point obstacles $p_1, \ldots, p_n \in \mathbb{R}^2$ and a preferred corridor width $w_{\max}$. Given two endpoints $s, g \in \mathbb{R}^2$, we show how to compute a (near-)optimal corridor that connects s and g.

3.1 A Single Point Obstacle

Let us assume we have a single point obstacle p. Without loss of generality we assume p is located at the origin. We start with computing an optimal corridor between two endpoints whose distance from p is smaller than or equal to $w_{\max}$. Note that the width of such a corridor at $\gamma(t)$ along its backbone is $\|\gamma(t)\|$.

We first approximate the optimal backbone by a polyline: for any $\Delta r > 0$, if we look at the circles of radii $\Delta r, 2\Delta r, 3\Delta r, \ldots$ that are centered at the origin, each two neighboring circles define an annulus; since Δr is small we assume that the distance from p of all points in the kth annulus is constant and equals $k\Delta r$. Consider the scenario depicted in Figure 1(a), where γ enters one of the annuli at some point A, where $\|A\| = r_1$, and leaves this annulus at B, where $\|B\| = r_2 = r_1 + \Delta r$. The angles that the backbone path forms with pA and pB are α_1 and β_1, respectively. When entering the annulus we have $w_1 = r_1$ and $w_2 = r_2$, so applying Equation (2) we can express the refracted angle α_2, using $\sin \alpha_2 = \frac{r_2}{r_1} \sin \alpha_1$. By applying the Law of Sines on the triangle $\triangle pAB$, we get $\frac{r_2}{\sin(\pi - \alpha_2)} = \frac{r_1}{\sin \beta_1}$, therefore:

$$\sin \beta_1 = \frac{r_1}{r_2} \sin(\pi - \alpha_2) = \frac{r_1}{r_2} \sin \alpha_2 = \sin \alpha_1 \ .$$

Thus $\beta_1 = \alpha_1$. Taking $\Delta r \longrightarrow 0$, we obtain a smooth curve γ, such that the angle that $\nabla \gamma(t)$ forms with $\overrightarrow{p\gamma(t)}$ is a constant ψ. It is possible to show that a curve that has this property must be segment of a *logarithmic spiral* (also named an *equiangular spiral*)[3] whose polar equation is given by $r(t) = ae^{b\theta(t)}$, where a is a constant and $b = \cot \psi$. See, e.g., [2] for a proof of this latter fact.

[2] See, e.g., http://scienceworld.wolfram.com/physics/SnellsLaw.html for the details and for a detailed proof. See also Mitchell and Papadimitriou [10], who used this observation in a similar setting of the problem.

[3] http://www-groups.dcs.st-and.ac.uk/~history/Curves/Equiangular.html

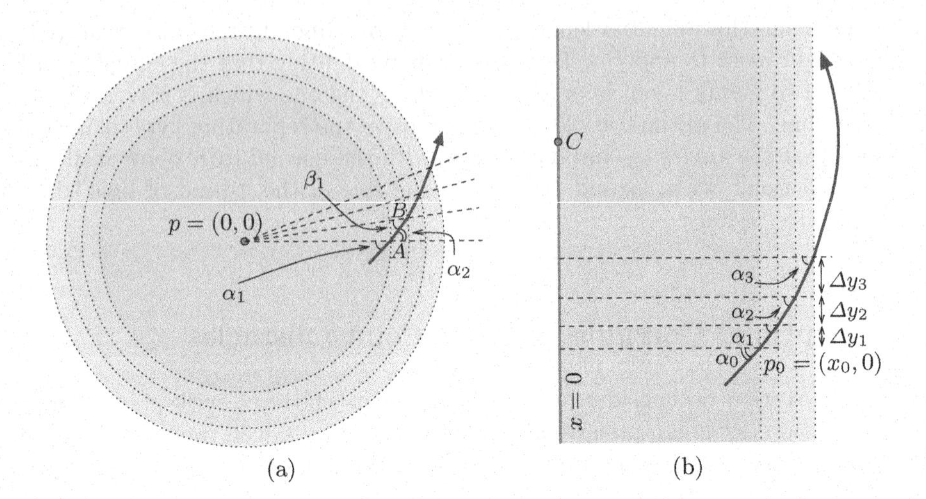

Fig. 1. Analysis of the optimal backbone path in the vicinity of a single obstacle: (a) near a point obstacle $p = (0, 0)$, (b) near a line segment supported by $x = 0$

Proposition 3. *Given a single point obstacle located at the origin, a start position* $s = r_s e^{i\theta_s}$ *and a goal position* $g = r_g e^{i\theta_g}$ *(in polar coordinates), where* $r_s, r_g \leq w_{\max}$*, the backbone of the optimal corridor connecting* s *and* g *is a spiral arc supported by the logarithmic spiral* $r = a^* e^{b^* \theta}$*. Since both* s *and* g *lie on this spiral, we have (assuming* $\theta_s \neq \theta_g$*, otherwise the optimal backbone path is simply a line segment):*

$$a^* = {r_g}^{\frac{\theta_s}{\theta_s - \theta_g}} \cdot {r_s}^{-\frac{\theta_g}{\theta_s - \theta_g}} \, , \qquad b^* = \frac{1}{\theta_g - \theta_s} \cdot \ln \frac{r_g}{r_s} \, . \tag{3}$$

We now consider the case where the clearance of the two endpoints exceeds $w_{\max}$, namely the two endpoints of our path lie outside the closure of the disc $B(p; w_{\max})$. There are two possible scenarios: (i) The straight line segment $\overline{sg}$ does not intersect $B(p; w_{\max})$; in this case, this segment is the backbone of the optimal corridor. (ii) $\overline{sg}$ intersects $B(p; w_{\max})$. In this latter case the optimal backbone path is a bit more involved. Consider some backbone path γ connecting s and g. It is clear that the intersection of γ with $B(p; w_{\max})$ comprises a single component, so we denote the point where the path enters the disc by s' and the point where it leaves the disc by g' (see the illustration to the right). As s' and g' lie on the disc boundary, their polar representation is $s' = w_{\max} e^{i\theta_{s'}}$ and $g' = w_{\max} e^{i\theta_{g'}}$, so we use Equation (3) and obtain $a^* = w_{\max}$ and $b^* = 0$. The optimal path between s' and g' therefore lies on the degenerate spiral $r = w_{\max}$, namely the circle that forms the boundary of $B(p; w_{\max})$. We conclude that the

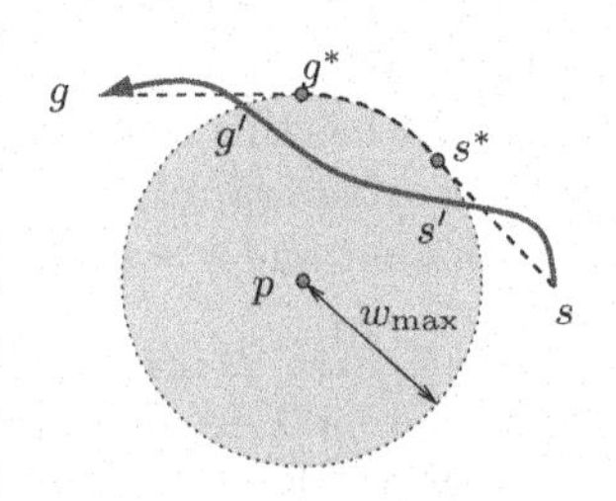

optimal backbone path between s and g must contain a circular arc on the boundary of $B(p; w_{\max})$. As according to Lemma 2 this path must be smooth, it should comprise two line segments ss^* and g^*g that are tangent to the disc and a circular arc that connects the two tangency points s^* and g^* (see the dashed path in the figure above). Note that as there are two possible smooth paths from s to g we select the shortest one.

3.2 Multiple Well-Separated Point Obstacles

Let us now go back to our original setting, where we are given a set of point obstacles $\mathcal{O} = \{p_1, \ldots, p_n\}$, along with a preferred width $w_{\max}$, and wish to compute the optimal corridor from s to g, where we assume that $c(s) = \min_i \|s - p_i\| \geq w_{\max}$ and $c(g) = \min_i \|g - p_i\| \geq w_{\max}$.

In case the points are well separated — that is, for each $i \neq j$ the discs $B(p_i; w_{\max})$ and $B(p_j; w_{\max})$ are disjoint in their interiors (implying that $\|p_i - p_j\| \geq 2w_{\max}$), we can follow the same arguments we used above for a single obstacle and conclude that the optimal backbone is either the straight line segment sg (in case it is *free*, namely its interior does not intersect the interior of any of the discs), or it comprises circular arcs and line segments that connect them.

We can therefore construct the visibility graph of the dilated obstacles and use it to construct optimal paths. The vertices of this graph are the endpoints of the free bitangents to two dilated obstacles, which in turn are represented as graph edges. In addition, each two neighboring tangency points on a disc $B(p_i; w_{\max})$ are connected by a circular arc. Given a path-planning query, namely two endpoints s and g, we treat s and g as vertices and add all free tangents from s and from g to the discs as graph edges. If the segment sg is free, we add it to the graph as well. We then perform Dijkstra's algorithm from s to find the shortest path to g in the resulting graph. The weight $\omega(e)$ given to each graph edge e is its weighted length, which simply equals its length in this case.

Proposition 4. *Given a set $\mathcal{O}$ of n point obstacles in the plane that are well-separated with respect to $w_{\max}$, and two endpoints s and g with clearance at least $w_{\max}$, it is possible to compute the optimal corridor connecting s and g in $O(E \log n)$ time using the visibility graph of the dilated obstacles, where E is the number of visibility edges in this graph.*

3.3 Corridors Amidst Point Obstacles: The General Case

We now consider the case where the endpoints s and g have arbitrary clearance, namely the dilated obstacles $B(p_1; w_{\max}), \ldots, B(p_n; w_{\max})$ are not necessarily pairwise disjoint in their interiors. The boundary of $\mathcal{M} = \bigcup_{i=1}^{n} B(p_i; w_{\max})$ comprises whole circles and circular arcs, such that a common endpoint of two arcs is a reflex vertex. We now construct $\mathcal{V}$, the Voronoi diagram of the points, and compute the intersection $\mathcal{V} \cap \mathcal{M}$, namely the portions of the Voronoi edges contained within the union of the dilated obstacles. Note that reflex vertices are equidistant to two point obstacles, so they serve as the connection points between the Voronoi edges and the boundary arcs of $\mathcal{M}$. We will refer to the Voronoi edges in $\mathcal{V} \cap \mathcal{M}$, together with the circular arcs that form the boundary of $\mathcal{M}$, as the *bounded Voronoi diagram* of the point set $\mathcal{O} = \{p_1, \ldots, p_n\}$, which we denote $\hat{\mathcal{V}}(\mathcal{O})$. The figure to the right shows the bounded Voronoi diagram of six points; the boundary of $\mathcal{M}$ is drawn is solid lines and the Voronoi edges are dotted.

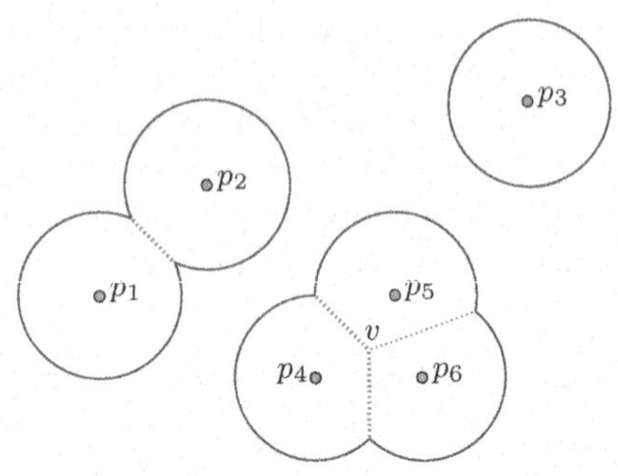

Note that $\hat{\mathcal{V}}(\mathcal{O})$ partitions the plane into two-dimensional cells of two types: Voronoi regions of the point obstacles, and regions where the clearance is larger than $w_{\max}$. Given two points s' and g' that belong to the same cell κ, we know that:

- If κ is a cell whose clearance is greater than $w_{\max}$, the optimal backbone path between $s' = (x_1, y_1)$ and $g' = (x_2, y_2)$ is the straight line segment σ that connects them, provided that σ does not intersect any feature of $\hat{\mathcal{V}}(\mathcal{O})$. The weighted length of this segment simply equals the Euclidean distance $\|g' - s'\| = \sqrt{(x_2 - x_1)^2 + (y_2 - y_1)^2}$.
- If κ is a Voronoi cell of a point obstacle p_i, the optimal backbone path between s' and g' is a spiral arc σ centered at p_i, provided that σ does not intersect any feature of $\hat{\mathcal{V}}(\mathcal{O})$. If $s' = r_1 e^{i\theta_1}$ and $g' = r_2 e^{i\theta_2}$ are the polar coordinates of the endpoints with respect to p_i, the weighted length of σ is given by (recall that from Equation (3) we have $b = \frac{1}{\theta_2 - \theta_1} \cdot \ln \frac{r_2}{r_1}$):

$$
\begin{aligned}
L^*(\sigma) &= \int_{\theta_1}^{\theta_2} \frac{w_{\max}}{r(\theta)} \sqrt{r^2(\theta) + \left(\frac{dr}{d\theta}\right)^2(\theta)}\, d\theta = \int_{\theta_1}^{\theta_2} \frac{w_{\max}}{ae^{b\theta}} \sqrt{1+b^2}\, ae^{b\theta}\, d\theta = \\
&= \int_{\theta_1}^{\theta_2} w_{\max}\sqrt{1+b^2}\, d\theta = w_{\max}\sqrt{1+b^2}(\theta_2 - \theta_1) = \\
&= w_{\max}\sqrt{(\theta_2 - \theta_1)^2 + (\ln r_2 - \ln r_1)^2}\ .
\end{aligned}
$$

In addition, the features of $\hat{\mathcal{V}}(\mathcal{O})$ are also *locally optimal*, namely they can serve as backbone paths of optimal corridors (see Figure 2(a)). We already know that portions of the circular arcs that form the boundary of $\mathcal{M}$ are locally optimal, and that the weighted length of such a circular arc simply equals its length. The Voronoi edges are also locally optimal: given s' and g' on the same Voronoi edge, the optimal backbone path that connects them is simply the straight line segment $s'g'$ which coincides with the Voronoi edge.

Following the construction of the visibility graph of the dilated point obstacles (Section 3.2), it is possible to add *visibility edges* to the bounded Voronoi diagram, namely to consider every free bitangent of two circular arcs, every free line segment from a reflex vertex tangent to a circular arc and every free line segment

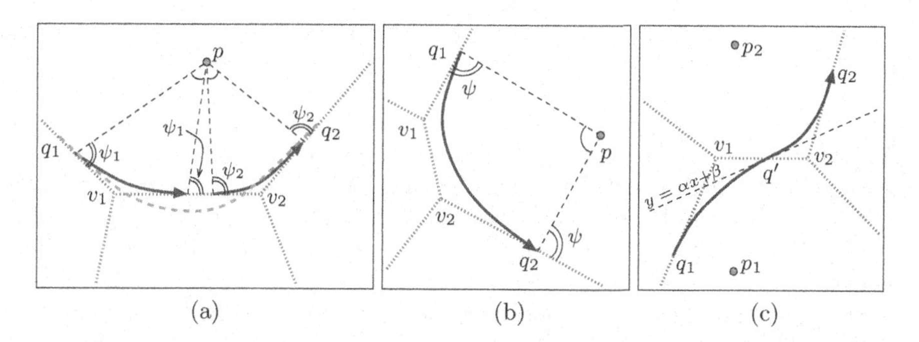

Fig. 2. (a) The spiral arc connecting q_1 and q_2 (dashed) crosses the Voronoi edge v_1v_2; the optimal backbone path between q_1 and q_2 therefore comprises two spiral arcs that shortcut v_1 and v_2 (solid arrows) and portions of Voronoi edges. (b) Shortcutting two adjacent Voronoi vertices v_1 and v_2 by a single spiral arc. (c) Shortcutting two Voronoi vertices by a cross-cell curve, which is a smooth concatenation of two spiral arcs. Both arcs have a common tangent $y = \alpha x + b$, which crosses the Voronoi edge v_1v_2 at q'.

between two reflex vertices.[4] However, a path extracted from such a graph may pass through Voronoi vertices and reflex vertices, thus it may contain sharp turns. According to Lemma 2, such a path cannot serve as a backbone to an optimal corridor. We can try and rectify this problem by introducing a *shortcut edge* between each pair of Voronoi edges that are incident to a common Voronoi vertex (see Figure 2(a) for an illustration), and between each pair consisting of a Voronoi edge and a visibility edge that are both incident to a common reflex vertex. However, this is not sufficient. We can show that it is sometimes possible to shortcut two Voronoi vertices v_1 and v_2 at once by connecting two Voronoi edges that are separated by another edge using a single curve. This curve may be contained in a single Voronoi cell, as in the example depicted in Figure 2(b), or it may cross the Voronoi edge v_1v_2 at some point q' (see Figure 2(c)). We should continue and examine the possibility of shortcutting $k > 2$ Voronoi vertices by considering sequences of $(k + 1)$ contiguous Voronoi edges and trying to locate an endpoint q_1 on the first edge and q_2 on the last edge that are connected by a smooth curve comprising spiral arcs. This operation is not trivial, and requires solving a system of low-degree polynomial equations with $2(N_c + 1)$ unknowns, where N_c is the number of crossings between the shortcut curve and the Voronoi diagram. In some scenarios it may be possible to construct shortcuts to $\Theta(n)$ Voronoi vertices by considering sequences of $\Theta(n)$ contiguous Voronoi edges, thus the size of the augmented diagram may blow up exponentially.

We therefore devise an approximation algorithm based on the structure of the bounded Voronoi diagram $\hat{\mathcal{V}}(\mathcal{O})$ and the planar partition it induces. Given $\varepsilon > 0$, we subdivide the line segments and the circular arcs that form the features of

[4] The resulting construct is the visibility–Voronoi diagram of the obstacles; see [19] for more details.

$\hat{\mathcal{V}}(\mathcal{O})$ into small intervals of length $\frac{c(I)}{w_{\max}}\varepsilon$ (as ε is small, we consider the clearance of an interval I to be constant and denote it $c(I)$). Notice that the intervals are shorter in regions where the clearance is smaller, and that each interval has weighted length ε. Hence, if Λ is the total weighted length of the features of $\hat{\mathcal{V}}(\mathcal{O})$, then there are $\frac{\Lambda}{\varepsilon}$ intervals in total. Let us now define a graph $\mathcal{D}$ whose set of nodes equals the set of intervals $\mathcal{I}$. Each interval is incident to two of the cells defined by the bounded Voronoi diagram, and we connect $I_1, I_2 \in \mathcal{I}$ by an edge if and only if they are incident to a common cell. This edge is a line segment in a cell where the clearance is larger than $w_{\max}$, a spiral segment in a Voronoi region of one of the point obstacles, a circular arc on the boundary of a dilated obstacle, or a straight line segment on a Voronoi edge. In addition, an edge should not cross any of the features of $\hat{\mathcal{V}}(\mathcal{O})$. Using a brute-force algorithm that checks each candidate edge versus the $O(n)$ diagram features, $\mathcal{D}$ can be constructed in $O\left(\frac{\Lambda^2}{\varepsilon^2}n\right)$ time.

Given two endpoints s and g, we can connect them to the graph and use Dijkstra's algorithm to compute a near-optimal backbone connecting s and g in $O\left(\frac{\Lambda^2}{\varepsilon^2}\right)$ time. Let γ^* be the backbone path of the optimal corridor between s and g, which comprises $k = O(n)$ segments $\gamma_1, \ldots, \gamma_k$ (a path segment may be a straight line segment, a spiral arc, a portion of a circular arc or a portion of a Voronoi edge). We next show that each such segment is approximated by an edge in the graph $\mathcal{D}$ we have constructed.

Lemma 5. *For each segment γ_i of the optimal backbone path γ^*, there exists an edge e in $\mathcal{D}$ such that $L^*(e) < L^*(\gamma_i) + 2\sqrt{2}\varepsilon$.*

Proof. Let us denote the endpoints of the path segment γ_i by q_1 and q_2, and let I_1 and I_2 be the intervals that contain these endpoints, respectively.

In case γ_i is a straight line segment in a cell κ whose clearance is greater than $w_{\max}$, then its weighted length simply equals $\|q_2 - q_1\|$, the Euclidean distance between its endpoints. In the graph $\mathcal{D}$ there exists an edge connecting I_1 and I_2, and we denote its endpoints by $\tilde{q}_1$ and $\tilde{q}_2$. By the construction of the intervals, we know that $\|q_j - \tilde{q}_j\| \leq \frac{c(I_j)}{w_{\max}}\varepsilon = \varepsilon$ (for $j = 1, 2$), hence:

$$\|\tilde{q}_2 - \tilde{q}_1\| < \|q_2 - q_1\| + 2\varepsilon \ .$$

Similar arguments hold when γ_i is a circular arc with clearance $w_{\max}$.

In case γ_i is a segment on a Voronoi edge, the graph $\mathcal{D}$ contains a segment $\tilde{q}_1\tilde{q}_2$ that in the worst case extends $\frac{c(q_1)}{w_{\max}}\varepsilon$ to one side of q_1 and $\frac{c(q_2)}{w_{\max}}\varepsilon$ to the other side of q_2. Since the contribution of each of these extensions is $\frac{w_{\max}}{c(q_j)}$ times its length (for $j = 1, 2$), the weighted length of $\tilde{q}_1\tilde{q}_2$ is at most 2ε more than $L^*(\gamma_i)$.

The case where γ_i is a spiral arc contained in a Voronoi cell of a point obstacle p_i is a bit more involved. Let $q_1 = r_1 e^{i\theta_1}$ and $q_2 = r_2 e^{i\theta_2}$ be the polar coordinates of γ_i's endpoints with respect to p_i, then we have $L^*(\gamma_i) = w_{\max}\sqrt{(\theta_2 - \theta_1)^2 + (\ln r_2 - \ln r_1)^2}$. $\mathcal{D}$ contains a spiral arc connecting I_1 and I_2, and we denote its endpoints by $\tilde{q}_j = \tilde{r}_j e^{i\tilde{\theta}_j} \in I_j$ (for $j = 1, 2$). As $c(q_j) = r_j$, we know that the length of each of these two intervals is $\|I_j\| = \frac{r_j}{w_{\max}}\varepsilon$. If we denote $\Delta\theta_j = \theta_j - \tilde{\theta}_j$, we can write:

$$\sin\left(\frac{\Delta\theta_j}{2}\right) < \frac{\frac{1}{2}\|I_j\|}{r_j} = \frac{\varepsilon}{2w_{\max}} .$$

As for small angles $\sin\phi \approx \phi$, we conclude that $|\Delta\theta_j| < \frac{\varepsilon}{w_{\max}}$. At the same time, $|\Delta r_j| = |r_j - \tilde{r}_j| < \frac{\varepsilon r_j}{w_{\max}}$, thus we have:

$$|\ln\tilde{r}_j - \ln r_j| < \left|\ln\left(r_j\left(1 + \frac{\varepsilon}{w_{\max}}\right)\right) - \ln r_j\right| = \ln\left(1 + \frac{\varepsilon}{w_{\max}}\right) .$$

As $\ln(1+x) \approx x$ for small x values, we conclude that $|\ln\tilde{r}_j - \ln r_j| < \frac{\varepsilon}{w_{\max}}$. The length of the approximated spiral arc contained in $\mathcal{D}$ can therefore be at most $L^*(\gamma_i) + 2\sqrt{2}\varepsilon$. □

Corollary 6. *For each two endpoints s and g, it is possible to use the graph $\mathcal{D}$ and compute a near-optimal backbone path $\tilde{\gamma}$ connecting s and g in $O\left(\frac{\Lambda^2}{\varepsilon^2}\right)$ time, such that $L^*(\tilde{\gamma}) < L^*(\gamma^*) + O(n)\varepsilon$.*

4 Optimal Corridors Amidst Polygonal Obstacles

In this section we generalize the data structures introduced in Section 3 to compute optimal corridors amidst polygonal obstacles. As we did in case of point obstacles, we first examine how an optimal backbone path looks like in the vicinity of a single obstacle. Note that the polygon P can be viewed as a collection of points (vertices) and line segments (edges), such that the distance of a point $q \in \mathbb{R}^2$ to P is attained on a polygon vertex or in the interior of an edge. We can thus subdivide the plane into regions, such that the identity of the closest polygon feature is the same for all points in any of the regions. Using the analysis we performed in Section 3.1 we already know that the optimal backbone path in a region closest to a polygon vertex is an arc of a logarithmic spiral. We now study the case of two points that lie in a region closest to a polygon edge.

Without loss of generality, we shall assume that the polygon edge we consider is an arbitrarily long segment of the vertical line $x = 0$, and analyze the optimal backbone path γ between two points s and g, whose distance from this line is less than $w_{\max}$ (see Figure 1(b) for an illustration). Note that the width of the corridor at $\gamma(t) = (x(t), y(t))$ simply equals $|x(t)|$.

We begin by approximating the backbone path by a polyline. Assume that $\gamma(t)$ passes through a point $p_0 = (x_0, 0)$ and forms an angle α_0 with the line $y = 0$ perpendicular to the obstacle. For any $\Delta x > 0$ we can define the lines $x = x_0, x = x_0 + \Delta x, x = x_0 + 2\Delta x, \ldots$, where each two neighboring lines define a vertical slab; since Δx is small we assume that the distance of all points in the slab from the obstacle is constant and equals $x_0 + k\Delta x$. We can now use Equation (2) and write: $\sin\alpha_1 = \frac{x_0+\Delta x}{x_0}\sin\alpha_0$, $\sin\alpha_2 = \frac{x_0+2\Delta x}{x_0+\Delta x}\sin\alpha_1 = \frac{x_0+2\Delta x}{x_0}\sin\alpha_0$, ..., $\sin\alpha_k = \frac{x_0+k\Delta x}{x_0}\sin\alpha_0$. If we examine the kth slab we can write $x = x_0 + k\Delta x$, so we have:

$$\Delta y_k = \Delta x \tan\alpha_k = \Delta x \cdot \frac{\sin\alpha_k}{\sqrt{1-\sin^2\alpha_k}} = \Delta x \cdot \frac{x\sin\alpha_0}{\sqrt{x_0^2 - x^2\sin^2\alpha_0}} . \tag{4}$$

Letting Δx tend to zero we obtain a smooth curve. We can use Equation (4) to express the derivative of the curve and we obtain:

$$y'(x) = \lim_{\Delta x \longrightarrow 0} \frac{\Delta y_k}{\Delta x} = \frac{x\sin\alpha_0}{\sqrt{x_0^2 - x^2\sin^2\alpha_0}} , \tag{5}$$

$$y(x) = -\frac{1}{\sin\alpha_0}\sqrt{x_0^2 - x^2\sin^2\alpha_0} + K . \tag{6}$$

As the point $(x_0, 0)$ lies on the curve, it is easy to see that the constant K equals $x_0\cot\alpha_0$.

Observe that $y(x)$ is defined only for $x < \frac{x_0}{\sin\alpha_0}$. When $x = \frac{x_0}{\sin\alpha_0}$ the path is reflected from the vertical wall and starts approaching the obstacle. We note that squaring and re-arranging Equation (6) we obtain that $x^2 + (y - x_0\cot\alpha_0)^2 = \left(\frac{x_0}{\sin\alpha_0}\right)^2$, thus we conclude that γ is a circular arc, whose supporting circle is centered at $C = (0, x_0\cot\alpha_0)$ and its radius is $\frac{x_0}{\sin\alpha_0}$.

Proposition 7. *Given a start position $s = (x_s, y_s)$ and a goal position $g = (x_g, y_g)$ in the vicinity of a segment supported by $x = 0$ and with $0 < x_s, x_g \leq w_{\max}$, the backbone of the optimal corridor between these two endpoints is a circular arc supported by a circle of radius r^* that is centered at $(0, y^*)$, where (we assume that $y_s \neq y_g$, otherwise the optimal backbone path is simply the line segment sg):*

$$y^* = \frac{y_s + y_g}{2} + \frac{x_g^2 - x_s^2}{2(y_g - y_s)} , \tag{7}$$

$$r^* = \sqrt{\frac{1}{2}(x_s^2 + x_g^2) + \frac{1}{4}(y_g - y_s)^2 + \frac{(x_g^2 - x_s^2)^2}{4(y_g - y_s)^2}} . \tag{8}$$

4.1 Moving Amidst Multiple Polygons

We are given a set $\mathcal{P} = \{P_1, \ldots, P_k\}$ of polygonal obstacles having n vertices in total, along with a preferred corridor width $w_{\max}$.

We first mention that if the polygons are well-separated, namely the distance between each P_i and P_j $(1 \leq i < j \leq k)$ is more than $2w_{\max}$, we can use the visibility graph of the dilated polygons to plan optimal backbone paths. The dilated obstacles in this case are Minkowski sums of the polygonal obstacles with a disc of radius $w_{\max}$ and their boundary comprises line segments, which correspond to dilated polygon edges, and circular arcs, which correspond to dilated vertices. Visibility edges in this case correspond to line segments tangent to two circular arcs. Proving that the visibility graph indeed contains optimal backbone paths is done exactly the same as we did in Section 3.2 for point obstacles.

In case there exist narrow passages between the obstacles, we generalize the construction detailed in Section 3.3 to polygons, and introduce the bounded Voronoi diagram of the set of polygons $\mathcal{P}$. Note that in this case we have *Voronoi chains* that are sequences of Voronoi edges. A Voronoi edge may be induced by two polygon vertices or by two polygon edges, in which case it is a line segment, or by a polygon vertex and an edge of another polygon, in which case it is a parabolic arc. Thus, the Voronoi chains are smooth curves that are piecewise linear or piecewise parabolic and are equidistant to two nearest polygons; see, e.g., [8] for more details. The bounded Voronoi diagram $\hat{\mathcal{V}}(\mathcal{P})$ also contains edges that separate the Voronoi cells of adjacent polygon features, namely a polygon edge and a vertex incident to this edge. These edges are line segments perpendicular to the obstacles (see Figure 3 for an illustration).

Observe that if we are given two points on the same Voronoi chain, then the locally optimal backbone path between them is simply the segment of the chain they define. This is clear in case of point obstacles, as the edges are straight line segments. In case of chains that separate Voronoi cells of polygons and may contain parabolic arcs this fact is less obvious. However, we are able to prove that parabolic arcs are also locally optimal — namely, it is not possible to shortcut such an arc by choosing a shorter route that is closer to one of the polygons, as such a route always has a larger weighted length. This proof is rather technical and we refer the reader to [18] for its details.

$\hat{\mathcal{V}}(\mathcal{P})$ subdivides the plane into cells of three types: regions where the clearance is larger than $w_{\max}$, Voronoi cells of polygon vertices, and Voronoi cells of polygon edges. We have already encountered cells of the first two types in the bounded Voronoi diagram of a set of points (Section 3.3). We also know from Proposition 7 that if we have two points in the Voronoi cell of a polygon edge, the optimal backbone path connecting them is a circular arc whose center lies on this edge. Assume, without loss of generality, that the obstacle edge lies on the line $y = 0$ and that the center of the circular arc a is the origin, and let $r^* e^{i\theta_1}$ and $r^* e^{i\theta_2}$ be the arc endpoints. The weighted length of the circular arc is therefore given by (note that $r(\theta) = r^*$):

$$L^*(a) = \int_{\theta_1}^{\theta_2} \frac{w_{\max}}{r^* \sin\theta} \sqrt{r^2(\theta) + \left(\frac{dr}{d\theta}\right)^2(\theta)}\, d\theta = \int_{\theta_1}^{\theta_2} \frac{w_{\max}}{\sin\theta}\, d\theta =$$

$$= w_{\max} \left(\ln \frac{1 - \cos\theta}{\sin\theta} \right) \Bigg|_{\theta_1}^{\theta_2} = w_{\max} \left(\ln \tan\frac{\theta_2}{2} - \ln \tan\frac{\theta_1}{2} \right) .$$

The approximation algorithm given in Section 3.3 can also be extended to handle polygonal obstacles. In this case we also consider intervals that lie on Voronoi edges that separate the Voronoi cell of each polygon into simple regions — thus, each region is induced by a polygon vertex, a polygon edge, or correspond to regions where the clearance is above $w_{\max}$. We can show that Lemma 5 also

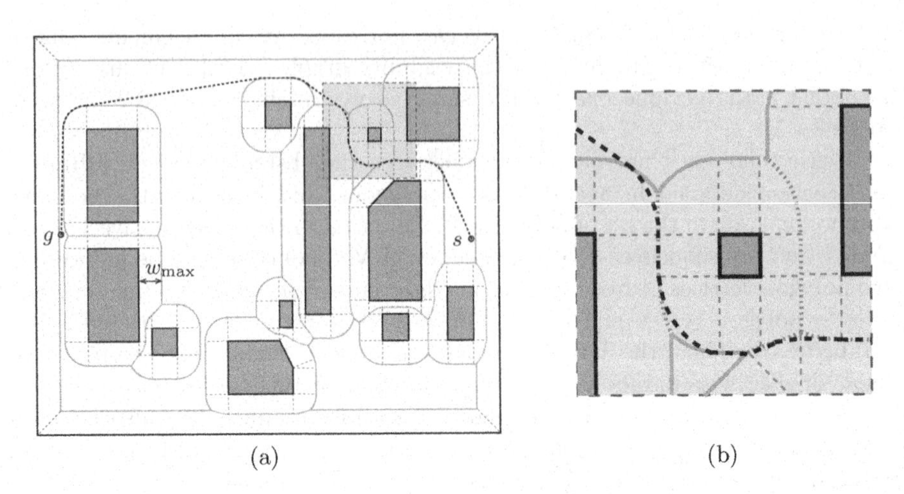

Fig. 3. (a) A near-optimal backbone path (dashed) amidst polygonal obstacles, overlayed on top of the the bounded Voronoi diagram of the obstacles. Boundary edges are drawn in light solid lines, Voronoi chains between polygons are dotted, and Voronoi edges that separate cells of adjacent polygon features are drawn in a light dashed line. The bounded Voronoi diagram was computed using the software described in [19]. The backbone path was computed using an A* algorithm on a fine grid discretizing the environment. (b) Zooming on a portion of the path; notice the shortcuts that the path takes.

applies for the circular arcs inside a Voronoi cell of a polygon edge: Let γ_i be such a circular arc and let I_1 and I_2 be the intervals containing its endpoints q_1 and q_2, receptively. $\mathcal{D}$ contains a circular arc σ connecting I_1 and I_2, and we denote its endpoints $\tilde{q}_j = \tilde{r}_j e^{i\tilde{\theta}_j} \in I_j$ (for $j = 1, 2$). As $c(q_j) = r^* \sin\theta_j$, we know that the length of each interval is $\|I_j\| = \frac{r^* \sin\theta_j}{w_{\max}}\varepsilon$. If we denote $\Delta\theta_j = \theta_j - \tilde{\theta}_j$, we can write:

$$\sin\left(\frac{\Delta\theta_j}{2}\right) < \frac{\frac{1}{2}\|I_j\|}{r^*} = \frac{\varepsilon}{2w_{\max}}\sin\theta_j \; .$$

As for small angles $\sin\phi \approx \phi$, we conclude that $|\Delta\theta_j| < \frac{\varepsilon}{w_{\max}}\sin\theta$. If we use the fact that $f(x + \Delta x) \approx f(x) + f'(x)\Delta x$ (for small Δx) with $f(x) = \ln\tan\frac{x}{2}$, we can bound the weighted length of σ (recall that $f'(x) = \frac{1}{\sin x}$ in our case):

$$L^*(\sigma) = w_{\max}\left(\ln\tan\frac{\theta_2 + \Delta\theta_2}{2} - \ln\tan\frac{\theta_1 + \Delta\theta_1}{2}\right) <$$

$$w_{\max}\left(\ln\tan\frac{\theta_2}{2} + \frac{\varepsilon\sin\theta_2}{w_{\max}}\cdot\frac{1}{\sin\theta_2} - \ln\tan\frac{\theta_1}{2} + \frac{\varepsilon\sin\theta_1}{w_{\max}}\cdot\frac{1}{\sin\theta_1}\right) = L^*(\gamma_i) + 2\varepsilon \; .$$

Corollary 8. *Given a set of polygonal obstacles $\mathcal{P}$ having n vertices in total, let Λ be the total weighted length of the bounded Voronoi diagram $\hat{\mathcal{V}}(\mathcal{P})$ with* *respect to a given $w_{\max}$ value. Given $\varepsilon > 0$, we can construct a graph $\mathcal{D}$ over the intervals of $\hat{\mathcal{V}}(\mathcal{P})$ in $O\left(\frac{\Lambda^2}{\varepsilon^2}n\right)$ time, such that for each two endpoints s and g it is possible to use $\mathcal{D}$ and compute a near-optimal backbone of a corridor C connecting s and g. $L^*(C)$ is at most $O(n)\varepsilon$ more than the weighted length of the optimal corridor connecting s and g.*

5 Conclusions and Future Work

In this paper we have introduced a measure for the quality of corridors and studied the structure of optimal corridors amidst point obstacles and polygonal obstacles in the plane. We have devised an approximation algorithm for computing near-optimal corridors amidst obstacles. We are also investigating methods to speed up our approximation algorithm, as well as design simple practical methods to compute good corridors. We are interested in extending our result to corridors in three dimensions as well.

In some applications having a winding backbone path decreases the quality of the corridor. We can therefore augment the weighted length function by considering the curvature of the backbone path γ as follows:

$$L^*_\mu(C) = \int_\gamma \left(\frac{w_{\max}}{w(t)}\right)^{d-1} dt + \mu \int_\gamma w(t)\kappa(t)dt\ , \tag{9}$$

where $\kappa(t)$ is the curvature of $\gamma(t)$, and $0 < \mu \leq 1$ is the weight we give to the curvature measure. We are able to show that in case of well-separated obstacles, optimal corridors under the L^*_μ measure are still contained in the visibility graph of the obstacles dilated by $w_{\max}$. We are still exploring methods of computing optimal corridors in the case of denser scenes.

References

1. Choset, H., Lynch, K.M., Hutchinson, S., Kantor, G., Burgard, W., Kavraki, L.E., Thrun, S.: Principles of Robot Motion: Theory, Algorithms, and Implementations, ch.7. MIT Press, Boston (2005)
2. Gray, A.: Modern Differential Geometry of Curves and Surfaces with Mathematica. In: Logarithmic Spirals, 2nd edn., pp. 40–42. CRC Press, Boca Raton (1997)
3. Kamphuis, A., Overmars, M.H.: Motion planning for coherent groups of entities. In: Proc. IEEE Int. Conf. Robotics and Automation, pp. 3815–3822 (2004)
4. Kamphuis, A., Pettre, J., Overmars, M.H., Laumond, J.-P.: Path finding for the animation of walking characters. In: Proc. Eurographics/ACM SIGGRAPH Sympos. Computer Animation, pp. 8–9 (2005)
5. Kavraki, L.E., Švestka, P., Latombe, J.-C., Overmars, M.H.: Probabilistic roadmaps for path planning in high-dimensional configuration spaces. IEEE Trans. Robotics and Automation 12, 566–580 (1996)
6. Khatib, O.: Real-time obstacle avoidance for manipulators and mobile robots. Int. J. Robotics Research 5(1), 90–98 (1986)

7. Latombe, J.-C.: Robot Motion Planning, ch. 7. Kluwer Academic Publishers, Boston (1991)
8. Lee, D.-T., Drysdale III, R.L.: Generalization of Voronoi diagrams in the plane. SIAM J. on Computing 10(1), 73–87 (1981)
9. Mitchell, J.S.B.: Shortest paths and networks. In: Goodman, J.E., O'Rourke, J. (eds.) Handbook of Discrete and Computational Geometry, ch.27, 2nd edn., pp. 607–642. Chapman & Hall/CRC, Boca Raton (2004)
10. Mitchell, J.S.B., Papadimitriou, C.H.: The weighted region problem: Finding shortest paths through a weighted planar subdivision. J. of the ACM 38(1), 18–73 (1991)
11. Nieuwenhuisen, D., Kamphuis, A., Mooijekind, M., Overmars, M.H.: Automatic construction of roadmaps for path planning in games. In: Proc. Int. Conf. Computer Games: Artif. Intell., Design and Education, pp. 285–292 (2004)
12. Nieuwenhuisen, D., Overmars, M.H.: Motion planning for camera movements. In: Proc. IEEE Int. Conf. Robotics and Automation, pp. 3870–3876 (2004)
13. Nieuwenhuisen, D., Overmars, M.H.: Useful cycles in probabilistic roadmap graphs. In: Proc. IEEE Int. Conf. Robotics and Automation, pp. 446–452 (2004)
14. Ó'Dúnlaing, C., Yap, C.K.: A "retraction" method for planning the motion of a disk. J. Algorithms 6, 104–111 (1985)
15. Overmars, M.H.: Path planning for games. In: Proc. 3rd Int. Game Design and Technology Workshop, pp. 29–33 (2005)
16. Reif, J., Wang, H.: Social potential fields: a distributed behavioral control for autonomous robots. In: Goldberg, K., Halperin, D., Latombe, J.-C., Wilson, R. (eds.) International Workshop on Algorithmic Foundations of Robotics, pp. 431–459. A. K. Peters (1995)
17. Russel, S., Norvig, P.: Artificial Intelligence: A Modern Approach, 2nd edn. Prentice Hall, Englewood Cliffs (2002)
18. Wein, R., van den Berg, J., Halperin, D.: Planning near-optimal corridors amidst obstacles. Technical report, Tel-Aviv University (2006), `http://www.cs.tau.ac.il/~wein/publications/pdfs/corridors.pdf`
19. Wein, R., van den Berg, J.P., Halperin, D.: The visibility-Voronoi complex and its applications. In: Proc. 21st Annu. ACM Sympos. Comput. Geom., pp. 63–72 (2005); In: Computational Geometry: Theory and Applications (to appear)

Using Motion Primitives in Probabilistic Sample-Based Planning for Humanoid Robots

Kris Hauser[1], Timothy Bretl[1], Kensuke Harada[2], and Jean-Claude Latombe[1]

[1] Computer Science Department, Stanford University
{khauser,tbretl}@stanford.edu and latombe@cs.stanford.edu
[2] National Institute of Advanced Industrial Science and Technology (AIST)
kensuke.harada@aist.go.jp

Abstract. This paper presents a method of computing efficient and natural-looking motions for humanoid robots walking on varied terrain. It uses a small set of high-quality motion primitives (such as a fixed gait on flat ground) that have been generated offline. But rather than restrict motion to these primitives, it uses them to derive a sampling strategy for a probabilistic, sample-based planner. Results in simulation on several different terrains demonstrate a reduction in planning time and a marked increase in motion quality.

1 Introduction

In this paper we present a method of planning efficient and natural-looking motions for humanoid robots on varied terrain. One thing that makes this problem difficult is that although humanoids have many degrees of freedom (DOF), we do not know in advance which of these DOF are actually useful, nor which contacts may be needed. On easy terrain like flat ground or stairs of fixed height, the motion of a humanoid is lightly constrained, most of its DOF are redundant, and only feet need contact the ground. On hard terrain like steep rock or urban rubble, the motion of a humanoid is highly constrained, most of its DOF are essential, and additional contacts (hands, knees, shoulders) might be required for balance. On varied terrain, the number of relevant DOF and the types of required contacts may change from step to step.

Consequently, planners that simplify the problem by considering a subset of the robot's DOF work well on easy terrain, but are not flexible enough to handle varied terrain. For example, one strategy for a humanoid on mostly flat ground is to precompute a library of feasible steps [22]. Each step is a continuous trajectory that places one foot in a new location relative to the other. Motions are constructed as a sequence of these steps. Because this only requires searching a graph, rather than a high-dimensional configuration space, it can be done quickly. More importantly, because the steps are precomputed, the resulting motion is efficient and robust, and looks natural. However, when the ground is not flat – in particular, when hands are required for balance – this approach may not be able to find a feasible motion.

S. Akella et al. (Eds.): Algorithmic Foundation of Robotics VII, STAR 47, pp. 507–522, 2008.
springerlink.com

Conversely, planners that consider all of the robot's DOF work well on hard terrain, but do not generate efficient or natural-looking motions (when this is possible) on varied terrain. For example, one strategy for a humanoid on severely uneven ground (based on earlier work for a free-climbing robot [5]) begins by identifying a number of potentially useful contacts [16]. Each mapping of hands or feet to contacts is a *stance*, associated with a (possibly empty) set of feasible configurations that satisfy all motion constraints. The robot can take a step from one stance to another if they differ by a single contact and if they share a feasible configuration, called a *transition*. The planner proceeds in two stages: first, it generates a candidate sequence of contacts by finding transitions between stances; then, it refines this sequence into a feasible, continuous trajectory by finding paths between subsequent transitions. Probabilistic, sample-based algorithms are used to find both transitions and paths. This approach is fast on irregular and steep terrain, because in this situation the robot's motion is most constrained just as it makes or breaks a contact. But when the ground is flat, this approach takes longer than the one of [22], and may generate needless motions of the arms or other DOF that are not required for balance. These motions are hard to eliminate in post-processing.

Rather than select one approach or the other, our planner combines the strengths of both. First, we generate a small set of high-quality *motion primitives* (similar to [22]), that might include a single step on flat ground, or an arm movement that places a hand on a wall for balance. Here, these primitives are produced by a lengthy off-line precomputation, but they might also be designed by hand or even captured or learned from examples of human motion. We record each motion primitive as a nominal path through the robot's configuration space (a joint-angle trajectory). Then, we use the two-stage strategy of [5, 16] to plan motions of the humanoid on-the-fly. But instead of sampling across all of configuration space to find transitions between stances and paths between transitions, we sample in a growing distribution around the nominal path associated with a chosen motion primitive. Although still preliminary, our simulation results demonstrate a reduction in planning time and a marked increase in motion quality[1] for a humanoid walking on varied terrain.

2 Related Work

Motion primitives and other types of maneuvers have been applied widely to robotics and digital animation. Four general strategies have been used:

Record and playback. This strategy restricts motion to a library of maneuvers. Natural-looking humanoid locomotion on mostly flat ground can be planned as a sequence of precomputed feasible steps [22]. Robust helicopter flight can be planned as a sequence of feedfoward control strategies (learned by observing

[1] Exactly how motion quality should be measured is an open question, beyond the scope of this paper. Here, we define quality as inversely proportional to a linear combination of path length and sum-squared distance from an upright posture.

skilled human operators) to move between trim states [10, 11, 12, 31]. Robotic juggling can be planned as a sequence of feedback control strategies [8]. The motion of peg-climbing robots can be planned as a sequence of actions like "grab the nearest peg" [3]. In these applications, a reasonably small library of maneuvers is sufficient to achieve most desired motions. For humanoid robots on varied terrain, such a library may grow to impractical size.

Warp, blend, or transform. Widely used for digital animation, this strategy also restricts motion to a library of maneuvers, but allows these maneuvers to be superimposed or transformed to better fit the task at hand. For example, captured motions of human actors can be "warped" to allow characters to reach different footfalls [40] or "retargetted" to control characters of different morphologies [13]. Of course, for a digital character it is most important to look good while for a humanoid robot it is most important to satisfy hard motion constraints. So although some techniques have been proposed to transform maneuvers while maintaining physical constraints [34,39], this strategy seems better suited for animation than robotics.

Model reduction. This strategy plans overall motion first, following this motion with a concatenation of primitives. For example, another way to generate natural-looking humanoid locomotion on flat ground is to approximate the robot as a cylinder, plan a 2-D collision-free path of this cylinder, and follow this path with a fixed gait [21, 19, 32, 20]. A similar method is used to plan the motion of nonholonomic wheeled vehicles [24, 23]. A related strategy plans the motion of key points on a robot or digital actor (such as the center of mass or related ground reference points [33]), tracking these points with an operational space controller [38]. These approaches work well when it does not matter much where a robot or digital actor contacts its environment. When the choice of contact location is critical, as is often the case for humanoids on varied terrain, it makes more sense to compute a sequence of footfalls first.

Bias inverse kinematic solutions. Like model reduction, this strategy first plans the motion of key points on a robot or digital actor, such as the location of hands or feet. But instead of a fixed controller, a search algorithm is used to compute a pose of the robot or actor at each instant that tracks these points (an inverse kinematic solution). One approach is to choose an inverse kinematic solution according to a probability density function learned from high-quality example motions [41, 15, 28, 29]. The set of examples give the resulting pose a particular "style." In fact, we take a similar approach in this paper, planning steps for a humanoid by sampling waypoints in a growing distribution around high-quality nominal paths.

3 Background

Our planner extends a similar one for humanoid robots [16], which was based on earlier work for a free-climbing robot [5]. Here, we summarize our basic approach and describe the limitations we address by using motion primitives.

(a) (b)

Fig. 1. (a) The humanoid robot HRP-2 [18]. (b) Example of varied terrain.

3.1 Motion Constraints

We consider the humanoid HRP-2 (Fig. 1(a)). A *configuration* q consists of 6 parameters defining the position and orientation of the torso and a list of 30 revolute joint angles. The set of all such q is the *configuration space*, denoted $\mathcal{Q}$, of dimensionality 36. We consider terrain that might include a mixture of flat, sloped, or rocky ground (Fig. 1(b)). We assume that this terrain and all robot links are perfectly rigid. We also assume that we are given in advance a set of links (such as hands, feet, or knees) that are allowed to touch the terrain. We call the placement of a link on the terrain a *contact*, and fix the position and orientation of the link while the contact is maintained. We call a set of simultaneous contacts a *stance*, denoted by σ. Consider a stance σ with $n \geq 1$ contacts. The *feasible space* $\mathcal{F}_\sigma$ is the set of all feasible configurations of the robot at σ. To be in $\mathcal{F}_\sigma$, a configuration q must satisfy several constraints:

Contact. The n contacts form a linkage with multiple closed-loop chains, so q must satisfy inverse kinematic equations. Let $\mathcal{Q}_\sigma \subset \mathcal{Q}$ be the set of all configurations q that satisfy these equations. This set is a possibly empty sub-manifold of $\mathcal{Q}$ of dimensionality $36 - 6n$, which we call the *stance manifold.*

Equilibrium. To balance at a fixed stance σ, HRP-2 must apply forces at contacts in σ that compensate for gravity without slip. For valid forces to exist, HRP-2's center of mass (CM) must lie above its *support polygon.* On varied terrain, this polygon does not always correspond to the base of HRP-2's feet [7, 5, 6]. So we model each contact as a set of frictional points, and compute the support polygon as in [16, 6]. When the CM lies above this polygon, we also check that joint torques achieving the required contact forces are within bounds.

Collision. In addition to satisfying joint angle limits, the robot must avoid collision with the environment (except at contacts) and with itself [14, 37].

3.2 Motion Planning

We assume HRP-2 moves from one place to another by taking a sequence of *steps.* Each step is a continuous motion at a fixed stance that ends by making or breaking a single contact. In particular, suppose the robot begins at a configuration $q \in \mathcal{F}_\sigma$ at a stance σ. A single step from q consists of three parts: first, a contact that is made or broken to move from σ to a new stance σ'; second, a configuration $q' \in \mathcal{F}_\sigma \cap \mathcal{F}_{\sigma'}$, which we call a *transition*, that is feasible at both σ and σ'; third, a feasible path in $\mathcal{F}_\sigma$ from q to q'.

Following the approach of [16, 5], we make these three choices hierarchically. To find a contact, we randomly sample potential placements of the robot's links in the terrain (or select a placement in σ to release). We use heuristics to decide which placement is most likely to lead toward the goal. To find a transition given σ', we randomly sample configurations in $\mathcal{Q}_\sigma$ (or in $\mathcal{Q}_{\sigma'}$ if $\sigma \subset \sigma'$) and reject them if they are not in $\mathcal{F}_\sigma \cap \mathcal{F}_{\sigma'}$. We use the combination of a bounding-volume technique similar to [9] and an iterative Newton-Raphson method to sample configurations in $\mathcal{Q}_\sigma$ (which has zero measure in $\mathcal{Q}$). To find a path given q', we use a variant of the probabilistic roadmap (PRM) algorithm called SBL [36]. This algorithm is bidirectional (growing trees, as in [25], from both q and q') and lazy (delaying the creation of local paths until a candidate sequence of milestones is found).

3.3 Current Limitations

Our search strategy postpones finding one-step paths (a costly computation) until after finding transitions and contacts [16, 5]. It works well for HRP-2 on irregular and steep terrain because in this situation, the robot's motion is most constrained just as it makes or breaks a contact. In particular, we have observed in our experiments that if $q \in \mathcal{F}_\sigma \cap \mathcal{F}_{\sigma'}$ and $q' \in \mathcal{F}_\sigma \cap \mathcal{F}_{\sigma''}$ exist, then a path between q and q' in $\mathcal{F}_\sigma$ likely also exists.

However, because we randomly sample each transition and use PRM to plan each one-step path, the motions we generate are feasible (given an accurate terrain model) but not necessarily high-quality. For example, when HRP-2 walks on terrain that is not irregular and steep, its motion is lightly constrained. Each step we generate might contain strange or erratic motions of the arms and legs. These motions are difficult to eliminate in post-processing.

Also, because we randomly sample each contact, we might end up trying difficult steps when simpler ones would have led to the goal as well. For example, the robot might reach a stance σ associated with a feasible space $\mathcal{F}_\sigma$ containing a narrow passage. With only a small perturbation of the contacts at σ, this narrow passage is likely to disappear [17]. So although additional steps might still be possible, they would be easier to compute if we had made a better choice of contacts at σ.

4 Generating Motion Primitives

We address the limitations of our planner by using a library of *motion primitives*. Each primitive is a single step of very high quality. In this section, we describe how we generate primitives. In the following section, we will describe how they guide our selection of paths, transitions, and contacts.

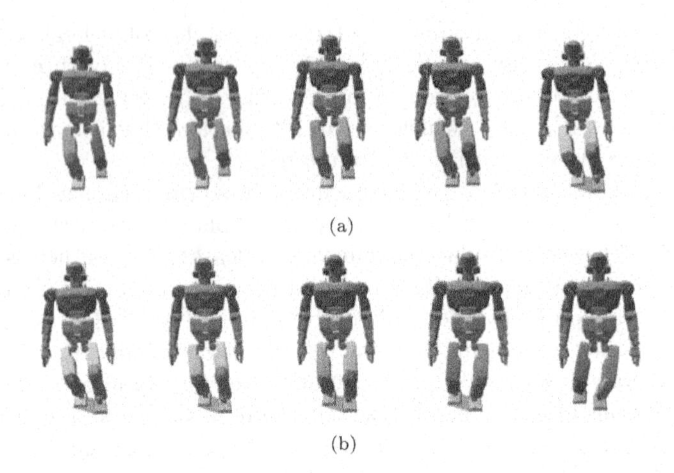

Fig. 2. Two primitives on flat ground, to (a) place a foot and (b) remove a foot. The support polygon – here, just the convex hull of supporting feet – is shaded blue.

Currently, it is the responsibility of the user to decide which primitives to include in the library. First, we need to identify a small but representative set of steps to be learned and to specify start and goal stances (differing by a single contact) for each one. These steps should be both important (often repeated) and broadly applicable (similar to a wide variety of other steps). For example, we might choose to include several consecutive steps on flat ground, each placing or removing a foot (Fig. 2). Next, we need to define a weighted set of criteria to judge the quality of each step. For example, we might choose to minimize path length, torque, energy, or the amount of deviation from an upright posture. Finally, we need to decide whether to accept or reject a candidate primitive, because we are not guaranteed that our optimization criteria correspond to our aesthetic notion of what is "natural."

It is the responsibility of the planner to actually compute each primitive. First, we generate an initial trajectory between the given start and goal stances by randomly sampling a feasible transition and creating a path to reach it using PRM, as in [16, 5]. Then, we optimize this trajectory with respect to the given objective function using a standard nonlinear optimization package [26]. This entire process is an off-line precomputation; several hours were required to generate the two example primitives in Fig. 2.

The generation of motion primitives has not been the main focus of our work (this paper concerns their application to planning), so many improvements may be possible. For example, we expect better results to be obtained by using the method of optimization proposed by [4]. Likewise, we might use a learned classifier to decide (without supervision) whether candidate primitives look natural, as in [35]. Finally, we might automate the selection of primitives to include in our library by learning a statistical model of importance (similar to location-based activity recognition [27]) or applicability after perturbation (similar to PRM planning with model uncertainty [30]).

We record each primitive in our library as a nominal path

$$u\colon t \in [0,1] \to u(t) \in \mathcal{Q}$$

in configuration space that does one of two things:

- *Adds a contact.* For some σ and σ' such that $\sigma \subset \sigma'$, u is a feasible path in $\mathcal{F}_\sigma$ from $u(0) \in \mathcal{F}_\sigma$ to $u(1) \in \mathcal{F}_\sigma \cap \mathcal{F}_{\sigma'}$.
- *Removes a contact.* For some σ and σ' such that $\sigma \supset \sigma'$, u is a feasible path in $\mathcal{F}_\sigma$ from $u(0) \in \mathcal{F}_\sigma$ to $u(1) \in \mathcal{F}_\sigma \cap \mathcal{F}_{\sigma'}$.

We will denote the start and goal stances for each primitive u by σ_u and σ'_u, respectively. In general, u will only define a feasible step between σ_u and σ'_u, but we will see in the next section that it can still be used to help guide our choice of path, transition, and contact to reach other stances.

5 Using Primitives for Planning

We use motion primitives to help our planner generate each step. We do this at three levels: finding a path (given a transition and a final stance), finding a transition (given only the final stance), and finding a contact (in order to define the final stance). In each case, first we transform the primitive to better match the step we are trying to plan, then we apply the transformed primitive to bias the sampling strategy used by our planner.

5.1 Finding Paths

Consider the robot at an initial configuration $q_{\text{initial}} \in \mathcal{F}_\sigma$ at an initial stance σ. Assume that we are given a final stance σ' and a transition $q_{\text{final}} \in \mathcal{F}_\sigma \cap \mathcal{F}_{\sigma'}$ (recall q_{final} is a configuration feasible at both σ and σ'). Also assume that we are given an appropriate primitive $u \subset \mathcal{Q}$ (as described in Section 4). We want to use u to guide our search for a path from q_{initial} to q_{final} in $\mathcal{F}_\sigma$. As before, we use SBL (a variant of PRM) to grow trees from root configurations [36]. But rather than root these trees only at q_{initial} and q_{final}, we root them at additional configurations (similar to [1]) sampled according to the primitive u.

Transforming the primitive to match $q_{initial}$ and q_{final}. Although we assume u is similar to the step we are trying to plan, it will not be identical. So first, we transform u so that it starts at q_{initial} and ends at q_{final}. We have chosen to use an affine transformation of the form

$$\hat{u}(t) = A\left(u(t) - u(0)\right) + q_{\text{initial}} \tag{1}$$

that maps the straight-line segment between $u(0)$ and $u(1)$ to the segment between q_{initial} and q_{final}. In other words,

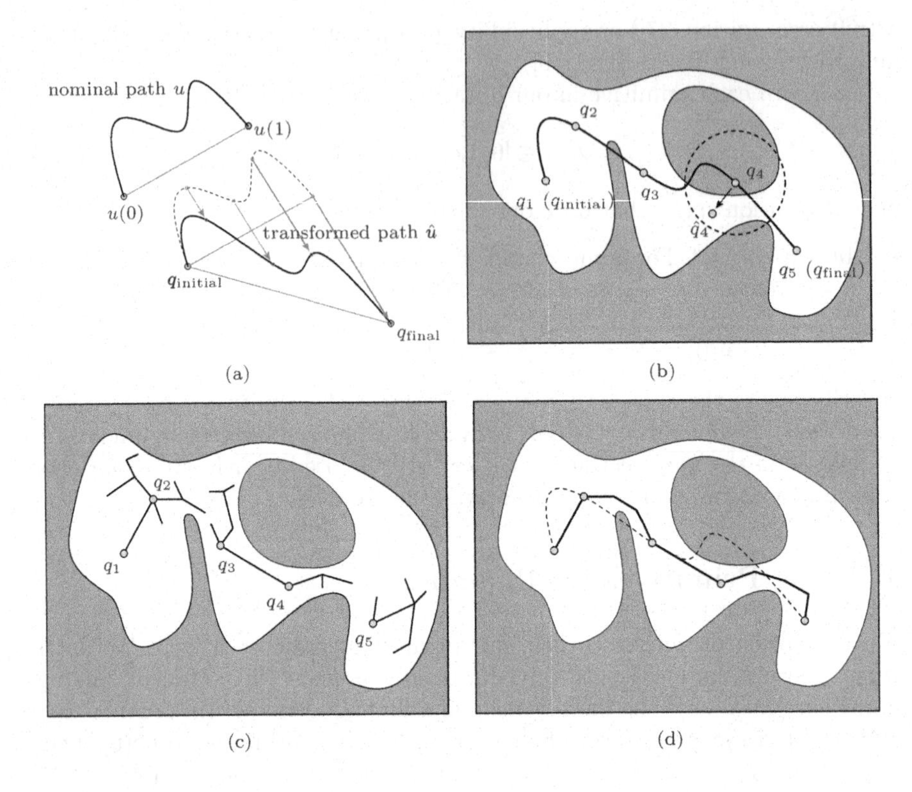

Fig. 3. Using a primitive to guide path planning. (a) Transforming a motion primitive to start at q_{initial} and end at q_{final}. (b) Sampling root milestones in $\mathcal{F}_\sigma$ near equally spaced waypoints along $\hat{u}$. (c) Growing trees to connect neighboring roots. (d) The resulting path, which if possible is close to $\hat{u}$ (dotted).

$$\begin{aligned}\hat{u}(0) &= A\,(u(0) - u(0)) + q_{\text{initial}} \\ &= 0 + q_{\text{initial}} \\ &= q_{\text{initial}}\end{aligned} \qquad \begin{aligned}\hat{u}(1) &= A\,(u(1) - u(0)) + q_{\text{initial}} \\ &= (q_{\text{final}} - q_{\text{initial}}) + q_{\text{initial}} \\ &= q_{\text{final}}\end{aligned}$$

In particular, we select A closest to the identity matrix, minimizing

$$\min_A \sum_{i,j} (A_{ij} - \delta_{i,j})^2 \text{ such that } A\,(u(1) - u(0)) = q_{\text{final}} - q_{\text{initial}}$$

where $\delta_{ij} = 1$ if $i = j$ and 0 otherwise. We compute A in closed form as

$$A = I + \frac{((q_{\text{final}} - q_{\text{initial}}) - (u(1) - u(0)))\,(u(1) - u(0))^T}{\|u(1) - u(0)\|_2^2}.$$

We can visualize this transformation as in Fig. 3(a). First, u is translated to start at q_{initial}. Then, the farther we move along u (the more we increase t), the closer $\hat{u}$ is pushed toward the segment from q_{initial} to q_{final}.

Sampling root milestones. Let $q_1, \ldots, q_n$ be configurations evenly distributed along $\hat{u}$ from q_{initial} to q_{final} (Fig. 3(b)). For each $i = 1, \ldots, n$, we test if $q_i \in \mathcal{F}_\sigma$. If so, we add q_i as a root milestone in our roadmap. If not, we repeatedly sample other configurations in a growing neighborhood of q_i until we find some feasible $q'_i \in \mathcal{F}_\sigma$, which we add as a root instead of q_i.

Connecting neighboring roots with sampled trees. For $i = 1, \ldots, n-1$, we check if the root milestone q_i can be connected to its neighbor q_{i+1} with a feasible local path (as in [16]). If not, we add the pair of roots (q_i, q_{i+1}) to a list $\mathcal{R}$. Then, we apply PRM to grow trees between every pair in $\mathcal{R}$. For example, in Fig. 3(c) we add (q_2, q_3) and (q_4, q_5) to $\mathcal{R}$ and grow trees to connect both q_2 with q_3 and q_4 with q_5. We process all trees in parallel. So at every iteration, for each pair $(q_i, q_{i+1}) \in \mathcal{R}$, we first add m milestones to the trees at both q_i and q_{i+1} (in our experiments, we set $m = 5$). Then, we find the configurations q connected to q_i and q' connected to q_{i+1} that are closest. If q and q' can be connected by a local path, we remove (q_i, q_{i+1}) from $\mathcal{R}$. When we connect all neighboring roots, we return the resulting path; if this does not happen after a fixed number of iterations, we return failure. Just like our original implementation, this approach will find a path between q_{initial} and q_{final} whenever one exists (given enough time). However, since we seed our roadmap with milestones that are close to u, we expect the resulting motion to be similar (and of similar quality) to this primitive whenever possible (Fig. 3(d)), deviating significantly from it only when necessary.

5.2 Finding Transitions

Again consider the robot at a configuration $q_{\text{initial}} \in \mathcal{F}_\sigma$ at a stance σ. But now, assume that we are only given a final stance σ', so we use a primitive u to guide our search for a transition before we plan a path to reach it.

Transforming the primitive to match σ and σ'. Since we do not know q_{final}, we can not use the same transformation (1) that we used for planning paths. Instead, we choose a rigid-body transformation of the form

$$\hat{u}(t) = Au(t) + b \tag{2}$$

that maps the nominal stances σ_u and σ'_u (associated with the primitive u) as closely as possible to the stances σ and σ'.

Recall that a stance consists of several contacts, each placing a link of the robot on the terrain. If we model the surface of the terrain and all robot links as a triangular mesh, then we can define the location of each placement by a finite number of points $r_i \in \mathbb{R}^3$. For example, the face-face contact between a foot and the ground might be defined by the vertices r_1, r_2, and r_3 of a triangle. We consider these points to be attached to the robot, so if the foot is placed against a different face in the terrain, the points r_1, r_2, and r_3 move in $\mathbb{R}^3$ but remain in the same location relative to the foot. We will use these points to define our mapping between stances.

In particular, let $r_i \in \mathbb{R}^3$ for $i = 1, \ldots, m$ be the set of all points defining the contacts in both σ_u and σ'_u, and let $s_i \in \mathbb{R}^3$ for $i = 1, \ldots, m$ be the set of all points defining the contacts in both σ and σ'. (We assume u has been chosen so that both sets have the same number of points.) Then we choose the rotation matrix A and translation b in (2) that minimize

$$\min_{A,b} \sum_i \|Ar_i + b - s_i\|_2^2.$$

We can compute A and b in closed form [2]. But, we only consider rotations A about the gravity vector to avoid tilting the robot into an unstable orientation.

Sampling a transition. As before, we sample configurations $q \in \mathcal{Q}_\sigma$, keeping them if $q \in \mathcal{F}_\sigma \cap \mathcal{F}_{\sigma'}$. But rather than sample configurations completely at random, we sample them in a growing neighborhood of $\hat{u}(1)$. We expect a well-chosen transition to further improve the quality of the path to reach it.

5.3 Finding Contacts

Once more, consider the robot at a configuration $q_{\text{initial}} \in \mathcal{F}_\sigma$ and a stance σ. But now, assume we are given neither a final stance nor a transition, but only a primitive u. If u removes a robot link from the terrain, we immediately generate a final stance σ' by removing the corresponding contact from σ. But if u places a link in the terrain, we use it to guide our search for a new contact.

Transforming the primitive to match σ. We use the same transformation (2) to construct $\hat{u}$ as for finding transitions. But here, we compute A and b to map only σ_u to σ, since we do not know σ'. We use this transformation to adjust the placement of the new contact given by u. Let $r_i \in \mathbb{R}^3$ for $i = 1, \ldots, m$ be the set of points defining this contact. Then the transformed contact is given by $\hat{r}_i = Ar_i + b$ for $i = 1, \ldots, m$.

Sampling a contact. We define a sphere of radius δ, centered at $(1/m)\sum_i \hat{r}_i$. We increase δ until the intersection of this sphere with the terrain is non-empty (initially, we set δ approximately the size of HRP-2's foot). We randomly sample a placement of the points $\hat{r}_i$ on the surface of the terrain inside the sphere, by first sampling a position of their centroid $s \in \mathbb{R}^3$ on the surface, then sampling a rotation of $\hat{r}_i$ about the surface normal at s. We check that the contact defined by this placement has similar properties (normal vector, friction coefficient) to the contact defined by u. If so, we add it to σ to form σ'. If not, we reject it and sample another placement.

5.4 Deciding Which Primitive to Use

It only remains to decide which primitive u should be used, given an initial stance σ and configuration q_{initial}. We have experimented with a variety of heuristics. For example, we might pick the primitive that most closely matches σ_u with σ (in other words, that minimizes the error in a transformation of the form (2)). Likewise, we might pick the primitive that most closely matches σ'_u with the actual terrain. However, the best approach is still not clear, and this issue remains an important area for future work.

6 Results

An example of climbing a single stair. With each additional part of a step that we compute using a primitive, we add to the quality of the result. For example, consider the motion of HRP-2 in Fig. 4 to climb a single stair of height 0.3m (just below the knee). This motion was planned from scratch, by randomly sampling contacts and transitions and by using PRM to generate paths. The robot does not look natural – its arm and leg motions are erratic, and its step over the stair is needlessly long. To improve this motion, we applied the two primitives shown in Fig. 2 (steps on flat ground). Fig. 5 shows the result of using these primitives to plan each path. Some erratic leg motions are eliminated, such as the backward movement of the leg in the second frame. The erratic arm motions remain, however, because the transition in the fourth frame is the same (still randomly sampled). Fig. 6 shows the result of using primitives to adjust this transition as well as to plan paths, eliminating most of the erratic arm motions. Finally, Fig. 7 shows the result of using primitives to select contacts well as plan transitions and paths. The chosen contact resulted in a much easier step, eliminating the extreme lean in the fifth frame.

Planning time and motion quality for stairs of different heights. In our experiments, we have observed that planning time remains low and motion quality remains high even when we use a primitive to plan a step that is quite different. For example, we adapted the same two primitives in Fig. 2 to stairs of height 0.2m and 0.4m as well as 0.3m. Fig. 8 shows the results, averaged over five runs. Quality is measured by an objective function that penalizes both path length and deviations from an upright posture (lower values indicate higher quality). For comparison, we report the minimum objective value achieved after a lengthy off-line optimization. These results demonstrate that our use of primitives provides a modest reduction in planning time but significantly improves motion quality. Note also that both time and quality degrade gracefully as the step we are planning deviates further from the primitive.

A variety of other examples. We have tested our planner in many other example environments. Fig. 9 shows HRP-2 on uneven terrain (using the primitives in Fig. 2), in which the highest and lowest point differ by 0.5m. Fig. 10 shows HRP-2 climbing a ladder with rungs that have non-uniform spacing and that deviate from horizontal by up to 15°. The primitives for this example were generated on a ladder with horizontal, uniformly spaced rungs. Fig. 11 shows HRP-2 making several sideways steps among boulders, using the hands for support. Here, the primitives were generated by stepping sideways on flat ground while pushing against a vertical wall. Fig. 12 shows HRP-2 traversing very rough terrain with slopes up to 40°. This motion was generated with a larger set of primitives (including steps of several heights, a pivot step, and a high step using the hand for support). In all of these examples, contacts were sampled on-the-fly (using motion primitives), not placed by hand. Planning for the first three examples took about one minute on a 1.8 GHz PC. The fourth took example about eight minutes.

Fig. 4. Stair step planned entirely from scratch

Fig. 5. Primitives guide path planning, reducing unnecessary leg motions

Fig. 6. Primitives guide transition sampling, reducing unnecessary arm motions

Fig. 7. Primitives guide the choice of contact, resulting in an easier step

Stair height	From scratch		Adapt primitive		Optimal objective
	Time	Objective	Time	Objective	
0.2m	8.61	5.03	5.42	3.04	2.19
0.3m	10.3	4.67	4.08	2.31	2.17
0.4m	12.2	5.15	10.8	3.27	2.55

Fig. 8. Planning time and objective function values for stair steps, averaged over 5 runs

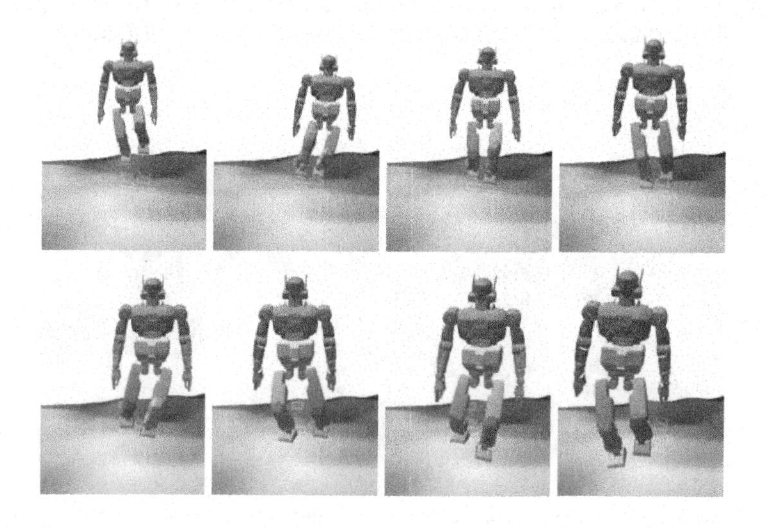

Fig. 9. A planar walking primitive adapted to slightly uneven terrain

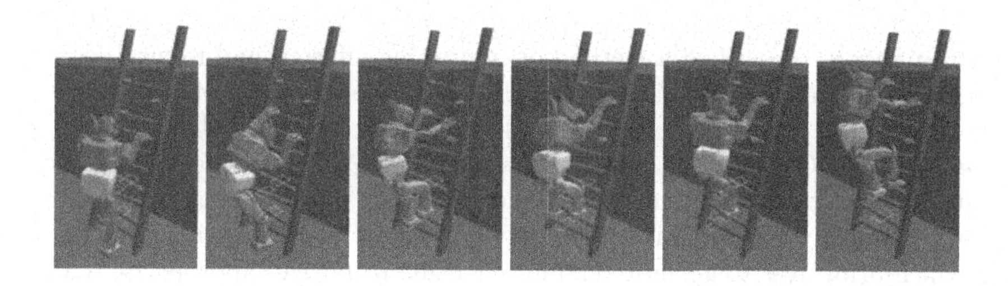

Fig. 10. A ladder climbing primitive adapted to a new ladder with uneven rungs

Fig. 11. A side-step primitive using the hands for support, adapted to a terrain with large boulders. Hand support is necessary because the robot must walk on a highly sloped boulder.

Fig. 12. A motion on steep and uneven terrain generated from a set of several primitives. A hand is being used for support in the third configuration.

7 Conclusion

In this paper we described a method of computing efficient and natural-looking motions for humanoids walking on varied terrain. We used a set of motion primitives, generated offline, to derive a sampling strategy for a probabilistic, sample-based planner. Our experimental results on several different examples demonstrated a reduction in planning time and a marked increase in motion quality. However, much work remains to be done. For example, our heuristics for deciding which primitives to generate and for choosing primitives appropriate to each step could be improved. One might even consider the use of several primitives concurrently, or the use of a primitive that encodes several steps rather than just a single step. Finally, even though primitives increase motion quality, a better method of post-processing would improve our results.

Acknowledgments. This work was partially supported by NSF grant 0412884. K. Hauser is supported by a Thomas V. Jones Stanford Graduate Fellowship.

References

1. Akinc, M., Bekris, K.E., Chen, B.Y., Ladd, A.M., Plaku, E., Kavraki, L.E.: Probabilistic roadmaps of trees for parallel computation of multiple query roadmaps. In: Int. Symp. Rob. Res., Siena, Italy (2003)
2. Arun, K., Huang, T., Blostein, S.: Least-squares fitting of two 3-d point sets. IEEE Trans. Pattern Anal. Machine Intell. 9(5), 698–700 (1987)
3. Bevly, D., Farritor, S., Dubowsky, S.: Action module planning and its application to an experimental climbing robot. In: IEEE Int. Conf. Rob. Aut., pp. 4009–4014 (2000)
4. Bobrow, J., Martin, B., Sohl, G., Wang, E., Park, F., Kim, J.: Optimal robot motions for physical criteria. J. of Robotic Systems 18(12), 785–795 (2001)
5. Bretl, T.: Motion planning of multi-limbed robots subject to equilibrium constraints: The free-climbing robot problem. Int. J. Rob. Res. 25(4), 317–342 (2006)
6. Bretl, T., Lall, S.: A fast and adaptive test of static equilibrium for legged robots. In: IEEE Int. Conf. Rob. Aut., Orlando (2006)
7. Bretl, T., Latombe, J.-C., Rock, S.: Toward autonomous free-climbing robots. In: Int. Symp. Rob. Res., Siena, Italy (2003)
8. Burridge, R., Rizzi, A., Koditschek, D.: Sequential composition of dynamically dexterous robot behaviors. Int. J. Rob. Res. 18(6), 534–555 (1999)
9. Cortés, J., Siméon, T., Laumond, J.-P.: A random loop generator for planning the motions of closed kinematic chains using prm methods. In: IEEE Int. Conf. Rob. Aut., Washington (2002)
10. Frazzoli, E., Dahleh, M.A., Feron, E.: Maneuver-based motion planning for nonlinear systems with symmetries. IEEE Trans. Robot. 25(1), 116–129 (2002)
11. Frazzoli, E., Dahleh, M.A., Feron, E.: Real-time motion planning for agile autonomous vehicles. AIAA J. of Guidance, Control, and Dynamics 25(1), 116–129 (2002)
12. Gavrilets, V., Frazzoli, E., Mettler, B., Peidmonte, M., Feron, E.: Aggressive maneuvering of small autonomous helicopters: A human-centered approach. Int. J. Rob. Res. 20(10), 795–807 (2001)
13. Gleicher, M.: Retargetting motion to new characters. In: SIGGRAPH, pp. 33–42 (1998)
14. Gottschalk, S., Lin, M., Manocha, D.: OBB-tree: A hierarchical structure for rapid interference detection. In: ACM SIGGRAPH, pp. 171–180 (1996)
15. Grochow, K., Martin, S.L., Hertzmann, A., Popović, Z.: Style-based inverse kinematics. ACM Trans. Graph. 23(3), 522–531 (2004)
16. Hauser, K., Bretl, T., Latombe, J.-C.: Non-gaited humanoid locomotion planning. In: Humanoids, Tsukuba, Japan (2005)
17. Hsu, D., Latombe, J., Kurniawati, H.: On the probabilistic foundations of probabilistic roadmap planning. In: Int. Symp. Rob. Res., San Francisco (2005)
18. Kaneko, K., Kanehiro, F., Kajita, S., Hirukawa, H., Kawasaki, T., Hirata, M., Akachi, K., Isozumi, T.: Humanoid robot HRP-2. In: IEEE Int. Conf. Rob. Aut., New Orleans, pp. 1083–1090 (2004)
19. Kovar, L., Gleicher, M., Pighin, F.: Motion graphs. In: SIGGRAPH, San Antonio, Texas, pp. 473–482 (2002)

20. Kron, T., Shin, S.Y.: Motion modeling for on-line locomotion synthesis. In: Eurographics/ACM SIGGRAPH Symposium on Computer Animation, Los Angeles, pp. 29–38 (2005)
21. Kuffner Jr., J.J.: Autonomous Agents for Real-Time Animation. PhD thesis, Stanford University (1999)
22. Kuffner Jr., J.J., Nishiwaki, K., Kagami, S., Inaba, M., Inoue, H.: Motion planning for humanoid robots. In: Int. Symp. Rob. Res., Siena, Italy (2003)
23. Laumond, J., Jacobs, P., Taix, M., Murray, R.: A motion planner for nonholonomic mobile robots. IEEE Trans. Robot. Automat. 10(5), 577–593 (1994)
24. Laumond, J.-P.: Finding collision-free smooth trajectories for a non-holonomic mobile robot. In: International Joint Conference on Artificial Intelligence (IJCAI), pp. 1120–1123 (1987)
25. LaValle, S.M., Kuffner Jr., J.J.: Rapidly-exploring random trees: progress and prospects. In: WAFR (2000)
26. Lawrence, C., Zhou, J., Tits, A.: User's guide for CFSQP version 2.5: A C code for solving (large scale) constrained nonlinear (minimax) optimization problems, generating iterates satisfying all inequality constraints. Technical Report TR-94-16r1, 20742, Institute for Systems Research, University of Maryland, College Park, MD (1997)
27. Liao, L., Fox, D., Kautz, H.: Location-based activity recognition. In: Advances in Neural Information Processing Systems (2005)
28. Liu, C.K., Hertzmann, A., Popović, Z.: Learning physics-based motion style with nonlinear inverse optimization. ACM Trans. Graph. 24(3), 1071–1081 (2005)
29. Meredith, M., Maddock, S.: Adapting motion capture data using weighted real-time inverse kinematics. Comput. Entertain. 3(1) (2005)
30. Missiuro, P.E., Roy, N.: Adapting probabilistic roadmaps to handle uncertain maps. In: IEEE Int. Conf. Rob. Aut., Orlando (2006)
31. Ng, A.Y., Kim, H.J., Jordan, M., Sastry, S.: Autonomous helicopter flight via reinforcement learning. In: Neural Information Processing Systems 16 (2004)
32. Pettré, J., Laumond, J.-P., Siméon, T.: A 2-stages locomotion planner for digital actors. In: Eurographics/SIGGRAPH Symp. Comp. Anim. (2003)
33. Popovic, M.B., Goswami, A., Herr, H.: Ground reference points in legged locomotion: Definitions, biological trajectories and control implications. Int. J. Rob. Res. 24(12), 1013–1032 (2005)
34. Popović, Z., Witkin, A.: Physically based motion transformation. In: SIGGRAPH, pp. 11–20 (1999)
35. Ren, L., Patrick, A., Efros, A.A., Hodgins, J.K., Rehg, J.M.: A data-driven approach to quantifying natural human motion. ACM Trans. Graph. 24(3), 1090–1097 (2005)
36. Sánchez, G., Latombe, J.-C.: On delaying collision checking in PRM planning: Application to multi-robot coordination. Int. J. of Rob. Res. 21(1), 5–26 (2002)
37. Schwarzer, F., Saha, M., Latombe, J.-C.: Exact collision checking of robot paths. In: WAFR, Nice, France (December 2002)
38. Sentis, L., Khatib, O.: Synthesis of whole-body behaviors through hierarchical control of behavioral primitives. Int. J. Humanoid Robotics 2(4), 505–518 (2005)
39. Shin, H.J., Lee, J., Shin, S.Y., Gleicher, M.: Computer puppetry: An importance-based approach. ACM Trans. Graph. 20(2), 67–94 (2001)
40. Witkin, A., Popović, Z.: Motion warping. In: SIGGRAPH, Los Angeles, CA, pp. 105–108 (1995)
41. Yamane, K., Kuffner, J.J., Hodgins, J.K.: Synthesizing animations of human manipulation tasks. ACM Trans. Graph. 23(3), 532–539 (2004)

Author Index

Springer Tracts in Advanced Robotics

Edited by B. Siciliano, O. Khatib and F. Groen

Vol. 47: Akella, S.; Amato, N.M.; Huang, W.H.; Mishra, B.; (Eds.)
Algorithmic Foundation of Robotics VII
524 p. 2008 [978-3-540-68404-6]

Vol. 46: Bessière, P.; Laugier, C.; Siegwart R. (Eds.)
Probabilistic Reasoning and Decision Making in Sensory-Motor Systems
375 p. 2008 [978-3-540-79006-8]

Vol. 45: Bicchi, A.; Buss, M.; Ernst, M.O.; Peer A. (Eds.)
The Sense of Touch and Its Rendering
281 p. 2008 [978-3-540-79034-1]

Vol. 44: Bruyninckx, H.; Přeučil, L.; Kulich, M. (Eds.)
European Robotics Symposium 2008
356 p. 2008 [978-3-540-78315-2]

Vol. 43: Lamon, P.
3D-Position Tracking and Control for All-Terrain Robots
105 p. 2008 [978-3-540-78286-5]

Vol. 42: Laugier, C.; Siegwart, R. (Eds.)
Field and Service Robotics
597 p. 2008 [978-3-540-75403-9]

Vol. 41: Milford, M.J.
Robot Navigation from Nature
194 p. 2008 [978-3-540-77519-5]

Vol. 40: Birglen, L.; Laliberté, T.; Gosselin, C.
Underactuated Robotic Hands
241 p. 2008 [978-3-540-77458-7]

Vol. 39: Khatib, O.; Kumar, V.; Rus, D. (Eds.)
Experimental Robotics
563 p. 2008 [978-3-540-77456-3]

Vol. 38: Jefferies, M.E.; Yeap, W.-K. (Eds.)
Robotics and Cognitive Approaches to Spatial Mapping
328 p. 2008 [978-3-540-75386-5]

Vol. 37: Ollero, A.; Maza, I. (Eds.)
Multiple Heterogeneous Unmanned Aerial Vehicles
233 p. 2007 [978-3-540-73957-9]

Vol. 36: Buehler, M.; Iagnemma, K.; Singh, S. (Eds.)
The 2005 DARPA Grand Challenge – The Great Robot Race
520 p. 2007 [978-3-540-73428-4]

Vol. 35: Laugier, C.; Chatila, R. (Eds.)
Autonomous Navigation in Dynamic Environments
169 p. 2007 [978-3-540-73421-5]

Vol. 34: Wisse, M.; van der Linde, R.Q.
Delft Pneumatic Bipeds
136 p. 2007 [978-3-540-72807-8]

Vol. 33: Kong, X.; Gosselin, C.
Type Synthesis of Parallel Mechanisms
272 p. 2007 [978-3-540-71989-2]

Vol. 32: Milutinović, D.; Lima, P.
Cells and Robots – Modeling and Control of Large-Size Agent Populations
130 p. 2007 [978-3-540-71981-6]

Vol. 31: Ferre, M.; Buss, M.; Aracil, R.; Melchiorri, C.; Balaguer C. (Eds.)
Advances in Telerobotics
500 p. 2007 [978-3-540-71363-0]

Vol. 30: Brugali, D. (Ed.)
Software Engineering for Experimental Robotics
490 p. 2007 [978-3-540-68949-2]

Vol. 29: Secchi, C.; Stramigioli, S.; Fantuzzi, C.
Control of Interactive Robotic Interfaces – A Port-Hamiltonian Approach
225 p. 2007 [978-3-540-49712-7]

Vol. 28: Thrun, S.; Brooks, R.; Durrant-Whyte, H. (Eds.)
Robotics Research – Results of the 12th International Symposium ISRR
602 p. 2007 [978-3-540-48110-2]

Vol. 27: Montemerlo, M.; Thrun, S.
FastSLAM – A Scalable Method for the Simultaneous Localization and Mapping Problem in Robotics
120 p. 2007 [978-3-540-46399-3]

Vol. 26: Taylor, G.; Kleeman, L.
Visual Perception and Robotic Manipulation – 3D Object Recognition, Tracking and Hand-Eye Coordination
218 p. 2007 [978-3-540-33454-5]

Vol. 25: Corke, P.; Sukkarieh, S. (Eds.)
Field and Service Robotics – Results of the 5th International Conference
580 p. 2006 [978-3-540-33452-1]

Vol. 24: Yuta, S.; Asama, H.; Thrun, S.; Prassler, E.; Tsubouchi, T. (Eds.)
Field and Service Robotics – Recent Advances in Research and Applications
550 p. 2006 [978-3-540-32801-8]

Vol. 23: Andrade-Cetto, J,; Sanfeliu, A.
Environment Learning for Indoor Mobile Robots – A Stochastic State Estimation Approach to Simultaneous Localization and Map Building
130 p. 2006 [978-3-540-32795-0]

Vol. 22: Christensen, H.I. (Ed.)
European Robotics Symposium 2006
209 p. 2006 [978-3-540-32688-5]

Vol. 21: Ang Jr., H.; Khatib, O. (Eds.)
Experimental Robotics IX – The 9th International Symposium on Experimental Robotics
618 p. 2006 [978-3-540-28816-9]

Vol. 20: Xu, Y.; Ou, Y.
Control of Single Wheel Robots
188 p. 2005 [978-3-540-28184-9]

Vol. 19: Lefebvre, T.; Bruyninckx, H.; De Schutter, J. Nonlinear Kalman Filtering for Force-Controlled Robot Tasks
280 p. 2005 [978-3-540-28023-1]

Vol. 18: Barbagli, F.; Prattichizzo, D.; Salisbury, K. (Eds.)
Multi-point Interaction with Real and Virtual Objects
281 p. 2005 [978-3-540-26036-3]

Vol. 17: Erdmann, M.; Hsu, D.; Overmars, M.; van der Stappen, F.A (Eds.)
Algorithmic Foundations of Robotics VI
472 p. 2005 [978-3-540-25728-8]

Vol. 16: Cuesta, F.; Ollero, A.
Intelligent Mobile Robot Navigation
224 p. 2005 [978-3-540-23956-7]

Vol. 15: Dario, P.; Chatila R. (Eds.)
Robotics Research – The Eleventh International Symposium
595 p. 2005 [978-3-540-23214-8]

Vol. 14: Prassler, E.; Lawitzky, G.; Stopp, A.; Grunwald, G.; Hägele, M.; Dillmann, R.; Iossifidis. I. (Eds.)
Advances in Human-Robot Interaction
414 p. 2005 [978-3-540-23211-7]

Vol. 13: Chung, W.
Nonholonomic Manipulators
115 p. 2004 [978-3-540-22108-1]

Vol. 12: Iagnemma K.; Dubowsky, S.
Mobile Robots in Rough Terrain – Estimation, Motion Planning, and Control with Application to Planetary Rovers
123 p. 2004 [978-3-540-21968-2]

Vol. 11: Kim, J.-H.; Kim, D.-H.; Kim, Y.-J.; Seow, K.-T.
Soccer Robotics
353 p. 2004 [978-3-540-21859-3]

Vol. 10: Siciliano, B.; De Luca, A.; Melchiorri, C.; Casalino, G. (Eds.)
Advances in Control of Articulated and Mobile Robots
259 p. 2004 [978-3-540-20783-2]

Vol. 9: Yamane, K.
Simulating and Generating Motions of Human Figures
176 p. 2004 [978-3-540-20317-9]

Vol. 8: Baeten, J.; De Schutter, J.
Integrated Visual Servoing and Force Control – The Task Frame Approach
198 p. 2004 [978-3-540-40475-0]

Vol. 7: Boissonnat, J.-D.; Burdick, J.; Goldberg, K.; Hutchinson, S. (Eds.)
Algorithmic Foundations of Robotics V
577 p. 2004 [978-3-540-40476-7]

Vol. 6: Jarvis, R.A.; Zelinsky, A. (Eds.)
Robotics Research – The Tenth International Symposium
580 p. 2003 [978-3-540-00550-6]

Vol. 5: Siciliano, B.; Dario, P. (Eds.)
Experimental Robotics VIII – Proceedings of the 8th International Symposium ISER02
685 p. 2003 [978-3-540-00305-2]

Vol. 4: Bicchi, A.; Christensen, H.I.; Prattichizzo, D. (Eds.)
Control Problems in Robotics
296 p. 2003 [978-3-540-00251-2]

Vol. 3: Natale, C.
Interaction Control of Robot Manipulators – Six-degrees-of-freedom Tasks
120 p. 2003 [978-3-540-00159-1]

Vol. 2: Antonelli, G.
Underwater Robots – Motion and Force Control of Vehicle-Manipulator Systems
268 p. 2006 [978-3-540-31752-4]

Vol. 1: Caccavale, F.; Villani, L. (Eds.)
Fault Diagnosis and Fault Tolerance for Mechatronic Systems – Recent Advances
191 p. 2003 [978-3-540-44159-5]